alkyl hydroperoxides in the presence of molybdenum, vanadium, tungsten, and titanium catalysts (eq 5).

The activation of dioxygen by transition metal complexes is discussed in Chapter 4, and it includes the reactions of the various types of peroxometal complexes involving both homolytic and heterolytic processes. The considerable effort devoted to this subject in the last decade was undoubtedly spurred by the hope that the activation of dioxygen by transition metal complexes would lead to selective oxidations of hydrocarbons under mild conditions.[22–24] The results of this research have been generally disappointing, except for the recent demonstration[25] that terminal olefins can be selectively oxidized to methyl ketones in the presence of rhodium complexes:

$$2\,RCH{=}CH_2 + O_2 \xrightarrow{[Rh]} 2\,RCOCH_3 \tag{29}$$

The direct interaction with organic compounds proceeding via homolytic one-electron processes is treated in Chapter 5. The initial steps in the cobalt-catalyzed aerial oxidation of *p*-xylene and cyclohexane are considered in this context.

The awareness of oxometal species as important participants in a variety of stoichiometric and catalytic oxidations has increased in recent years and is covered in Chapter 6. Many important commercial processes for the vapor-phase oxidation of olefins over metal oxide catalysts, such as bismuth molybdate for the ammoxidation of propylene to acrylonitrile,

$$CH_2{=}CHCH_3 + NH_3 + \tfrac{3}{2}\,O_2 \xrightarrow{[Bi_2O_3/MoO_3]} CH_2{=}CHCN + 3\,H_2O \tag{30}$$

fall into this category. Other examples include several well-known reagents used in organic synthesis, e.g., the OsO_4-catalyzed hydroxylation of olefins,[26]

$$\text{cyclohexene} + H_2O_2 \xrightarrow{[OsO_4]} \textit{cis}\text{-cyclohexane-1,2-diol (OH, OH)} \tag{31}$$

and the SeO_2-catalyzed allylic oxidation of olefins.[10]

$$n\text{-}C_8H_{17}CH{=}CH_2 + (CH_3)_3COOH \xrightarrow{[SeO_2]} n\text{-}C_7H_{15}\underset{\displaystyle OH}{\underset{|}{C}H}{-}CH{=}CH_2 + (CH_3)_3COH \tag{32}$$

Activation of organic substrates by coordination to transition metals is included in Chapter 7. The palladium-catalyzed oxidation of ethylene to vinyl acetate,

$$CH_2{=}CH_2 + \tfrac{1}{2}\,O_2 \xrightarrow[HOAc]{[Pd^{II}/Cu^{II}]} CH_2{=}CHOAc + H_2O \tag{33}$$

and acetaldehyde (eq 3) are such examples.

V. HOMOGENEOUS AND HETEROGENEOUS CATALYSIS

Homogeneous and heterogeneous catalysis in liquid- and vapor-phase processes can be considered on the basis of common mechanistic principles. Although these two areas of catalysis have tended to develop independently in the past, it has become increasingly evident in recent years that the fundamental chemical steps are essentially the same whether the oxidation occurs in the coordination sphere of a soluble metal complex or on the adsorbed metal-containing surface. Similarities in behavior have often been overlooked in the past because heterogeneous catalysts are generally employed under quite different conditions (at higher temperatures in the gas phase) than are their homogeneous counterparts.

The qualitative principle of soft (heterolytic) and hard (homolytic) catalysts developed for homogeneous catalysts is equally applicable to heterogeneous catalysts. Soft surfaces are typified by transition metals and their alloys, as well as transition metal oxides having metal-type conduction bands. Hard surfaces are typified by most transition metal oxides. In the first case, chemisorption can best be compared to coordination by metal complexes in low oxidation states, whereas coordination by metal complexes in high oxidation states pertains to hard catalysts. The same mechanistic pathways are available to both heterogeneous and homogeneous catalysts. Thus, homolytic and heterolytic mechanisms involve interactions of metal centers with hydrocarbon substrates, with molecular oxygen, and with intermediate hydroperoxides. Moreover, the concepts of electron and ligand transfer (Chapter 3) and activation by coordination and oxidative addition (Chapter 7) provide a mechanistic basis for discussing both homogeneous and heterogeneous oxidations. The local molecular structure of the active site is more pertinent than a superficial examination of the macroscopic features of the reacting system, such as its physical state. The activity and selectivity of heterogeneous catalysts can often be changed by varying the support, an effect that is directly analogous to ligand effects in homogeneous catalysis.

The frenetic activity devoted in recent years to "heterogenizing" homogeneous catalysts[27–32] is a further illustration of the general trend toward a dissolution of the traditional disciplinary barriers between homogeneous and heterogeneous catalysis. Although immobilization of homogeneous catalysts has generally been applied to hydrogenation and carbonylation reactions, the technique is, in principle, applicable to oxidations. The recent report[33] of the use of immobilized Schiff base complexes of cobalt as oxidation catalysts is interesting in this context.

Much of the decrease in energy consumption in petrochemical processes which has been recently achieved is directly attributable to the development of new, more efficient catalytic processes. Homogeneous, liquid-phase processes generally allow for a better control of oxidation conditions and hydrocarbon conversions. These advantages must be weighed against those derived from heterogeneous catalysts, the most important being the ease of product separation from the catalyst and continuous processing.

VI. BIOCHEMICAL OXIDATIONS

Not only is catalysis by metal complexes of considerable technological interest, but it also plays a vital role in enzymatic oxidation processes. Although enzymatic pathways are influenced to a large extent by the protein structure associated with the catalyst, the fundamental chemical steps are essentially the same as those occurring with simpler metal complexes. Thus, if there is one basic underlying theme in our approach to oxidation, it is that the same fundamental mechanistic concepts can be applied equally to the areas of *homogeneous*, *heterogeneous*, and *enzymatic* catalysis. Indeed, progress in the understanding of enzymatic oxidations has probably been hampered by the myth that these biochemical reactions involve fundamentally different mechanisms. The treatment of biochemical oxidations in Chapter 8 is designed to emphasize the unifying features in all of these processes.

The advantage of using biochemical oxidations for the synthesis of industrial chemicals is that the reaction step is carried out at ambient temperature and hence consumes significantly less energy than conventional technology. There are problems associated with the dilute aqueous solutions involved (which means that considerable energy may be expended in product recovery) and the often low rates of reaction. However, these problems may well be solved through utilization of the latest developments in recombinant DNA techniques and membrane separation technology.

Such biochemical oxidations will probably be carried out with whole cells, possibly immobilized, rather than enzymes. The former (fermentations), unlike the latter, can be carried out without any added cosubstrates (cofactors). Microbial epoxidation of olefins[34,35] may be one of the first processes to attain commercial importance, in the manufacture of propylene oxide, for example. It is worth noting that the synthesis of low-molecular-weight products, such as ethylene and propylene oxides, could make separation from an aqueous solution much easier. In principle, propylene could be used as the sole substrate to give a theoretical yield of propylene oxide of 90%:

$$10\ CH_3CH{=}CH_2 + 9\ O_2 \longrightarrow 9\ CH_3\overset{\overset{O}{\diagup\ \diagdown}}{CH{-}CH_2} + 3\ H_2O + 3\ CO_2 \qquad (34)$$

Biochemical oxidations may also be exploited for the asymmetric oxidation of prochiral substrates, e.g., asymmetric epoxidation.

VII. SYNTHETIC METHODOLOGY FOR ORGANIC OXIDATIONS

Mechanistic studies are clearly a means of developing better synthetic methods. Therefore, the application of the various oxidative transformations involving metal catalysis to specific transformations of different organic substrates (olefins, aromatic hydrocarbons, alkanes, and oxygen-containing compounds) is considered separately in Part II of this book. Stoichiometric oxidations with metal oxidants such as permanganate and chromic acid are generally not discussed unless they are particularly relevant. We have focused our attention on all those catalytic oxidative processes that are potentially amenable to the commercial production of organic compounds. These oxidations are also of interest to a wide range of synthetic organic chemists engaged in laboratory-scale preparations. Emphasis has been placed on processes utilizing dioxygen, hydrogen peroxide, and alkyl hydroperoxides as the primary oxidants. Several reactions employing other cheap primary oxidants, such as sodium hypochlorite, are also included. Reactions of peroxy acids and ozone are not discussed, since they do not generally involve metal catalysis. Their reactions and synthetic applications have been adequately covered elsewhere.

REFERENCES

1. For historical background, see C. Moureu and C. Dufraisse, *Chem. Rev.* **7**, 113 (1926); N. A. Milas, *ibid.* **10**, 295 (1932).
2. J. L. Bolland, *Q. Rev., Chem. Soc.* **3**, 1 (1949); L. Bateman, *ibid.* **8**, 147 (1954).
3. H. M. Stanley, *Chem. Ind. (London)* p. 681 (1979).
4. J. D. Idol, *Chem. Ind. (London)* p. 272 (1979).
5. M. Sittig, "Combine Hydrocarbons and Oxygen for Profit," Chem. Process Rev. No. 11. Noyes Dev. Co., Park Ridge, New Jersey, 1968.
6. T. Dumas and W. Bulani, "Oxidation of Petrochemicals: Chemistry and Technology." Appl. Sci. Publ., London, 1974.
7. D. J. Hucknall, "Selective Oxidation of Hydrocarbons." Academic Press, New York, 1974.
8. E. Davy, *Philos. Trans. R. Soc. London* p. 108 (1820) (quoted in Ref. 3).
9. E. Rideal, *Trans. Faraday Soc.* **42**, 100 (1946).
10. K. B. Sharpless and T. R. Verhoeven, *Aldrichimica Acta* **12**, 63 (1979).
11. J. Tsuji, *Pure Appl. Chem.* **51**, 1235 (1979).
12. G. W. Parshall, *J. Mol. Catal.* **4**, 243 (1970).
13. L. F. Hatch, *Hydrocarbon Process.* **49**(3), 101(1970).
14. H. W. Prengle and N. Barona, *Hydrocarbon Process.* **49**(3), 106(1970).

To
Jetty and Marion

Metal-Catalyzed Oxidations of Organic Compounds

Mechanistic Principles and Synthetic Methodology Including Biochemical Processes

Metal-Catalyzed Oxidations of Organic Compounds

Mechanistic Principles and Synthetic Methodology Including Biochemical Processes

ROGER A. SHELDON

Research and Development Department
Océ-Andeno BV
Venlo, The Netherlands

JAY K. KOCHI

Department of Chemistry
Indiana University
Bloomington, Indiana

ACADEMIC PRESS

New York London Toronto Sydney San Francisco 1981

A Subsidiary of Harcourt Brace Jovanovich, Publishers

ACADEMIC PRESS, INC.
111 Fifth Avenue, New York, New York 10003

United Kingdom Edition published by
ACADEMIC PRESS, INC. (LONDON) LTD.
24/28 Oval Road, London NW1 7DX

Library of Congress Cataloging in Publication Date

Sheldon, Roger A.
Metal-catalyzed oxidations of organic compounds.

Includes bibliographical references and index.
1. Oxidation. 2. Chemistry, Organic. 3. Metal catalysts. I. Kochi, Jay K. II. Title.
QD281.09S46 547'.23 81-10883
ISBN 0-12-639380-X AACR2

PRINTED IN THE UNITED STATES OF AMERICA

83 84 9 8 7 6 5 4 3 2

Contents

Preface

Oxidative transformations of functional groups are basic to organic chemistry, oxidation being extensively used in the laboratory synthesis of fine organic chemicals as well as in the manufacture of large-volume petrochemicals. The majority of the processes employed industrially involve catalysis by metal complexes, and an increasing variety of catalytic processes are being developed for laboratory-scale synthesis. Catalytic processes enjoy the advantages over their noncatalytic counterparts of proceeding efficiently under milder conditions, thus leading to more energy-efficient processes. Furthermore, catalytic processes are generally more selective, and capable of leading to optimal utilization of raw materials. Unlike the stoichiometric oxidations with traditional oxidants, such as permanganate and dichromate, the catalytic processes do not produce vast amounts of inorganic effluents which are difficult to dispose of. As a result, a greater emphasis must be placed on the development of new, improved catalytic processes.

The subject of catalytic oxidation actually embraces the three traditionally different fields of homogeneous liquid-phase oxidations, heterogeneous gas-phase oxidations, and biochemical enzymatic oxidations. These three areas have generally developed as virtually separate disciplines, with research in one area having little apparent relevance to that in the other two. Thus, it is not surprising that there are no books that treat catalytic oxidation in these three fields in a single volume.

A primary aim of this book is to show how a consideration of the essential chemistry, both organic and inorganic, can lead to a common mechanistic basis for unifying these disparate fields. Accordingly, the first part of the book emphasizes the mechanistic principles of oxidation–reduction in organic, inorganic, and biochemical systems. We have focused our attention on the use of molecular oxygen (dioxygen), hydrogen peroxide, and alkyl hydroperoxides as the primary oxidants. Reactions of peroxy acids and ozone are not discussed, since they do not generally involve metal catalysis and since their chemistry has been adequately covered elsewhere. Stoichiometric oxidations with metal oxidants such as permanganate and chromic acid are discussed only when pertinent to the mechanistic discussions. No rigorous distinction is made between homogeneous liquid-phase and heterogeneous gas-phase processes, since it is desirable for the mechanistic analysis to consider both at the molecular level. By similar reasoning, we consider the mechanisms of biochemical oxidations in terms of oxygen activation, oxometal interactions, etc. of the substrate with the enzyme by keying on the metal prosthetic group. Differences between enzymatic oxidations and simple metal-catalyzed oxidations, such as the high regioselectivities and stereospecificities observed in biochemical systems, can generally be attributed to the influence of the protein apoenzyme and not to fundamental differences in mechanism.

The synthetic applications of catalytic oxidations are emphasized in the second part of this book. The material is divided into substrate and reaction types, by which we have attempted to include most of the oxidative transformations involving metal catalysis, especially those potentially amenable to commercial production of organic compounds. In so doing, we hope to stimulate the application of what may be described as the technology of petrochemical oxidation to the broader field of organic synthesis. By providing a sound mechanistic basis, we also hope that the search for new catalytic processes to replace the stoichiometric ones will be less empirical than it has been.

This book is directed toward chemists engaged in organic synthesis, organometallic chemistry, and catalysis, both in industrial laboratories and in academic institutions. Although our primary concern is to develop a deeper understanding and a greater utilization of metal catalysis, this book may be used to augment a more traditional textbook in organic synthesis. A chemical background beyond the first-year courses in organic and inorganic chemistry is desirable, since we assume a rudimentary knowledge of the kinetics and mechanism of organic and inorganic reactions.

An extensive bibliography of additional material is included at the end of each chapter to provide for additional reading. The index has been thoroughly cross-referenced, and the reader is encouraged to seek the various concepts and correlations under more than one entry. Metal catalysis of organic reactions is a burgeoning area of research, and we welcome additions to this literature as well as errors and oversights in this book. The principal literature has been covered through the end of 1980.

Part of this book was written while one of us (R. A. S.) was employed by the Koninklijke Shell Laboratorium, Amsterdam. Financial support from the National Science Foundation is gratefully acknowledged (J. K. K.). We thank Terry O'Donnell-Moore for help with the proofreading of the manuscript and the preparation of the index.

R. A. Sheldon
J. K. Kochi

Glossary

Autoxidation: Oxidation of an organic compound with molecular oxygen, usually via a free radical chain process.

Catalyst: An additive that increases the rate of a chemical process. The *catalytic species* actually involved in the mechanism (i.e., the individual steps) of the catalytic process often differ from that employed in the addition (i.e., the *catalyst precursor*).

Conversion: Amount of reactant consumed in a chemical process, expressed as a percentage relative to the original charge.

Coordination: The complement of ligands directly bonded to the metal center. Full coordination about the metal will not always be included in this book, unless required for the discussion.

Electron transfer: Addition or removal of an electron from a chemical species leading to its reduction or oxidation, respectively. Reference to *electron transfer* may be made either in the context of the stoichiometry of a redox process or with reference to the mechanism of a particular step (usually outersphere).

β-Elimination: Loss of a ligand usually induced by the removal of a β-proton, e.g.,

$$CH_3CH_2OPt^{II} \longrightarrow CH_3CH{=}O + HPt^{II}$$
$$AcOCH_2CH_2Pd^{II}OAc \longrightarrow AcOCH{=}CH_2 + HPd^{II}OAc$$

Heterolysis: Fragmentation of a neutral diamagnetic compound into a cation and anion pair. In a nucleophilic or electrophilic substitution, *heterolysis* is merely inferred from the stoichiometry, e.g.,

$$I^- + HOMo(O)_2OOH \longrightarrow HOMo(O)_2O^- + IOH$$

Homolysis: Fragmentation of a diamagnetic molecule into two paramagnetic species, e.g.,

$$CH_3OOH \longrightarrow CH_3O\cdot + HO\cdot$$

Furthermore, *homolysis* is implicit in oxidation–reduction reactions in which radicals are produced as a result of one-equivalent changes, e.g.,

$$CH_3OOH + Cu^I \longrightarrow CH_3O\cdot + HOCu^{II}$$

Homolytic substitution: Innersphere reactions of metal complexes with radicals involving either a *displacement* (S_H2) as in

$$CH_3OO\cdot + PhCo^{III} \longrightarrow CH_3OOCo^{III} + Ph\cdot$$

or *atom* (*ligand*) *transfer* as in

$$CH_3\cdot + Cu^{II}Cl_2 \longrightarrow CH_3Cl + Cu^{I}Cl$$

The distinction between S_H2 and atom transfer is largely made on whether attack occurs on the metal or on the ligand, respectively. Atom transfer involves a one-equivalent change in the formal oxidation oxidation state of the metal whereas S_H2 does not.

Inhibition: Prevention of the propagation cycle in a radical chain or catalytic process by destruction of the reactive intermediates or potential initiators. The processes of *inhibition* and *initiation* are opposed.

Initiator: An additive that starts a (radical) chain process, usually by homolysis.

Insertion: The interposition of a molecule into a ligand–metal bond, e.g.,

$$CH_2{=}CH_2 + HOPd^{II} \longrightarrow HOCH_2CH_2Pd^{II}$$

The term is usually synonymous with *addition*—focus being primarily placed on the addend in insertion.

Ion-radical: Species formed from a neutral diamagnetic precursor by either the addition (*anion-radical*) or the removal (*cation-radical*) of an electron, e.g.,

$$C_6H_6^{\overset{\cdot}{-}},\quad (CH_3)_4Sn^{\overset{\cdot}{+}},\quad \overset{+}{\dot{C}}H_2\dot{C}H_2$$

Kinetic chain length or turnover number: In a chain or catalytic process, the number of reaction cycles carried out per initiating or catalytic species. The kinetic chain length is equal to the rate of one of the propagation steps divided by the rate of termination.

Organometal: Chemical species containing at least one ligand in which carbon is bonded to the metal.

Oxidation: Electron removal from a chemical species. It is also used in a qualitative context to portray the increase in the *oxidation number* of a specific nucleus in a chemical compound. It is expressed in different ways in organic chemistry and inorganic chemistry, as elaborated in Chapter 1, p. 6.

Oxidation number: The formal oxidation state of the metal center in a complex determined as:

$$\text{metal charge} = \text{charge on complex ion} - \sum \text{ligand charges}$$

Some examples are $CH_3Tl^{III}Cl_2$, $C_3H_5Mo^{VI}O_2(OH)$, $(NC)_5Co^{III}OO\cdot^{3-}$. The designation is completely arbitrary. It is useful as a bookkeeping device to keep track of electron changes in oxidation–reduction processes.

Oxidative addition: Reaction in which the oxidation of a metal complex by an electrophile is accompanied by an increase in its coordination number, e.g.,

$$O_2 + Mo^{IV} \longrightarrow \begin{matrix} O \\ | \\ O \end{matrix}\!\!>Mo^{VI}$$

usually applied to two-equivalent changes of the metal center without regard to the mechanism. When oxidative addition involves an overall one-equivalent change, a radical is implicated, e.g.,

$$CH_3OO\cdot + Co^{II}(DMG)_2 \longrightarrow CH_3OOCo^{III}(DMG)_2$$

Oxidizability: Susceptibility of an organic substrate RH to autoxidation and given by the ratio: $k_p/(2k_t)^{1/2}$, expressed in units of ($M^{-1/2}\,sec^{-1/2}$) where k_p is the second-order rate constant for propagation ($RO_2\cdot + RH \rightarrow$) and k_t is the second-order rate constant for termination ($2\ RO_2\cdot \rightarrow$).

Oxygen activation: Enhancement of dioxygen reactivity in autoxidation, usually by coordination to a metal center.

Peroxymetallocycle: Cyclic organometallic intermediate containing a peroxometal bond, as postulated in a variety of oxygen-transfer reactions with dioxygen and peroxides. Formally described as a 1,2-dioxa-3-metallacycloalkane, $\overline{M(CH_2)_nOO}$, where n is most commonly 2. Analogous to *oxymetallocycles*, $\overline{M(CH_2)_nO}$ proposed as intermediates in the reactions of oxometals with alkenes with $n = 2$.

Promoter: An additive that increases the rate of a (radical) chain process.

Reductive elimination: The reverse of oxidative addition, e.g.,

$$CH_2{=}CHCH_2{-}OMo^{VI}O_2(OH) \longrightarrow CH_2{=}CHCH{=}O + Mo^{IV}O_2 + H_2O$$

See also *homolysis*.

Retardation: The slowing up of a chain or catalytic process, usually due to inefficient inhibition.

Selectivity: The relative rates of two or more simultaneous processes occurring on the same substrate. Commonly measured by product distributions. For example, it is the amount of a particular product formed divided by the amount of the reactant consumed, expressed as a percentage.

Terminal oxidant: In a catalytic process, the reagent consumed and expressed in the overall stoichiometry.

Yield: The amount of a particular product formed divided by the amount of reactant charged, expressed as a percentage. The *yield* equals the *conversion* multiplied by the *selectivity*.

Abbreviations

acacen 1,2-bis(4-oxopentyl-2-imino)ethane
AIBN azobisisobutyronitrile
Ar aryl
B axial base in metalloporphyrin complexes
bipy α,α′-bipyridyl
Bu butyl
CHP cumene hydroperoxide
cyt cytochrome
D oxidized form of cosubstrate in enzymatic oxidations
DH_2 reduced form of cosubstrate in enzymatic oxidations
DMAC *N*,*N*-dimethylacetamide
DMF *N*,*N*-dimethylformamide
DMK acetone (dimethyl ketone)
DMSO dimethyl sulfoxide
Diphos 1,2-bis(diphenylphosphino)ethane
E^+ electrophile
EBHP ethylbenzene hydroperoxide
EC Enzyme Commission
ee enantiomeric excess
Et ethyl
FAD flavin adenine dinucleotide coenzyme
FADH reduced form of FAD
FMN flavin mononucleotide coenzyme
FMNH reduced form of FMN
HMPA hexamethyl phosphoramide
Hb hemoglobin
HRP horseradish peroxidase
KCL kinetic chain length
L neutral ligand (usually phosphine)
M metal
Mb myoglobin
MCPBA *meta*-chloroperbenzoic acid
Me methyl
MEK methyl ethyl ketone
NAD nicotinamide adenine dinucleotide coenzyme
NADH reduced form of NAD
NADP nicotinamide adenine dinucleotide phosphate coenzyme
NADPH reduced form of NADP
OAc acetate
OEP octaethylporphyrinato
OxyHb oxyhemoglobin
Ⓟ polymeric group
P porphyrinato
PBA perbenzoic acid
Ph phenyl
Pr propyl
PTC phase-transfer catalysis
py pyridine
pyO pyridine *N*-oxide
R alkyl
RS alkylmercapto
S substrate or solvent
SH_2 reduced form of substrate S
SO/SO_2 oxidized forms of substrate S
Salen *N*,*N*-bis(salicylidene)-ethylenediamino
Saloph *N*,*N*-bis(salicyclidene)-*o*-phenylenediamino
Salpr bis(3-salicylideneimino-propyl)amino
SCE standard calomel electrode
TAHP *tert*-amyl hydroperoxide
TBHP *tert*-butyl hydroperoxide
TPP *meso*-tetraphenylporphyrinato
X anionic ligand (usually halide)

Chapter 1

Introduction to Metal-Catalyzed Oxidations

I. HISTORICAL DEVELOPMENT

The oxidation of organic compounds by molecular oxygen (dioxygen) has a long history[1]—stemming as it does from Lavoisier's explanation of combustion, which also marked the beginning of the modern era of chemistry. Observations made in the nineteenth century linked the deterioration of many organic materials, such as rubber and natural oils and fats, to the absorption of dioxygen. Early investigators were mainly concerned with the *inhibition* of such oxidative degradations.

Around the turn of the century, it was recognized that these oxidative processes involved the formation of organic peroxides. Subsequent mechanistic studies of the interaction of simple hydrocarbons with dioxygen, carried out in the 1940s, provided the basic concepts for the development of the free radical chain theory of autoxidation.[2] Although autoxidation (which refers to oxidations with dioxygen) can be spontaneously initiated, it is more commonly promoted by metal species, often in trace quantities.

The control of autoxidation is desirable not only from the standpoint of inhibiting the oxidative deterioration of plastics, gasoline, lubricating oils, and rubber, but also for *promoting* the selective oxidation of petroleum hydrocarbon feedstocks to a variety of industrial organic chemicals. Although autoxidations of organic compounds are generally highly exothermic reactions, they do not readily undergo spontaneous combustion in air, largely owing to the relatively high activation energies for the initiation process. In other words, aerial oxidations are usually difficult to initiate, but, once underway, they are often difficult to interrupt short of the most thermodynamically stable products, namely, carbon dioxide and water.

Catalysis by metal complexes plays an important role in the control of selective, partial oxidation of alkanes, olefins, and aromatic hydrocarbons to useful products. If we are in the age of petroleum today, with 90% of organic chemicals derived from this raw material, it is also the age of catalysis.[3,4] Indeed, the majority of petrochemical processes are catalytic, the most important ones being catalytic oxidations.[5–7]

The first observation of a catalytic oxidation can probably be attributed to Davy,[8] who showed in 1820 that ethanol is oxidized to acetic acid in the presence of platinum.*

$$CH_3CH_2OH + O_2 \xrightarrow{[Pt]} CH_3CO_2H + H_2O \tag{1}$$

Although there were some industrial processes, such as the silver-catalyzed oxidation of ethylene to ethylene oxide discovered in 1933, the field of metal-catalyzed oxidation assumed importance only in the 1940s; that is, it coincided with the development of the theory of free radical autoxidation. The following observation, made in 1945 by Sir Eric Rideal in his introductory remarks to the Faraday Society meeting devoted to the subject of oxidation,[9] is indicative of the state of the art only 35 years ago:

> We know a little about the mechanism of hydrogenation, but in oxidation we still sail on uncharted seas.

Several important industrial processes were developed for the selective partial oxidation of hydrocarbon feedstocks during the 1940s and 1950s. Initially, the majority of these processes involved vapor-phase oxidations over heterogeneous catalysts (e.g., ethylene to ethylene oxide). The serious study and application of homogeneous catalysis of liquid-phase oxidation

* Hereinafter, the catalyst will be indicated in [brackets] without necessarily specifying either the oxidation state or the coordination.

originates from the late 1950s. For example, the Mid-Century process, aptly named, for the production of terephthalic acid,

$$p\text{-}CH_3C_6H_4CH_3 + 3\,O_2 \xrightarrow[\text{HOAc}]{[Co(OAc)_2/Br^-]} p\text{-}HO_2CC_6H_4CO_2H + 2\,H_2O \tag{2}$$

and the Wacker process for acetaldehyde,

$$2\,CH_2{=}CH_2 + O_2 \xrightarrow[H_2O]{[PdCl_2/CuCl_2]} 2\,CH_3CHO \tag{3}$$

were both reported in 1959. The Celanese process for the liquid-phase oxidation of *n*-butane to acetic acid was also developed in the 1950s and is still in use today. In the last decade, an efficient liquid-phase process for the epoxidation of olefins (such as propylene), which cannot be carried out with silver, has been developed in two steps.

$$(CH_3)_3CH + O_2 \longrightarrow (CH_3)_3COOH \tag{4}$$

$$(CH_3)_3COOH + CH_2{=}CHCH_3 \xrightarrow{[Mo]} (CH_3)_3COH + H_2\overset{O}{C{-}C}HCH_3 \tag{5}$$

A variety of metal-catalyzed oxidations has also been developed for selective oxidations that can be carried out in laboratory-scale synthesis. The importance of metal catalysts is illustrated in the following transformations of olefins and acetylenes with *tert*-butyl hydroperoxide.[10]

$$RCH_2{-}CH{=}CH_2 + (CH_3)_3COOH \xrightarrow{[Mo^{VI}]} RCH_2\overset{O}{CH{-}CH_2} \tag{6}$$

$$RCH_2{-}CH{=}CH_2 + (CH_3)_3COOH \xrightarrow{[Os^{VIII}]} RCH_2{-}CH(OH){-}CH_2(OH) \tag{7}$$

$$RCH_2{-}CH{=}CH_2 + (CH_3)_3COOH \xrightarrow{[Se^{IV}]} RCH(OH){-}CH{=}CH_2 \tag{8}$$

$$RCH(OH){-}CH{=}CH_2 + (CH_3)_3COOH \xrightarrow{[V^{V}]} RCH(OH){-}\overset{O}{CH{-}CH_2} \tag{9}$$

$$RCH_2{-}C{\equiv}CH + (CH_3)_3COOH \xrightarrow{[Se^{IV}]} RCH(OH){-}C{\equiv}CH \tag{10}$$

The application of the Wacker oxidation in eq 3 to a variety of higher terminal olefins affords methyl ketones selectively.

$$RCH{=}CH_2 + O_2 \xrightarrow{[PdCl_2/CuCl_2]} RCOCH_3 + H_2O \tag{11}$$

The catalytic oxidation can be effectively employed in the synthesis of a variety of natural products.[11]

II. SOURCES OF HYDROCARBONS

The success of the Wacker process has led to a burgeoning interest in the chemistry of organometallic compounds and in homogeneous catalysis. In the last two decades, there has been a tendency for more petrochemical processes to employ homogeneous catalysts in the liquid phase, stimulated by certain process advantages.[12–15] Thus, homogeneous catalysts are generally more active and selective than their heterogeneous counterparts. In practice, this means less severe reaction conditions leading to lower energy requirements, and cleaner processes affording savings of raw materials and reduced pollution. In today's climate of spiraling prices for hydrocarbon feedstocks and fuels, emphasis is firmly placed on the conservation of energy and raw materials.

TABLE I Product Distributions from Ethane and Naphtha Cracking[a]

Product (wt%)	Feedstock	
	Ethane[b]	Naphtha[c]
Methane/hydrogen	12.9	16.0
Ethylene	80.9	31.3
Ethane	Starting material	3.4
Propylene	1.8	12.1
Propane	0.3	0.5
Butadiene	1.9	4.2
Butenes	0.4	2.8
Butane	0.4	0.2
C_5^+	1.4	9.0
BTX		13.0
Fuel oil		6.0
Others		1.5
	100	100

[a] Data taken from L. F. Hatch and S. Matar, *Hydrocarbon Process.*, **57**(1), 135(1978) and **57**(3), 129(1978).

[b] 750°–850°C; 1–1.25 bar; steam diluent (0–50%); no catalyst; 60% conversion.

[c] 840°C; 1 bar; 0.6 kg steam per kilogram naphtha; no catalyst.

The petrochemicals industry was founded, in the 1920s in the United States, in order to utilize the lower olefins formed as by-products in the manufacture of gasoline by the cracking of gas oil.[16] Synthetic ethanol and isopropanol were produced by absorbing ethylene and propylene in sulfuric acid followed by dilution with water. Subsequently, many petrochemicals were produced by the selective oxidation of olefins, alkanes, and aromatic hydrocarbons available from oil conversion processes (*vide infra*).

The three most important olefins used for the production of petrochemicals are ethylene, propylene, and butadiene. These base chemicals are coproduced in various ratios by the thermal cracking, at 750°–850°C in the presence of steam, of hydrocarbon feedstocks ranging from ethane (derived from natural gas liquids) to crude oil.

Ethylene is the largest volume organic chemical. In the United States it is produced mainly by ethane cracking, but in Europe more than 90% of the ethylene is produced from naphtha cracking. The product distributions obtained from ethane and naphtha cracking are compared in Table I.

The major aromatic feedstocks—benzene, toluene, and xylenes (BTX)—are obtained as by-products of olefin production by naphtha cracking. They are also obtained from the catalytic reforming of naphtha, which involves the dehydrocyclization of alkanes over supported platinum or rhenium catalysts. The sources of the major hydrocarbon base chemicals are summarized in Fig. 1.

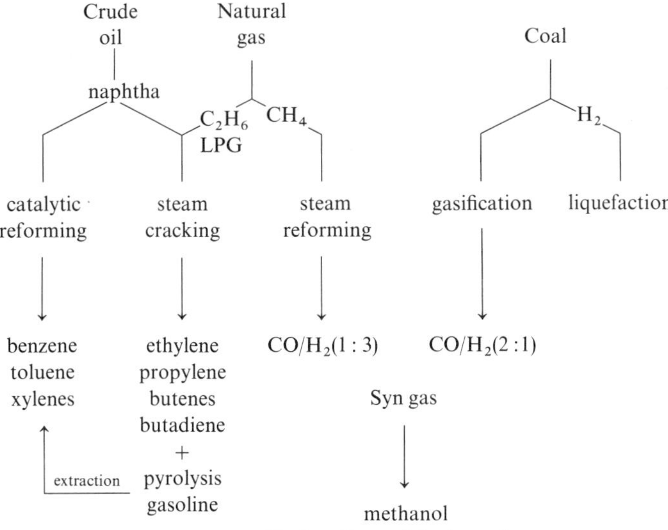

FIG. 1. The availability of the major source of hydrocarbon feedstocks from crude oil, natural gas, and coal.

III. DEFINITION OF OXIDATION

The concept of oxidation in organic chemistry differs from that in inorganic chemistry. Historically, oxidation in organic chemistry[17] usually refers to either (1) the elimination of hydrogen, as in the sequential dehydrogenation of ethane,

$$CH_3CH_3 \longrightarrow CH_2{=}CH_2 \longrightarrow HC{\equiv}CH \tag{12}$$

or (2) the replacement of a hydrogen atom bonded to carbon with another, more electronegative element such as oxygen in the following series of oxidative transformations of methane,

$$CH_4 \longrightarrow CH_3OH \longrightarrow CH_2O \longrightarrow HCO_2H \longrightarrow CO_2 \tag{13}$$

By contrast, oxidation in inorganic chemistry usually refers to the loss of one or more electrons from a metal ion, e.g., $Fe^{2+} \rightarrow Fe^{3+} + e^-$; $Tl^+ \rightarrow Tl^{3+} + 2e^-$; etc. When this concept is applied to a metal complex, it is necessary to assign an oxidation state formally to the metal. In this book we have arbitrarily adopted the formalism that, in a metal complex, the bonding electrons are assigned to the ligand. The formal oxidation state of the metal is indicated by a roman numeral, as illustrated in the examples below:

$$Mo^{VI}O_4^{2-} \qquad (Ph_3P)_2Pt^{II}(CH_3)_2 \qquad (CH_3)_3Sn^{IV}H \qquad (Ph_3P)_2(CO)ClIr^{III}\!\left\langle\begin{matrix}O\\|\\O\end{matrix}\right.$$

The oxidation state formalism is especially useful in the discussion of oxidative processes involving organic compounds and metal complexes, since the electron changes are not always obvious. This formalism, which allows a convenient method for electron "bookkeeping," should not be construed as denoting the actual oxidation states of the various species.[18] Its utility is best shown by considering the following redox reactions between organic compounds and metal complexes (only the formal oxidation state of the metal center is indicated).

$$PhCH_3 + Co^{III} \longrightarrow PhCH_2\cdot + Co^{II} + H^+ \tag{14}$$

$$(CH_3)_3COOH + Fe^{II} \longrightarrow (CH_3)_3CO\cdot + HOFe^{III} \tag{15}$$

$$(Ph_3P)_4Pt^0 + CH_3I \longrightarrow (Ph_3P)_2Pt^{II}(CH_3)I + 2\,Ph_3P \tag{16}$$

$$AcOCH_2CH_2Te^{IV}Br_3 + HOAc \longrightarrow AcOCH_2CH_2OAc + Te^{II}Br_2 + HBr \tag{17}$$

$$CH_2{=}CHCH_3 + Mo^{VI}O_4^{2-} \longrightarrow CH_2{=}CHCH_2\cdot + Mo^{V}O_4^{3-} + H^+ \tag{18}$$

$$CH_3CH{=}CHCH_2Mo^{VI}(OH)O_2 \longrightarrow CH_2{=}CHCH{=}CH_2 + Mo^{IV}O_2 + H_2O \tag{19}$$

Note that the change in the formal oxidation state of the carbon center in eqs 14 and 18 is from C(IV) to C(III).

IV. MECHANISTIC PRINCIPLES OF METAL-CATALYZED OXIDATION

Metal-catalyzed oxidations may be conveniently divided into two types, which we arbitrarily designate as *homolytic* and *heterolytic*. The first type of catalysis usually involves soluble transition metal salts (homogeneous), such as the acetates or naphthenates of Co, Mn, Fe, Cu, etc., or the metal oxides (heterogeneous). Furthermore, homolytic catalysis necessitates the recycling of the metal species between several oxidation states by one-equivalent changes. Free radicals are formed as intermediates from the organic substrate. Heterolytic catalysis involves reactions of organic substrates coordinated to transition metals. It is characterized by the metal complex acting as a Lewis acid or formally undergoing two-equivalent changes. Free radicals are not intermediates. For example, consider the acyloxylation of benzene by trifluoroacetate,

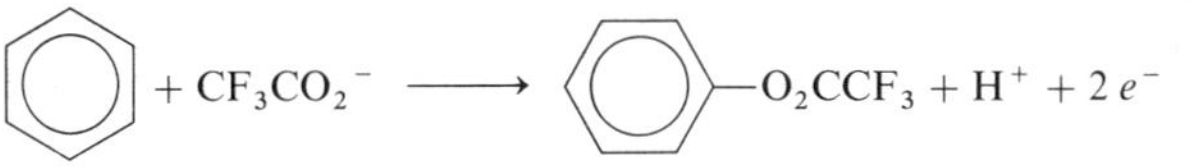

$$C_6H_6 + CF_3CO_2^- \longrightarrow C_6H_5{-}O_2CCF_3 + H^+ + 2\,e^- \qquad (20)$$

which overall corresponds to a two-electron oxidation.[19] This oxidative substitution may be effected by lead tetrakistrifluoroacetate,

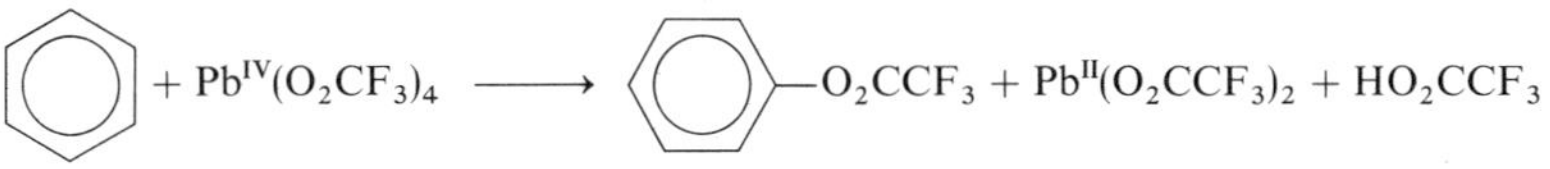

$$C_6H_6 + Pb^{IV}(O_2CF_3)_4 \longrightarrow C_6H_5{-}O_2CCF_3 + Pb^{II}(O_2CCF_3)_2 + HO_2CCF_3 \qquad (21)$$

leading to the concomitant reduction of Pb(IV) to Pb(II). The same transformation can be effected by cobalt tristrifluoroacetate,

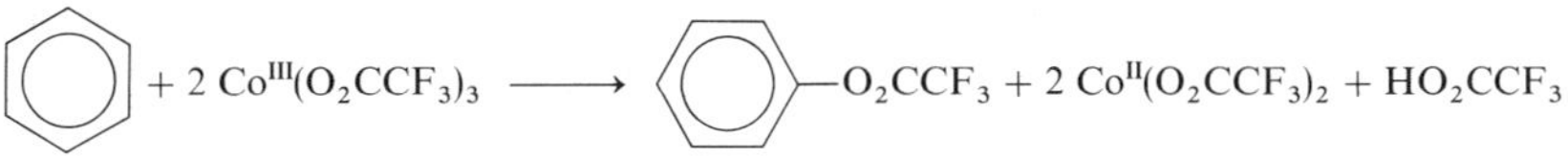

$$C_6H_6 + 2\,Co^{III}(O_2CCF_3)_3 \longrightarrow C_6H_5{-}O_2CCF_3 + 2\,Co^{II}(O_2CCF_3)_2 + HO_2CCF_3 \qquad (22)$$

leading to the concomitant reduction of 2 Co(III) to 2 Co(II). It can be shown that the oxidative substitution promoted by Pb(IV) proceeds via the heterolytic process in Scheme I.

Scheme I:

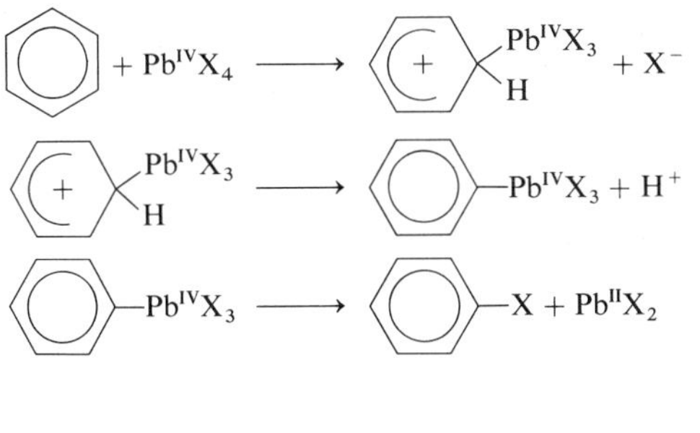

$$C_6H_6 + Pb^{IV}X_4 \longrightarrow [C_6H_6{-}Pb^{IV}X_3]^+ + X^- \qquad (23)$$

$$[C_6H_6{-}Pb^{IV}X_3]^+ \longrightarrow C_6H_5{-}Pb^{IV}X_3 + H^+ \qquad (24)$$

$$C_6H_5{-}Pb^{IV}X_3 \longrightarrow C_6H_5{-}X + Pb^{II}X_2 \qquad (25)$$

According to this mechanism, electrophilic attack by Pb(IV) in eq 23 is rate limiting and followed by reductive elimination of Pb(II) in eq 25. On the other hand, the oxidative substitution of benzene promoted by Co(III) proceeds via a homolytic mechanism in Scheme II.[20]

Scheme II:

$$C_6H_6 + Co^{III}X_3 \longrightarrow C_6H_6^{+\cdot} + Co^{II}X_3^{-} \quad (26)$$

$$C_6H_6^{+\cdot} + X^{-} \longrightarrow \cdot C_6H_6X \text{ (X, H)} \quad (27)$$

$$\cdot C_6H_6X + Co^{III}X_3 \longrightarrow C_6H_5\text{—}X + Co^{II}X_2 + HX \quad (28)$$

Electron transfer in eq 26 is rate limiting, and all the subsequent reactions are fast. The successive one-electron transfers in eqs 26 and 28 correspond to the formation of benzene cation-radicals and the destruction of cyclohexadienyl radicals, respectively.

Catalytic processes may also be devised which correspond to the recycling of the metal species by heterolytic and homolytic mechanisms. These two types of catalytic processes will be treated separately in the ensuing discussion, although the distinction is not always clear since there are transition metal complexes that are capable of participating in both types of catalysis. Homolytic catalysis and heterolytic catalysis also fall into the categories that have been described as "hard" and "soft" processes, respectively.[21]

Historically, the homolytic type of catalysis has been known and studied for a long time. The heterolytic catalysts represent a relatively recent innovation but nevertheless include such important developments as the Wacker process for the oxidation of olefins (eq 3). Regardless of the mechanism involved, the most important characteristics of metal catalysts for effecting oxidation are the accessibility of several oxidation states as well as the accommodation of various coordination numbers, both of which are properties of transition metal complexes.

The mechanistic discussion in this book has been divided according to the basic types of reagents and reactions involved. For example, the metal-catalyzed oxidations with hydrogen peroxide and alkyl hydroperoxides via homolytic and heterolytic processes are included in Chapter 3. Homolytic processes are intimately involved in classical autoxidation processes catalyzed by cobalt, manganese, iron, and copper complexes, as in the commercially important conversion of *p*-xylene to terephthalic acid (eq 2). Heterolytic processes include the industrial epoxidation of olefins with

15. (a) G. Szonyi, *Adv. Chem. Series* **70**, 53 (1968).
    (b) See also *Adv. Chem. Series* **132** (1974).
16. B. T. Brooks, *Ind. Eng. Chem.* **27**, 278 (1935).
17. See, for example, A. Streitwieser, Jr. and C. H. Heathcock, "Introduction to Organic Chemistry," p. 232. Macmillan, New York, 1976.
18. K. F. Purcell and J. C. Kotz, "Inorganic Chemistry," p. 519. Saunders, Philadelphia, Pennsylvania, 1977.
19. R. A. Sheldon and J. K. Kochi, *Adv. Catal.* **25**, 272 (1976).
20. J. K. Kochi, R. T. Tang, and T. Bernath, *J. Am. Chem. Soc.* **95**, 7114 (1973).
21. R. Ugo, *Chim. Ind.* (*Milan*) **51**, 1319 (1969).
22. J. Lyons, *in* "Fundamental Research in Homogeneous Catalysis ", Vol. 3, (M. Tsutsui and R. Ugo, eds.), p. 1. Plenum, New York, 1977.
23. J. Lyons, *Aspects Homogeneous Catal.* **3**, 1 (1977).
24. H. Mimoun, *Rev. Inst. Fr. Pet.* **33**, 259 (1978).
25. H. Mimoun, M. M. Perez Machirant, and I. Sérée de Roch, *J. Am. Chem. Soc.* **100**, 5437 (1978).
26. M. Schröder, *Chem. Rev.* **80**, 187 (1980).
27. J. C. Bailar, *Catal. Rev.* **10**, 17 (1974).
28. H. Heinemann, *Chemtech.* **1**, 286 (1971).
29. C. U. Pittman and G. O. Evans, *Chemtech.* **3**, 560 (1973).
30. Z. M. Michalska and D. E. Webster, *Platinum Met Rev.* **18**, 65 (1974); *Chemtech* **5**, 117 (1975).
31. R. H. Grubbs, *Chemtech* **7**, 512 (1977).
32. D. Commereuc and G. Martino, *Rev. Inst. Fr. Pet.* **30**, 89 (1975).
33. R. S. Drago, J. Gaul, A. Zombeck, and D. K. Straub, *J. Am. Chem. Soc.* **102**, 1033 (1980).
34. H. Ohta and H. Tetsukawa, *J. Chem. Soc., Chem. Commun.* p. 849 (1978).
35. S. W. May and R. D. Schwartz, *J. Am. Chem. Soc.* **96**, 4031 (1974).

ADDITIONAL READING

F. R. Mayo, ed., "Oxidation of Organic Compounds," Vols. I, II, III, Adv. Chem. Series 75, 76, 77. Am. Chem. Soc., Washington, D.C., 1968. The proceedings of the International Oxidation Symposium includes liquid-phase oxidations, base-catalyzed and heteroatom compound oxidations, free radical initiations, interactions and inhibitions, gas-phase oxidations, heterogeneous catalysis, applied oxidations and synthetic processes, ozone chemistry, photooxidation and singlet oxygen oxidation, and biochemical oxidations.

K. B. Wiberg, ed., "Oxidation in Organic Chemistry." Academic Press, New York, 1965. Includes oxidations with permanganate, chromic acid and chromyl compounds, vanadium(V), cobalt(III), manganese(III), cerium(IV), and lead(IV) tetraacetate and glycol cleavages.

W. S. Trahanovsky, ed., "Oxidation in Organic Chemistry." Academic Press, New York, 1973. Includes oxidations with cupric ion, thallium(III), and ruthenium tetroxide as well as the mechanism of phenolic oxidative coupling.

R. L. Augustine, ed., "Oxidation," Vol. I. Dekker, New York, 1969. Includes oxidations of hydrocarbons and heteroatom compounds using transition metal compounds, selenium dioxide oxidation, glycol cleavage, peracid and peroxide reactions, and ozonization. Volume II, coedited with D. J. Trecker, includes sulfoxide–carbodiimide oxidations, photosensitized oxygenations epoxidation with hydroperoxides, and metal-catalyzed peroxide reactions.

D. J. Hucknall, "Selective Oxidation of Hydrocarbons." Academic Press, New York, 1974.

E. K. Fields, ed., "Selective Oxidation Process," Adv. Chem. Series 51, Am. Chem. Soc., Washington, D.C., 1965. Includes autoxidations.

B. J. Luberoff, ed., "Homogeneous Catalysis," Adv. Chem. Series 70, Am. Chem. Soc., Washington, D.C., 1968. Includes metal-catalyzed oxidations.

W. A. Waters, "Mechanism of Oxidation of Organic Compounds." Methuen, London, 1964.

R. Stewart, "Oxidation Mechanisms. Application to Organic Chemistry." Benjamin, New York, 1964.

T. A. Turney, "Oxidation Mechanisms." Butterworth, London, 1965. Includes organic and inorganic oxidations.

G. W. Parshall, "Homogeneous Catalysis. The Applications and Chemistry of Catalysis by Soluble Transition Metal Complexes." Wiley, New York, 1980. Includes industrial oxidations.

B. C. Gates, J. R. Katzer, and G. C. A. Schuit, "Chemistry of Catalytic Processes." McGraw-Hill, New York 1979. Mainly concerned with heterogeneous processes including hydrocarbon cracking and reforming, catalysis by transition metal complexes of the Wacker oxidation, vinyl acetate, oxo, methanol carbonylation, and Ziegler–Natta polymerization and ammoxidation of propylene.

C. H. Bamford and C. H. Tipper, eds., "Comprehensive Chemical Kinetics, Section 6: Oxidation and Combustion Reactions," Vol. 16, Liquid Phase Oxidations, Elsevier, Amsterdam, 1980.

Houben Weyl, "Methoden der Organische Chemie," Vol. 4/*1*b, Oxidation, Georg Thieme, Stuttgart, 1975.

Catalytica Associates, Inc., "Selective Catalytic Oxidation of Hydrocarbons, A Critical Analysis", Multiclient Study No. 1077, October, 1979, Santa Clara, California.

Part One

Mechanistic Principles of Metal-Catalyzed Oxidations

Chapter 2

Oxidations with Molecular Oxygen

Many liquid-phase oxidations known as *autoxidations* proceed under relatively mild conditions of temperature and oxygen pressure.[1] They are often subject to autocatalysis by products. The pioneering work of Bäckström[2] demonstrated that liquid-phase autoxidations are chain reactions. Criegee[3] made an important contribution in 1939 when he showed that the primary product of autoxidation of cyclohexene is the allylic hydroperoxide.

$$\text{cyclohexene} \xrightarrow{O_2} \text{3-cyclohexenyl-}O_2H \qquad (1)$$

Subsequent studies of the interaction of a variety of simple hydrocarbons with molecular oxygen provided the basis for the development of the free radical chain theory of autoxidation.[4–8] Although initial studies were concerned mainly with finding ways of preventing autoxidation, it was soon recognized that the controlled oxidation of hydrocarbons can be a useful method for preparing a wide range of oxygenated derivatives of hydrocarbons. The most well known commercial process involves the conversion of cumene to phenol and acetone, discovered by Hock and Lang in 1944.[9]

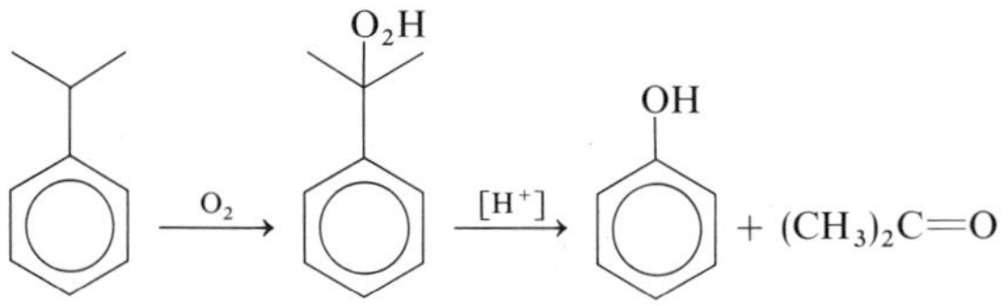

$$\text{PhCH(CH}_3)_2 \xrightarrow{O_2} \text{PhC(CH}_3)_2O_2H \xrightarrow{[H^+]} \text{PhOH} + (CH_3)_2C{=}O \tag{2}$$

I. FUNDAMENTALS OF RADICAL CHAIN PROCESSES

The liquid-phase autoxidation of hydrocarbons has been studied extensively and is the subject of several monographs and reviews.[10–25] Autoxidations proceed via a free radical chain mechanism described by the following general Scheme I.

Scheme I:

Initiation

$$In_2 \xrightarrow{R_i} 2\ In\cdot \tag{3}$$

$$In\cdot + RH \longrightarrow InH + R\cdot \tag{4}$$

Propagation

$$R\cdot + O_2 \longrightarrow RO_2\cdot \tag{5}$$

$$RO_2\cdot + RH \xrightarrow{k_p} RO_2H + R\cdot \tag{6}$$

Termination

$$R\cdot + RO_2\cdot \longrightarrow RO_2R \tag{7}$$

$$2\ RO_2\cdot \xrightarrow{2k_t} RO_4R \longrightarrow \text{nonradical products} + O_2 \tag{8}$$

Alkylperoxy radicals play vital roles in both the propagation and the termination processes. Hydroperoxides RO_2H are usually the primary products and in some cases may be isolated in high yields. Much of the present knowledge of autoxidation mechanisms has been acquired from studies of the reactions of alkylperoxy radicals and the parent alkyl hydroperoxide in the absence of oxygen.[26–29] The various modes of reaction of organic peroxides are now well characterized.[30–35]

At partial pressures of oxygen greater than approximately 100 torr, chain termination proceeds exclusively via the mutual destruction of two alkylperoxy radicals in reaction 8. The cross-termination reaction in eq 7 may be neglected. The predicted rate expression under steady-state conditions is given by

$$-\frac{d[\mathrm{RH}]}{dt} = -\frac{d[\mathrm{O_2}]}{dt} = k_\mathrm{p}[\mathrm{RH}]\left(\frac{R_\mathrm{i}}{2k_\mathrm{t}}\right)^{1/2} \tag{9}$$

These kinetics are often observed in practice. The susceptibility of any particular substrate to autoxidation is determined by the ratio $k_\mathrm{p}/(2k_\mathrm{t})^{1/2}$, which is usually referred to as its *oxidizability*. The oxidizabilities of some typical organic compounds are listed in Table I.

TABLE I Oxidizability of Various Organic Compounds

Substrate	$k_\mathrm{p}/(2k_\mathrm{t})^{1/2} \times 10^3(M^{-1/2}\ \mathrm{sec}^{-1/2})$
2,3-Dimethyl-2-butene	3.2
Cyclohexene	2.3
1-Octene	0.06
Cumene	1.5
Ethylbenzene	0.21
Toluene	0.01
p-Xylene	0.05
Benzaldehyde	290
Benzyl alcohol	0.85
2,4,6-Trimethylheptane	0.09

[a] From J. A. Howard, *Adv. Free-Radical Chem.* **4**, 49 (1972).

Scheme I becomes considerably more complicated at high conversions of the substrate owing to the accumulation of secondary products, such as the easily oxidized aldehydes and ketones, formed by the thermal decomposition of the alkyl hydroperoxide. Autoxidative syntheses are thus usually carried out at low conversions (<20%), and the excess unreacted substrate is recycled.

A. Initiation

Chain initiation is readily accomplished by the deliberate addition of initiators, that is, compounds that yield free radicals on thermal decomposition. In practice, initiators should have substantial rates of decomposition in the temperature range between 50° and 150°C. The rate of chain initiation R_i is given by

$$R_\mathrm{i} = 2ek_\mathrm{i}[\mathrm{In_2}] \tag{10}$$

where e is the efficiency of radical production (i.e., the fraction that escapes from the solvent cage) and k_i is the unimolecular rate constant for decomposition of the initiator In_2. Typical initiators are aliphatic azo compounds, dialkyl peroxides, diacyl peroxides, and peroxy esters. Table II gives the bond energies of several common initators. Kinetic studies have been greatly simplified by the use of initiators, which circumvents the long and often irreproducible induction periods observed earlier.

TABLE II Some Common Initiators for Autoxidation

Name	Structure	Activation energy ($kcal\ mol^{-1}$)	Temp (°C) for 1 hr half-life
Hydrogen peroxide	HO—OH	48	—
t-Butyl hydroperoxide	*t*-BuO—OH	42	—
t-Butyl peroxide	*t*-BuO—O—*t*-Bu	37	150
t-Butyl perbenzoate	*t*-BuO—OC(=O)Ph	34	125
Benzoyl peroxide	PhC(=O)O—OC(=O)Ph	30	95
Acetyl peroxide	$CH_3C(=O)O—OC(=O)CH_3$	30–32	85
Azoisobutyronitrile	$(CH_3)_2C(CN)—N{=}N—C(CN)(CH_3)_2$	30	85
t-Butyl hyponitrite	*t*-BuO—N=N—O—*t*-Bu	28	60
t-Butyl peroxalate	*t*-BuO—OC(=O)C(=O)O—O—*t*-Bu	25.5	40

Initiation by the direct reaction of most hydrocarbons with molecular oxygen, namely,

$$RH + O_2 \longrightarrow R\cdot + HO_2\cdot \quad (11)$$

is thermodynamically and kinetically unfavorable, although it has been observed in a few cases.[36–39] For example, spontaneous initiation has been observed in the autoxidation of indene.

$$\text{indene} + O_2 \longrightarrow \text{indenyl}\cdot + HO_2\cdot \quad (12)$$

Chain initiation in the absence of added initiators is usually attributed to radicals formed by the decomposition of adventitious peroxidic impurities present in the substrate. Direct attack can be favorable when it involves compounds in which hydrogen is bonded to elements other than carbon, as illustrated by the facile air oxidation of thiols, phosphines, and a variety of organometallic compounds.[40]

A third mechanism for initiation is the reaction of carbanions with molecular oxygen:[41–44]

$$R^- + O_2 \longrightarrow R\cdot + O_2^-\cdot \tag{13}$$

Except for highly acidic hydrocarbons, this pathway is not very common. However, the complexation of aralkanes such as fluorene, diphenylmethane, and cumene with the tricarbonylchromium moiety $Cr(CO)_3$ enhances the acidities of the benzylic protons and allows ready autoxidation to fluorenone, benzophenone, and cumyl alcohol.[45] An interesting development in base-catalyzed autoxidations is the use of phase transfer catalysts, as described in Chapter 6. This technique allows for efficient oxidation at low temperatures in two-phase mixtures of hydrocarbon and water.[46]

Thermal decomposition of alkyl hydroperoxides represents a major source of free radicals in autoxidation reactions. Unless hydrocarbons are rigorously purified prior to use, trace amounts of hydroperoxides can lead to erroneous results in kinetic studies, especially in the absence of added initiators. If initiation involves simple unimolecular homolysis of the alkyl hydroperoxide,

$$ROOH \longrightarrow RO\cdot + \cdot OH \tag{14}$$

and the autoxidation is carried out at sufficiently high temperatures so that it does not accumulate, the limiting rate is given by[47]

$$-\frac{d[RH]}{dt} = \frac{3k_p^2}{k_t}[RH]^2 \tag{15}$$

Only one-third of the RH is consumed by alkylperoxy radicals under these conditions. Thus, the chain lengths are short, and substantial amounts of RH are consumed in stoichiometric reactions with alkoxy and hydroxy radicals generated from the thermolysis of the hydroperoxide.

B. Propagation

The addition of the alkyl radical (R·) to oxygen is extremely rapid, in most cases being diffusion-controlled (i.e., $k_2 > 10^9\ M^{-1}\ sec^{-1}$). At partial pressures of oxygen above 100 torr, the rate-controlling step in autoxidations is hydrogen transfer from substrate to the alkylperoxy radical described in eq 6. The rate constants for hydrogen transfer from similar compounds can be roughly correlated with the exothermicity of reaction 6. Oxidations are likely to be rapid if the bond that is formed (ROO—H) is at least as strong as that which is broken (R—H). Some pertinent bond dissociation energies are listed in Table III. The ROO—H bond strength is estimated[48] to be about 90 kcal mol^{-1}, which is larger than that for either a benzylic or an allylic C—H bond (~85 kcal mol^{-1}) or an aldehydic C—H bond (86 kcal mol^{-1}). It is comparable to a tertiary C—H bond in a saturated hydrocarbon. The relatively weak O—H, S—H, N—H, and P—H bonds of phenols, thiols, aromatic amines, and phosphines, respectively, also provide readily abstractable hydrogens.

Alkylperoxy radicals, being relatively stable and persistent, are quite selective and preferentially abstract only the most weakly bound hydrogen. The selectivity of alkylperoxy radicals is similar to that of bromine atoms [D(H—Br) = 87 kcal mol^{-1}]. The relative rates of attack at the primary, secondary, and tertiary C—H bonds of 2-methylpentane increase in the order 1 : 30 : 300.[29]

TABLE III X—H Bond Energies[a]

Compound	Energy (kcal mol^{-1})	Compound	Energy (kcal mol^{-1})
CH_3—H	103	$PhCH_2$—H	85
n-C_3H_7—H	99	RCO—H	86
i-C_3H_7—H	94	CH_3S—H	88
t-C_4H_9—H	90	CH_3PH—H	85
CH_2=CH—H	105	PhO—H	88
C_6H_5—H	103	PhNH—H	80
CH_2=CH—CH_2—H	85	ROO—H	90

[a] Data from J. A. Kerr, *Chem. Rev.* **66**, 465 (1966); S. W. Benson and R. Shaw, *in* "Organic Peroxides" (D. Swern, ed.), p. 105. Wiley (Interscience), New York, 1970.

The propagation rate constants are also dependent on the nature of the attacking alkylperoxy radical. In order to obtain a meaningful correlation of the propagation rate constants with the C—H bond energies, it is necessary to compare the rate constants for the reaction of a series of substrates RH with the same alkylperoxy radical. These rate constants can be measured experimentally by carrying out the autoxidations of various substrates RH in the presence of moderate concentrations of an alkyl hydroperoxide $R'O_2H$. Under these conditions, all of the alkylperoxy radicals derived from

RH undergo chain transfer with the added hydroperoxide,

$$RO_2\cdot + R'O_2H \longrightarrow RO_2H + R'O_2\cdot \tag{16}$$

and the rate-controlling propagation and termination steps are represented by

$$R'O_2\cdot + RH \xrightarrow{k'_p} R'O_2H + R\cdot \tag{17}$$

$$2\ R'O_2\cdot \xrightarrow{2k'_t} \text{nonradical products} \tag{18}$$

The overall rate of oxidation is given by

$$-\frac{d[O_2]}{dt} = k'_p[RH]\left(\frac{R_i}{2k'_t}\right)^{1/2} \tag{19}$$

and the determination of the absolute rate constants gives the crossed propagation rate constant k'_p. In Table IV the rate constants k'_p for the reaction of several substrates with *tert*-butylperoxy radicals are compared with the rate constants k_p for the reaction with their own peroxy radicals.

Table IV shows that the reactivities of peroxy radicals are strongly dependent on their structure. Reactivities are influenced by both steric and polar effects[21,26–28] and, in general, increase as the electron-withdrawing

TABLE IV Rate Constants per Labile Hydrogen for Reaction of Substrates with Their Own Peroxy Radicals (k_p) and with *tert*- Butylperoxy Radicals (k'_p) at 30°C[a]

Substrate	k_p (M^{-1} sec^{-1})	k'_p (M^{-1} sec^{-1})	k_p/k'_p
1-Octene	0.5	0.084	6.0
Cyclohexene	1.5	0.80	1.9
Cyclopentene	1.7	0.85	2.0
2,3-Dimethyl-2-butene	0.14	0.14	1.0
Toluene	0.08	0.012	6.7
Ethylbenzene	0.65	0.10	6.5
Cumene	0.18	0.22	0.9
Tetralin	1.6	0.5	3.2
Benzyl ether	7.5	0.3	25.0
Benzyl alcohol	2.4	0.065	37.0
Benzyl acetate	2.3	0.0075	307
Benzyl chloride	1.50	0.008	190
Benzyl bromide	0.6	0.006	100
Benzyl cyanide	1.56	0.01	156
Benzaldehyde	33,000	0.85	~40,000

[a] See Howard.[21,28]

capacity of the α substituent increases. Acylperoxy radicals, which possess a strong electron-withdrawing carbonyl group, are considerably more reactive than alkylperoxy radicals. For example, the benzoylperoxy radical is 4×10^4 times more reactive than the *tert*-butylperoxy radical toward benzaldehyde as substrate. This difference explains, in large part, the very high rates and the long chain lengths observed in the autoxidation of aldehydes. On the basis of bond dissociation energies (see Table III) alone, one would expect aldehydes and alkylaromatic compounds to be autoxidized at roughly the same rates, whereas the former actually exhibit a much higher rate of autoxidation.

C. Termination

At reasonable oxygen pressures, the termination step occurs exclusively by the self-reaction of two alkylperoxy radicals.

$$2\, RO_2\cdot \rightleftharpoons RO_4R \tag{20}$$

The modes of decomposition of tetroxides are dependent on the structure of the alkyl group.[26–28,49,50] For example, the tetroxides derived from secondary and primary alkylperoxy radicals undergo disproportionation to an alcohol and a carbonyl compound via a cyclic mechanism:[49]

$$R_2C(H)O{-}O{-}O{-}O{-}CHR_2 \text{ (cyclic)} \longrightarrow R_2C{=}O + R_2CHOH + O_2 \tag{21}$$

Such a pathway is unavailable to tetroxides derived from *tert*-alkylperoxy radicals, which undergo decomposition to dialkyl peroxides and molecular oxygen. The further thermolysis of dialkyl peroxides explains in part, the much higher rates of termination of primary and secondary alkylperoxy radicals compared to tertiary alkylperoxy radicals shown in Table V.

TABLE V Approximate Rate Constants for Various Alkylperoxy Radical Terminations at 30°C[a]

Alkylperoxy radical	$2k_t$ (M^{-1} sec^{-1})
$HO_2\cdot$	7.6×10^5
Primary, $RCH_2O_2\cdot$	10^7
Secondary, $R_2CHO_2\cdot$	10^6
Tertiary, $R_3CO_2\cdot$	10^3

[a] Data from Howard.[28]

The overall rate of autoxidation of a substrate is determined not only by the propagation rate constant k_p, but also by the termination rate constant k_t, as given by eq 9. Examination of the rate constants in Table V reveals that the lower rates of autoxidation of primary and secondary C—H functions compared to their tertiary counterparts are due not only to the lower reactivity of the C—H bonds (see Table IV) in the former, but also to the significantly higher rates of termination of primary and secondary alkylperoxy radicals. This explains why such a fairly reactive hydrocarbon as toluene [$D(PhCH_2—H) = 85$ kcal mol^{-1}] has a rather slow rate of autoxidation.

II. ALDEHYDE AUTOXIDATION

The autoxidation of aldehydes is analogous to that of hydrocarbons. Acylperoxy radicals are involved as principal chain carriers, and peracids are the primary products.

$$RCHO \xrightarrow{\text{initiation}} R\dot{C}O \quad (22)$$

$$R\dot{C}O + O_2 \longrightarrow RCO_3\cdot \quad (23)$$

$$RCO_3\cdot + RCHO \longrightarrow RCO_3H + R\dot{C}O \quad (24)$$

III. OLEFIN AUTOXIDATION

In the autoxidation of olefins, chain propagation can occur via the usual abstraction mechanism to produce the allylic radical in reaction 25 or via the addition of the alkylperoxy radical to the double bond in reaction 26.

$$RO_2\cdot + -\overset{|}{\underset{H}{C}}-\overset{|}{C}=C\langle \xrightarrow{\text{abstraction}} RO_2H + -\overset{|}{\dot{C}}-\overset{|}{C}=C\langle \quad (25)$$

$$RO_2\cdot + -\overset{|}{\underset{H}{C}}-\overset{|}{C}=C\langle \xrightarrow{\text{addition}} RO_2-\overset{|}{\underset{|}{C}}-\overset{|}{\underset{|}{C}}\cdot \quad (26)$$

Addition can be followed by the unimolecular decomposition of the β-alkylperoxyalkyl radical, affording epoxide and an alkoxy radical,

$$RO_2-\overset{|}{\underset{|}{C}}-\overset{|}{\underset{|}{C}}\cdot \longrightarrow RO\cdot + \rangle\overset{O}{C-C}\langle \quad (27)$$

or by the reaction with oxygen, giving polyperoxides.

$$RO_2-\overset{|}{\underset{|}{C}}-\overset{|}{\underset{|}{C}}\cdot + O_2 \longrightarrow RO_2-\overset{|}{\underset{|}{C}}-\overset{|}{\underset{|}{C}}-O_2\cdot, \text{ etc.} \quad (28)$$

TABLE VI Abstraction and Addition Ratios for Various Olefins

Olefin	Temp (°C)	Method	Abstraction (%)	Addition (%)
Propene	120	t-$BuO_2\cdot$[a]	60	40
	70	Autoxidation[b]	50	50
1-Hexene	120	t-$BuO_2\cdot$	85	15
	70	Autoxidation	68	32
2,3-Dimethyl-2-butene	120	t-$BuO_2\cdot$	61	39
	70	Autoxidation	56	44
α-Methylstyrene	120	t-$BuO_2\cdot$	7	93
	70	Autoxidation	0	100
Cyclopentene	70	Autoxidation	88	12
Cyclohexene	70	Autoxidation	95	5
Cycloheptene	70	Autoxidation	78	22
Cyclooctene	70	Autoxidation	30	70
Isobutene	70	Autoxidation	17	83
2-Butene	70	Autoxidation	38	62
1-butene	70	Autoxidation	73	27
3-Methyl-1-butene	70	Autoxidation	93	7

[a] From Koelewijn.[56]
[b] From Mayo.[19]

Much of the present knowledge of the addition mechanism for olefin autoxidation resulted from the studies of Mayo,[19,51–53] Brill,[54,55] and co-workers. Abstraction and addition are competing processes in olefin autoxidations. The ratio of addition to abstraction is strongly dependent on the structure of the olefin[19,56] and, of course, the oxygen pressure. The ratios of abstraction to addition have been measured for various olefins in autoxidation studies[51] and in the model reaction of *tert*-butylperoxy radical with olefins.[56] The differences with some olefins in Table VI, such as 1-hexene and propene, arise from primary or secondary alkylperoxy radicals (formed in autoxidation) which exhibit significantly different reactivities from that of the *tert*-butylperoxy radical.

IV. COOXIDATIONS

Studies of *cooxidation*, in which a mixture of two substrates is autoxidized, can provide quantitative information on the relative reactivities of peroxy radicals toward various hydrocarbons.[57–59] Cooxidations are also important from a practical viewpoint, since it is possible to utilize the peroxidic intermediates for additional oxidation processes rather than wasting this active form of oxygen. For example, in the cooxidation of aldehydes and olefins,[60,61] the active oxygen of the acylperoxy radical and the peracid are utilized for the epoxidation of the olefin.

$$\mathrm{RCHO} \xrightarrow{\text{initiation}} \mathrm{R\dot{C}O} \tag{29}$$

$$\mathrm{R\dot{C}O} + \mathrm{O_2} \longrightarrow \mathrm{RCO_3\cdot} \tag{30}$$

$$\mathrm{RCO_3\cdot} + \mathrm{RCHO} \longrightarrow \mathrm{RCO_3H} + \mathrm{R\dot{C}O} \tag{31}$$

$$\mathrm{RCO_3\cdot} + \;\rangle\mathrm{C{=}C}\langle \longrightarrow \mathrm{RCO_3{-}\overset{|}{\underset{|}{C}}{-}\overset{|}{\underset{|}{C}}\cdot} \tag{32}$$

$$\mathrm{RCO_3{-}\overset{|}{\underset{|}{C}}{-}\overset{|}{\underset{|}{C}}\cdot} \longrightarrow \mathrm{RCO_2\cdot} + \;\rangle\overset{\;O}{\mathrm{C{-}C}}\langle \tag{33}$$

$$\mathrm{RCO_2\cdot} + \mathrm{RCHO} \longrightarrow \mathrm{RCO_2H} + \mathrm{R\dot{C}O} \tag{34}$$

$$\mathrm{RCO_3H} + \;\rangle\mathrm{C{=}C}\langle \longrightarrow \mathrm{RCO_2H} + \;\rangle\overset{\;O}{\mathrm{C{-}C}}\langle \tag{35}$$

This reaction affords yields of epoxides that are much higher than those obtained from the autoxidation of the olefin alone, since acylperoxy radicals are more selective than alkylperoxy radicals in favoring addition relative to abstraction.

The addition of a second substrate to an autoxidation reaction can produce dramatic effects, as illustrated by the observation[62] that the presence of 3 mol % of tetralin reduces the rate of cumene autoxidation by two-thirds, despite the fact that tetralin itself is oxidized 10 times faster than cumene. The retardation is due to the higher rate of termination of the secondary tetralylperoxy radicals compared to the tertiary cumylperoxy radicals (see above). Thus, the small amount of tetralin in cumene is readily converted to tetralylperoxy radicals, which leads to an increase in the rate of chain termination and a consequent decrease in the rate of oxidation.

The kinetics of the autoxidation of mixtures of substrates have been discussed by Walling.[47] A large increase in the rate of oxidation of the unreactive component is observed in the presence of a small amount of a substrate readily attacked by alkylperoxy radicals, provided that the rate of termination remains essentially constant. An interesting example in which cooxidation produces a dramatic result is observed with thiols and olefins.[63] Thiyl radicals (which are inert to oxygen) participate in the addition process below, taking advantage of the rapid addition of thiyl radicals to alkenes.

$$\mathrm{RS\cdot} + \mathrm{R'CH{=}CH_2} \longrightarrow \mathrm{RSCH_2\dot{C}HR'} \tag{36}$$

$$\mathrm{RSCH_2\dot{C}HR'} + \mathrm{O_2} \longrightarrow \mathrm{RSCH_2CH(R')O_2\cdot} \tag{37}$$

$$\mathrm{RSCH_2CH(R')O_2\cdot} + \mathrm{RSH} \longrightarrow \mathrm{RSCH_2CH(R')O_2H} + \mathrm{RS\cdot} \tag{38}$$

The importance of cooxidation in relation to autoxidation cannot be overemphasized. Subsequent to the initial stages of reaction, most liquid-phase autoxidations are effectively cooxidations in the presence of secondary products such as alcohols, aldehydes, and ketones. In many commercial oxidations, small amounts of reactive substrates, such as aldehydes and ketones, are often added to provide for high initial rates of reaction.

V. GAS-PHASE VERSUS LIQUID-PHASE PROCESSES

Although the primary products of liquid-phase autoxidations of hydrocarbons are alkyl hydroperoxides, which in some cases can be isolated in high yields, gas-phase oxidations generally afford carbonyl compounds and/or dehydrogenation products. The differences are not due to changes in mechanism, but rather to the availability of alternate pathways for reaction of the alkylperoxy radical intermediates under gas-phase conditions (i.e., high temperature and low substrate concentration). A classic example lies in the gas- and liquid-phase oxidations of isobutane, which have been studied in detail by Mayo[19] and co-workers. The liquid-phase autoxidation of isobutane at 125°C affords approximately 75% *tert*-butyl hydroperoxide (TBHP), 21% *tert*-butyl alcohol, 2% acetone, and 1% isobutyl derivatives.[64] The kinetic chains are rather long, and TBHP is formed via reactions 40–41.

$$t\text{-BuH} \xrightarrow{\text{initiation}} t\text{-Bu}\cdot \tag{39}$$

$$t\text{-Bu}\cdot + O_2 \longrightarrow t\text{-BuO}_2\cdot \tag{40}$$

$$t\text{-BuO}_2\cdot + t\text{-BuH} \longrightarrow t\text{-BuO}_2\text{H} + t\text{-Bu}\cdot \tag{41}$$

$$t\text{-BuO}_2\cdot + t\text{-BuO}_2\cdot \xrightarrow{\text{termination}} t\text{-BuOO-}t\text{-Bu} + O_2 \tag{42}$$

$$t\text{-BuO}_2\cdot + t\text{-BuO}_2\cdot \longrightarrow t\text{-BuO}\cdot + O_2 + t\text{-BuO}\cdot \tag{43}$$

$$t\text{-BuO}\cdot + t\text{-BuH} \longrightarrow t\text{-BuOH} + t\text{-Bu}\cdot \tag{44}$$

Termination proceeds mainly by reaction 42, and the amount of alcohol formed is dependent on the rate of chain initiation. High rates of initiation afford high concentration of *tert*-butylperoxy radicals (cf. eqs 39–40). In the gas phase at 155°C and at low total pressures, there is insufficient isobutane to provide for efficient chain transfer via reaction 41. Consequently, the self-reaction of *tert*-butylperoxy radicals produces *tert*-butoxy radicals via reaction 43. Since there is also insufficient isobutane to sustain reaction 44, most of the *tert*-butoxy radicals undergo unimolecular fragmentation to acetone and methyl radicals. The latter react with oxygen to form methylperoxy radicals, which are more reactive in chain termination than are *tert*-butylperoxy radicals. As a result, the principal products are acetone, methanol, and *tert*-butyl alcohol.

$$t\text{-BuO}\cdot \longrightarrow \text{Me}_2\text{CO} + \text{Me}\cdot \quad (45)$$

$$\text{Me}\cdot + \text{O}_2 \longrightarrow \text{MeO}_2\cdot \quad (46)$$

$$\text{MeO}_2\cdot + t\text{-BuO}_2\cdot \longrightarrow \text{MeO}\cdot + t\text{-BuO}\cdot + \text{O}_2 \quad (47)$$

$$\text{MeO}_2\cdot + t\text{-BuO}_2\cdot \xrightarrow{\text{termination}} t\text{-BuOH} + \text{CH}_2\text{O} + \text{O}_2 \quad (48)$$

Interestingly, the transition from a liquid-phase to a gas-phase reaction is fairly smooth. Thus, an increase in the concentration of isobutane in the gas-phase oxidation at 100°C leads to an increase in the yield of *tert*-butyl hydroperoxide at the expense of *tert*-butyl alcohol, acetone, and methanol. At low rates of initiation and 13 atm of isobutane, a 92% yield of TBHP can be obtained. Similarly, the dilution of the liquid-phase reaction at 100°C with carbon tetrachloride favors reactions 42 and 43 at the expense of the chain propagation in eq 41, and the yields of *tert*-butyl alcohol and acetone increase at the expense of TBHP. At even higher temperatures (300°C) in the gas phase, the major primary product is isobutene, formed via reaction 49.

$$t\text{-Bu}\cdot + \text{O}_2 \longrightarrow \text{Me}_2\text{C}{=}\text{CH}_2 + \text{HO}_2\cdot \quad (49)$$

Although reaction 40 is faster than reaction 49 at all temperatures, it is reversible at high temperatures, whereas reaction 49 is not.

The scheme outlined above for isobutane applies to all autoxidations. In the liquid phase at relatively low temperatures, the kinetic chain lengths are long and the major products are hydroperoxides. In the gas phase, kinetic chain lengths are short and the major products are carbonyl compounds formed by thermal cleavage of alkylperoxy radical intermediates. By a suitable choice of conditions, it is possible to direct gas-phase oxidations toward the efficient production of alkyl hydroperoxides.

VI. INHIBITION OF AUTOXIDATIONS

The inhibition of autoxidations generally involves the addition of substances (inhibitors) that scavenge alkylperoxy radicals and/or destroy alkyl hydroperoxides.[65,66] For example, in the oxidation of alkylaromatic compounds it is often essential to remove traces of phenolic inhibitors in order to ensure a smooth autoxidation, with no induction period.

The most commonly used inhibitors are the substituted phenols, such as Ionol® (2,6-di-*tert*-butyl-4-methylphenol), which interrupt the autoxidation chain by forming stable phenoxyl radicals.

$$\text{RO}_2\cdot + \text{ArOH} \longrightarrow \text{RO}_2\text{H} + \text{ArO}\cdot \quad (50)$$

Hiatt[66] has discussed in detail the various types of reagents for the destruction of hydroperoxides and their mode of action. Small amounts of strong acids or strong bases are often used to effect the heterolytic decomposition of alkyl hydroperoxides. Divalent sulfur and trivalent phosphorus compounds cause inhibition by reducing alkyl hydroperoxides to the corresponding alcohols. Certain sulfur-containing metal chelates, such as zinc dithiocarbamates and dithiophosphates, are very effective in the neutralization of hydroperoxides. The detailed mechanisms of the inhibiting action of these metal complexes are not very well understood, although their inhibiting action may arise from the formation of sulfur dioxide (which causes heterolytic decomposition of the hydroperoxide) by oxidation of the ligands. In practice, the desired synergistic effect can be obtained by using a zinc dithiophosphate in combination with a peroxy radical scavenger, such as Ionol.

VII. SOLVENT EFFECTS

Autoxidations of hydrocarbons have generally been carried out in neat hydrocarbon as solvent, and few systematic studies have been made of solvent effects. In a study of the autoxidation of cyclohexane in different solvents, the dielectric constant was found to have no effect on the rate constant for propagation.[67] However, the medium strongly influenced the rate constant for termination ($RO_2\cdot + RO_2\cdot$), which can be considered an interaction of two dipoles. The rate of termination decreased with increasing dielectric constant. In this context the reported high selectivities to form hydroperoxides in the autoxidations of cyclohexane, *p*-xylene, ethylbenzene, and cumene in the presence of nitriles are an interesting development.[68]

REFERENCES

1. For reviews of early work see C. Moureu and C. Dufraisse, *Chem. Rev.* **7**, 113 (1926); N. A. Milas, *ibid.* **10**, 295 (1932).
2. H. L. J. Bäckström, *J. Am. Chem. Soc.* **49**, 1460 (1927).
3. R. Criegee, H. Pilz, and H. Flygare, *Chem. Ber.* **72**, 1799 (1939).
4. E. H. Farmer and A. Sundralingam, *J. Chem. Soc.* 121 (1942).
5. D. Swern and J. E. Coleman, *J. Am. Oil Chem. Soc.* **32**, 700 (1955).
6. J. H. Skellon, *Chem. Ind.* (*London*) p. 629 (1951); p. 1047 (1953).
7. J. L. Bolland, *Q. Rev., Chem. Soc.* **3**, 1 (1949).
8. L. Bateman, *Q. Rev., Chem. Soc.* **8**, 147 (1954).
9. H. Hock and B. Lang, *Chem. Ber.* **77**, 257 (1944).
10. L. Reich and S. S. Stivala, "Autoxidation of Hydrocarbons and Polyolefins." Dekker, New York, 1969.
11. N. M. Emanuel, E. T. Denisov, and Z. K. Maizus, "Liquid Phase Oxidation of Hydrocarbons" (B. J. Hazzard, transl.). Plenum, New York, 1967.

12. N. M. Emanuel, ed., "Oxidation of Hydrocarbons in the Liquid Phase." Pergamon, Oxford, 1965.
13. E. T. Denisov and N. M. Emanuel, *Russ. Chem. Rev.* **29**, 645 (1960).
14. W. O. Lundberg, ed., "Autoxidation and Antioxidants," Vols. 1 and 2. Wiley, New York, 1962.
15. G. Scott, "Atmospheric Oxidation and Antioxidants." Elsevier, Amsterdam, 1965.
16. I. V. Berezin, E. T. Denisov, and N. M. Emanuel, "The Oxidation of Cyclohexane" (K. A. Allen, transl.). Pergamon, Oxford, 1965.
17. E. T. Denisov, N. I. Mitskevich, and V. E. Agabekov, "Liquid Phase Oxidation of Oxygen-Containing Compounds" (D. A. Paterson, Engl. transl.). Consultants Bureau (Plenum), New York, 1977.
18. F. R. Mayo, ed., "Oxidation of Organic Compounds," Adv. Chem. Series 75, 76, 77. Am. Chem. Soc., Washington, D.C., 1968.
19. F. R. Mayo, *Accts. Chem. Rsch.* **1**, 193 (1968).
20. G. A. Russell, *J. Chem. Educ.* **36**, 111 (1959).
21. J. A. Howard, *in* "Free Radicals" (J. K. Kochi, ed.), Vol. 2, p. 3, Wiley, New York, 1973.
22. E. S. Huyser, "Free Radical Chain Reactions," p. 39. Wiley (Interscience), New York, 1970.
23. J. Betts, *Q. Rev., Chem. Soc.* **25**, 265 (1971).
24. W. G. Lloyd, *Methods Free-Radical Chem.* **4**, 1 (1973).
25. I. Sérée de Roch, *Ind. Chim. Belge* **33**, 994 (1968).
26. K. U. Ingold, *Accts. Chem. Rsch.* **2**, 1 (1969).
27. K. U. Ingold, *Pure Appl. Chem.* **15**, 49 (1967).
28. J. A. Howard, *Adv. Free-Radical Chem.* **4**, 55 (1972).
29. J. E. Bennett, D. M. Brown, and B. Mile, *Trans. Faraday Soc.* **66**, 386, 397 (1970).
30. D. Swern, ed., "Organic Peroxides," Vol. 1, Wiley, New York, 1970; Vol. 2, 1971; Vol. 3, 1973.
31. E. G. E. Hawkins, "Organic Peroxides." Van Nostrand-Reinhold, Princeton, New Jersey, 1961.
32. L. F. Martin, "Organic Peroxide Technology," Noyes Data Corp., Park Ridge, New Jersey, 1973.
33. A. G. Davies, "Organic Peroxides." Butterworth, London, 1961.
34. A. V. Tobolsky and R. B. Mesrobian, "Organic Peroxides." Wiley (Interscience), New York, 1954.
35. J. O. Edwards, ed., "Peroxide Reaction Mechanisms." Wiley (Interscience), New York, 1962.
36. G. A. Russell, *J. Am. Chem. Soc.* **78**, 1035, 1041 (1956).
37. D. J. Carlson and J. C. Robb, *Trans. Faraday Soc.* **62**, 3403 (1966).
38. L. Dulog, *Makromol. Chem.* **76**, 119 (1964).
39. A. Bromberg and K. A. Muzzket, *J. Am. Chem. Soc.* **91**, 2860 (1969).
40. T. G. Brilkina and V. A. Shushunov, "Reactions of Organometallic Compounds with Oxygen and Peroxides" (Engl. transl., A. G. Davies, ed.). Iliffe, London, 1969.
41. G. A. Russell, A. G. Bemis, E. G. Janzen, E. J. Geels, and A. J. Moye, *Adv. Chem. Series* **75**, 174 (1968).
42. G. A. Russell, A. G. Bemis, E. G. Janzen, E. J. Geels, A. J. Moye, A. J. Mak, and E. T. Strom, *Adv. Chem. Series* **51**, 112 (1965).
43. G. Sosnovsky, *in* "Organic Peroxides" (D. Swern, ed.), Vol. 3, p. 354. Wiley, New York, 1971.
44. (a) K. U. Ingold and B. P. Roberts, "Free Radical Substitution Reactions." Wiley (Interscience), New York, 1971.
    (b) J. F. Garst, *in* "Free Radicals" (J. K. Kochi, ed.), Vol. 1, p. 503, Wiley, New York, 1973.

45. S. Top, G. Jaouen, and M. McGlinchey, *J. Chem. Soc., Chem. Commun.* p. 643 (1980).
46. J. Yamashita, S. Ishikawa, and H. Hashimoto, *Prepr. Org. Chem, Div., Am. Chem. Soc.* Paper No. 76 (1979).
47. C. Walling, *J. Am. Chem. Soc.* **91**, 7590 (1969).
48. S. W. Benson, *J. Am. Chem. Soc.* **87**, 972 (1965).
49. G. A. Russell, *J. Am. Chem. Soc.* **79**, 3871 (1957).
50. P. D. Bartlett and T. G. Traylor, *J. Am. Chem. Soc.* **85**, 2407 (1963).
51. D. E. van Sickle, F. R. Mayo, and R. M. Arluck, *J. Am. Chem. Soc.* **87**, 4824 (1965).
52. D. E. van Sickle, F. R. Mayo, E. S. Gould, and R. M. Arluck, *J. Am. Chem. Soc.* **89**, 977 (1967).
53. D. E. van Sickle, F. R. Mayo, R. M. Arluck, and M. G. Syz, *J. Am. Chem. Soc.* **89**, 967 (1967).
54. W. F. Brill, *J. Am. Chem. Soc.* **85**, 141 (1963).
55. W. F. Brill and B. J. Barone, *J. Org. Chem.* **29**, 140 (1964).
56. M. Koelewijn, *Recl. Trav. Chim. Pays-Bas* **91**, 759 (1972).
57. E. Niki, Y. Kamiya, and N. Ohta, *Bul. Chem. Soc. Japan* **42**, 512 (1969).
58. F. R. Mayo, M. G. Syz, T. Mill, and J. K. Castleman, *Adv. Chem. Series* **75**, 38 (1968).
59. D. G. Hendry, *Adv. Chem. Series* **75**, 24 (1968).
60. F. Tsuchiya and T. Ikawa, *Canad. J. Chem.* **47**, 3191 (1969).
61. A. D. Vreugdenhil and H. Reit, *Recl. Trav. Chim. Pays-Bas* **91**, 237 (1972).
62. G. A. Russell, *J. Am. Chem. Soc.* **77**, 4583 (1955).
63. M. S. Kharasch, W. Nudenberg, and G. J. Mantell, *J. Org. Chem.* **16**, 524 (1951).
64. D. E. Winkler and G. W. Hearne, *Ind. Eng. Chem.* **41**, 2597 (1949).
65. W. G. Lloyd, *Chemtech.* **2**, 182 (1972).
66. R. R. Hiatt, *CRC Crit. Rev. Food Sci. Nutr.* **7**, 1 (1975).
67. G. E. Zaikov and Z. K. Maizus, *Bul. Acad. Sci. USSR, Div. Chem. Sci.* p. 267 (1969).
68. K. Tanaka and J. Imamura, *Chem. Lett.* p. 1347 (1974).

ADDITIONAL READING

L. Reich and S. S. Stivala, "Autoxidation of Hydrocarbons and Polyolefins." Dekker, New York, 1969.

W. O. Lundberg, ed., "Autoxidation and Antioxidants," Vols. I, II. Wiley (Interscience), New York, 1961.

E. S. Huyser, Kinetics of free-radical reactions. In "Physical Chemistry" H. Eyring, D. Henderson, and W. Jost, eds.), Vol. 7, Ch. 6. Academic Press, New York, 1965.

W. G. Lloyd, Autoxidation, *in* "Methods in Free-Radical Chemistry" (E. S. Huyser, Ed.), Vol. 4, p. 2ff. Marcel Dekker, 1973.

Chapter 3

Metal Catalysis in Peroxide Reactions

Reactions of metal complexes with peroxides are important in two ways: first, in the interaction with alkyl hydroperoxides, which are the important intermediates in autoxidation as described in Chapter 2, and, second, in organic synthesis, for which various types of peroxides represent useful forms of "active oxygen" reagents for oxidative transformations.

Alkyl hydroperoxides can be produced in high selectivity via autoxidation of tertiary aliphatic hydrocarbons and alkylaromatic hydrocarbons. The alkyl hydroperoxides are generally available with *tert*-butyl (TBHP), *tert*-amyl (TAHP), α-cumyl (CHP), and α-phenethyl groups. When the corresponding alcohol is readily available, the *tert*-alkyl hydroperoxide can also be prepared by the acid-catalyzed reaction of the alcohol with hydrogen peroxide.[1]

$$ROH + H_2O_2 \xrightleftharpoons{[H^+]} RO_2H + H_2O \qquad (1)$$

Hydrogen peroxide is also a readily available oxidant.[2] Organic peracids, which are available from the autoxidation of aldehydes or by the reaction of carboxylic acids with hydrogen peroxide, form another group of oxidants with wide synthetic applications.[3,4] However, they are generally used without metal catalysts and will not be dealt with here.

Reactions of peroxides mediated by metal ions can be divided into two groups involving fundamentally different mechanisms. The first class involves *homolytic*, one-electron processes in which free radicals are intermediates. The second group involves *heterolytic*, two-electron processes, and the function of the metal catalyst is to increase the electrophilicity of the hydroperoxide.

I. HOMOLYTIC CATALYSIS

Oxidation–reduction reactions of organic substrates with metal species proceeding by a one-equivalent change in the oxidation state of the metal generate free radicals as intermediates, e.g.,

$$C_6H_5{-}CH_3 + Co^{III} \longrightarrow C_6H_5{-}CH_2\cdot + Co^{II} + H^+ \qquad (2)$$

The subsequent reactions of free radicals with metal complexes can be classified as either *electron transfer* or *ligand transfer* processes, as schematically represented by the following formulations.[5–10]

Electron transfer

$$R\cdot + M^{n+} \rightleftharpoons R^+ + M^{(n-1)+} \qquad (3)$$

Ligand transfer

$$R\cdot + M^{n+}X \rightleftharpoons RX + M^{(n-1)+} \qquad (4)$$

Competition between electron transfer and ligand transfer depends to a large extent on the nature of the ligand X. A unified theory of the mechanisms of oxidation of alkyl radicals by copper(II) complexes, relating to the hard and soft acid–base classification, has been proposed.[7] Hard ligands, such as acetate, tend to favor electron transfer processes, whereas soft ligands, such as bromide, favor ligand transfer. Attack on the ligand and attack on the metal are competitive in the case of chloride, which is a borderline case.

A multitude of apparently different types of redox reactions may be classified within the general scheme contained in eqs 3 and 4. Since these reactions are influenced by changes in the redox potential of the metal complex, it is possible to change from one process to the microscopic reverse process by altering the ligands. For example, with acetate as a ligand, cobalt(II) is stable relative to cobalt(III), and in the presence of bromide ions cobalt(III) is reduced by alkyl radicals in a ligand transfer oxidation.

$$Co^{III}(OAc)_2Br + R\cdot \longrightarrow Co^{II}(OAc)_2 + RBr \tag{5}$$

On the other hand, with cyanide as a ligand, cobalt(III) is stable with respect to cobalt(II), and the microscopic reverse process obtains, i.e.,

$$[Co^{II}(CN)_5]^{3-} + RBr \longrightarrow [Co^{III}(CN)_5Br]^{3-} + R\cdot \tag{6}$$

These concepts can be applied to the catalysis of oxidation reactions by defining the role of metal ions in homolytic processes. Thus, metal oxidants such as cobalt(III) can catalyze autoxidations by generating alkylperoxy radicals from alkyl hydroperoxides, i.e.,

$$RO_2H + Co^{III} \rightleftharpoons RO_2\cdot + Co^{II} + H^+ \tag{7}$$

On the other hand, cobalt(II) species can inhibit autoxidations via the microscopic reverse process of eq 7 under certain conditions (*vide infra*).

A. Hydrogen Peroxide

The mild oxidizing action of hydrogen peroxide is considerably enhanced in the presence of certain metal catalysts.[11–23] The best known of these reagents is Fenton's reagent, which consists of ferrous salts and H_2O_2. The iron(II)-catalyzed decomposition of hydrogen peroxide proceeds via a free radical chain process involving hydroxyl radicals as transient intermediates.

Scheme I:

$$Fe^{II} + H_2O_2 \longrightarrow Fe^{III}OH + HO\cdot \tag{8}$$

$$Fe^{III} + H_2O_2 \longrightarrow Fe^{II} + HO_2\cdot + H^+ \tag{9}$$

$$Fe^{II} + HO\cdot \longrightarrow Fe^{III}OH \tag{10}$$

$$Fe^{III} + HO_2\cdot \longrightarrow Fe^{II} + O_2 + H^+ \tag{11}$$

$$HO\cdot + H_2O_2 \longrightarrow H_2O + HO_2\cdot \tag{12}$$

Since reaction 11 is energetically more favorable than reaction 9, the catalytic decomposition proceeds via the sequence of reactions in eqs 8, 11, and 12. In the reaction of metal ions with alkyl hydroperoxides (see Section II,B,1), the reaction analogous to that in eq 11 is energetically unfavorable.

In the presence of organic substrates, the hydroxyl radicals produce organic free radicals as in eq 13, which can undergo dimerization, oxidation by Fe(III), or reduction by Fe(II).

$$RH + HO\cdot \longrightarrow R\cdot + H_2O \tag{13}$$

$$R\cdot + Fe^{III} \longrightarrow [R^+] + Fe^{II} \longrightarrow \text{products} \tag{14}$$

$$R\cdot + Fe^{II} \longrightarrow [R^-] + Fe^{III} \xrightarrow{H_2O} RH \tag{15}$$

Reaction 13 is competitive with reactions 10 and 12, the latter leading to the nonproductive decomposition of H_2O_2. Thus, the yields of oxidized substrates are often low, which severely limits the synthetic utility of Fenton's reagent. This is especially true for water-immiscible substrates, since the production of hydroxyl radical in reactions 10 and 12 occurs in the aqueous phase. The hydroxyl radical is rather indiscriminate in its reactions with organic substrates. Walling[20] has measured the reactivities of various substrates (see Table I) with hydroxyl radicals relative to reaction 10 ($k_{10} = 3 \times 10^8\ M^{-1}\ \text{sec}^{-1}$). It is noteworthy that all the substrates are water soluble.

The classical redox reactions of iron(II) and iron(III) with hydrogen peroxide in Scheme I have been augmented recently by the realization that ferryl species can be formed by the further reaction of hydroxyl radicals, i.e.,

$$\text{Fe}^{\text{III}} + \text{HO}\cdot \longrightarrow \text{O}{=}\text{Fe}^{\text{IV}} + \text{H}^+$$

The high formal oxidation state of iron in these transient species (either IV or V) leads to reactive oxidants capable of many interesting reactions, as described further in Chapter 6, Section V, and in Chapter 8.

Substrate radicals (R·), which yield relatively stable carbenium ions (e.g., those with α-OH or α-OR substituents and *tert*-alkyl radicals), are efficiently oxidized by Fe(III). In contrast, allylic and benzylic radicals show marginal reactivity, and primary and secondary alkyl radicals are rather inert to Fe(III). With the latter, a small amount of Cu(II) added to the system enhances oxidation, since Cu(II) is much more effective than Fe(III) in the oxidation of alkyl radicals via electron transfer. The various pathways for reaction are illustrated by the oxidation of *tert*-butyl alcohol.[20] With

TABLE I Relative Reactivities (k_{13}/k_{10}) of Substrates toward Hydroxyl Radicals[a]

Substrate (RH)	k_{13}/k_{10}
Methanol	4.0
Ethanol	7.0
n-Propanol	11.3
Isopropyl alcohol	6.7
Isobutyl alcohol	14.0
t-Butyl alcohol	1.9
Neopentyl alcohol	12.0
Tetrahydrofuran	6.8
Acetone	0.3
Acetic acid	0.1
N,*N*-Dimethylformamide	6.0

[a] From Walling.[20]

Fenton's reagent, 2,5-dimethyl-2,5-hexanediol is produced in 84% yield by the dimerization of the intermediate primary alkyl radicals.

$$(CH_3)_3COH + HO\cdot \longrightarrow (CH_3)_2\underset{\displaystyle OH}{\underset{|}{C}}CH_2\cdot + H_2O \tag{16}$$

$$2\,(CH_3)_2\underset{\displaystyle OH}{\underset{|}{C}}CH_2\cdot \xrightarrow{\text{dimerization}} (CH_3)_2\underset{\displaystyle OH}{\underset{|}{C}}CH_2CH_2\underset{\displaystyle OH}{\underset{|}{C}}(CH_3)_2 \tag{17}$$

However, in the presence of a small amount of Cu(II), the intermediate radical is intercepted by Cu(II) to afford the diol in 86% yield.[20,23]

$$(CH_3)_2\underset{\displaystyle OH}{\underset{|}{C}}CH_2\cdot + Cu^{II} \xrightarrow{H_2O} (CH_3)_2\underset{\displaystyle OH}{\underset{|}{C}}CH_2OH + Cu^{I} + H^+ \tag{18}$$

Oxidative coupling reactions analogous to reaction 17 are also observed with Fenton's reagent and aliphatic esters, ethers, nitriles, and carboxylic acids.[16,17]

Fenton's reagent has been used for the conversion of hydrocarbons to carboxylic acids in the presence of carbon monoxide.[24]

$$RH + CO + H_2O_2 \xrightarrow{[Fe^{II}]} RCO_2H + H_2O \tag{19}$$

When *tert*-butyl alcohol is used as the substrate, β-hydroxyvaleric acid is formed in 23% yield together with 33% of 2,5-dimethyl-2,5-hexanediol.[24,25]

$$(CH_3)_3COH + CO + H_2O_2 \xrightarrow{[Fe^{II}]} (CH_3)_2\underset{\displaystyle OH}{\underset{|}{C}}CH_2CO_2H + H_2O \tag{20}$$

The hydroxyl radical can add to unsaturated substrates such as maleic acid to afford the α-carboxyalkyl radical, which is readily reduced by Fe(II) to malic acid.[20,21]

$$HO\cdot + \begin{matrix} CHCO_2H \\ \| \\ CHCO_2H \end{matrix} \longrightarrow \begin{matrix} HOCHCO_2H \\ | \\ \cdot CHCO_2H \end{matrix} \xrightarrow[H^+]{Fe^{II}} \begin{matrix} HOCHCO_2H \\ | \\ CH_2CO_2H \end{matrix} \tag{21}$$

Other radicals derived from ketones afford relatively stable anions and are reduced by Fe(II). Interestingly, when reaction 21 is carried out in the presence of methanol, a γ-lactone is formed via the following steps.[20,21]

$$HO\cdot + CH_3OH \longrightarrow H_2O + \cdot CH_2OH \tag{22}$$

$$\cdot CH_2OH + \begin{matrix} CHCO_2H \\ \| \\ CHCO_2H \end{matrix} \longrightarrow \xrightarrow[H^+]{Fe^{II}} \text{(γ-butyrolactone-4-carboxylic acid: five-membered ring O=C–O with } CO_2H\text{)} \tag{23}$$

Fenton's reagent has also been used for the hydroxylation of aromatic hydrocarbons to the corresponding phenols.[14]

$$HO\cdot + C_6H_6 \longrightarrow HO(H)C_6H_6\cdot \xrightarrow{Fe^{III}} HO-C_6H_5 + Fe^{II} + H^+ \quad (24)$$

However, yields are generally low (<20%), and biaryls are formed as byproducts via dimerization of the intermediate hydroxycyclohexadienyl radicals.

A number of other metal complexes, including those of cerium, copper, cobalt, manganese, and silver,[11–17] also decompose hydrogen peroxide via reactions analogous to eqs 8–12.

B. Alkyl Hydroperoxides

The metal-catalyzed homolytic decomposition of alkyl hydroperoxidic intermediates is the most common pathway for the catalysis of liquid-phase autoxidations. Information concerning the roles of metal complexes in oxidations has been gained from the separate studies of their interactions with alkyl hydroperoxides under nonautoxidizing conditions.[26–30] The rapid decomposition of alkyl hydroperoxides in hydrocarbon solutions in the presence of trace amounts of iron, manganese, cobalt, and copper compounds is well known. The two principal reactions of alkyl hydroperoxides with metal complexes are

Reduction

$$RO_2H + M^{(n-1)+} \longrightarrow RO\cdot + M^{n+} + HO^- \quad (25)$$

Oxidation

$$RO_2H + M^{n+} \longrightarrow RO_2\cdot + M^{(n-1)+} + H^+ \quad (26)$$

Reduction and oxidation in this manner are analogous to reactions 8 and 9, respectively, previously described for hydrogen peroxide. In inert solvents, reaction 25 or 26 can be followed by the radical chain decomposition of the hydroperoxide, i.e.,

$$2\,RO_2\cdot \longrightarrow 2\,RO\cdot + O_2 \quad (27)$$

$$RO\cdot + RO_2H \longrightarrow RO_2\cdot + ROH \quad (28)$$

Under these circumstances, the metal ion acts as an *initiator* rather than as a *catalyst*. In general, metal complexes catalyze autoxidations by generating chain-initiating radicals via reaction 25 or 26, the relative rates of which are roughly correlated with the redox potential of the $M^{n+}/M^{(n-1)+}$ couple. Table II lists the redox potentials of some metal ions that are known to react with hydroperoxides. These values pertain to aqueous solutions, since redox potentials of metal complexes in organic solvents are not generally known. Redox potentials are influenced by the nature of the ligands and the solvent.

TABLE II Redox Potentials (Aqueous Solution)

$M^{n+} + e \rightarrow M^{(n-1)+}$	E_0(V)	$M^{n+} + 2e \rightarrow M^{(n-2)+}$	E_0(V)
$Ag(II) + e \rightarrow Ag(I)$	1.98	$Pb(IV) + 2e \rightarrow Pb(II)$	1.69
$Co(III) + e \rightarrow Co(II)$	1.82	$Tl(III) + 2e \rightarrow Tl(I)$	1.25
$Ce(IV) + e \rightarrow Ce(III)$	1.61	$2\,Hg(II) + 2e \rightarrow Hg(I)_2$	0.92
$Mn(III) + e \rightarrow Mn(II)$	1.51		
$Fe(III) + e \rightarrow Fe(II)$	0.77		
$Cu(II) + e \rightarrow Cu(I)$	0.15		
$Mo(VI) + e \rightarrow Mo(V)$	~0.2[a]	$HO_2\cdot + e + H^+ \rightarrow H_2O_2$	1.68[b]
$W(VI) + e \rightarrow W(V)$	−0.03	$H_2O_2 + 2e + 2H^+ \rightarrow 2H_2O$	1.77[c]
$V(III) + e \rightarrow V(II)$	−0.20	$O_2 + e + H^+ \rightarrow HO_2\cdot$	−0.32[b]
$Ti(IV) + e \rightarrow Ti(III)$	−0.37	$O_2 + 2e + 2H^+ \rightarrow H_2O_2$	0.68[c]
$Cr(III) + e \rightarrow Cr(II)$	−0.41	$O_2 + e \rightarrow O_2^{\cdot-}$	−0.45[b]

[a] R. J. P. Williams, *Adv. Chem. Coord. Compd.* p. 279 (1961).
[b] See P. George, "Oxidases," Vol. I. Wiley, New York, 1965.
[c] See M. M. Jones, "Ligand Reactivity and Catalysis." Academic Press, New York, 1968.

Other routes that do not involve a change in the oxidation state of the metal can be envisaged[31] and are applicable to the catalytic decomposition of hydroperoxides by main group elements, such as boron and sulfur. For example, the following catalytic mechanism has been proposed to account for the catalytic effects of sulfonium compounds in cumene autoxidation.[32]

$$R_3SX + R'O_2H \rightleftharpoons R_3SOOR' + HX \qquad (29)$$

$$R_3SOOR' \longrightarrow R_3SO\cdot + R'O\cdot \qquad (30)$$

$$R_3SO\cdot + \underset{(R'O_2H)}{R'H} \longrightarrow R_3SOH + \underset{(R'O_2\cdot)}{R'\cdot} \qquad (31)$$

$$R_3SOH + HX \longrightarrow R_3SX + H_2O \qquad (32)$$

Catalysis via such a mechanism is dependent on the weakening of the peroxidic linkage by coordination to the metal.

Since alkyl hydroperoxides are fairly strong oxidants but weak reducing agents, reaction 25 is generally faster than reaction 26. When the metal ion is a strong reducing agent, reaction 25 predominates. For example, chromous ion Cr(II) reduces alkyl hydroperoxides to the corresponding alcohols in a stoichiometric reaction.[5]

$$RO_2H + Cr^{II} \longrightarrow RO\cdot + Cr^{III}OH \qquad (33)$$

$$RO\cdot + Cr^{II} \longrightarrow ROCr^{III} \qquad (34)$$

The unavailability of a convenient route for regeneration of the Cr(II) precludes a catalytic process. Copper(I) complexes similarly reduce alkyl hydroperoxides to the corresponding alcohols, but a catalytic process is possible since there are several routes available for regenerating copper(I),[5,7] including the electron transfer oxidation of alkyl radicals produced by hydrogen transfer from the solvent by alkoxy radicals, i.e.,

$$RO\cdot + R'H \longrightarrow ROH + R'\cdot \tag{35}$$

$$R'\cdot + Cu^{II} \longrightarrow R'^{+} + Cu^{I} \tag{36}$$

Alkyl radicals can also be formed by fragmentation of the *tert*-alkoxy radicals, e.g.,

$$RC(CH_3)_2O\cdot \longrightarrow R\cdot + (CH_3)_2C{=}O \tag{37}$$

In view of the low redox potential of the Cu(II)/Cu(I) couple, regeneration of Cu(I) with the hydroperoxide,

$$Cu^{II} + RO_2H \longrightarrow Cu^{I} + RO_2\cdot + H^{+} \tag{38}$$

is less likely. Hiatt has reported that alkyl hydroperoxides at room temperature are inert to cupric acetate alone.[29] Iron(II) complexes also reduce alkyl hydroperoxides (compare eq 25), leading to a series of synthetically useful reactions.

When the metal has two oxidation states of comparable stability, reactions 25 and 26 can occur concurrently. Thus, cobalt and manganese compounds are the most effective catalysts for autoxidations, since they are able to induce efficient catalytic (as opposed to stoichiometric) decomposition of alkyl hydroperoxides, e.g.,

$$RO_2H + Co^{II} \longrightarrow RO\cdot + Co^{III}OH \tag{39}$$

$$RO_2H + Co^{III} \longrightarrow RO_2\cdot + Co^{II} + H^{+} \tag{40}$$

The net transformation constitutes a catalytic decomposition of the hydroperoxide into alkoxy and alkylperoxy radicals, i.e.,

$$2\,RO_2H \xrightarrow{[Co^{II}/Co^{III}]} RO_2\cdot + RO\cdot + H_2O \tag{41}$$

Cobalt complexes are generally effective catalysts and have received the most attention.[26,33,34] Hiatt and co-workers studied the decomposition of *tert*-butyl hydroperoxide in chlorobenzene at 25°C in the presence of catalytic amounts of cobalt, iron, cerium, vanadium, and lead complexes.[29] The time required for complete decomposition of the TBHP varied from a few minutes for cobalt carboxylates to several days for lead naphthenates. In all cases, the products consisted of approximately 86% *tert*-butyl alcohol, 12% di-*tert*-butyl peroxide, and 93% oxygen. A radical-induced chain decomposition via reactions 27 and 28, initiated by a redox decomposition of the hydroperoxide, was postulated. In alkane solvents, shorter kinetic chain lengths and lower oxygen yields are observed owing to competition from hydrogen transfer with the solvent (reaction 35). Competition between the metal-ion-induced and radical-induced decompositions of hydroperoxides is influenced by several factors, such as the relative concentrations of the metal complex as well as the hydroperoxide, the nature of the solvent (see above), and the presence of radical scavengers.[31] The contribution from the metal-induced decomposition can be determined by carrying out the reaction in the presence of scavengers to inhibit the radical-induced decomposition. Thus, Kamiya *et al.*[33] showed that the initial rate of the cobalt-catalyzed decomposition of tetralin hydroperoxide, when corrected for the contribution from radical-induced decomposition by the addition of an inhibitor, was equal to the limiting rate of cobalt-catalyzed autoxidation of tetralin under the same conditions. This result demonstrates that chain initiation occurs exclusively via cobalt-catalyzed decomposition of the hydroperoxide under autoxidizing conditions.

The relative rates of reactions 39 and 40 are markedly dependent on the solvent.[31,34] In nonpolar solvents, reaction 40 constitutes the slower, rate-determining step, and the cobalt catalyst is present mainly in the trivalent oxidation state. Other workers[26] have reported that the commencement of the cobalt-catalyzed autoxidations of hydrocarbons is accompanied by the oxidation of Co(II) to Co(III). The transformation is easily observed by the change in color from pale violet or pink of Co(II) species to the dark green of Co(III) intermediates. Similarly, the commencement of manganese-catalyzed autoxidations coincides with the conversion of Mn(II) to Mn(III). The concentrations of Co(III) and Mn(III) reach a maximum during the autoxidation and then decrease. The maximum coincides with the appearance of aldehydes in the reaction mixture. The subsequent decrease in the concentrations of Co(III) and Mn(III) is attributed to the reduction by aldehydes by the reaction in eq 42, which is more facile than reduction by alkyl hydroperoxide (see eq 40).

$$Co^{III} + RCHO \longrightarrow Co^{II} + R\dot{C}O + H^+ \tag{42}$$

Rhodium and iridium complexes also rapidly decompose alkyl hydroperoxides in a catalytic sequence analogous to reactions 39 and 40.[35–42]

The metal-ion-promoted decomposition of alkyl hydroperoxides can be utilized for the introduction of the alkylperoxy group into olefins, ethers, and alkylaromatic compounds.[43,44] Copper salts are the catalysts of choice for this reaction.

$$2\,R'O_2H + RH \xrightarrow{[Cu^I/Cu^{II}]} RO_2R' + H_2O + R'OH \tag{43}$$

Oxidative substitution in this manner is analogous to the peroxy ester reaction.[45,46]

$$RH + t\text{-}BuO_2Ac \xrightarrow{[Cu^{II}/Cu^I]} ROAc + t\text{-}BuOH \tag{44}$$

Both reactions proceed via the following redox steps.[47]

Scheme II:

$$t\text{-BuO}_2\text{Ac} + \text{Cu}^{\text{I}} \longrightarrow t\text{-BuO}\cdot + \text{Cu}^{\text{II}}\text{OAc} \quad (45)$$

$$t\text{-BuO}_2\text{H} + \text{Cu}^{\text{I}} \longrightarrow t\text{-BuO}\cdot + \text{Cu}^{\text{II}}\text{OH} \quad (46)$$

$$t\text{-BuO}\cdot + \text{RH} \longrightarrow t\text{-BuOH} + \text{R}\cdot \quad (47)$$

$$\text{R}\cdot + \text{Cu}^{\text{II}} \xrightarrow{t\text{-BuO}_2\text{H}} \text{RO}_2\text{-}t\text{-Bu} + \text{H}^+ + \text{Cu}^{\text{I}} \quad (48)$$

$$\text{R}\cdot + \text{Cu}^{\text{II}} \xrightarrow{\text{HOAc}} \text{ROAc} + \text{H}^+ + \text{Cu}^{\text{I}} \quad (49)$$

These reactions have been utilized for the allylic substitution of olefins.[45,46,48]

Iron(II) complexes reduce alkyl hydroperoxides in the presence of butadiene to afford oligomeric products by the following sequence of reations.[49]

$$\text{RO}_2\text{H} + \text{Fe}^{\text{II}} \longrightarrow \text{RO}\cdot + \text{Fe}^{\text{III}}\text{OH} \quad (50)$$

RO· + (butadiene) ⟶ RO–CH₂–ĊH–CH=CH₂ (allylic radical) (51)

2 RO(allylic radical) ⟶ RO–CH₂CH=CHCH₂CH₂CH=CHCH₂–OR (52)

2 RO(allylic radical) ⟶ RO–CH₂CH=CHCH₂CH₂CH(OR)CH=CH₂ (53)

The hydrogen peroxide adduct of cyclohexanone reacts with ferrous sulfate in acidic solutions to produce the 5-carboxypentyl radical, which dimerizes to dodecanedioic acid.[50]

1-hydroxycyclohexyl hydroperoxide (HO, O$_2$H) + Fe^{II} ⟶ 1-hydroxycyclohexyloxy radical (HO, O·) + Fe^{III}OH (54)

1-hydroxycyclohexyloxy radical (HO, O·) ⟶ $\text{HO}_2\text{C(CH}_2)_4\text{CH}_2\cdot \longrightarrow \text{HO}_2\text{C(CH}_2)_{10}\text{CO}_2\text{H}$ (55)

In the presence of butadiene, the 5-carboxypentyl radical generates an allylic adduct, which dimerizes to a mixture of C_{20} dicarboxylic acids:[50]

$\text{HO}_2\text{C(CH}_2)_4\text{CH}_2\cdot$ + (butadiene) ⟶ $\text{HO}_2\text{C(CH}_2)_5$–CH₂–ĊH–CH=CH₂ (56)

$\text{HO}_2\text{C(CH}_2)_5$–CH₂–ĊH–CH=CH₂ $\xrightarrow{\text{dimerization}}$ C_{20} diacids (57)

A further modification[51] is achieved by intercepting the allylic radicals with Cu(II) to afford ω-hexenoic acid (eq 58). The facile ligand transfer oxidation of alkyl radicals may be employed to produce the corresponding ω-halo acids (eq 59).

$$HO_2C(CH_2)_4CH_2\cdot + Cu^{II} \longrightarrow HO_2C(CH_2)_3CH{=}CH_2 + Cu^{I} + H^{+} \quad (58)$$

$$HO_2C(CH_2)_4CH_2\cdot + CuCl_2 \longrightarrow HO_2C(CH_2)_5Cl + CuCl \quad (59)$$

These reactions have been utilized for the synthesis of a wide variety of polyfunctional long-chain molecules.[52]

The introduction of remote double bonds is possible in the catalytic decomposition of alkyl hydroperoxides,[53] e.g.,

$$R{-}O_2H + Fe^{II} \longrightarrow R{-}O\cdot + Fe^{III}OH \quad (60)$$

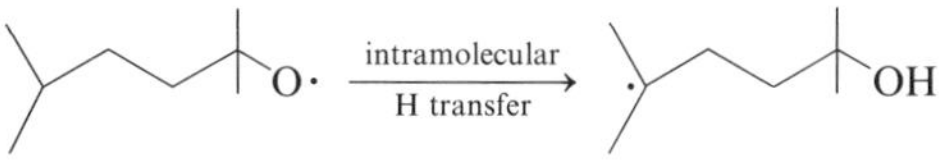

$$R{-}O\cdot \xrightarrow[\text{H transfer}]{\text{intramolecular}} \cdot R{-}OH \quad (61)$$

$$\cdot R{-}OH \xrightarrow{Cu^{II}} {-}OH + {-}OH \quad (62)$$

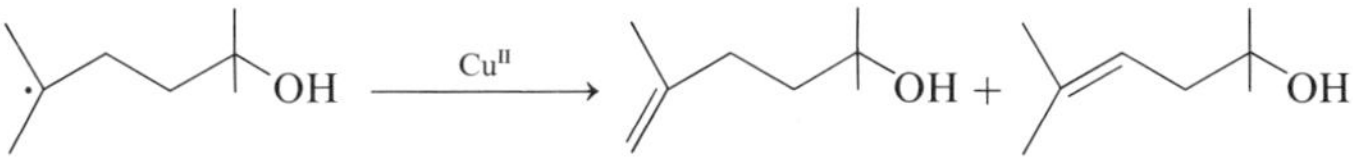

The copper-catalyzed decomposition of α-alkoxy hydroperoxides has been used in the efficient synthesis of the naturally occurring macrolide recifeiolide,[53a]

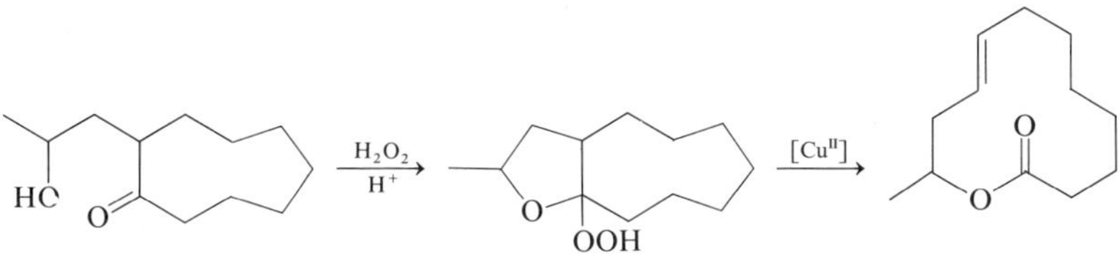

$$\xrightarrow[H^{+}]{H_2O_2} \quad \text{OOH} \quad \xrightarrow{[Cu^{II}]}$$

in which the key step is the β-scission of the alkoxy radical:

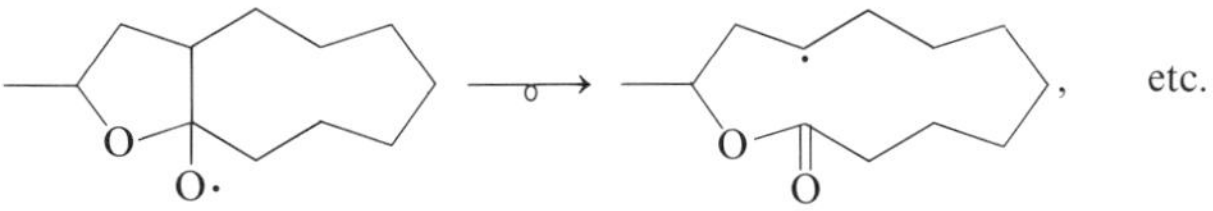

$$\text{O}\cdot \longrightarrow \quad , \quad \text{etc.}$$

C. Peroxy Acids

Metal ions can catalyze the decomposition of peracids via redox reactions analogous to those observed with alkyl hydroperoxides.[34] There is, however, some ambiguity as to whether the reductive cleavage takes place by eq 63 or eq 64.

$$RCO_2OH + M^{(n-1)+} \longrightarrow RCO_2M^{n+} + HO\cdot \quad (63a)$$

$$RCO_2OH + M^{(n-1)+} \longrightarrow RCO_2\cdot + M^{n+}OH \quad (63b)$$

$$RCO_2OH + M^{n+} \longrightarrow M^{(n-1)+} + RCO_3\cdot + H^+ \quad (64)$$

It is likely that reactions 63a and 63b are competing pathways, their relative contributions being solvent dependent.

Oxidation of Co(II) or Mn(II) complexes by peracids is facile,[26] but reduction by reaction 64 proceeds much more slowly, if at all. During the metal-catalyzed autoxidations of aldehydes, the reduced form of the catalyst is preferentially formed by direct oxidation of the aldehyde in eq 42. Reaction 43 is the slower, rate-determining step in aldehyde oxidations, since the cobalt catalyst is present largely as Co(III). In contrast to the reactions with hydrogen peroxide and alkyl hydroperoxides, the metal-catalyzed reactions of peracids are of limited synthetic utility.

A recent spectroscopic study of cobalt(II) acetate and peracetic acid in 95% acetic acid–water solutions has indicated the presence of two distinct phases of color changes.[54] The initial stage from pink to apple green is complete within a few seconds at 2°C and is kinetically first order in peracid and second order in cobalt(II). Coupled with the rapid monomer–dimer equilibration of cobalt(II) acetate, which favors the monomer,[55] the following sequence of changes was proposed:

$$2\ Co^{II} \rightleftharpoons (Co^{II})_2$$

$$(Co^{II})_2 + ClC_6H_4CO_3H \xrightarrow{2H^+} (Co^{III})_2 + ClC_6H_4CO_2H + H_2O \quad (65)$$

At this juncture, the bonding (and ligands) in the dimeric cobalt(III) species formed in eq 65 is undetermined; however, it is slowly converted to a more stable olive green form, which is considered to be similar to cobaltic acetate previously prepared by a variety of methods.[56] The corresponding peracid oxidation of manganese(II) acetate to the manganese(III) species shows an unusual autocatalytic behavior, which has been attributed to the reaction sequence:

$$(Mn^{III})_2 + ClC_6H_4CO_3H \xrightarrow{2H^+} 2\ Mn^{IV} + ClC_6H_4CO_2H + H_2O$$

$$Mn^{IV} + Mn^{II} \longrightarrow (Mn^{III})_2$$

and not to the direct oxidation of manganese(II), i.e.,

$$Mn^{II} + ClC_6H_4CO_3H \longrightarrow Mn^{IV} + ClC_6H_4CO_2H$$

$$Mn^{IV} + Mn^{II} \longrightarrow (Mn^{III})_2$$

According to this study, the peracid oxidation of both cobalt(II) and manganese(II) proceeds preferentially via the oxidation of binuclear metal acetates and not by the sequence of one-electron changes in eqs 63 and 64.[57]

D. Kinetics of Autoxidation Involving Redox Initiation

Chain initiation during metal-catalyzed autoxidations of hydrocarbons generates alkoxy and alkylperoxy radicals from the reaction of alkyl hydroperoxides with the metal catalyst (see eqs 39 and 40). Since alkylperoxy radicals are strong oxidants, they are capable of oxidizing the reduced form of the metal catalyst, e.g.,[26,58]

$$RO_2\cdot + Co^{II} \underset{}{\overset{k_t'}{\rightleftharpoons}} RO_2Co^{III} \tag{66}$$

Transition metal complexes, especially in media of low polarity such as neat hydrocarbons, often behave as catalysts at low concentrations but as inhibitors at high concentrations. This phenomenon, referred to as catalyst–inhibitor conversion, manifests itself in the long induction periods often observed in metal-catalyzed autoxidations in nonpolar media.[59,60] The induction period is effectively eliminated when the alkyl hydroperoxide concentration is greater than that of the metal.[61] In practice, a small amount of an alkyl hydroperoxide such as TBHP is generally added to metal-catalyzed autoxidations in order to eliminate any induction period.

At high cobalt concentrations, Co(II) competes effectively with the substrate RH for the alkylperoxy radicals and obviates chain propagation by

$$RO_2\cdot + RH \longrightarrow RO_2H + R\cdot \tag{67}$$

Under these conditions, termination proceeds virtually exclusively via reaction 66, rather than by the self-reaction of alkylperoxy radicals. The expression for the chain length in a process involving termination via the self-reaction of $RO_2\cdot$ is

$$\text{Chain length} = k_p[RH]/(2k_tR_i)^{1/2} \tag{68}$$

However, if termination occurs via reaction 66, it becomes

$$\text{Chain length} = k_p[RH]/k_t'[Co^{II}] \tag{69}$$

In other words, if reaction 66 predominates, catalyst–inhibitor conversion is observed when the chain length becomes less than unity, i.e., $[Co(II)] > k_p[RH]/k_t'$. In practice, an abrupt transition from catalysis to inhibition is generally observed for hydrocarbon autoxidations catalyzed by Co, Mn, Fe, Cu, etc.[59,61] For example, in the cobalt-catalyzed autoxidation of neat tetralin at 65°C, an abrupt transition from rapid reaction to inhibition was observed[61] at a catalyst concentration of approximately 0.1 *M*. Such puzzling kinetic phenomena associated with metal-catalyzed autoxidations have been recently analyzed by Black[62] on the basis of metal–hydroperoxide complexes. Chain initiation via the unimolecular decomposition of an iron–hydroperoxide complex was originally proposed by Banks *et al.*[63] and later generalized to other transition metals by Chalk and Smith.[64] There is now more experimental evidence in support of such complexes in the initiation of metal-catalyzed autoxidations.[62,65] Using cobalt as an example, initiation is assumed to involve the following steps:[62]

Scheme III:

$$Co^{II} + RO_2H \rightleftharpoons [Co^{II}\,RO_2H] \tag{70}$$

$$[Co^{II}\,RO_2H] \longrightarrow RO\cdot + Co^{III}OH \tag{71}$$

$$Co^{III} + RO_2H \rightleftharpoons [Co^{III}\,RO_2H] \tag{72}$$

$$[Co^{III}\,RO_2H] \longrightarrow RO_2\cdot + Co^{II} + H^+ \tag{73}$$

In media of low polarity, the cobalt is tied up as the complex with the alkyl hydroperoxide in excess. As the concentration of cobalt is increased, however, a point is reached at which $[Co] > [RO_2H]$, and the equilibrium concentration of the uncomplexed Co(II) increases suddenly to the point at which the oxidation can no longer proceed. In polar protic solvents such as acetic acid, the catalyst is mainly associated with the solvent molecules present in much higher concentrations than RO_2H, e.g.,

$$[Co^{II}\,RO_2H] + HOAc \longrightarrow [Co^{II}\,HOAc] + RO_2H \tag{74}$$

Consequently, the phenomena of abrupt catalyst–inhibitor conversion and long induction periods are generally restricted to media of low polarity.

Another kinetic observation associated with metal-catalyzed autoxidations is the phenomenon of the *limiting rate*, whereby reaction rates ultimately level off at some limiting values. This behavior was attributed by Tobolsky[66] to the attainment of a steady-state concentration of alkyl hydroperoxides, in which the rate of decomposition is counterbalanced by its rate of formation. Indeed, Woodward and Mesrobian[67,68] showed that the concentration of hydroperoxide in the cobalt-catalyzed autoxidation of tetralin in acetic acid ceased to increase once the limiting rate was attained. However, other investigators have found that, in media of low polarity, the hydroperoxide concentration increases steadily throughout the region of maximum rate.[33,61,69] This apparent anomaly has also been rationalized by Black[62] to be a consequence of the formation of metal–hydroperoxide complexes.

Walling[70] has derived the following general expression for the limiting rate:

$$-\frac{d[\mathrm{RH}]}{dt} = \frac{n}{f}\frac{k_\mathrm{p}^2}{2k_\mathrm{t}}[\mathrm{RH}]^2 \tag{75}$$

where n is the number of initiating radicals produced for each RO_2H decomposed and f is the fraction of RH consumed by $RO_2\cdot$ attack. When initiation proceeds by the reactions in eqs 71 and 73, $n = 1$ and $f = \frac{2}{3}$, giving

$$-\frac{d[\mathrm{RH}]}{dt} = \frac{3k_\mathrm{p}^2[\mathrm{RH}]^2}{4k_\mathrm{t}} \tag{76}$$

When metal-catalyzed autoxidations are carried out in nonpolar media, the phenomenon of *catalyst deactivation* is often observed, since the catalyst is extremely sensitive to polar substances formed during the reaction, especially in hydrocarbon solvents. For example, in the oxidation of cyclohexane catalyzed by cobalt stearate, an insoluble precipitate of cobalt adipate is formed.[26] The catalyst may also precipitate as an insoluble oxide or hydroxide during the reaction. These problems can be minimized in a polar protic medium, usually acetic acid.

Metal catalysts are usually added in the form of readily available, hydrocarbon-soluble salts such as stearates, naphthenates, octanoates, and acetylacetonates (acac). When reactions are carried out in acetic acid, the metal acetate is commonly employed. Other types of complexes, such as metal phthalocyanines and organometallic complexes, have been used. It is doubtful that the ligand has more than a minor effect in most autoxidations, especially when one considers that the ligand is in many cases destroyed during the early stages of reaction.[31] Sometimes these catalysts are used in a heterogeneous form, e.g., cobalt on silica, but the fundamental processes involved remain the same.

Transition metal carbonyls show interesting properties as homolytic catalysts in the oxidation of ketones and alcohols.[71] Cyclohexanone is oxidized to adipic acid under autoxidizing conditions with either $Rh_6(CO)_{16}$ or $Re_2(CO)_{10}$.

$$\text{cyclohexanone} + O_2 \xrightarrow{[M_n(CO)_m]} \text{HO}_2\text{C}(\text{CH}_2)_4\text{CO}_2\text{H}, \quad \text{etc.}$$

Since there is no evidence for complexation of dioxygen to either $Rh_6(CO)_{12}$ or $Re_2(CO)_{10}$, it was deemed unlikely that autoxidation involves oxygen activation. Instead, it was proposed that the metal carbonyls are involved in peroxide decomposition via a homolytic process. It is noteworthy that the dimeric $Re_2(CO)_{10}$ remains intact throughout the catalytic cycle since it can be recovered quantitatively upon completion of the autoxidation, even in the absence of a carbon monoxide atmosphere. The sequence of intermediates proposed for the catalytic oxidation is shown below.

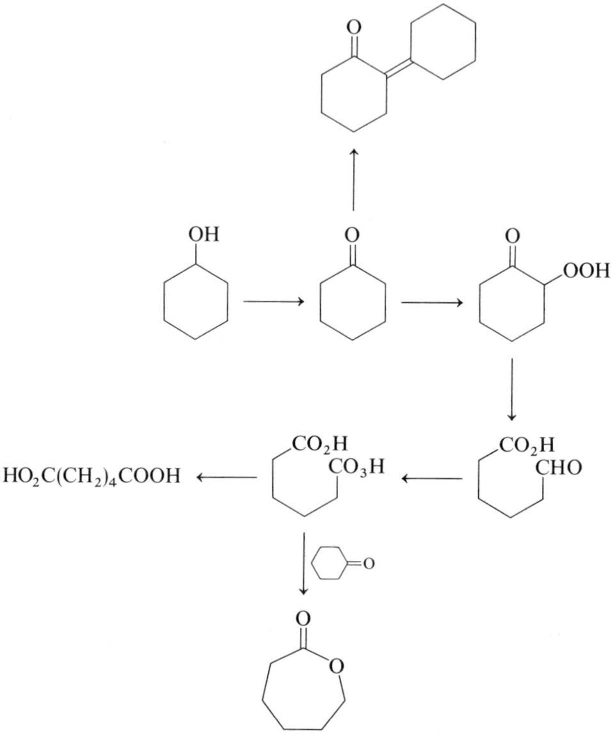

Cyclohexanol is similarly autoxidized to adipic acid in the presence of these metal carbonyls via the cyclohexanone intermediate.

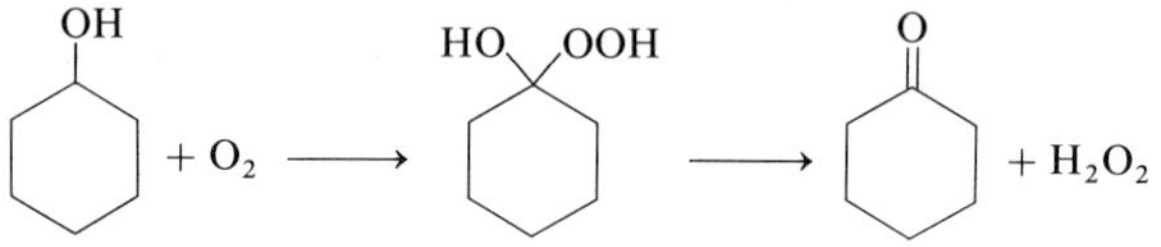

II. HETEROLYTIC CATALYSIS

Although the transition metal complexes of Co, Mn, Fe, and Cu facilitate homolysis of the peroxidic bonds, a second group of catalysts, which includes transition metals such as V, Mo, W, and Ti, can promote the heterolysis of the O—O bond in hydrogen peroxide and alkyl hydroperoxides. These metal–hydroperoxide combinations form an important class of reagents for a variety of selective oxidations.

A. Hydrogen Peroxide

Many acidic metal oxides, such as MoO_3, WO_3, V_2O_5, and SeO_2, catalyze the reactions of hydrogen peroxide through the formation of inorganic peracids.[72] These reagents closely resemble organic peracids and readily undergo heterolysis at the O—O bond with nucleophiles. Although hydrogen peroxide is a relatively weak electrophile, substitution of hydrogen by an electron-withdrawing carbonyl or oxometal (M=O) group renders the peroxidic oxygens more electrophilic.

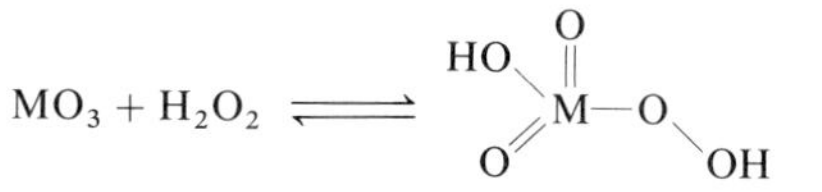

$$MO_3 + H_2O_2 \rightleftharpoons (HO)(O{=})M({=}O)\text{—O—OH} \tag{77}$$

Compare:

$$RCO_2H + H_2O_2 \rightleftharpoons R\overset{\overset{\displaystyle O}{\|}}{C}OOH + H_2O \tag{78}$$

The examples shown below include persulfuric acid (Caro's acid), permolybdic acid, pertungstic acid, perselenous acid, and perboric acid, respectively. In some cases, e.g., perboric acid, the peracid has not been isolated but only postulated as a reactive intermediate.

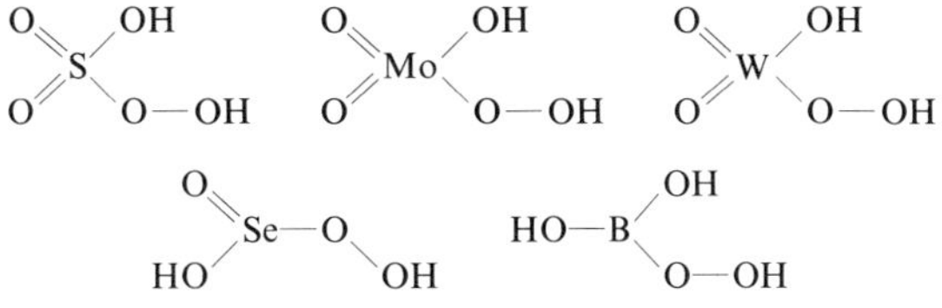

$$(O{=})_2S(OH)\text{—O—OH} \qquad (O{=})_2Mo(OH)\text{—O—OH} \qquad (O{=})_2W(OH)\text{—O—OH}$$

$$(O{=})Se(HO)\text{—O—OH} \qquad HO\text{—}B(OH)\text{—O—OH}$$

These peracids are all hydroxy hydroperoxides that are formed via a general reaction involving the addition of H_2O_2 to an M=O group.

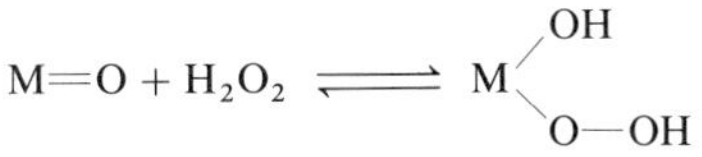

$$M{=}O + H_2O_2 \rightleftharpoons M(OH)\text{—O—OH} \tag{79}$$

In each of these inorganic peracids, the conjugate base of the acid provides an excellent leaving group for nucleophilic displacement. For example, the oxidation of iodide ion by hydrogen peroxide is catalyzed by, *inter alia*, molybdenum compounds through the intermediacy of permolybdic acid.

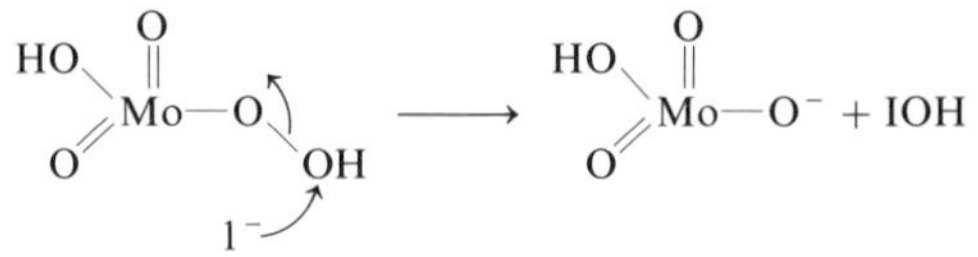

$$(HO)(O{=})Mo(=O){-}O{-}OH + I^- \longrightarrow (HO)(O{=})Mo(=O){-}O^- + IOH \quad (80)$$

These reagents were first used in organic chemistry for the hydroxylation of olefins to 1,2-glycols, a reaction discovered by Milas.[73]

$$RCH{=}CHR' + H_2O_2 \longrightarrow RCH(OH){-}CH(OH)R' \quad (81)$$

The preferred catalysts were CrO_3, RuO_4, MoO_3, V_2O_5, and OsO_4, the last being the catalyst of choice. It was later shown by Mugdan and Young[74] that the OsO_4-catalyzed hydroxylation (and that catalyzed by RuO_4) involves a fundamentally different mechanism from that catalyzed by MoO_3 and WO_3. Thus, the *cis*-glycol is formed via a cyclic osmate ester.

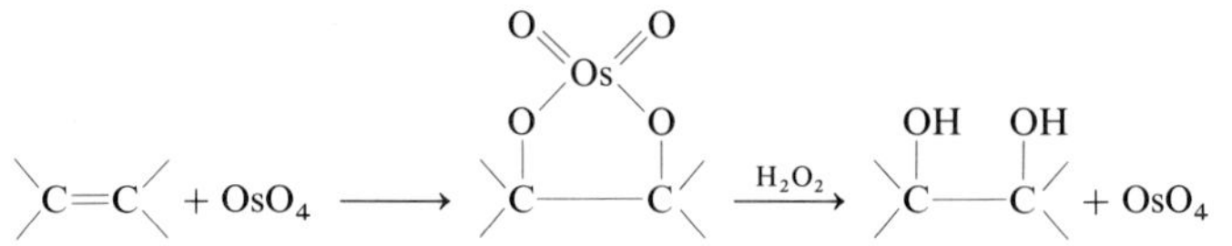

$$>C{=}C< + OsO_4 \longrightarrow >C(O){-}C(O)< \text{ [cyclic osmate ester, } O_2Os \text{]} \xrightarrow{H_2O_2} >C(OH){-}C(OH)< + OsO_4 \quad (82)$$

Equation 82 represents an example of an olefin oxidation by an oxometal reagent, in which H_2O_2 is employed as the terminal oxidant (see Chapter 6).

Reactions with WO_3, MoO_3, SeO_2, etc., on the other hand, involve inorganic peracids and afford the *trans*-glycol via an epoxide intermediate, which is hydrolyzed under the acidic reaction conditions. These reagents can be used under basic or neutral conditions for the epoxidation of olefins.[75–79] Similar mechanisms probably pertain to the epoxidation of olefins by organic and inorganic peracids via transition states (a) and (b), respectively.

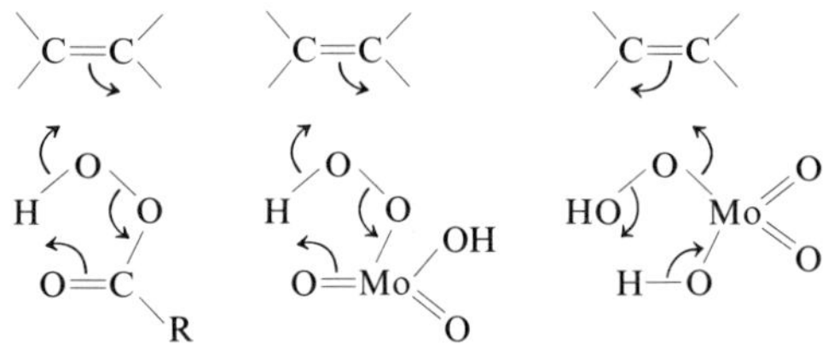

(a) $>C{=}C<$ / H–O–O–C(=O)R (b) $>C{=}C<$ / H–O–O–Mo(=O)(=O)OH (c) $>C{=}C<$ / HO–O–Mo(=O)(=O)–O–H

(a) (b) (c)

Alternatively, a transition state (c) can be envisaged in which a hydroxyhydroperoxide structure is involved.[80] Interestingly, an analog of transition state (c) has been described in the epoxidation with H_2O_2 in the presence of hexafluoroacetone:[81]

$$\begin{array}{c} \text{H} \\ \text{O}^{\nearrow} \\ \text{F}_3\text{C}\diagdown\ | \\ \text{C}\diagdown\ \ \text{OH} \\ \text{F}_3\text{C}\diagup\ \ \text{O}\diagup \\ \diagdown\text{C}=\text{C}\diagup \end{array} \longrightarrow (CF_3)_2C{=}O + \underset{\diagup\quad\diagdown}{\overset{\diagdown\ \overset{O}{\diagup\diagdown}\ \diagup}{C{-}C}} + H_2O \tag{83}$$

This reagent has also been utilized recently for the oxidation of aldehydes to carboxylic acids, amines to amine oxides, and dialkyl sulfides to the corresponding sulfoxides.[81b]

The metal oxides that form inorganic peracids exhibit the following order of reactivity: W > Mo, Se ≫ Ti, Ta.[74] The latter two show only faint activity owing to the low solubility in the reaction medium. The oxides V_2O_5 and CrO_3 were nonselective, causing much nonproductive decomposition of the H_2O_2 via homolytic pathways (see the previous section). On the basis of their electrophilic properties, one would not expect any large difference to exist between H_2O_2 and RO_2H (see following section) as epoxidizing agents in conjunction with metal catalysts. However, in practice, the combination of RO_2H and metal catalyst is generally far superior to that of H_2O_2 and metal catalyst. This apparent anomaly can be attributed to the presence of water in H_2O_2 systems.[80,82] Hydrogen peroxide is generally employed as an aqueous solution, and water is known to retard the epoxidation reaction seriously by complexing with the catalyst (see following section). Unfortunately, the complete exclusion of water is not possible since it is a product of the reaction. In addition to retarding the rate of reaction, water can also have a deleterious effect on the selectivity. The epoxide is often readily converted to the corresponding glycol by reaction with H_2O and/or H_2O_2. These reactions are catalyzed by the same oxometal catalysts that catalyze the epoxidation step, e.g.,[83]

$$(CH_3)_2C{=}CH_2 + H_2O_2 \xrightarrow{[Mo^{VI}]} (CH_3)_2\overset{\overset{O}{\diagup\ \diagdown}}{C{-}CH_2} + H_2O \tag{84}$$

$$(CH_3)_2\overset{\overset{O}{\diagup\ \diagdown}}{C{-}CH_2} + H_2O_2 \xrightarrow{[Mo^{VI}]} (CH_3)_2\underset{\underset{OH}{|}}{C}{-}CH_2O_2H \tag{85}$$

$$(CH_3)_2\underset{\underset{OH}{|}}{C}{-}CH_2O_2H + (CH_3)_2C{=}CH_2 \xrightarrow{[Mo^{VI}]} (CH_3)_2\overset{\overset{O}{\diagup\ \diagdown}}{C{-}CH_2} + (CH_3)_2\underset{\underset{OH}{|}}{C}{-}CH_2OH \tag{86}$$

In the original investigations[75–79] of metal-catalyzed epoxidations with H_2O_2, high rates and selectivities were generally obtained with water-soluble olefins, such as the allylic alcohols. These substrates represent special cases, since they form complexes with the catalyst (see following section), thus facilitating oxygen transfer as depicted in the transition state below.

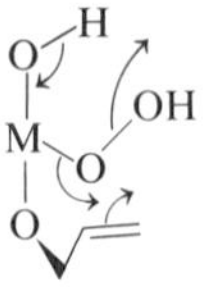

In contrast, simple olefins such as propylene and cyclohexene generally afford poor yields of epoxides with these reagents, mainly owing to the strong inhibiting effect of water noted above. For a valid comparison of RO_2H and H_2O_2, the reactions of the latter should be carried out in the absence of water. This was achieved by Schirmann and co-workers, who showed that simple epoxides can be obtained in reasonable yields with commercial 30–70% H_2O_2 when the water is continuously removed during the reaction by azeotropic distillation.[82,84–87] Good results were obtained in polyether solvents (e.g., dioxane) using Mo and W compounds,[84] As_2O_3,[85] and B_2O_3[86] as catalysts. Propylene oxide, for example, was obtained in 85% yield using 70% H_2O_2 and $MoO_2(acac)_2$ as the catalyst.[84] Interestingly, metaboric acid (HBO_2) is not only a good epoxidation catalyst, which is very soluble in polyether solvents, but also a good dehydrating agent. Thus, when HBO_2 was used in stoichiometric amounts, the results obtained were the same as those obtained with catalytic amounts of H_2O_2 in conjunction with continuous water removal.[82] The *o*-boric acid by-product is easily dehydrated to metaboric acid.

$$\rangle C{=}C\langle + H_2O_2 + HBO_2 \longrightarrow \rangle C\overset{O}{-}C\langle + H_3BO_3 \qquad (87)$$

$$H_3BO_3 \longrightarrow HBO_2 + H_2O \qquad (88)$$

When this system was used in dioxane at 90°C, propylene was epoxidized in 99% selectivity at 31% conversion.[82] Another apparent anomaly in these systems is the reversal of activity observed with Mo and W catalysts, the former being the more active with RO_2H and the latter being the more active with H_2O_2.[75–79] This difference arises from the retarding effect of water, which is more serious with Mo catalysts than with W.[80] When water is removed from the system, the expected order of reactivity, Mo > W, is observed.[84]

The catalytic effect of arsenic compounds has also been reported by McMullen,[88] who used highly concentrated H_2O_2 without water removal. Jacobson and co-workers[89] described the use of arsonated polystyrene resins Ⓟ—$AsO(OH)_2$ as catalysts for olefin epoxidation. Reactions were carried out in both biphase (water-immiscible solvent) and triphase systems (water-immiscible solvent) using 90% H_2O_2. The following mechanism, equivalent to the mechanism in the transition state (b) above, was suggested.

$$\text{(P)—As(=O)(OH)OH} + H_2O_2 \rightleftharpoons \text{(P)—As(=O)(OH)O—OH} + H_2O$$

$$\text{(P)—As(=O)(OH)O—OH} \xrightarrow{\;>C=C<\;} \left[\text{(P)—As(=O)(OH)—O—O(H)}\cdots\text{C=C}\right] \longrightarrow \text{>C—C<(O)} + \text{(P)—As(=O)(OH)OH}$$

The same workers found[90] that these arsonated polystyrenes also catalyzed the Baeyer–Villiger oxidation of ketones by aqueous hydrogen peroxide, another reaction characteristic of organic peracids. In contrast to the epoxidation and hydroxylation of olefins, this reaction is not generally achieved with the combination of H_2O_2 and metal catalyst. It was reasoned[90] that arylarsonic acids have the same pK_a (3.82) as *m*-chlorobenzoic acid, the peroxy form of which is the most common reagent used in the Baeyer–Villiger oxidation of ketones. In addition, the C—As bond in benzenearsonic acid ($PhAsO_3H$), unlike the C—Se bond in the analogous benzeneseleninic acid ($PhSeO_2H$), is very stable toward nucleophilic and electrophilic reagents. It was found that benzenearsonic acid is an efficient catalyst for the oxidation of cyclic ketones to lactones. Immobilization of the catalyst on a polystyrene support provided the additional advantage of allowing the facile separation of the toxic catalyst. The following mechanism was suggested for the reaction.

$$\text{(P)—As(=O)(OH)OOH} \xrightarrow{\text{RC(=O)R}'} \text{(P)(OH)As(=O)—O—O—C(R)(R)—O—H} \longrightarrow \text{R}'\text{C(=O)OR} + \text{(P)—As(=O)(OH)OH}$$

Molybdenum–peroxo complexes stabilized by picolinate and pyridine-2,6-dicarboxylate ligands have also been found to catalyze the Baeyer–Villiger oxidation of cyclic ketones to lactones with 90% H_2O_2.[91] The following mechanism was suggested (the chemistry of peroxometal complexes is discussed in more detail in Chapter 6).

$$\text{Mo=O} + H_2O_2 \longrightarrow \text{Mo}\langle\text{O—O}\rangle + H_2O$$

$$\text{Mo}\langle\text{O—O}\rangle \xrightarrow{(CH_2)_nC=O} \text{Mo(—O—O—C(CH}_2)_n\text{—O—)} \longrightarrow \text{Mo=O} + (CH_2)_nC(=O)O$$

Arylseleninic acids have been shown[92] to catalyze the epoxidation of olefins with H_2O_2, presumably via formation of the perseleninic acid,

$$\mathrm{ArSe}\begin{matrix}{\scriptstyle\diagup\!\!\!\diagup O}\\{\scriptstyle\diagdown OH}\end{matrix} + H_2O_2 \rightleftharpoons \mathrm{ArSe}\begin{matrix}{\scriptstyle\diagup\!\!\!\diagup O}\\{\scriptstyle\diagdown OOH}\end{matrix} + H_2O \tag{89}$$

which compares with

$$SeO_2 + H_2O_2 \rightleftharpoons HO{-}\mathrm{Se}\begin{matrix}{\scriptstyle\diagup\!\!\!\diagup O}\\{\scriptstyle\diagdown OOH}\end{matrix} + H_2O \tag{90}$$

The same inorganic-peracid-forming metal oxides and related compounds which catalyze the epoxidation and hydroxylation of olefins also catalyze other reactions that are characteristic of organic peracids, such as the oxidation of tertiary amine oxides and sulfides to sulfoxides.

$$R_3N + H_2O_2 \longrightarrow R_3NO + H_2O \tag{91}$$

$$R_2S + H_2O_2 \longrightarrow R_2SO + H_2O \tag{92}$$

Thus, tungsten[93] and vanadium[94] compounds catalyze the oxidation of amines (reaction 91) and sulfides (reaction 92) by H_2O_2. Selenium dioxide[95] and arylseleninic acids[92a] (*vide supra*) are also effective catalysts for the H_2O_2 oxidation of sulfides to sulfoxides.

DiFuria and Modena[96,97] carried out a detailed study of the vanadium(V)-catalyzed oxidation of sulfides in alcohol solvents. Under these conditions, hydrogen peroxide was calculated to be at least 100 times as effective as alkyl hydroperoxides for sulfide oxidation. The catalyst $VO(OR)_3$ associates much more strongly with H_2O_2 to form the peroxyvanadate [see $VO(OR)_2OOH$ in the reactions below] than with RO_2H. A diperoxovanadate is also formed, but it is a less effective oxidant, since it is a strong acid and exists largely in its anionic form, which is a poor electrophile.

$$(RO)_2V(=O){-}OR + H_2O_2 \underset{}{\overset{-ROH}{\rightleftharpoons}} (RO)_2V(=O){-}O{-}OH \overset{H_2O_2}{\rightleftharpoons} (RO)(HO{-}O)V(=O){-}O{-}OH + ROH$$

$$(RO)(HO{-}O)V(=O){-}O{-}OH \overset{-H^+}{\rightleftharpoons} (RO)(HO{-}O)V(=O){-}O{-}O^- \rightleftharpoons \left[O{=}V(O_2)_2\right]^- + ROH$$

This catalytic system was not effective for the epoxidation of olefins, mainly owing to a high rate of autodecomposition of H_2O_2.

Cyclohexanone oxime is produced by the reaction of cyclohexanone with ammonia and hydrogen peroxide in the presence of tungstic acid as catalyst.[98] The key step in this reaction is probably a W(VI)-catalyzed epoxidation of cyclohexanone imine, i.e.,

$$\text{cyclohexanone} + NH_3 \xrightarrow[-H_2O]{H^+} \text{cyclohexanone imine (C=NH)} \xrightarrow[\text{[W}^{VI}\text{]}]{H_2O_2} \text{oxaziridine (HN–O)} \longrightarrow \text{cyclohexanone oxime (C=NOH)} \tag{93}$$

Primary aliphatic amines containing an α-hydrogen atom are oxidized to ketoximes (60–80% yield) with H_2O_2 in the presence of tungstate, molybdate, or vanadate catalysts.[93,99,100]

$$R_2CH\text{—}NH_2 \xrightarrow[\text{[MoO}_4^{2-}\text{ or WO}_4^{2-}\text{]}]{H_2O_2} R_2C\text{=}NOH \tag{94}$$

In the presence of organic bases, the inorganic peracids undergo cyclization to afford stable cyclic peroxides in many cases.

$$\text{(O=)M(OOH)(OH)} \xrightarrow{\text{[base]}} \text{(O=)M(O}_2\text{)} + H_2O \tag{95}$$

These complexes have also been used as selective oxidizing agents, e.g., olefin epoxidation. Since they bear structural resemblance to other metal peroxides formed by reaction of metal complexes with molecular oxygen, they are described in the following chapter.

Finally, organometallic hydroperoxides of Sn, Pb, Ge, etc., formed *in situ* by the facile reaction of the corresponding hydroxy derivative with H_2O_2, have been used for the catalytic epoxidation of olefins.[72,101–103] Since the hydroxy derivative is regenerated, it can be used in catalytic amounts:

$$R_3MOH + H_2O_2 \rightleftharpoons R_3MO_2H + H_2O. \tag{96}$$

$$R_3MO_2H + \text{>C=C<} \xrightarrow[\text{or W}^{VI}\text{]}]{\text{[Mo}^{VI}} \text{>C—C< (epoxide, O)} + R_3MOH \tag{97}$$

where M = Sn, Pb, Ge. As would be expected, these reagents are less electrophilic than are inorganic peracids, such as phenylperarsonic acid (*vide supra*), and require the presence of an epoxidation catalyst such as Mo(VI) and W(VI). These systems are less effective and less practical than those employing Mo, B, or As with water removal, discussed above. In fact, they constitute a special case of the metal-catalyzed epoxidations of olefins with alkyl hydroperoxides described in the following section.

Metal hydroperoxides are also putative intermediates in catalytic reactions of palladium(II) trifluoroacetate using hydrogen peroxide,

$$(CF_3CO_2)_2Pd + H_2O_2 \longrightarrow CF_3CO_2PdOOH + CF_3CO_2H$$

or stoichiometric reactions involving the protonation of peroxopalladium(II),

$$(Ph_3P)_2PdO_2 + CH_3SO_3H \longrightarrow (Ph_3P)_2PdOOH^+ \ O_3SCH_3^-$$

as described further in Chapter 4, p. 88.

B. Alkyl Hydroperoxides

Metal complexes in high oxidation states, such as those of Mo(VI), W(VI), V(V), and Ti(IV), can facilitate the heterolysis of alkyl hydroperoxides by forming complexes analogous to the inorganic peracids formed with hydrogen peroxide. In practice, these metal–RO_2H reagents have greater synthetic utility than the analogous metal–H_2O_2 reagents owing to their solubility in nonpolar solvents such as hydrocarbons.

Indictor and Brill[104] reported in 1965 the selective epoxidation of olefins with *tert*-butyl hydroperoxide in the presence of catalytic amounts of soluble Mo, V, and Cr complexes at 25°C.

$$>C=C< + RO_2H \xrightarrow{[\text{catalyst}]} >\overset{O}{C-C}< + ROH \qquad (98)$$

Predictably, the usual autoxidation catalysts derived from Co, Cu, Mn, Fe, etc., gave poor yields of epoxides owing to the rapid catalytic decomposition of the hydroperoxide into free radicals. The production of epoxides by the reaction of olefins with alkyl hydroperoxides in the presence of soluble compounds of Mo, V, W, Ti, and other metals[105–107] has been developed into a commercial process for the manufacture of propylene oxide.[108] A titanium-based, heterogeneous catalyst for the epoxidation of olefins with alkyl hydroperoxides gives very high epoxide selectivities.[109] This catalyst has the added advantage of being completely insoluble in the reaction medium and suitable for a continuous, fixed-bed operation. These metal-catalyzed epoxidations exhibit the following characteristics:[110–113]

1. Metals with low oxidation potentials and high Lewis acidity are superior catalysts. The following order of activity is generally observed for soluble transition metal catalysts: Mo > W > V, Ti. Certain main group elements, e.g., Se, B, and Sn, are also catalysts but show a lower activity than the best transition metal catalysts.

2. The active catalyst contains the metal in its highest oxidation state, e.g., Mo(VI), W(VI), V(V), Ti(IV). The induction period, sometimes observed, arises from the oxidation of the metal to its highest oxidation state by the hydroperoxide.

3. Polar solvents, particularly alcohols and water, greatly retard the reaction by competing with the hydroperoxide for coordination sites on the metal. Consequently, autoretardation is generally observed from the alcoholic coproduct, and it increases in significance with metals in the order W < Mo < Ti < V.

4. The high yields of epoxides and the stereospecificity of the reaction accord only with a heterolytic mechanism.

5. Substituent effects indicate that the active epoxidizing agent is an electrophilic species.

6. By-products generally result from the competing metal-catalyzed, homolytic decomposition of the hydroperoxide.

The reaction scheme is outlined below for molybdenum catalysis.[113–115]

Scheme IV:

$$Mo^{VI} + RO_2H \rightleftharpoons [Mo^{VI}\ RO_2H]$$

$$Mo^{VI} + ROH \rightleftharpoons [Mo^{VI}\ ROH]$$

$$[Mo^{VI}\ RO_2H] + \gt C{=}C\lt \xrightarrow{k_e} \gt C\overset{O}{—}C\lt + Mo^{VI} + ROH$$

$$Mo^{VI} + RO_2H \xrightarrow{k_d} Mo^{V} + RO_2\cdot + H^+$$

$$Mo^{V} + RO_2H \xrightarrow{\text{fast}} Mo^{VI}OH + RO\cdot$$

1. Metal Catalysts

The epoxide selectivity is directly related to the ratio of rate constants k_e/k_d. The rate of oxidation (k_d) generally increases with the redox potential of the metal, but the rate of heterolysis (k_e) is related to the Lewis acidity of the catalyst. Significantly, all of the active epoxidation catalysts [with the exception of vanadium(V)] are weak oxidants with the metal in its highest oxidation state. This trend explains why vanadium catalysts are generally less selective than those derived from molybdenum, tungsten, and titanium. The Lewis acidity of metal complexes generally increases in parallel with the oxidation state of the metal. Active epoxidation catalysts are thus found among compounds in a high oxidation state. The Lewis acidity of transition metal oxides decreases in the order CrO_3, $MoO_3 \gg WO_3 > TiO_2$, V_2O_5, etc., which explains the high activity observed for MoO_3.

The Lewis acidity of the catalyst is also influenced by the electronic properties of the coordinated ligands. In practice, however, a ligand effect may be observable only during the initial stages of reaction because of the rapid destruction of the original ligands under the oxidizing conditions.[112] Thus, the observed rates of the molybdenum-catalyzed epoxidations of olefins were found to be independent of the structure of the added molybdenum catalyst. The Mo(VI)–glycol complex has been isolated.[112]

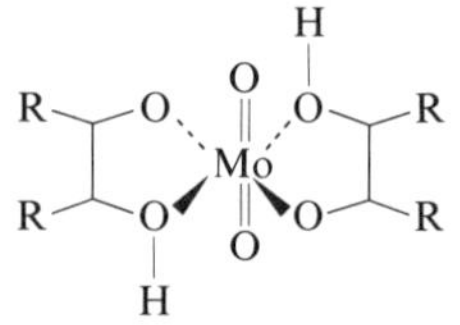

Complexes with very strongly bound ligands show low activity, presumably due to hindrance of complex formation between the catalyst and the hydroperoxide.[116–118] Catalysts with loosely bound ligands, such as $MoO_2(acac)_2$, are active but less selective than those with ligands of intermediate stability such as $MoO_2(oxine)_2$. Although insoluble catalyst precursors (e.g., molybdenum metal or MoO_3) could be used in the early studies, their catalytic activity results from dissolution in the presence of hydroperoxide.[105,106] Thus, metal oxides such as MoO_3 are generally given a pretreatment by heating with peroxide to effect dissolution,[105,116–118] which is reminiscent of the procedure for the preparation of the soluble Milas reagents from H_2O_2.[73] Of the metal oxides, only MoO_3 and V_2O_5 show any significant activity for epoxidation. Other oxides (WO_3, TiO_2, ZrO_2, Ta_2O_5, and UO_2) are essentially inert or cause rapid, nonproductive destruction of the hydroperoxide (CrO_3, Re_2O_7).[109,119] The use of MoO_3, WO_3, and V_2O_5 on a silica support leads to an increase in catalytic activity,[120] but it is due to the rapid leaching of the dispersed metal from the surface.[119] However, the supported Ti(IV) catalyst is unique in that it is highly active, and its action is truly heterogeneous. The chemical combination of Ti(IV) with SiO_2 provides the right stereochemical and electronic environment for coordination of the hydroperoxide and subsequent oxygen transfer to the olefin. A possible structure for the active site is illustrated below.[80,113]

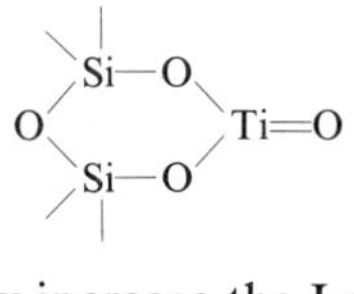

The silicate ligands apparently increase the Lewis acidity of the Ti(IV), and they may also stabilize the titanyl (Ti=O) group by retarding polymerization, which is a characteristic of titanyl complexes.

2. *Mechanism of Oxygen Transfer to Olefins*

Since the oxometal (M=O) functionality is common to all catalysts, it is included in the two types of mechanisms generally proposed for epoxidation. In Scheme V, the hydroperoxide is converted to a peroxometal intermediate:

Scheme V:[121]

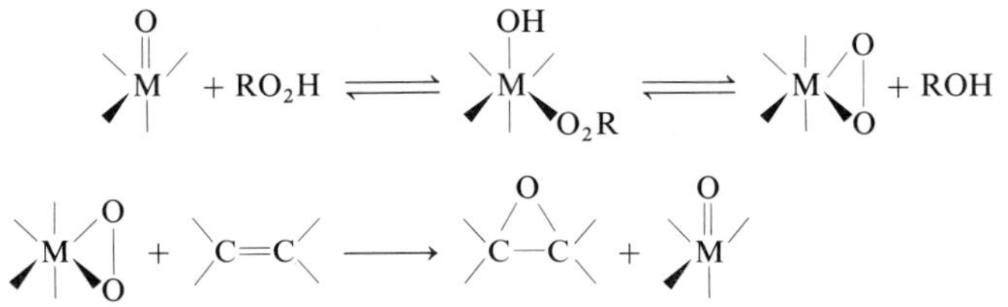

Alternatively, the alkyl hydroperoxide is directly involved in oxygen transfer in Scheme VI, by one of the three activated complexes depicted.

Scheme VI:[110,112,122]

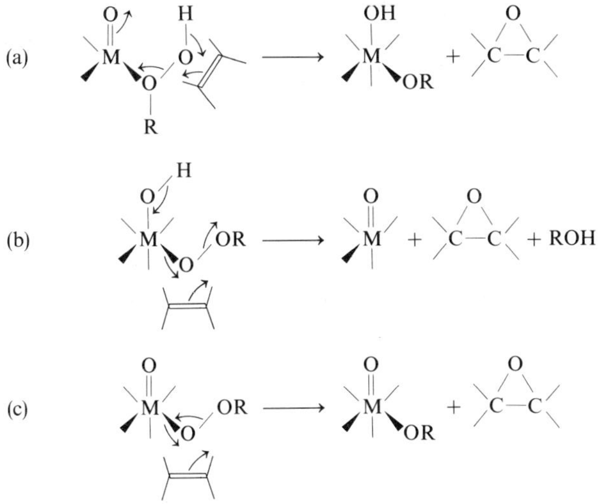

(a)

(b)

(c)

A third type of mechanism involving an intramolecular nucleophilic attack of an alkylperoxo ligand on a coordinated olefin has been proposed by Mimoun[123] to proceed via the quasi-peroxymetallocyclic intermediate depicted in Scheme VII.

Scheme VII:

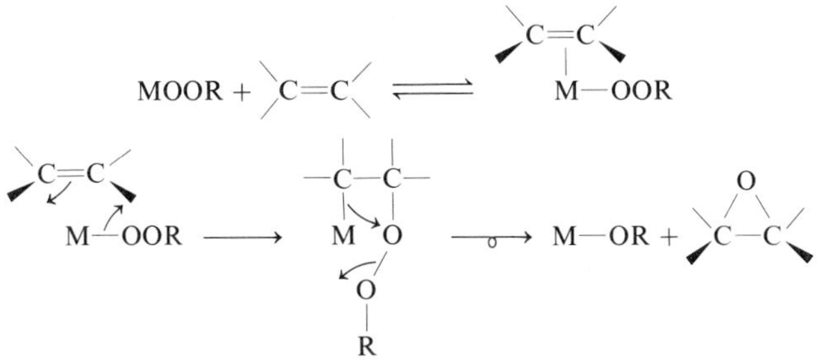

MOOR + C=C ⇌ M—OOR (olefin-coordinated)

M—OOR (olefin-coordinated) ⟶ M—O (metallocyclic intermediate, O—R) ⟶ M—OR + epoxide

It should be noted that, in this mechanism, olefin coordination to the metal constitutes the rate-determining step, as also is the case in Scheme V.

The results of an ^{18}O-labeling study support a mechanism involving an intact alkyl hydroperoxide complex.[122] In mechanism (a), the alkyl hydroperoxide is coordinated to the metal through the oxygen attached to the alkyl group. In mechanisms (b) and (c), on the other hand, the hydroperoxide is coordinated to the oxygen adjacent to hydrogen. Sharpless[122] has provided steric arguments in favor of mechanism (b) and/or (c). Thus, the exceptional reactivities and the high syn selectivities observed in the epoxidation of allylic alcohols with these reagents (*vide infra*) are consistent with a mechanism involving oxygen transfer within a ternary metal–alkyl hydroperoxide–allylic alcohol complex. The direction of approach required by mechanisms (b) and (c) is ideally suited for epoxidation of an allylic alcohol coordinated through its hydroxyl group as shown:

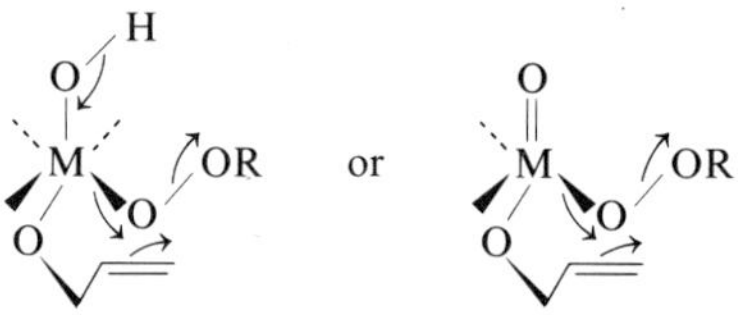

H O M OR O O or O M OR O O

The complex invoked in mechanisms (b) and (c) is formed by the addition of RO_2H to the M=O function,

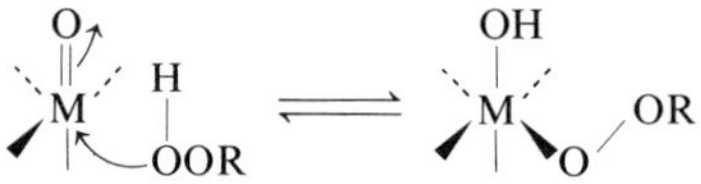

O M H OOR ⇌ OH M O OR

or by displacement of another ligand on the oxometal group.

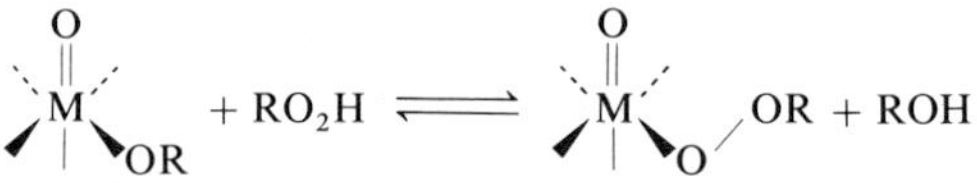

$$\text{O=M–OR} + RO_2H \rightleftharpoons \text{O=M–O–OR} + ROH \qquad (99)$$

Modena and co-workers[124] have recently shown in the vanadium-catalyzed epoxidation of sulfides with TBHP in alcohol that the catalyst is present as $VO(OR)_3$, which is formed by rapid ligand exchange with the solvent.

Similar mechanisms can be envisaged for epoxidations with hydrogen peroxide, e.g.,

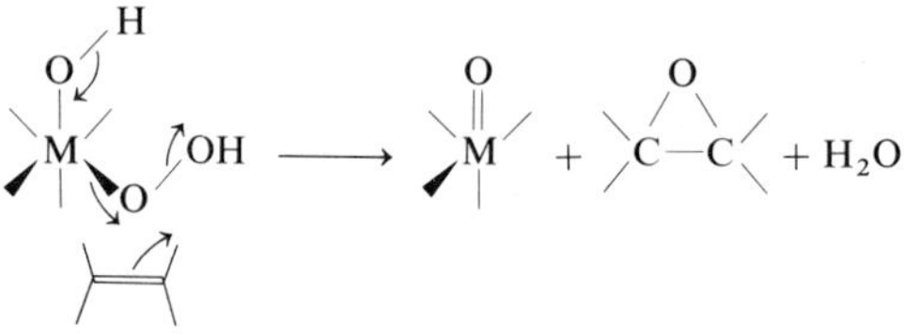

$$\text{H–O–M–O–OH (olefin)} \longrightarrow \text{O=M} + \text{C–C (epoxide, O)} + H_2O \qquad (100)$$

(The wholly organic analog of this epoxidation was given earlier in eq 83.[81]) Other workers[125] have presented evidence in favor of peroxometal intermediates in the presence of organic bases such as hexamethylphosphoric triamide (HMPA), which suggested that the active intermediate is pH dependent,[80] e.g.,

Basic pH

$$\text{M}{=}\text{O} + \text{RO}_2^- \longrightarrow \text{M}(\text{O}^-)(\text{O}{-}\text{OR}) \longrightarrow \text{M}(\text{O}_2) + \text{RO}^- \qquad (101)$$

Acidic pH

$$\text{M}{=}\text{O} + \text{RO}_2\text{H} \longrightarrow \text{M}(\text{OH})(\text{O}{-}\text{OR}) \qquad (102)$$

In this context, it is interesting that the addition of bases to these systems usually has an advantageous effect on the selectivity.[126]

The retarding effect of alcohols on the rate of epoxidation is manifested in autoretardation by the alcohol coproduct. According to Scheme IV, the extent of autoretardation is related to the ratio of the equilibrium constants for the formation of catalyst–RO_2H and catalyst–ROH complexes. This ratio varies with the metal (W < Mo < Ti < V).[110] The formation of strong catalyst–alcohol complexes accords with the observation that vanadium is a better catalyst than molybdenum for the epoxidation of allylic alcohols, whereas with simple olefins molybdenum is a much better catalyst by a factor of 10^2. This difference is also reflected in the regioselectivities observed in the epoxidation of 1,5-hexadien-3-ol.[127]

$$(\text{CH}_3)_3\text{COOH} + \text{1,5-hexadien-3-ol} \xrightarrow{[\text{catalyst}]} [\text{1,2-epoxy-5-hexen-3-ol} + \text{5,6-epoxy-1-hexen-3-ol}] + (\text{CH}_3)_3\text{COH} \qquad (103)$$

$[\text{VO(acac)}_2]$	4	:	1
$[\text{Mo(CO)}_6]$	1	:	1

The facile vanadium-catalyzed epoxidation of allylic alcohols can be rationalized on the basis of intramolecular oxygen transfer from a coordinated hydroperoxide molecule to the double bond of the coordinated allylic alcohol. Teranishi and co-workers[128] have recently reported a detailed study of the epoxidation of cyclic allylic alcohols of various ring sizes. With medium-ring alcohols, $VO(acac)_2$–TBHP affords predominantly the *syn*-epoxide, in contrast to *m*-chloroperbenzoic acid, which affords the *anti*-epoxide. The different stereochemistries were rationalized on the basis of transition state geometries, epoxidation with $VO(acac)_2$–TBHP, and a peracid involving 5,5-membered and 6,5-membered transition states, respectively.

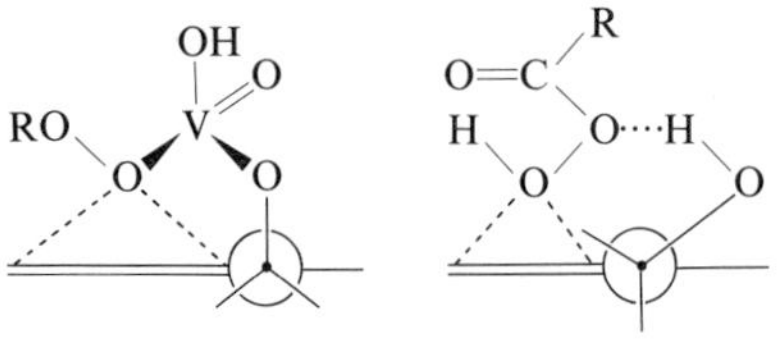

With conformationally rigid allylic alcohols, in which the attainment of the desirable geometry is difficult or impossible, oxidation of the alcohol by the vanadyl species affords the corresponding ketone.

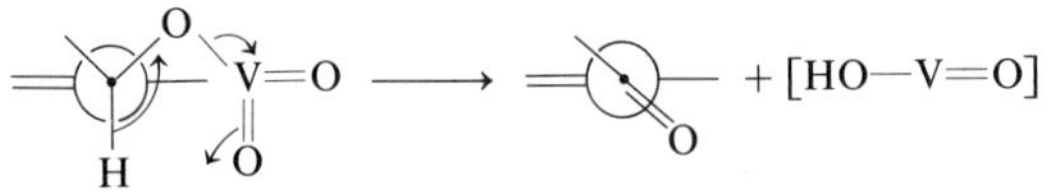

$$+ [HO{-}V{=}O] \qquad (104)$$

The coordinating ability of the hydroxyl group in allylic alcohols has been utilized in effecting asymmetric induction. For example, asymmetric epoxidations of allylic alcohols can be achieved in high enantiomeric excess ($>90\%$) by the use of a catalyst generated *in situ* from titanium tetraisopropoxide, (+)- or (−)-diethyl tartrate, and *tert*-butyl hydroperoxide.[128a] This chiral catalyst for epoxidation affords uniformly high asymmetric inductions for a variety of allylic alcohols with a wide range of substitution patterns. Oxygen is always delivered from the same enantioface of the olefin with a particular tartrate, as shown in Scheme VIII.

Scheme VIII:

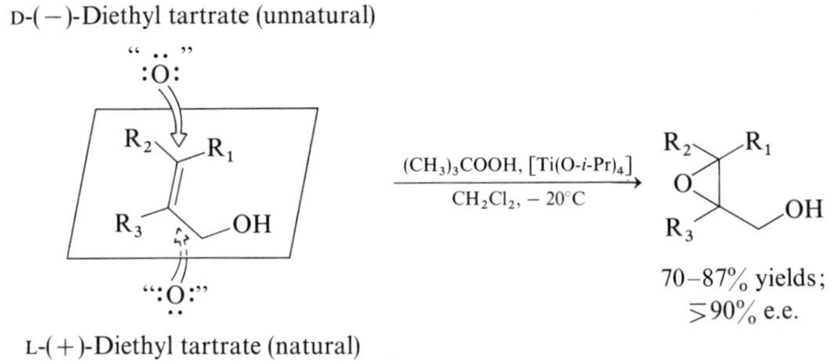

D-(−)-Diethyl tartrate (unnatural)

"$\ddot{O}$:"

R_2 R_1 R_3 OH

$\xrightarrow[CH_2Cl_2,\ -20°C]{(CH_3)_3COOH,\ [Ti(O\text{-}i\text{-}Pr)_4]}$

R_2 R_1 O R_3 OH

70–87% yields; $\gtrsim 90\%$ e.e.

"$:\ddot{O}:$"

L-(+)-Diethyl tartrate (natural)

The epoxidation of olefins with alkyl hydroperoxides as described above involves metal catalysts such as molybdenum(VI), vanadium(V), and titanium(IV) in rather high formal oxidation states. By contrast, metal complexes such as palladium(II) have recently been shown to be catalysts in the reaction of terminal olefins with alkyl hydroperoxides to produce methyl ketones, as described further in Chapter 4, Section III.

3. Oxidation of Other Substrates

In addition to olefins, other nucleophilic reagents undergo oxygen transfer reactions which closely parallel the reactions of the same substrates with organic peracids.[129] They almost certainly involve oxygen transfer from a metal–hydroperoxide complex to the substrate via a cyclic transition state analogous to that described above for epoxidations with these reagents. Tertiary amines are smoothly oxidized to the corresponding *N*-oxides in the presence of vanadium or molybdenum catalysts.[130–133]

$$R_3N + t\text{-}BuO_2H \xrightarrow{[VO(acac)_2]} R_3NO + t\text{-}BuOH \tag{105}$$

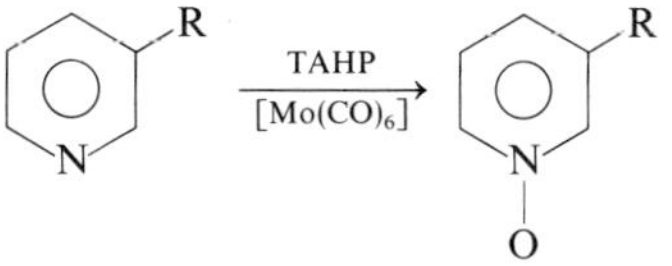

$$\text{3-R-pyridine} \xrightarrow[\text{[Mo(CO)}_6]]{\text{TAHP}} \text{3-R-pyridine N-oxide} \tag{106}$$

Nitrosamines are oxidized to the corresponding nitramines, and Schiff bases to oxaziridines.[131–133]

$$R_2N{-}NO \xrightarrow[\text{[Mo(CO)}_6]]{\text{TAHP}} R_2N{-}NO_2 \tag{107}$$

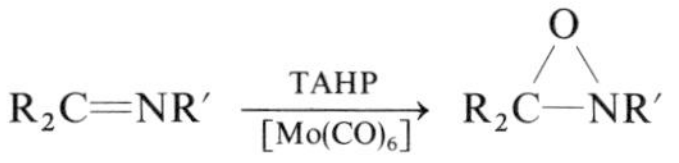

$$R_2C{=}NR' \xrightarrow[\text{[Mo(CO)}_6]]{\text{TAHP}} R_2\overset{O}{C{-}}NR' \ \text{(oxaziridine)} \tag{108}$$

The product of oxidation of aniline by TBHP depends on the metal catalyst used. Molybdenum and vanadium catalysts give nitrobenzene,[134] but the titanium catalyst affords azoxybenzene.[135]

$$ArNH_2 + t\text{-}BuO_2H \xrightarrow{[Mo^{VI} \text{ or } V^{V}]} ArNO_2, \quad \text{etc.} \tag{109}$$

$$ArNH_2 + t\text{-}BuO_2H \xrightarrow[\text{[Ti}^{IV}]]{} ArN{=}\overset{O}{\overset{\uparrow}{N}}Ar, \quad \text{etc.} \tag{110}$$

Oxidation of primary aliphatic amines produces oximes. Titanium compounds are the superior catalysts for this reaction. A process has been envisaged[136,137] for the coproduction of cyclohexanone oxime and styrene, as shown below.

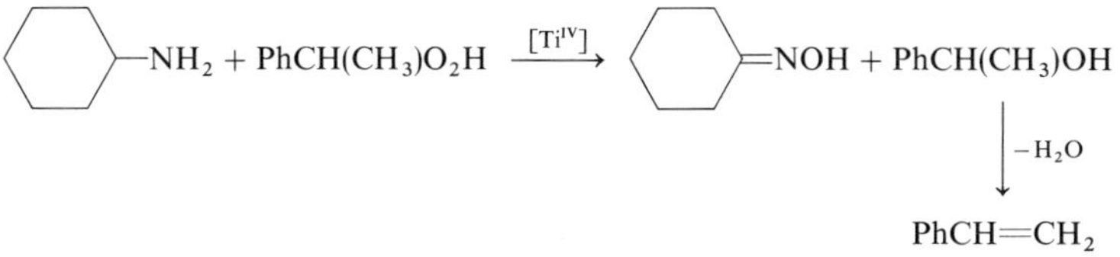

$$C_6H_{11}{-}NH_2 + PhCH(CH_3)O_2H \xrightarrow{[Ti^{IV}]} C_6H_{10}{=}NOH + PhCH(CH_3)OH \tag{111}$$

$$PhCH(CH_3)OH \xrightarrow{-H_2O} PhCH{=}CH_2$$

The molybdenum- and vanadium-catalyzed oxidation of sulfides to sulfoxides has been extensively studied.[138–143] In the presence of excess hydroperoxide, the slower further oxidation to the sulfone occurs, e.g.,

$$Bu_2S \xrightarrow[\text{[VO(acac)}_2]]{\text{TBHP}} Bu_2SO \xrightarrow[\text{[VO(acac)}_2]]{\text{TBHP}} Bu_2SO_2 \qquad (112)$$

Modena and co-workers[143] studied the vanadium(V)-catalyzed oxidation of sulfides with TBHP in alcohol solvents. Under these conditions, the catalyst is present as $VO(OR)_3$, which is formed by rapid ligand exchange with the solvent. The active oxidant is then formed via addition of TBHP to the vanadyl group or via ligand exchange, as shown:

$$(RO)_2V(=O){-}OR + R'O_2H \rightleftharpoons (RO)_2V(OH)(OR)(O_2R') \qquad (113a)$$

$$(RO)_2V(=O){-}OR + R'O_2H \rightleftharpoons (RO)_2V(=O){-}O_2R' + ROH \qquad (113b)$$

Amines and sulfides are generally oxidized much faster than olefins, as reflected in the selective oxidation of unsaturated sulfides.[129]

$$CH_3(CH_2)_3SCH_2CH{=}CH_2 \xrightarrow[\text{[MoO}_2\text{(acac)}_2]]{\text{TBHP}} CH_3(CH_2)_3SO_2CH_2CH{=}CH_2 \qquad (114)$$

Molybdenum and vanadium compounds also catalyze the oxidation of trivalent phosphorus compounds with TBHP.[144,145]

4. *In Situ Generation of Hydroperoxides*

In reactions such as the metal-catalyzed epoxidations, the alkyl hydroperoxide reagent is generally prepared in a separate step by autoxidation of the corresponding hydrocarbon. However, it is possible to generate the hydroperoxide *in situ*. For example, the preparation of propylene oxide and cyclohexanol by the cooxidation of cyclohexane and propylene in the presence of a molybdenum catalyst has been reported.[146]

$$C_6H_{12} + O_2 \longrightarrow C_6H_{11}{-}O_2H \qquad (115)$$

$$C_6H_{11}{-}O_2H + CH_3CH{=}CH_2 \xrightarrow{\text{[Mo]}} C_6H_{11}{-}OH + CH_3\overset{O}{CH{-}CH_2} \qquad (116)$$

However, the epoxidation catalyst tends to inhibit the autoxidation step by removing the chain-initiating hydroperoxide from the reaction mixture.

The oxidation of cyclohexene with molecular oxygen in the presence of a mixture of low-valent Group VIII metal complexes and molybdenum complexes has been examined.[147,148] The former act as autoxidation catalysts and the latter as epoxidation catalysts, to give a mixture of cyclohexen-3-ol and cyclohexene oxide.

$$+ O_2 \xrightarrow{\text{[autoxidation catalyst]}} \text{—}O_2H \qquad (117)$$

$$\text{—}O_2H + \xrightarrow{\text{[epoxidation catalyst]}} \text{—OH} + \qquad (118)$$

Owing to the exceptional reactivity of the vanadium reagent toward allylic alcohols, the stereoselective oxidation of cyclohexene to *syn*-1,2-epoxycyclohexan-3-ol can be effected.[149–151]

$$+ O_2 \xrightarrow{[C_5H_5V(CO)_4]} \text{OH}, \text{O}, \quad \text{etc.} \qquad (119)$$

REFERENCES

1. R. Hiatt, *in* "Organic Peroxides" (D. Swern, ed.), Vol. 2, p. 7. Wiley (Interscience), New York, 1971.
2. T. D. Manly, *Chem. Ind. (London)* p. 12 (1962); J. P. Schirmann and S. Y. Delavarenne, "Hydrogen Peroxide in Organic Chemistry," Informations Chimie, Paris, 1979.
3. D. Swern, *in* "Organic Peroxides" (D. Swern, ed.), Vol. 2, p. 355. Wiley (Interscience), New York, 1971.
4. B. Plesnicar, *in* "Oxidation in Organic Chemistry" Part C. (W. S. Trahavanosky, ed.), p. 211 Academic Press, New York, 1978.
5. J. K. Kochi, *Rec. Chem. Prog.* **27**, 207 (1966).
6. J. K. Kochi, *Science* **155**, 415 (1967).
7. J. K. Kochi, *Pure Appl. Chem.* **4**, 377 (1971).
8. W. A. Waters, *Pure Appl. Chem.* **4**, 307 (1971).
9. J. K. Kochi, *in* "Free Radicals" (J. K. Kochi, ed.), Vol. 1, p. 591. Wiley, New York, 1973.
10. J. K. Kochi, *Accts. Chem. Rsch.* **7**, 351 (1974).
11. J. H. Baxendale, *Adv. Catal.* **4**, 31 (1952).
12. J. Weiss, *Adv. Catal.* **4**, 343 (1952).
13. N. Uri, *Chem. Rev.* **50**, 375 (1952).
14. R. O. C. Norman and J. R. Lindsay Smith, *in* "Oxidases and Related Redox Systems" (T. E. King, H. S. Mason, and M. Morrison, eds.), p. 131. Wiley, New York, 1965.
15. L. S. Boguslavskaya, *Russ. Chem. Rev.* **34**, 503 (1965).
16. G. Sosnovsky, *in* "Organic Peroxides" (D. Swern, ed.), Vol. 2, p. 269. Wiley (Interscience), New York, 1971.

17. D. J. Rawlinson and G. Sosnovsky, *Synthesis* p. 1 (1972).
18. C. Walling and S. Kato, *J. Am. Chem. Soc.* **93**, 4275 (1971).
19. C. Walling, G. M. El-Taliawi, and R. A. Johnson, *J. Am. Chem. Soc.* **96**, 133 (1974).
20. C. Walling, *Accts. Chem. Rsch.* **8**, 125 (1975).
21. C. Walling and G. M. El-Taliawi, *J. Am. Chem. Soc.* **95**, 844, 848 (1973).
22. M. E. Snook and G. A. Hamilton, *J. Am. Chem. Soc.* **96**, 860 (1974).
23. H. E. de la Mare, J. K. Kochi, and F. F. Rust, *J. Am. Chem. Soc.* **85**, 1437 (1963).
24. D.D. Coffmann, E. L. Jenner, and R. D. Lipscomb, *J. Am. Chem. Soc.* **80**, 2864 (1958).
25. D. D. Coffman, U.S. Patent 2,687,432 (1954).
26. E. T. Denisov and N. M. Emanuel, *Russ. Chem. Rev.* **29**, 645 (1960).
27. R. Hiatt, T. Mill, and F. R. Mayo, *J. Org. Chem.* **33**, 1416 (1968).
28. R. Hiatt, T. Mill, K. C. Irwin, and J. K. Castleman, *J. Org. Chem.* **33**, 1421, 1428 (1968).
29. R. Hiatt, K. C. Irwin, and C. W. Gould, *J. Org. Chem.* **33**, 1430 (1968).
30. R. Hiatt and K. C. Irwin, *J. Org. Chem.* **33**, 1436 (1968).
31. R. A. Sheldon and J. K. Kochi, *Adv. Catal.* **25**, 272 (1976).
32. W. J. M. van Tilborg, *Tetrahedron* **31**, 2841 (1975); cf. S. K. Chung and K. Sasamoto, *J. Chem. Soc., Chem. Commun.* p. 346 (1981).
33. Y. Kamiya, S. Beaton, A. Lafortune, and K. U. Ingold, *Canad. J. Chem.* **41**, 2020 (1963); see also C. F. Hendriks, P. M. Heertjes, and H. C. A. van Beek, *Ind. Eng. Chem. Prod. Rsch. Dev.* **18**, 212 (1979).
34. I. I. Chuev, V. A. Shushunov, M. K. Shchennikova, and G. A. Abakumov, *Kinet. Katal. USSR* **10**, 75 (1969); **11**, 426 (1970), and references cited therein.
35. J. E. Lyons, *Adv. Chem. Series* **132**, 64 (1974).
36. A. Fusi, R. Ugo, F. Fox, A. Pasini, and S. Cenini, *J. Organometal. Chem.* **26**, 417 (1970).
37. V. P. Kurkov, J. Z. Pasky, and J. B. Lavigne, *J. Am. Chem. Soc.* **90**, 4743 (1968).
38. L. W. Fine, M. Grayson, and V. H. Suggs, *J. Organometal. Chem.* **8**, 219 (1970).
39. B. L. Booth, R. N. Haszeldine, and G. R. H. Neuss, *J. Chem. Soc., Chem. Commun.* p. 1074 (1972).
40. J. E. Lyons, *J. Chem. Soc., Chem. Commun.* p. 562 (1971).
41. J. E. Lyons and J. O. Turner, *Tetrahedron Lett.* p. 2903 (1972); *J. Org. Chem.* **37**, 2881 (1972).
42. H. Arzoumanian, A. A. Blanc, J. Metzger, and J. E. Vincent, *J. Organometal Chem.* **82**, 261 (1974).
43. M. S. Kharasch and A. Fono, *J. Org. Chem.* **23**, 325, 1322 (1958).
44. G. Sosnovsky and D. J. Rawlinson, *in* "Organic Peroxides" (D. Swern, ed.), Vol. 2, p. 153. Wiley, New York, 1971.
45. G. Sosnovsky and S. O. Lawesson, *Angew. Chem. Int. Ed.* **3**, 269 (1964).
46. G. Sosnovsky, *Tetrahedron* **21**, 871 (1965).
47. (a) J. K. Kochi, *J. Am. Chem. Soc.* **84**, 774, 2785 (1962).
    (b) J. K. Kochi and H. E. Mains, *J. Org. Chem.* **30**, 1862 (1965).
    (c) C. Walling and A. A. Zavitsas, *J. Am. Chem. Soc.* **85**, 2084 (1963).
48. A. R. Doumaux, *in* "Oxidation" (R. Augustine, ed.), p. 141. Dekker, New York, 1971.
49. (a) M. S. Kharasch, F. S. Arimoto, and W. Nudenberg, *J. Org. Chem.* **16**, 1556 (1951).
    (b) M. S. Kharasch and A. Fono, *J. Org. Chem.* **24**, 72 (1959).
50. M. S. Kharasch and W. Nudenberg, *J. Org. Chem.* **19**, 1921 (1954).
51. J. K. Kochi and F. F. Rust, *J. Am. Chem. Soc.* **84**, 3946 (1962).
52. (a) F. Minisci, *Accts. Chem. Rsch.* **8**, 165 (1975).
    (b) F. Minisci, M. Cecere, R. Galli, and A. Selva, *Org. Prep. Proced.* **1**, 11 (1969).
53. Ž. Čeković and M. M. Green, *J. Am. Chem. Soc.* **96**, 3000 (1974).

53a. S. L. Schreiber, *J. Am. Chem. Soc.* **102**, 6163 (1980).

54. G. H. Jones, *J. Chem. Soc., Chem. Commun.* p. 536 (1979).
55. D. Benson, P. J. Proll, L. H. Sutcliffe, and J. Walkley, *Discuss. Faraday Soc.* **29**, 60 (1960).
56. S. S. Lande, C. D. Falk, and J. K. Kochi, *J. Inorg. Nucl. Chem.* **33**, 4101 (1971), and references therein.
57. See however, C. F. Hendriks, H. C. A. van Beek, and P. M. Heertjes, *Ind. Eng. Chem. Prod. Rsch. Dev.* **18**(1), 38 (1979).
58. E. T. Denisov, *Russ. Chem. Rev.* **40**, 24 (1971).
59. A. T. Betts and N. Uri, *Adv. Chem. Series* **76**, 160 (1968); *Makromol. Chem.* **95**, 22 (1966).
60. D. G. Hendry and D. Schuetzle, *Prepr. Div. Pet. Chem., Am. Chem. Soc.* **14**(4), A31 (1969).
61. Y. Kamiya and K. U. Ingold, *Canad. J. Chem.* **42**, 1027, 2424 (1964).
62. J. F. Black, *J. Am. Chem Soc.* **100**, 527 (1978).
63. G. L. Banks, A. J. Chalk, J. E. Dawson, and J. F. Smith, *Nature* (*London*) **174**, 274 (1954).
64. A. J. Chalk and J. F. Smith, *Trans. Faraday Soc.* **53**, 1214, 1235 (1957).
65. G. M. Bulgakova, A. N. Shupik, I. P. Skibida, K. I. Zamaraev, and Z. K. Maizus, *Dokl. Akad. Nauk SSSR* **199**, 376 (1971).
66. A. V. Tobolsky, *India Rubber World* **118**, 363 (1948).
67. A. E. Woodward and R. B. Mesrobian, *J. Am. Chem. Soc.* **75**, 6189 (1953).
68. A. V. Tobolsky, D. J. Metz, and R. B. Mesrobian, *J. Am. Chem. Soc.* **72**, 1942 (1950).
69. Y. Kamiya, S. Beaton, A. Lafortune, and K. U. Ingold, *Canad. J. Chem.* **41**, 2034 (1963).
70. C. Walling, *J. Am. Chem. Soc.* **91**, 7590 (1969).
71. D. M. Roundhill, M. K. Dickson, N. S. Dixit, and B. P. Sudha-Dixit, *J. Am. Chem. Soc.* **102**, 5538 (1980).
72. (a) J. A. Connor and E. A. V. Ebsworth, *Adv. Inorg. Chem. Radiochem.* **6**, 279 (1964).
(b) J. P. Schirmann and S. Y. Delavarenne, "Hydrogen Peroxide in Organic Chemistry," Informations Chimie, Paris, 1979.
73. N. A. Milas and S. Sussman, *J. Am. Chem. Soc.* **58**, 1302 (1936); **59**, 2345 (1937; see also L. F. Fieser and M. Fieser, "Organic Reagents," Vol. 1, p. 472. Wiley, New York, 1967.
74. M. Mugdan and D. P. Young, *J. Chem. Soc.* p. 2988 (1949).
75. G. B. Payne and P. H. Williams, *J. Org. Chem.* **24**, 54 (1959); see also T. M. Shryne and L. Kim, U.S. Patent 4,024,165 (1977) to Shell.
76. G. B. Payne and C. W. Smith, *J. Org. Chem.* **22**, 1682 (1957).
77. H. C. Stevens and A. J. Kaman, *J. Am. Chem. Soc.* **87**, 734 (1965).
78. Z. Raciszewski, *J. Am. Chem. Soc.* **82**, 1267 (1960).
79. J. Itakura, H. Tanaka, and H. Ito, *Bul. Chem. Soc. Japan* **42**, 1604 (1969).
80. R. A. Sheldon, *J. Mol. Catal.* **7**, 107 (1980).
81a. L. Kim, Br. Patent 1,399,639 (1975) to Shell; see also R. D. Chambers and M. Clark, *Tetrahedron Lett.* p. 2741 (1970).
81b. B. Ganem, R. P. Heggs, A. J. Biloski, and D. R. Schwartz, *Tetrahedron Lett.* **21**, 685, 689 (1980); Synthesis, 810 (1980).
81c. Compare also J. Rebek, Jr. and R. McCready, *J. Am. Chem. Soc.* **102**, 5602 (1980); J. Rebek, Jr., R. McCready, and R. Wolak, *J. Chem. Soc., Chem. Commun.* p. 705 (1980). *Tetrahedron Lett.*, 4337 (1979).
82. M. Pralus, J. C. Lecoq, and J. P. Schirmann, *in* "Fundamental Research in Homogeneous Catalysis" (M. Tsutsui, ed.), Vol. 3, p. 327. Plenum, New York, 1979.
83. A. M. Matucci, E. Perrotti, and A. Santambrogio, *J. Chem. Soc. Chem. Commun.* p. 1198 (1970).
84. J. P. Schirmann and S. Y. Delavarenne, Ger. Patent 2,752,626 (1978) to Ugine Kuhlmann.
85. J. P. Schirmann, M. Pralus, and S. Y. Delavarenne, Ger. Patent 2,803,757 (1977) to Ugine Kuhlmann.
86. J. P. Schirmann and S. Y. Delavarenne, Ger. Patent 2,803,791 (1977) to Ugine Kuhlmann.

87. *Chem. & Eng. News* **56** (Dec. 11), p. 24 (1978).
88. C. H. McMullen, U.S. Patent 3,993,673 (1976) to Union Carbide.
89. S. E. Jacobson, F. Mares, and P. M. Zambri, *J. Am. Chem. Soc.* **101**, 6946 (1979).
90. S. E. Jacobson, F. Mares, and P. M. Zambri, *J. Am. Chem. Soc.* **101**, 6938 (1979).
91. S. E. Jacobson, R. Tang, and F. Mares, *J. Chem. Soc., Chem. Commun.* p. 888 (1978).
92. P. A. Grieco, Y. Yokoyama, S. Gilman, and M. Nishizawa, *J. Org. Chem.* **42**, 2034 (1977); see also ref. 92a and 96.

92a. H. J. Reich, F. Chow, and S. L. Peake, *Synthesis* p. 299 (1978).

93. K. Kahr and C. Berther, *Chem. Berichte* **93**, 132 (1960).
94. H. S. Schultz, H. B. Freyermuth, and S. R. Buc, *J. Org. Chem.* **28**, 1140 (1963).
95. J. Drabowicz and M. Mikolajczyk, *Synthesis* p. 758, (1978).
96. F. DiFuria and G. Modena, *Recl. Trav. Chim. Pays-Bas* **98**, 181 (1979).
97. F. DiFuria and G. Modena, *in* "Fundamental Research in Homogeneous Catalysis" (M. Tsutsui, ed.), Vol. 3, p. 433 Plenum, New York, 1979.
98. S. Tsuda, *Chem. Econ. & Eng. Rev.* **2**, 39 (1970); see also O. L. Lebedev and S. N. Kazarnovskii, *Zh. Obshch. Khim.* **30**, 1631 (1960).
99. K. Kahr, *Angew. Chem.* **72**, 135 (1960).
100. P. Burckard, J. P. Fleury, and F. Weiss, *Bul. Soc. Chim. France* p. 2730 (1965).
101. French Patent 1,332,004 to Ugine Kuhlmann [*Chem. Abstr.* **60**, 423e (1900)].
102. French Patent 2,038,948 (1970) to Sumitomo.
103. S. Y. Delavarenne, F. Weiss, and J. P. Schirmann, U.S. Patent 3,952,480 (1976) to Ugine Kuhlmann; see also M. Pralus, J. P. Schirmann, and S. Y. Delavarenne, U.S. Patent, 4,026,908 (1977) and French Patent, 2,300,765, both to Ugine Kuhlmann.
104. N. Indictor and W. F. Brill, *J. Org. Chem.* **30**, 2074 (1965).
105. J. Kollar, U.S. Patents 3,350,422 (1967), 3,351,635 (1967), 3,507,809 (1970), and 3,625,981 (1971) to Halcon International; see also J. Kollar, *Pap., Div. Pet. Chem., Am. Chem. Soc. Meet., 1978.*
106. M. N. Sheng and J. G. Zajacek, *Adv. Chem. Series* **76**, 418 (1968).
107. K. Allison, P. Johnson, G. Foster, and M. B. Sparke, *Ind. Eng. Chem. Prod. Rsch. Dev.* **5**, 166 (1966).
108. R. Landau, *Hydrocarbon Process.* **46**, 141 (1967); see also R. Landau, G. A. Sullivan, and D. Brown, *Chemtech* p. 602 (1979).
109. H. P. Wulff, U.S. Patent 3,923,843 (1975) to Shell Oil; Br. Patent 1,249,079 (1971) to Shell Oil.
110. R. A. Sheldon and J. A. van Doorn, *J. Catal.* **31**, 427, 438 (1973).
111. R. A. Sheldon and J. A. van Doorn, *J. Catal.* **34**, 242 (1974).
112. R. A. Sheldon, *Recl. Trav. Chim. Pays-Bas* **92**, 253, 367 (1973).
113. R. A. Sheldon, in "Aspects of Homogeneous Catalysis", Vol. 4, (R. Ugo, ed.), D. Reidel, Dordrecht. p.3, 1981.
114. T. N. Baker, G. J. Mains, M. N. Sheng, and J. G. Zajacek, *J. Org. Chem.* **38**, 1145 (1973).
115. M. I. Farberov, G. A. Stozhkova, A. V. Bondarenko, and T. M. Kirik, *Int. Chem. Eng.* **12**, 634 (1972); *Kinet. Katal.* (Eng. Trans.) **13**, 263 (1972).
116. P. Forzatti and F. Trifiro, *React. Kinet. Catal. Lett.* **1**, 367 (1974).
117. F. Trifiro, P. Forzatti, S. Preite, and I. Pasquon, *J. Less-Common Met.* **36**, 319 (1974).
118. J. Sobczak and J. J. Ziolkowski, *Inorg. Chim. Acta* **19**, 15 (1976).
119. F. Mashio and S. Kato, *Mem. Fac. Ind. Arts, Kyoto Tech. Univ., Sci. Technol.* 16, 79 (1967) [*Chem. Abstr.* **69**, 68762e (1968)].
120. F. Trifiro, P. Forzatti, and I. Pasquon, *in* "Catalysis, Heterogeneous and Homogeneous" (B. Delmon and G. Jannes, eds.), p. 509. Elsevier, Amsterdam, 1975.
121. H. Mimoun, I. Sérée de Roch, and L. Sajus, *Tetrahedron* **26**, 37 (1970).

122. A. O. Chong and K. B. Sharpless, *J. Org. Chem.* **42**, 1587 (1977). See also E. D. Mihelich, *Tetrahedron Lett.* p. 4729 (1979).
123. H. Mimoum, *J. Mol. Catal.* **7**, 1 (1980).
124. S. Cenci, F. DiFuria, G. Modena, R. Curci, and J. O. Edwards, *J. Chem. Soc., Perkin II* p. 979 (1978).
125. H. Arakawa and A. Ozaki, *Chem. Lett.* p. 1245 (1975).
126. C. Y. Wu and H. E. Swift, *J. Catal.* **43**, 380 (1976).
127. M. N. Sheng and J. G. Zajacek, *J. Org. Chem.* **35**, 1839 (1970).
128. T. Itoh, K. Jitsukawa, K. Kaneda, and S. Teranishi, *J. Am. Chem. Soc.* **101**, 159 (1979).
128a T. Katsuki and K. B. Sharpless, *J. Am. Chem. Soc.* **102**, 5974 (1980).
129. F. List and L. Kuhnen, *Erdoel Kohle, Erdgas, Petrochem. Brennst.-Chem.* **20**, 192 (1967).
130. M. N. Sheng and J. G. Zajacek, *J. Org. Chem.* **33**, 588 (1968).
131. G. A. Tolstikov, U. M. Jemilev, V. P. Yur'ev F. B. Gershanov, and S. F. Rafikov, *Tetrahedron Lett.* p. 2807 (1971).
132. G. A. Tolstikov, U. M. Dzhemilev, and V. P. Yur'ev, *J. Org. Chem. USSR* **8**, 1200 (1971).
133. G. A. Tolstikov, U. M. Dzhemilev, V. P. Yur'ev A. A. Pozdeeva, and F. G. Gerchikova, *J. Gen. Chem. USSR* **43**, 1350 (1973).
134. G. R. Howe and R. R. Hiatt, *J. Org. Chem.* **35**, 4007 (1970).
135. K. Kosswig, *Justus Liebigs Ann. Chem.* **749**, 206 (1971).
136. J. L. Russell and J. Kollar, Br. Patents 1,100,672 and 1,111,892 (1968) to Halcon.
137. G. N. Koshel, M. I. Farberov, L. L. Zalygin, and G. A. Krushinskaya, *J. Appl. Chem. USSR* **44**, 885 (1971).
138. G. A. Tolstikov, U. M. Dzhemilev, N. N. Novitskaya, and V. P. Yur'ev, *Bul. Acad. Sci. USSR, Div. Chem. Sci.* p. 2675 (1972).
139. G. A. Tolstikov, U. M. Dzhemilev, N. N. Novitskaya, V. P. Yur'ev and R. G. Kantyukova, *J. Gen. Chem. USSR* **41**, 1896 (1971).
140. R. Curci, F. DiFuria, R. Testi, and G. Modena, *J. Chem. Soc., Perkin II* p. 752 (1974).
141. R. Curci, F. DiFuria, and G. Modena *J. Chem. Soc., Perkin II* p. 576 (1977).
142. F. DiFuria, G. Modena, and R. Curci, *Tetrahedron Lett.* p. 4637 (1976).
143. F. DiFuria, G. Modena, R. Curci, and J. O. Edwards, *Gazz. Chim. Ital.* **109**, 571 (1979); see also Curci *et al.*[124]
144. R. Hiatt and C. McColeman, *Canad. J. Chem.* **49**, 1712 (1971).
145. D. G. Pobedimskii, E. G. Chebotareva, S. A. Nasybullin, P. A. Kirpchnikov, and A. L. Buchachenko, *Proc. Acad. Sci. USSR, Phys. Chem. Sect.* **220**, 59 (1975).
146. E. de Ruiter, *Erdoel Kohle, Erdgas, Petrochem. Brennst.-Chem.* **25**, 653 (1972).
147. A. Fusi, R. Ugo, and G. M. Zanderighi, *J. Catal.* **34**, 175 (1974).
148. H. Arzoumanian, A. Blanc, U. Hartig, and J. Metzger, *Tetrahedron Lett.* p. 1011 (1974); see also M. Baccouche, J. Ernst, J. H. Fuhrhop, R. Schlozer, and H. Arzoumanian, *J. Chem. Soc., Chem. Commun.* p. 821 (1977).
149. J. E. Lyons, *Tetrahedron Lett.* p. 2737 (1974).
150. J. E. Lyons, *Adv. Chem. Series* **132**, 64 (1974).
151. J. E. Lyons, *in* "Catalysis in Organic Synthesis" (P. N. Rylander and H. Greenfield, eds.), p. 235. Academic Press, New York, 1976.

ADDITIONAL READING

D. Swern, ed., "Organic Peroxides," Vols. I, II, III. Wiley (Interscience), New York, 1970, 1971, 1972. Includes metal-catalyzed reactions of various types of peroxides.

K. U. Ingold and B. P. Roberts, "Free Radical Substitution Reactions." Wiley (Interscience), New York, 1971. Includes homolytic substitution (S_H2) at metal centers.

J. K. Kochi, ed., Oxidation-reduction reactions of free radicals and metal complexes. *In* "Free Radicals," Vol. I. Wiley, New York, 1973.

K. B. Sharpless and T. R. Verhoeven, Metal-catalyzed highly selective oxygenations of olefins and acetylenes with *tert*-butyl hydroperoxide. Practical considerations and mechanisms. *Alrichimica Acta* **12**, 63 (1979).

R. A. Sheldon, Metal-catalyzed epoxidations of olefins with hydroperoxides, *in* "Aspects of Homogeneous Catalysis" (R. Ugo, Ed.), Vol. 4, D. Reidel, Dordrecht. 1981, p. 3ff.

J. P. Schirmann and S. Y. Delavarenne, "Hydrogen Peroxide in Organic Chemistry." Informations Chimie, Paris, 1979.

J. O. Edwards, ed., "Peroxide Reaction Mechanisms." Wiley (Interscience), New York, 1962.

Chapter 4

Activation of Molecular Oxygen by Metal Complexes

The direct participation of dioxygen in the oxidation of organic compounds, as effected by transition metal complexes, is referred to as *oxygen activation*. The best known example of a metal-mediated transfer of oxygen to an organic substrate is the commercially important gas-phase oxidation of ethylene to ethylene oxide over a supported silver catalyst,

$$CH_2{=}CH_2 + O_2 \xrightarrow[250^\circ C]{[Ag/Al_2O_3]} H_2C\overset{O}{—}CH_2, \quad \text{etc.} \tag{1}$$

in which a silver–dioxygen complex of ethylene has been implicated.[1]

Reactions of organic compounds with dioxygen in the ground state (triplet) are restricted by spin conservation, which may, in principle, be overcome by forming metal complexes. Indeed, a wide variety of transition metal complexes are now known to form stable dioxygen adducts.[2–19] However, the same metal complexes react directly with alkyl hydroperoxides to generate chain-initiating radicals. Since alkyl hydroperoxides are ubiquitous in most hydrocarbon mixtures, it is often very difficult to distinguish between the metal–hydroperoxide interactions and those involving oxygen activation. In general, the former tend to be energetically more favorable and mask any contribution from the latter, except in those cases in which hydroperoxide formation is difficult or impossible (e.g., with ethylene). Incorrect mechanistic conclusions have been drawn when this complication has not been taken into account.

I. OXYGEN FIXATION: HISTORICAL DEVELOPMENT

The reversible complexation of molecular oxygen by ammine cobalt(II) salts was already known in the nineteenth century.[20] The first example of a synthetic dioxygen chelate is the Co(II) complex of bis(salicylaldehyde ethyleneimine) (Salen), the progenitor of a series of Schiff base complexes that reversibly combine with oxygen.

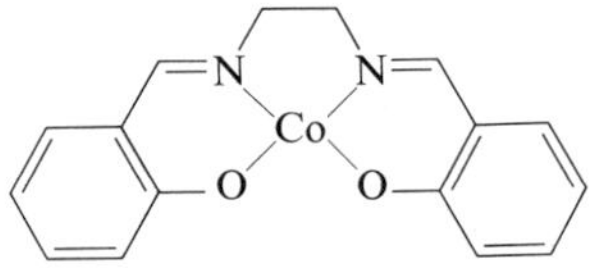

(Salen)Co(II) (Salcomine)

Since the original discovery of the oxygen-carrying properties of (Salen) Co(II),[21] a wide variety of analogous dioxygen complexes have been characterized,[14–18] generally as 2 : 1 μ-peroxo complexes, e.g.,

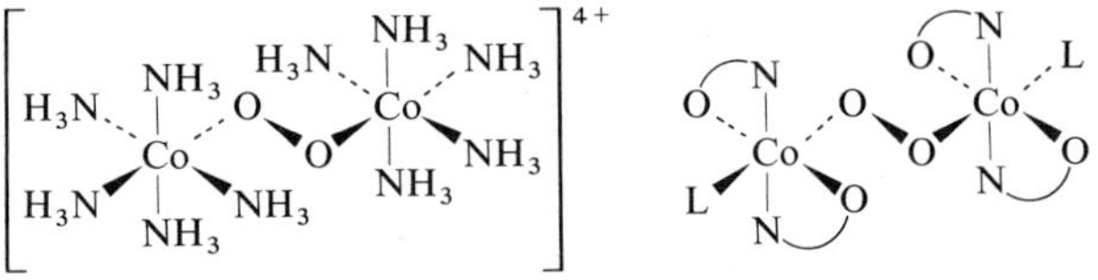

$$\left[(H_3N)_5Co{-}O{-}O{-}Co(NH_3)_5\right]^{4+}$$

where L = DMF, py, etc. Mononuclear, paramagnetic dioxygen complexes of Co(II) are also known, such as the 1:1 dioxygen complex with (3-methoxy-Salen)Co(II)[22] and the analogous (Acacen)Co(II)–O_2 complex.[23]

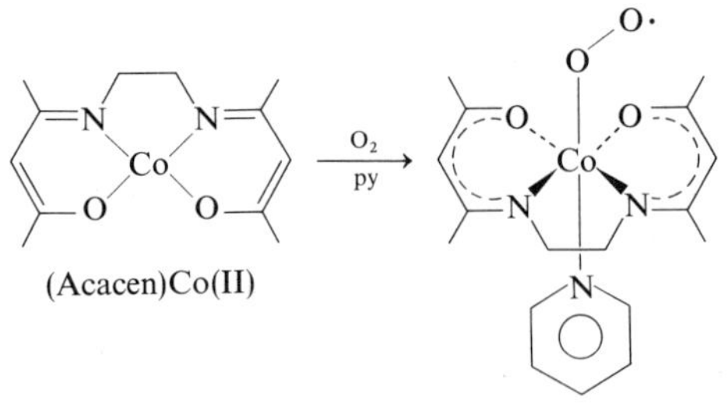

$$\text{(Acacen)Co(II)} \xrightarrow[\text{py}]{O_2} \text{(Acacen)(py)Co–O–O}\cdot \tag{2}$$

(Acacen)Co(II)

In 1963 Vaska[24] reported the first example of a reversible mononuclear dioxygen complex(**I**). Since then numerous stable, diamagnetic, mononuclear dioxygen complexes of Group VIII elements (Co, Ir, Rh, Ru, Os, Ni, Pd, Pt) have been prepared and characterized.[2–14] A few examples are shown below.

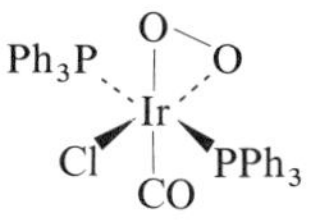

I 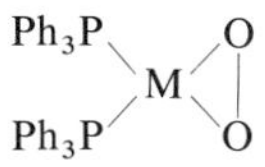 **II** (M = Ni, Pd, Pt) 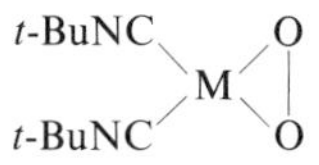 **III** (M = Ni, Pd)

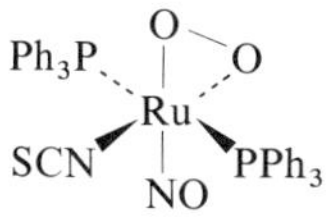

IV 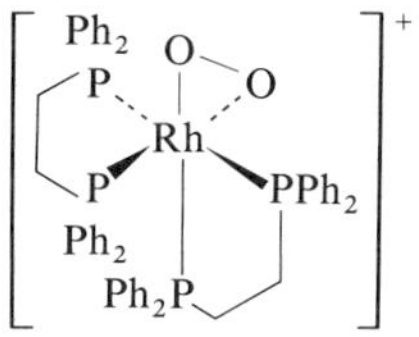 **V**

Metal–dioxygen complexes are important for two principal reasons. First, studies of model metal–dioxygen complexes should provide a better understanding of the natural oxygen carriers such as hemoglobin.[15,16,19] Second, the reactions of metal–dioxygen complexes could lead to the discovery of novel processes for the metal-mediated oxygenations of organic substrates.

II. PEROXOMETAL COMPLEXES: STRUCTURAL TYPES

The interaction of metal complexes with dioxygen can lead to successive one-electron transfers from the metal species, giving *superoxo* and *peroxo* complexes, MO_2 and MOOM, respectively, Vaska[2] reserves the terms "superoxo" and "peroxo" for the ligands, i.e., monohapto- and dihapto-coordinated O_2, respectively. "Superoxide" and "peroxide" refer to the free ions $O_2^{\cdot-}$ and O_2^{2-}, respectively (Scheme I).

Scheme I:

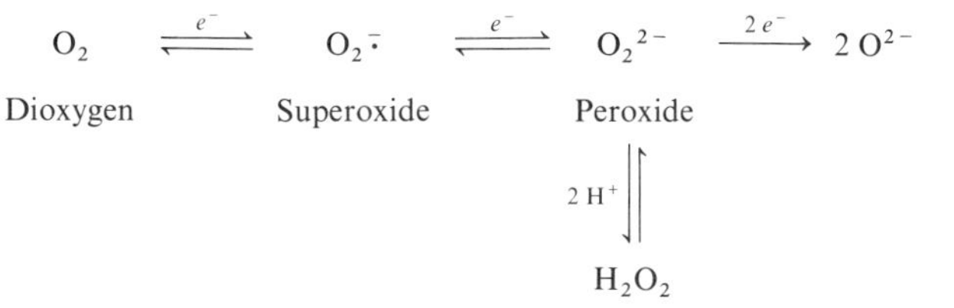

$$\underset{\text{Dioxygen}}{O_2} \overset{e^-}{\rightleftharpoons} \underset{\text{Superoxide}}{O_2^{\cdot-}} \overset{e^-}{\rightleftharpoons} \underset{\text{Peroxide}}{O_2^{2-}} \xrightarrow{2e^-} 2\,O^{2-}$$

$$O_2^{2-} \underset{}{\overset{2H^+}{\rightleftharpoons}} H_2O_2$$

Metal species in low oxidation states can often reduce dioxygen, the simplest example being the irreversible reduction of dioxygen to superoxide by alkali metals. The subsequent reduction of superoxide to peroxide has a higher reduction potential, and it does not proceed as readily.[25]

$$O_2 + K \longrightarrow KO_2 \tag{3}$$

Analogously, transition metal complexes can reduce dioxygen to superoxo complexes, many of which are formed reversibly at ambient temperatures. For example, superoxocobalt(III) complexes are formed by one-electron reduction of dioxygen by Co(II). In many cases, the superoxocobalt(III) species is only a transient intermediate and is rapidly reduced by a second Co(II) to a μ-peroxocobalt(III) complex.[25a]

$$Co^{II} + O_2 \rightleftharpoons Co^{III}{-}O{-}O\cdot \tag{4}$$

Superoxocobalt(III)

$$Co^{III}{-}O{-}O\cdot + Co^{II} \rightleftharpoons Co^{III}{-}O{-}O{-}Co^{III} \tag{5}$$

μ-Peroxocobalt(III)

When the next higher oxidation state of the metal is accessible, the dihapto peroxo complex is formed. Thus, a variety of Co(I), Ir(I), Pd(0), and Pt(0) complexes afford mononuclear peroxo complexes, in which the ligand is coordinated side-on.

$$M^{I} + O_2 \rightleftharpoons M^{II}{-}O{-}O\cdot \rightleftharpoons M^{III}\langle{}^{O}_{O}\rangle \tag{6}$$

Superoxo Peroxo

(See Luow *et al.*[25b] for kinetic evidence showing the sequential transformation of a superoxo- to a peroxometal complex.) These peroxo complexes generally afford hydrogen peroxide on treatment with a proton donor, e.g.,[26]

$$(Ph_3P)_2Pt^{II}O_2 + 2\,RCO_2H \longrightarrow (Ph_3P)_2Pt^{II}(O_2CR)_2 + H_2O_2 \tag{7}$$

$$(Ph_3P)_2Pt^{II}O_2 + C_6H_4(OH)_2 \longrightarrow C_6H_4O_2Pt^{II}(PPh_3)_2 + H_2O_2 \tag{8}$$

The reduction of dioxygen to hydrogen peroxide may also proceed by a direct process without the intermediacy of a dihapto peroxo complex. For example, the coordinatively saturated ruthenium(II) complex in eq 9 reacts with dioxygen via a ruthenium(IV) intermediate to afford hydrogen peroxide by the overall transformation.[27]

$$\text{(2-(1-hydroxyethyl)pyridine)N—Ru}^{\text{II}}(\text{NH}_3)_4 + \text{O}_2 \longrightarrow \text{(2-acetylpyridine)N—Ru}^{\text{II}}(\text{NH}_3)_4 + \text{H}_2\text{O}_2 \qquad (9)$$

accompanied by the oxidation of the alcoholic side chain to a ketonic function. The reduction of dioxygen probably proceeds via an outer-sphere electron transfer to yield a superoxide intermediate, as recently shown to occur in the kinetically inert osmium–porphyrin complex,

$$(\text{OEP})\text{Os}(\text{Py})_2 + \text{O}_2 \longrightarrow (\text{OEP})\text{Os}(\text{Py})_2^{\,+} + \text{O}_2^{\,\overset{\cdot}{-}}$$

where OEP = octaethylporphyrin[27a] (compare eq 3).

The peroxometal complexes formed from dioxygen in eq 6 are structurally analogous to the peroxometal complexes formed by reaction of various types of inorganic peracids[28] with different organic bases. For example, Mimoun and co-workers[29] prepared a series of Mo(VI) and W(VI) complexes by the reaction of MoO_3 or WO_3 with H_2O_2 in the presence of organic bases,

$$\text{MO}_3 + 2\,\text{H}_2\text{O}_2 \longrightarrow \text{O=M(OH)}_2(\text{O}_2\text{H})_2 \xrightarrow{\text{L}} \text{O=M(O}_2)_2\text{L} + 2\,\text{H}_2\text{O}$$

where M = Mo, W and L = DMF, DMAC, HMPA, py, etc. The O—O bond lengths of various peroxo- and superoxometal complexes are listed in Table I.

A distinction between a metal–dioxygen complex and a metal–peroxide complex has been made on the basis that the former is generated from molecular oxygen, whereas the latter is derived from hydrogen peroxide. However, this distinction is rather arbitrary since they are structurally equivalent and may, in some instances, actually be the same compound. For example, the peroxocobalt(III) complex **VI** has been synthesized from both Co(I) and O_2 as well as Co(III) and H_2O_2.[30]

$$\text{L}_4\text{Co}^{\text{I}} \xrightleftharpoons{\text{O}_2} \underset{\textbf{VI}}{\text{L}_4\text{Co}^{\text{III}}(\text{O}_2)} \xleftarrow[(-2\,\text{H}_3\text{O}^+)]{\text{H}_2\text{O}_2} cis\text{-}[\text{Co}^{\text{III}}(\text{H}_2\text{O})_2\text{L}_4]^{3+} \qquad (10)$$

where L_4 = two chelating diarsines. Similarly, a superoxocobalt(III) species is formed by the air oxidation of Co(II) or by reaction of Co(III) with superoxide anion.[31]

$$\text{Co}^{\text{II}} + \text{O}_2 \longrightarrow \text{Co}^{\text{III}}\text{—O—O}\cdot \longleftarrow \text{Co}^{\text{III}} + \text{O}_2^{\,\overset{\cdot}{-}} \qquad (11)$$

TABLE I O—O Bond Lengths in Peroxometal Complexes

Structure[a]		O—O (Å)	Reference[b]
O_2		1.21	1, 2
$O_2^{\overline{\cdot}}$		1.28	2
H_2O_2 (O_2^{2-})		1.47	2
Peroxo complexes			
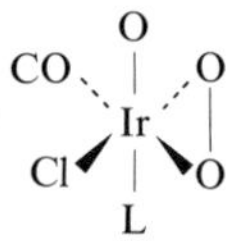 Ir(CO)(Cl)(L)(O)(O—O)	L = Ph_3P	1.30	3
	L = Ph_2EtP	1.46	4
 M^+(L)$_4$(O—O)	M — L–L		
	Ir $Ph_2PCH_2CH_2PPh_2$	1.52	5
	Rh $Ph_2PCH_2CH_2PPh_2$	1.42	6
	Co $Ph_2PCH{=}CHPPh_2$	1.42	7
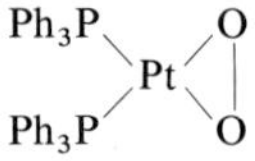 $(Ph_3P)_2Pt(O—O)$	1.5 C_6H_6	1.45	8
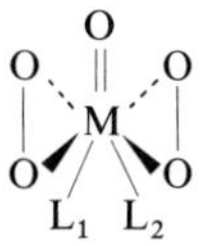 M(O)(O—O)$_2$(L_1)(L_2)	M — L_1, L_2		
	Cr py, DMF		
	Mo pyO, HMPA	1.40–1.50	9
	W Ph_3PO		
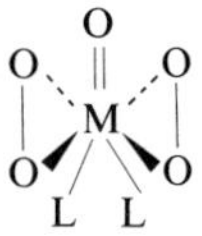 M(O)(O—O)$_2$(L–L)	M — L–L		
	Cr Bipy	1.40	10
	Mo $C_2O_4^{2-}$	1.55	11
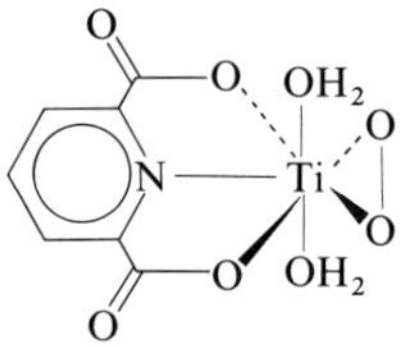 Ti(pyridine-2,6-dicarboxylato)(OH_2)$_2$(O—O)		1.46	12
μ-Peroxo complexes			
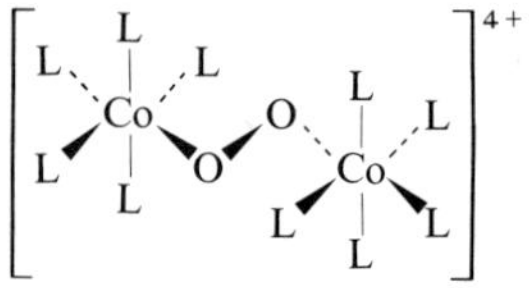 $[L_5Co—O—O—CoL_5]^{4+}$	L = NH_3	1.47	13

(continued)

TABLE I (*continued*)

Structure[a]	O—O (Å)	Reference[b]
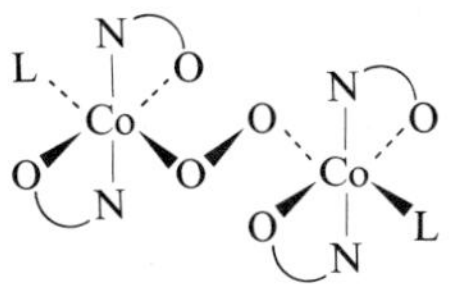 = [Co(Salen), DMF]$_2$O$_2$	1.35	14
Superoxo complexes		
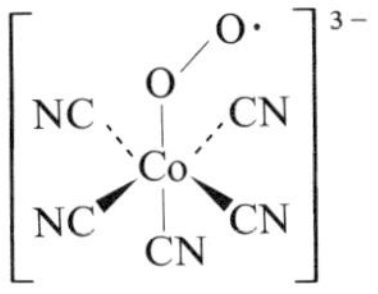	1.24	15
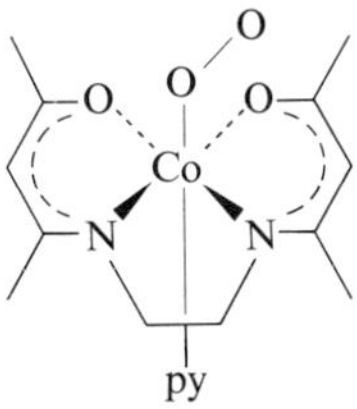	1.26	16

[a] Abbreviations: py, pyridine; pyO, pyridine oxide; DMF, dimethylformamide; HMPA, hexamethylphosphorus triamide; Bipy, bipyridine; Salen, salicylaldehyde ethylenimine.

[b] Key to references: 1, J. A. Connor and E. A. V. Ebsworth, *Adv. Inorg. Chem. Radiochem.* **6**, 279 (1964); 2, S. C. Abraham, *Q. Rev. Chem. Soc.* **10**, 407 (1956); 3, S. J. LaPlaca and J. A. Ibers, *J. Am. Chem. Soc.* **87**, 2581 (1965); 4, M. S. Weininger, I. F. Taylor, and E. L. Amma, *J. Chem. Soc., Chem. Commun.* p. 1172 (1971); 5, M. J. Nolte, E. Singleton, and M. Laing, *J. Am. Chem. Soc.* **97**, 6396 (1975); 6, J. A. McGinnety, N. C. Payne, and J. A. Ibers, *J. Am. Chem. Soc.* **91**, 6301 (1969); 7, N. W. Terry, E. L. Amma, and L. Vaska, *J. Am. Chem. Soc.* **94**, 653 (1972); 8, T. Kashiwagi, N. Yasuoka, N. Kasai, M. Kakudo, S. Takahashi, and N. Hagihara, *J. Chem. Soc., Chem. Commun.* p. 743 (1969); 9, H. Mimoun, *Rev. Inst. Fr. Pet.* **33**, 259 (1978); 10, R. Stomberg and I. Ainalem, *Acta Chem. Scand.* **22**, 1439 (1968); 11, R. Stomberg, *Acta Chem. Scand.* **23**, 2755 (1969); 12, D. Schwarzenbach, *Helv. Chim. Acta* **55**, 2990 (1972); 13, W. P. Schaeffer, *Inorg. Chem.* **7**, 725 (1968); 14, M. Calligans, G. Nardis, L. Randaccio, and A. Ripamonti, *J. Chem. Soc. A* p. 1069 (1970); 15, L. O. Brown and K. N. Raymond, *Inorg. Chem.* **14**, 2595 (1975); 16, G. A. Rodley and W. T. Robinson, *Nature* (*London*) **235**, 438 (1972).

The observation that dioxygen in oxyhemoglobin can be displaced by univalent anions to yield authentic ferric derivatives as in reaction 12 accords with its formulation as a superoxoferric complex.[32]

$$HbFe^{III}O_2\cdot + CN^- \longrightarrow HbFe^{III}CN + O_2^{\cdot -} \qquad (12)$$

Indeed, the microscopic reverse process, i.e., the formation of a superoxo complex by reaction of an Fe(III)–porphyrin complex with superoxide ion has been reported.[33]

Superoxide ion has also been used as the oxygen source in the preparation of dioxygen complexes of palladium.[34] Rhodium–olefin complexes have also been prepared by this method, in which the dioxygen serves as a bridging ligand,

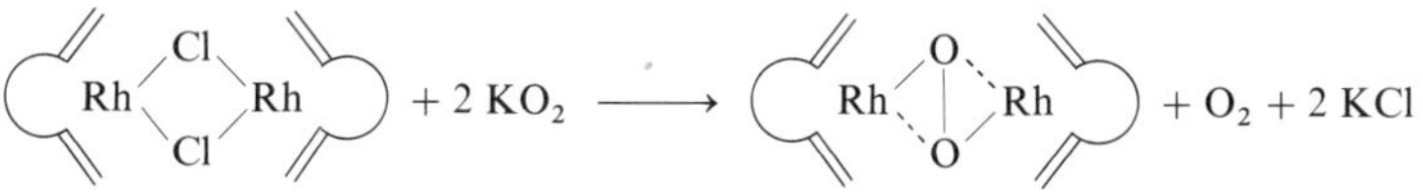

$$(\text{diene})Rh(\mu\text{-}Cl)_2Rh(\text{diene}) + 2\,KO_2 \longrightarrow (\text{diene})Rh(\mu\text{-}O_2)Rh(\text{diene}) + O_2 + 2\,KCl$$

The tentative structure above differs from the more common mononuclear dioxygen complexes of rhodium, and it is somewhat analogous to the complex,

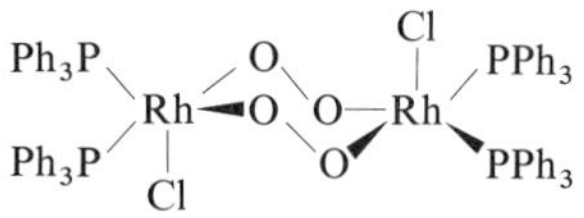

$$(Ph_3P)_2ClRh(O_2)_2RhCl(PPh_3)_2$$

the structure of which was determined earlier.[34]

In principle, peroxomolybdenum(VI) complexes should be available from the reaction of an appropriate Mo(IV) species with molecular oxygen. However, such reactions tend to afford oxometal species by the subsequent cleavage of a transient μ-peroxometal intermediate, e.g.,

$$2\,Mo^{IV} + O_2 \longrightarrow Mo^{V}{-}O{-}O{-}Mo^{V} \longrightarrow 2\,O{=}Mo^{VI} \qquad (13)$$

A peroxotitanium(IV)–porphyrin complex has been prepared both by the reaction of H_2O_2 with the oxotitanium(IV) porphyrin and by the reaction of the titanium(III) porphyrin with dioxygen,[33a]

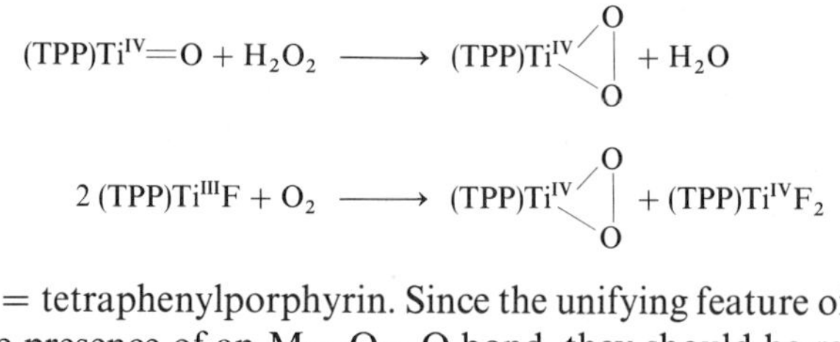

$$(TPP)Ti^{IV}{=}O + H_2O_2 \longrightarrow (TPP)Ti^{IV}(O_2) + H_2O$$

$$2\,(TPP)Ti^{III}F + O_2 \longrightarrow (TPP)Ti^{IV}(O_2) + (TPP)Ti^{IV}F_2$$

where TPP = tetraphenylporphyrin. Since the unifying feature of these complexes is the presence of an M—O—O bond, they should be referred to as peroxometal (or superoxometal) complexes, irrespective of their origins.

It is more reasonable to classify peroxometal complexes according to the nucleophilic and electrophilic properties of the coordinated dioxygen. Usually, peroxometal complexes containing metals in high oxidation states [e.g., Mo(VI), W(VI)] are electrophilic, whereas those containing metals in lower oxidation states are generally nucleophilic.

The ease with which lithium *n*-butoxide is formed in the reaction of *n*-butyllithium with a peroxo complex has been taken as a measure of the electrophilicity of the peroxidic linkage.[35] Typical high-valent peroxometal complexes such as MoO_5(HMPA) afford lithium butoxide readily at $-78°C$. The low-valent peroxometal complex $(Ph_3P)_2PtO_2$, on the other hand, resembles sodium peroxide in that it does not afford lithium *n*-butoxide, but $(Ph_3P)_2Pt(Bu)_2$ via the nucleophilic displacement of peroxide.

The various oxygenated metal species that can be formed in the interaction of metal complexes with dioxygen are summarized in Scheme II.

In addition to forming peroxo complexes, dioxygen can also insert into M—H or M—R bonds to form hydroperoxo- and alkylperoxometal complexes, respectively, in Scheme II. The best known example of the former is the reaction of pentacyanocobalt(III) hydride with oxygen.[36]

$$[(NC)_5Co^{III}H]^{3-} + O_2 \longrightarrow [(NC)_5Co^{III}—O—OH]^{3-} \tag{14}$$

Hydroperoxo complexes can also be formed by hydrolysis of peroxo or μ-peroxo complexes, e.g.,[36]

$$[(NC)_5Co^{III}—O—O—Co^{III}(CN)_5]^{6-} + H_2O \longrightarrow [(NC)_5Co^{III}—O—OH]^{3-} \tag{15}$$

Bis(triphenylphosphine)peroxoplatinum(II) has recently been reported[37] to form a dimeric μ-peroxo complex by hydrolysis, presumably via the hydroperoxo complex.

$$L_2Pt(O_2) + H_2O \longrightarrow L_2Pt(O—OH)(OH) \longrightarrow L_2Pt^+(OH) + HO_2^- \xrightarrow{L_2PtO_2} \left[L_2Pt(\mu\text{-}O—O)(\mu\text{-}OH)PtL_2\right]^+ \tag{16}$$

Hydroperoxometal complexes are probably intermediates in the hydrogenation of oxygen catalyzed by noble metals and their complexes.[38]

$$M—H + O_2 \longrightarrow M—O—OH \xrightarrow{H_2} MH + H_2O_2 \tag{17}$$

Peroxo- and superoxometal complexes usually have limited existence in dipolar, protic solvents owing to the rapid displacement of the peroxo or superoxo ligand by the solvent (or a protic substrate), e.g.,

$$M^{n+}—O—O\cdot + ROH \rightleftharpoons M^{n+}—OR + HO_2\cdot \tag{18}$$

Scheme II:

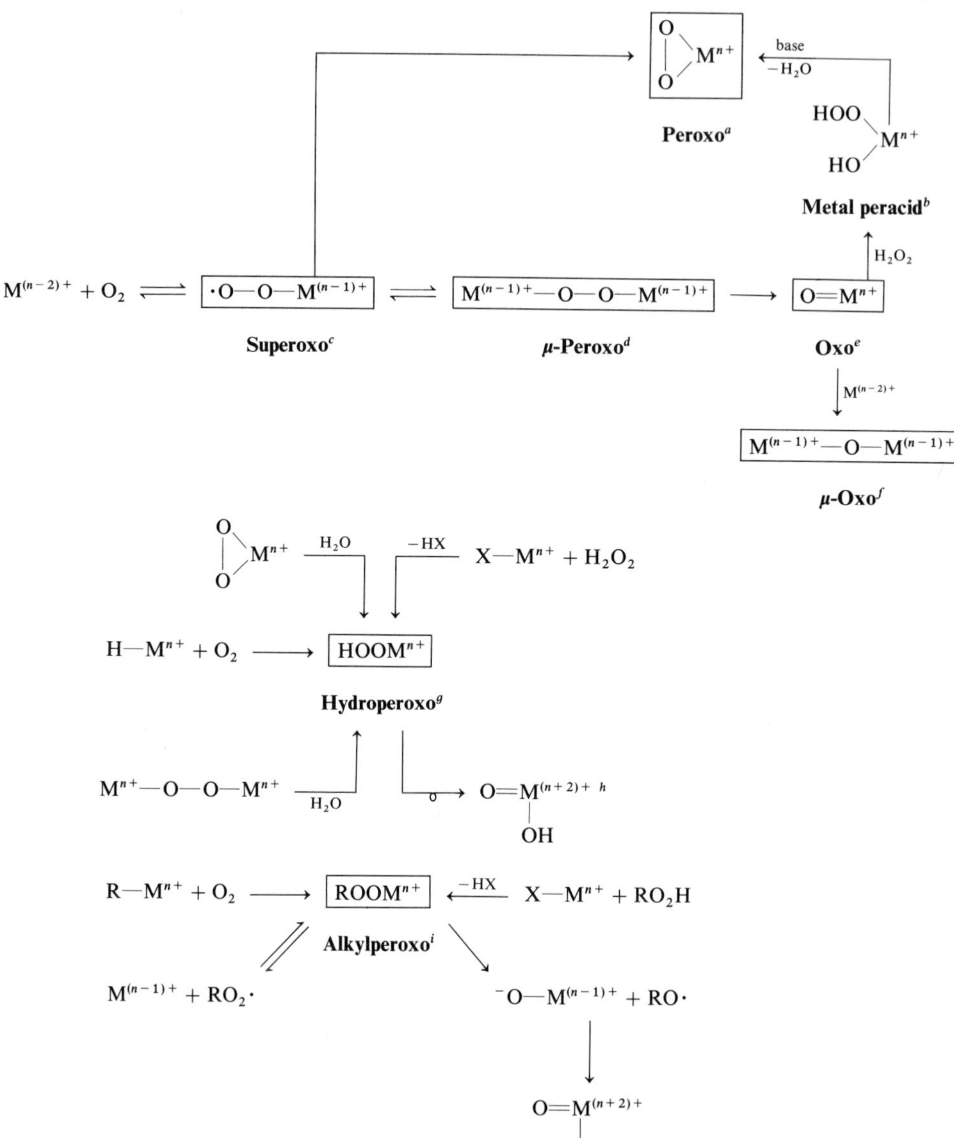

[a] Co^{III}, Rh^{III}, Ir^{III}, Ni^{II}, Pd^{II}, Pt^{II}, Ti^{IV}, V^{V}, Nb^{V}, Ta^{V}, Cr^{VI}, Mo^{VI}, W^{VI}, $Ru^{III,IV}$, $Os^{II,IV}$.
[b] Ti^{IV}, V^{V}, Nb^{V}, Ta^{V}, Cr^{VI}, Mo^{VI}, W^{VI}.
[c] Fe^{III}, Co^{III}, (Cu^{II}, Mn^{III}).
[d] Fe^{III}, Co^{III}, (Cu^{II}, Mn^{III}).
[e] Ti^{IV}, Zr^{IV}, Hf^{IV}, V^{V}, Nb^{V}, Ta^{V}, Cr^{VI}, Mo^{VI}, W^{VI}, $Fe^{IV,V,VI}$.
[f] V^{V}, Mo^{VI}, Cr^{VI}, Mn^{III}, Fe^{III}.
[g] Co^{III}, Rh^{III}, (Fe^{III}).
[h] (Fe^{IV}, Fe^{V}).
[i] Co^{III}, Pd^{II}.

This substitution has an important bearing on any mechanistic discussions of the reactions of peroxometal complexes (*vide infra*).

Alkylperoxometal complexes M—O—OR are intermediates in the autoxidations of organometallic compounds.[39] Insertion of oxygen into alkylcobalt(III) bonds has been suggested as a key step in the oxidation of various compounds catalyzed by cobalt(II)–Schiff base complexes (see Section V). It should be noted, however, that the insertion of ground-state, triplet oxygen into an M—R or an M—H bond is almost certainly not concerted, but a multistep process involving several one-electron changes. For example, it has been recently reported[40] that the autoxidation of isopropylchromium(III) cation proceeds via the following radical chain mechanism.

$$Cr^{III}\text{—}CH(CH_3)_2{}^{2+} \longrightarrow Cr^{2+} + (CH_3)_2CH\cdot \tag{19}$$

$$(CH_3)_2CH\cdot + O_2 \longrightarrow (CH_3)_2CHO_2\cdot \tag{20}$$

$$(CH_3)_2CHO_2\cdot + Cr^{III}\text{—}CH(CH_3)_2 \longrightarrow Cr^{III}\text{—}O\text{—}O\text{—}CH(CH_3)_2 + (CH_3)_2CH\cdot \tag{21}$$

Alkylperoxometal complexes are also, in principle, available from the reaction of alkyl hydroperoxides with metal complexes (see Scheme II).

III. NUCLEOPHILIC PEROXOMETAL COMPLEXES

A. Stoichiometric Oxidations

Mononuclear peroxo complexes of Group VIII metals oxidize a variety of inorganic oxides in stoichiometric reactions.[3–5,11,12] A few examples for the most widely studied complex, $(Ph_3P)_2PtO_2$, are shown below.

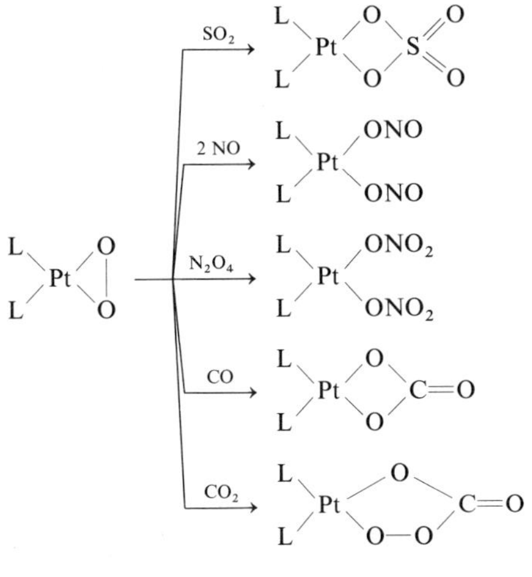

$$L_2Pt(O_2) \xrightarrow{SO_2} L_2Pt(O_2)SO_2 \text{ (cyclic } L_2Pt\text{–O–S(=O)}_2\text{–O)}$$
$$L_2Pt(O_2) \xrightarrow{2\,NO} L_2Pt(ONO)_2$$
$$L_2Pt(O_2) \xrightarrow{N_2O_4} L_2Pt(ONO_2)_2 \tag{22}$$
$$L_2Pt(O_2) \xrightarrow{CO} L_2Pt(O)_2C{=}O \text{ (cyclic carbonate)}$$
$$L_2Pt(O_2) \xrightarrow{CO_2} L_2Pt\text{–O–C(=O)–O–O–} \text{ (cyclic peroxocarbonate)}$$

where L = Ph_3P. These reactions are related to the catalytic oxidations of the oxides over noble metal catalysts; e.g., the contact process for the manufacture of sulfuric acid involves the aerial oxidation of SO_2 to SO_3 over a platinum catalyst.

The nucleophilic character of the coordinated dioxygen in $(Ph_3P)_2PtO_2$ is manifested in the addition to the carbonyl groups of aldehydes[41,41a] and ketones[41,41a] and to the double bond of electrophilic olefins.[42]

$$L_2Pt(O_2) \xrightarrow{PhCHO} L_2Pt(\text{O—O—CHPh—O}) \quad (23)$$

$$L_2Pt(O_2) \xrightarrow{(CH_3)_2CO} L_2Pt(\text{O—O—C(CH}_3)_2\text{—O}) \quad (24)$$

$$L_2Pt(O_2) \xrightarrow{(H_3C)_2C{=}C(CN)_2} L_2Pt(\text{O—O—C(CH}_3)_2\text{—C(CN)}_2) \quad (25)$$

A kinetic study is in agreement with a dual reaction mechanism in which the major pathway involves precoordination of the carbonyl group to the vacant axial site of platinum followed by insertion.[41a] The minor pathway is zero order in ketone and suggests a prior activation of the dioxygen moiety such as an isomerization to an end-on bonded dioxygen, e.g.,

$$L_2Pt(O_2) \longrightarrow L_2Pt^+\text{—O—O}^-$$

followed by a nucleophilic attack on the unsaturated center, i.e.,

$$L_2Pt^+\text{—O—O}^- + \;{>}C{=}C(CN)_2 \longrightarrow {>}C(\text{—O—O—Pt}^+)\text{—C}^-(CN)_2, \quad \text{etc.} \quad (26)$$

As expected, simple olefins do not react with $(Ph_3P)_2PtO_2$. Oxygen transfer to alkenes can be effected, however, in the presence of stoichiometric amounts of benzoyl chloride.[43]

$$L_2Pt(O_2) + PhCOCl \longrightarrow L_2Pt(\text{O—O}_2\text{CPh})(Cl) \quad (27)$$

$$L_2Pt(\text{O—O}_2\text{CPh})(Cl) + \text{cyclohexene} \longrightarrow L_2Pt(\text{O}_2\text{CPh})(Cl) + \text{cyclohexene oxide} \quad (28)$$

The benzoylperoxoplatinum(II) intermediate has been identified.

B. Catalytic Oxidations

For an efficient catalytic process, the product must be readily released from the metal complex. In other words, the coordination of the substrate must exceed that of the product. A few catalytic oxidations of easily oxidizable substrates are known with the Group VIII mononuclear peroxo complexes.[3–5,11,12] For example, numerous transition metal complexes catalyze the oxidation of organic phosphines to phosphine oxides. Wilke and co-workers[45] showed that complexes of the type $M(PPh_3)_4$, where M = Ni, Pd, Pt, catalyze the oxidation of triphenylphosphine in toluene. Sen and Halpern[46] demonstrated that the initially formed peroxoplatinum(II) complex $(Ph_3P)_2PtO_2$ reacts with a molecule of Ph_3P to form the active intermediate. It seems likely that triphenylphosphine promotes the dissociation to the active 1,3-dipolar form.

$$L_2Pt(O_2) + L \longrightarrow L_2Pt^+(L)(O{-}O^-) \quad (29a)$$

$$L_2Pt^+(L)(O{-}O^-) + L \longrightarrow L_2Pt^+(L)(O^-) + LO \quad (29b)$$

$$L_2Pt^+(L)(O^-) + L \longrightarrow L_2Pt(L) + LO \quad (29c)$$

where L = triphenylphosphine. The last step in eq 29c results in the regeneration of the Pt(0) catalyst and represents an example of an oxygen transfer from an oxometal species (compare the Rh(I) promoted oxygenation of olefins discussed later). In protic solvents such as ethanol, solvolysis could produce the hydroperoxide anion as the active oxidant.[46]

$$L_2Pt^+(L)(O{-}O^-) + EtOH \rightleftharpoons L_2Pt^+(L)(OEt) + HO_2^-$$

Peroxometal complexes of Ru, Rh, Ir, and Co also catalyze the autoxidation of phosphines, presumably via analogous mechanisms.[3–5,11,12] Since all of these reactions involve oxidations by the peroxide anion (whether it is coordinated or free), they apply only to easily oxidizable substrates.

The oxygenation of triphenylphosphine to triphenylphosphine oxide is also catalyzed by the iron porphyrin (TPP)Fe(II) by a mechanism that differs from the $(Ph_3P)_3Pt(0)$-catalyzed process.[47] Thus, it could be shown independently that the oxoiron(IV) complexes (TPP)O=FeB, where B = *N*-methylimidazole, pyridine, etc., formed in the autoxidation of (TPP)Fe(II), are capable of reacting directly with triphenylphosphine at $-80°C$ in a stoichiometric reaction.

$$(TPP)O{=}Fe^{IV}B + Ph_3P \xrightarrow{B} (TPP)Fe^{II}B_2 + Ph_3P{=}O$$

Such an oxygen atom transfer from oxoiron(IV) to phosphine is represented in the catalytic process outlined below,

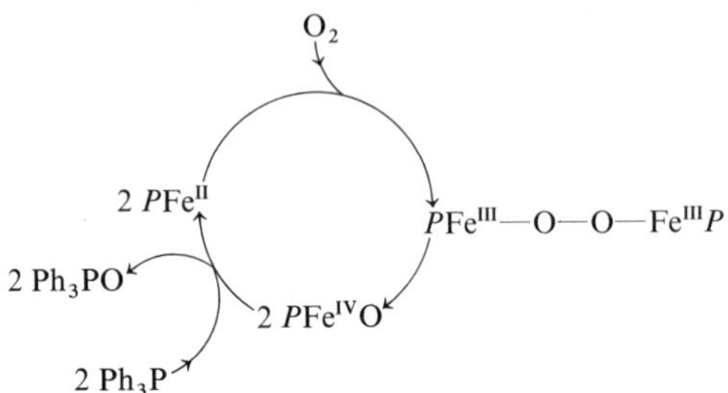

where P is TPP.

Of greater interest is the possibility of effecting selective, nonradical oxidations of hydrocarbons under mild conditions. For example, the selective epoxidation of olefins analogous to the known Ag-catalyzed epoxidation of ethylene in eq 1 is desirable, since the application of this reaction to the higher homologs results in nonselective, homolytic oxidation.

The oxidation of cyclohexene[48] and alkylbenzenes[49,50] in the presence of Group VIII metal–peroxo complexes was originally considered to proceed by nonradical pathways. However, more detailed studies[51–59] have invariably revealed that these reactions involve initiation via the metal-catalyzed decomposition of hydroperoxides into chain-initiating radicals (see Chapter 3). Thus, the oxidation of cyclohexene in the presence of various low-valent phosphine complexes of Group VIII transition metals [Pd(0), Pt(0), Ir(I), Rh(I)] involves chain initiation via the usual redox reactions of the metal complexes with adventitious hydroperoxides,[54] since long induction periods are observed with peroxide-free hydrocarbons.[53,54] Low-valent Group VIII metal complexes are known to catalyze the rapid decomposition of alkyl hydroperoxides under anaerobic conditions.[55–60]

The first authentic example of a selective nonradical oxidation of an olefin is the $RhCl(PPh_3)_3$-promoted cooxidation of terminal olefins and triphenylphosphine to methyl ketones and triphenylphosphine oxide.[61] Oxygen-labeling studies show that the oxygen atoms in both products are derived from dioxygen when the reaction is carried out in the presence of a small amount of water.[62,63] The following mechanism was proposed by Read and Walker.[64]

Scheme III:

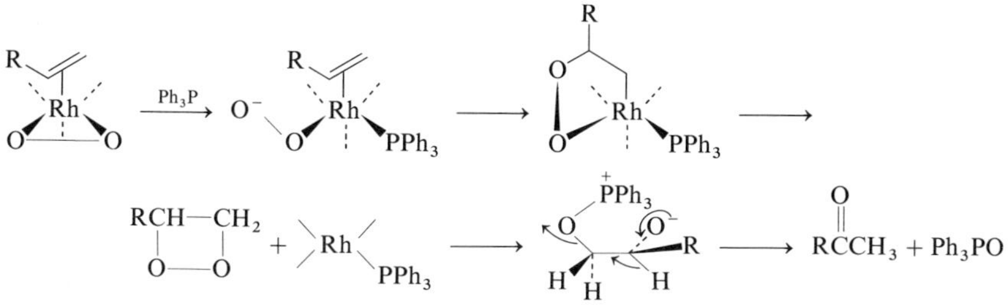

$$\text{R-CH=CH}_2\cdot\text{Rh}(\text{O}_2) \xrightarrow{\text{Ph}_3\text{P}} \text{O}^-\text{-O-Rh(olefin)PPh}_3 \longrightarrow \text{peroxymetallocycle (Rh, PPh}_3) \longrightarrow$$

$$\text{RCH-CH}_2\text{(O-O)} + \text{Rh-PPh}_3 \longrightarrow [\text{PPh}_3^+ \ldots] \longrightarrow \text{RCCH}_3(=\text{O}) + \text{Ph}_3\text{PO}$$

This mechanism exhibits several interesting features. First, the triphenylphosphine appears to promote dissociation of the peroxometal group to the 1,3-dipolar form (compare eq 29). Second, a peroxymetallocycle is formed by an intramolecular attack of a peroxo ligand on a coordinated olefin. As we have already noted, such a nucleophilic addition is not to be expected for an intermolecular process, but the coordination of an olefin to rhodium could render the double bond electrophilic as in the Pd(II)-catalyzed nucleophilic additions to simple olefins in the well-known Wacker and related processes.[65] Although Read and co-workers favor initial decomposition of the peroxymetallocycle to a dioxetane, an alternative decomposition can be envisioned in which β-elimination from the Rh-alkyl bond affords the enol and an oxorhodium(III) species.

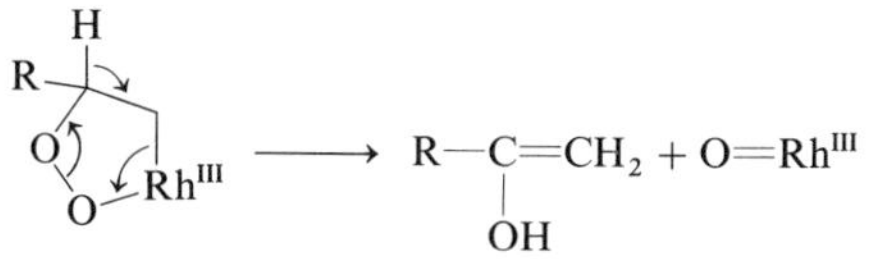

$$\text{R(H)C-CH}_2\text{-Rh}^{\text{III}}\text{(O-O)} \longrightarrow \text{R-C(OH)=CH}_2 + \text{O=Rh}^{\text{III}} \tag{30}$$

Reduction of the latter by Ph_3P regenerates the Rh(I) catalyst. Despite the unattractive use of stoichiometric amounts of triphenylphosphine, this reaction illustrates the important principle that oxygen transfer to simple olefins is feasible when accompanied by the simultaneous transfer of the second oxygen atom to an electron acceptor. Other workers[66–68] have reported similar Rh(I)-catalyzed oxidations of styrene and cyclooctene to ketones, but products derived from oxidative cleavage of the double bond were also observed.

An interesting example of a metal-catalyzed oxygen transfer reaction is provided by the rhodium-catalyzed oxidation of a variety of olefins reported by Mimoun and co-workers.[63] Terminal olefins are selectively converted to methyl ketones in ethanol at ambient temperatures in the presence of $RhCl_3$ and cupric perchlorate or nitrate and dioxygen. The first steps in the catalytic sequence involve the generation of a cationic Rh(I)–olefin complex,

$$2\,RhCl_3 + 4\,RCH{=}CH_2 + 2\,C_2H_5OH \longrightarrow [(RCH{=}CH_2)_2RhCl]_2 + 2\,CH_3CHO + 4\,HCl$$

$$[(RCH{=}CH_2)_2RhCl]_2 + 2\,CuClO_4 \longrightarrow 2\,(RCH{=}CH_2)_2Rh^+ClO_4^- + 2\,CuCl$$

which reacts with molecular oxygen via the following sequence.

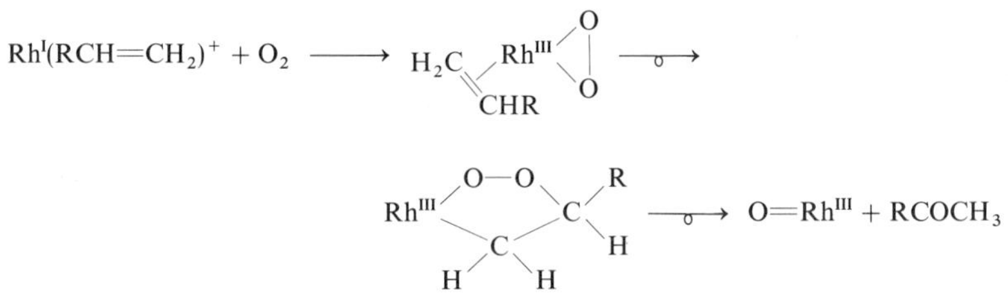

$$Rh^I(RCH{=}CH_2)^+ + O_2 \longrightarrow (H_2C{=}CHR)Rh^{III}(O_2) \longrightarrow$$

$$Rh^{III}(O{-}O{-}CHR{-}CH_2) \longrightarrow O{=}Rh^{III} + RCOCH_3 \quad (31)$$

Up to this point, the mechanism is similar to the cooxygenation of olefins with Ph_3P in Scheme III. However, in the absence of a coreductant, Mimoun proposed that the catalyst is regenerated from the oxorhodium(III) species and a strong acid via reduction by the olefin in a Wacker-type process which generates an additional molecule of ketone (Scheme IV).[63]

Scheme IV:

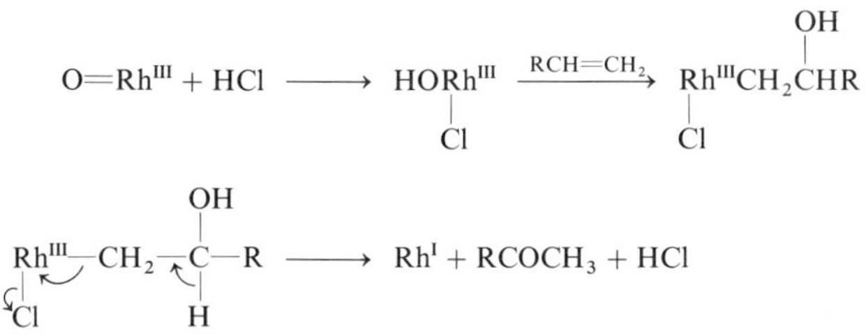

$$O{=}Rh^{III} + HCl \longrightarrow HORh^{III}Cl \xrightarrow{RCH{=}CH_2} Cl{-}Rh^{III}CH_2CH(OH)R$$

$$Cl{-}Rh^{III}{-}CH_2{-}C(OH)(H){-}R \longrightarrow Rh^I + RCOCH_3 + HCl$$

The overall process constitutes a Rh(I)-catalyzed conversion of two molecules of olefin and one of oyxgen into two molecules of ketone. The reaction is strongly inhibited by water, in contrast to a normal Wacker oxidation described in Chapter 7. Furthermore, copper does not play a direct role in the oxidation and is not essential for catalytic activity.

Cycloheptene and cyclooctene are oxidized to the corresponding ketones. Straight-chain internal olefins are oxidized much more slowly than their terminal counterparts. Small amounts of cleavage products are often observed in these reactions; e.g., 1-octene affords some *n*-heptanal,[61–64] possibly by an alternative decomposition of the peroxymetallocycle.

$$\mathrm{RHC{-}O{-}O{-}Rh^{III}{-}CH_2} \text{ (peroxymetallocycle)} \longrightarrow \mathrm{O{-}Rh^{III}{-}CH_2} + \mathrm{RCHO}$$

$$\mathrm{O{-}Rh^{III}{-}CH_2} \longrightarrow \mathrm{Rh^{I}} + \mathrm{CH_2O} \qquad (32)$$

Such a reaction is analogous to the thermal decomposition of the peroxyplatinum(II) and-palladium(II) compounds.[42]

$$\mathrm{(Ph_3P)_2M[O{-}O{-}C(CH_3)_2{-}C(CN)_2]} \xrightarrow{\Delta} \mathrm{(Ph_3P)_2Pt[O{-}C(CN)_2]} + \mathrm{(CH_3)_2CO} \qquad (33)$$

Mimoun[9,63] favors an alternative mechanism involving oxidation by the oxorhodium(III) species via a [2 + 2]-cycloaddition (see Chapter 6).

$$\mathrm{O{=}Rh^{III}} + \mathrm{RCH{=}CH_2} \longrightarrow \mathrm{Rh{-}O{-}CHR{-}CH_2}$$

$$\mathrm{Rh{-}O{-}CHR{-}CH_2} \longrightarrow \mathrm{H_2C{=}Rh^{III}} + \mathrm{RCHO} \qquad (34)$$

$$\mathrm{H_2C{=}Rh^{III}} + \mathrm{O_2} \longrightarrow \mathrm{Rh{-}O{-}O{-}CH_2} \longrightarrow \mathrm{O{=}Rh^{III}} + \mathrm{CH_2O}$$

In accord with this mechanism, oxidative cleavage seems to occur when no acid is present in the reaction mixture to effect reduction of the oxorhodium(III) via the Wacker pathway described in Scheme IV. The series of reactions above (eq 34) are particularly noteworthy, since they constitute the catalytic cleavage of an olefinic double bond by molecular oxygen.

Hexanol is another minor product formed in the $RhCl(PPh_3)_3$-catalyzed oxidation of 1-octene with O_2.[69] The cyclic mechanism in eq 35 evokes a migration of the R = hexyl group in the peroxymetallocycle intermediate,

$$\mathrm{Rh^{III}[O{-}O{-}CH(R){-}CH_2]} \longrightarrow \mathrm{O{=}Rh^{III}} + \mathrm{ROCH{=}CH_2}$$

$$\mathrm{ROCH{=}CH_2} \xrightarrow{H_2O} \mathrm{ROH} + \mathrm{CH_3CHO} \qquad (35)$$

reminiscent of the Baeyer–Villiger oxidation with peroxides.

The Rh-catalyzed autoxidation of terminal olefins to methyl ketones described above compares with a similar Pd-catalyzed transformation using *tert*-alkyl hydroperoxides as the terminal oxidant, e.g.,

$$n\text{-}C_4H_9CH{=}CH_2 + BuOOH \xrightarrow{[Pd]} n\text{-}C_4H_9\overset{\displaystyle O}{\overset{\|}{C}}CH_3 + BuOH \qquad (36)$$

In this catalytic process, an alkylperoxopalladium(II) species formed as a tetrameric complex by the metathesis,

$$4\,(RCO_2)_2Pd + 4\,BuOOH \longrightarrow [RCO_2PdOOBu]_4 + 4\,RCO_2H$$

(where R = CH_3, Cl_3C, CF_3, and C_5F_{11}), reacts with terminal olefins to form methyl ketones, e.g.,[44]

$$CF_3CO_2PdOOBu + n\text{-Bu}\diagup\!\!\diagdown\!\!\!= \longrightarrow n\text{-Bu}\overset{\displaystyle O}{\overset{\|}{C}}CH_3 + CF_3CO_2PdOBu \qquad (37)$$

Mimoun and co-workers have suggested that the oxidative ketonization of terminal olefins proceeds by a two-step mechanism involving the formation of a β-peroxyalkyl adduct,[44]

$$CF_3CO_2PdOOBu + RCH{=}CH_2 \longrightarrow CF_3CO_2PdCH_2\overset{\displaystyle R}{\overset{|}{C}}HOOBu \qquad (38)$$

followed by a β-hydride migration,

$$CF_3CO_2Pd(CH_2)(O{-}O{-}Bu)C(H)(R) \longrightarrow CF_3CO_2PdOBu + CH_3\underset{\underset{\displaystyle O}{\|}}{C}R \qquad (39)$$

Indeed the support for this formulation derives from the formation of acetophenone in 60% yield when palladium(II) trifluoroacetate is treated with $CF_3CO_2HgCH_2CH(Ph)OO\text{-}t\text{-}Bu$.

The similarity of the Pd-catalyzed process in eq 36 to the Rh-catalyzed one in Scheme III has been shown to be even greater by the recent observation that the peroxopalladium(II) complex is capable of stoichiometrically converting terminal olefins to methyl ketones in the presence of acids, i.e.,[144]

$$(Ph_3P)_2PdO_2 + CH_3SO_3H \longrightarrow [(Ph_3P)_2PdOOH^+O_3SCH_3^-] \xrightarrow{RCH=CH_2} RCOCH_3, \quad \text{etc.} \qquad (40)$$

Olefin insertion into a palladium–hydroperoxide bond is the presumed route by which such a transformation occurs, since it is related to the catalytic process with hydrogen peroxide, i.e.,[145]

$$RCH{=}CH_2 + H_2O_2 \xrightarrow{[Pd]} RCOCH_3 + H_2O \tag{41}$$

This process occurs with 85–95% selectivity. The mechanism of this reaction is to be distinguished from that pertaining to the Wacker process, since the ketonic oxygen is derived from hydrogen peroxide, not water.

$$n\text{-}C_4H_9CH{=}CH_2 + H_2O_2{}^* \xrightarrow{[Pd]} n\text{-}C_4H_9\overset{O^*}{\overset{\|}{C}}CH_3 + H_2O$$

The key steps are:

$$\begin{array}{l}CH_2{=}CHR\\ \swarrow\quad\nearrow\\ Pd{-}OOH\end{array} \longrightarrow \begin{array}{l}CH_2{-}CHR\\ \;|\qquad\;|\\ Pd\quad\;\; O\\ HO\diagup\end{array} \longrightarrow Pd^{II}OH + RCOCH_3 \tag{42}$$

Evidence for the β-hydroperoxyalkylpalladium(II) intermediate was obtained from the observation that ligand transfer from the corresponding β-hydroperoxyalkylmercury(II) derivative also results in the formation of methyl ketone.[145]

$$CF_3CO_2HgCH_2CH(R)OOH \xrightarrow{[Pd]} R\overset{O}{\overset{\|}{C}}CH_3 + CF_3CO_2HgOH$$

The decomposition of the quasi-peroxymetallocycle in eqs 39 and 42 is akin to the decomposition of the peroxymetallocycle in eq 30. Coupled with the similarity in the steps leading to olefin insertion (eqs 38 and 42), these processes bear striking resemblance to olefin insertions into metal–alkyl and metal–hydride bonds,

$$M{-}R + \rangle C{=}C\langle \longrightarrow M{-}\overset{|}{\underset{|}{C}}{-}\overset{|}{\underset{|}{C}}{-}R$$

$$M{-}H + \rangle C{=}C\langle \longrightarrow M{-}\overset{|}{\underset{|}{C}}{-}\overset{|}{\underset{|}{C}}{-}H$$

which are so important in olefin oligomerization, polymerization, and reduction, respectively.[39]

The postulation of peroxymetallocycles as intermediates allows all the different types of products to be rationally formulated in the metal-mediated oxygen transfer to olefinic substrates. At least four different pathways can be envisaged, as delineated in Scheme V.

Scheme V:

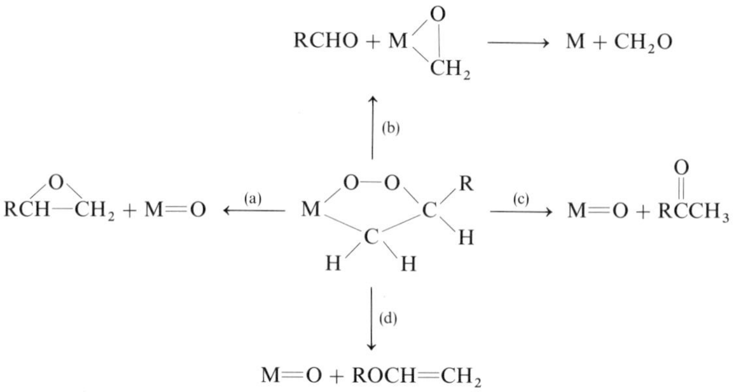

$$\mathrm{RCHO} + \mathrm{M}\!\left\langle{}^{\mathrm{O}}_{\mathrm{CH_2}}\right| \longrightarrow \mathrm{M} + \mathrm{CH_2O}$$

$$\uparrow \text{(b)}$$

$$\mathrm{RCH}\overset{\mathrm{O}}{\frown}\mathrm{CH_2} + \mathrm{M{=}O} \xleftarrow{\text{(a)}} \mathrm{M}\langle\mathrm{O{-}O{-}C(R)(H){-}C(H)(H)}\rangle \xrightarrow{\text{(c)}} \mathrm{M{=}O} + \mathrm{R\overset{\overset{\displaystyle O}{\|}}{C}CH_3}$$

$$\downarrow \text{(d)}$$

$$\mathrm{M{=}O} + \mathrm{ROCH{=}CH_2}$$

With Group VIII metal complexes such as Rh(I) complexes, products derived from paths (b), (c), and (d) are found, but epoxides via (a) are not. Thus, it is interesting to compare these Rh(I)-catalyzed oxidations with the Ag-catalyzed epoxidation of ethylene, in the gas phase, which affords ethylene oxide in 75% selectivity.[1] Mimoun has proposed a mechanism involving a peroxyargentocycle (Scheme VI).[9]

Scheme VI:

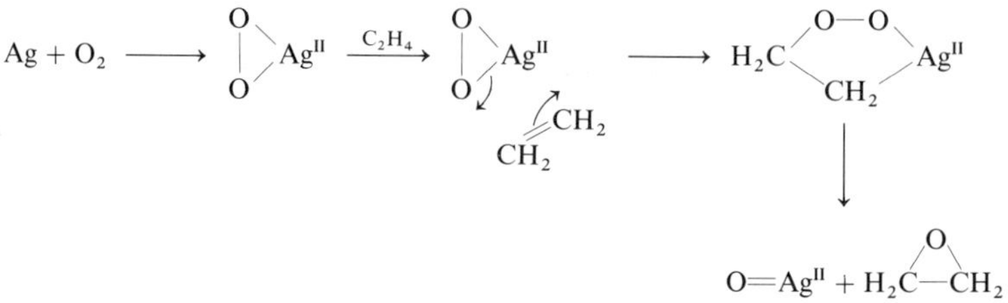

$$\mathrm{Ag} + \mathrm{O_2} \longrightarrow \left|{}^{\mathrm{O}}_{\mathrm{O}}\right\rangle\mathrm{Ag^{II}} \xrightarrow{\mathrm{C_2H_4}} \left|{}^{\mathrm{O}}_{\mathrm{O}}\right\rangle\mathrm{Ag^{II}}\ (\mathrm{CH_2{=}CH_2}) \longrightarrow \mathrm{H_2C}\langle\mathrm{O{-}O}\rangle\mathrm{Ag^{II}}\ (\text{ring via }\mathrm{CH_2})$$

$$\downarrow$$

$$\mathrm{O{=}Ag^{II}} + \mathrm{H_2C}\overset{\mathrm{O}}{\frown}\mathrm{CH_2}$$

Coordination of the ethylene to the strong Ag(II) electrophile renders the double bond susceptible to nucleophilic attack by coordinated peroxide. Oxosilver(II) is reduced by another molecule of ethylene to yield formaldehyde and a carbene–silver(II) complex (eq 43),

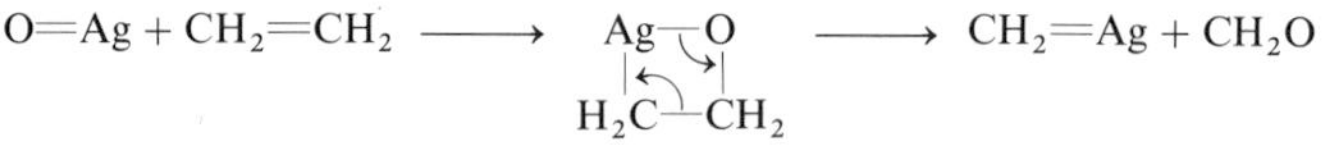

$$\mathrm{O{=}Ag} + \mathrm{CH_2{=}CH_2} \longrightarrow \begin{array}{c}\mathrm{Ag{-}O}\\ |\quad\ |\\ \mathrm{H_2C{-}CH_2}\end{array} \longrightarrow \mathrm{CH_2{=}Ag} + \mathrm{CH_2O} \qquad (43)$$

which affords another molecule of formaldehyde by reaction with oxygen (eq 44).

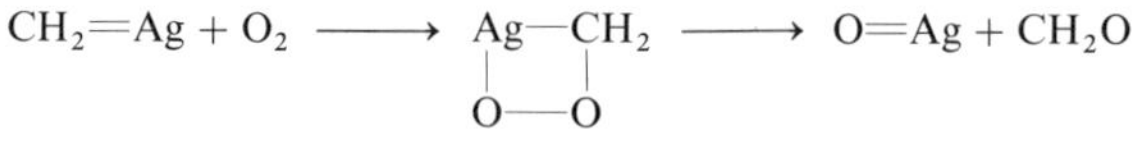

$$\mathrm{CH_2{=}Ag} + \mathrm{O_2} \longrightarrow \begin{array}{c}\mathrm{Ag{-}CH_2}\\ |\qquad |\\ \mathrm{O{-}O}\end{array} \longrightarrow \mathrm{O{=}Ag} + \mathrm{CH_2O} \qquad (44)$$

Formaldehyde and its oxidation product, formic acid, are both strong reducing agents that are capable of regenerating the Ag(0) catalyst from argentic oxide. Such a reaction scheme leads to the stoichiometry in eq 45, which predicts a maximal selectivity of 85% to ethylene oxide.

$$7\,C_2H_4 + 6\,O_2 \longrightarrow 6\,C_2H_4O + 2\,CO_2 + 2\,H_2O \qquad (45)$$

The Ag-catalyzed epoxidation constitutes an example of the reaction of an electrophilic peroxometal species formed from molecular oxygen, in which coordination of ethylene to Ag(II) renders the double bond susceptible to attack by the coordinated peroxy group. It is analogous to the stoichiometric epoxidations observed with the high-valent, electrophilic peroxometal complexes discussed in the following section.

Iridium(I) and rhodium(I) complexes catalyze the autoxidation of butadiene to polyperoxides at 55°C,[70] presumably by the reaction of dioxygen with the metal–butadiene complex to form a metal-bonded alkylperoxy radical.

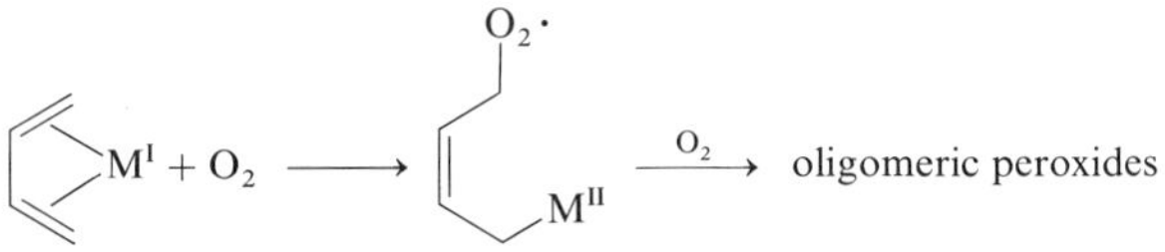

$$(C_4H_6)M^{I} + O_2 \longrightarrow \cdot O_2CH_2CH{=}CHCH_2M^{II} \xrightarrow{O_2} \text{oligomeric peroxides} \qquad (46)$$

The trapping of dioxygen by carbenes generated in the presence of palladium(II) acetate leads to intermediates that can transfer an oxygen atom to olefins to afford epoxides under mild conditions, e.g.,[71]

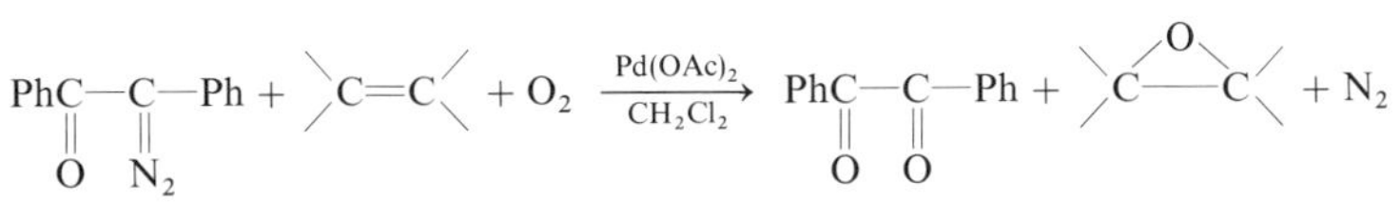

$$PhC(=O){-}C(=N_2){-}Ph + {>}C{=}C{<} + O_2 \xrightarrow[CH_2Cl_2]{Pd(OAc)_2} PhC(=O){-}C(=O){-}Ph + {>}C\overset{O}{\frown}C{<} + N_2$$

Peroxopalladium species such as:

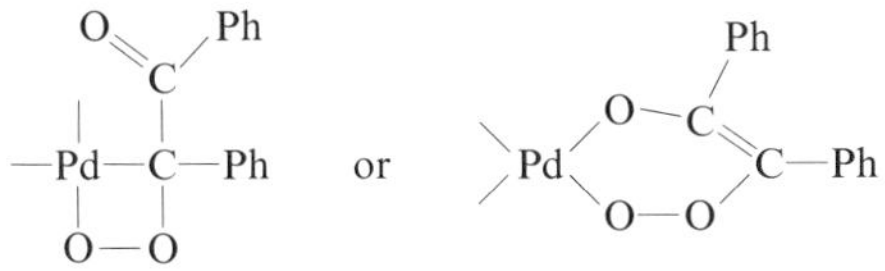

$$-Pd(-O-O-)C(Ph)(C(=O)Ph) \quad \text{or} \quad {>}Pd(-O-C(Ph){=}C(Ph)-O-O-)$$

have been suggested as the intermediates responsible for oxygen atom transfer.

IV. ELECTROPHILIC PEROXOMETAL COMPLEXES

It has been known for many years that Ti(IV), V(V), Cr(VI), Mo(VI), and W(VI) oxides react with hydrogen peroxide to form stable compounds containing the O=M—O—OH group, designated inorganic peracids.[28]

In the presence of bases, peroxometal adducts containing one or more organic ligands can be isolated,[29,72–77] as illustrated by a few examples below.

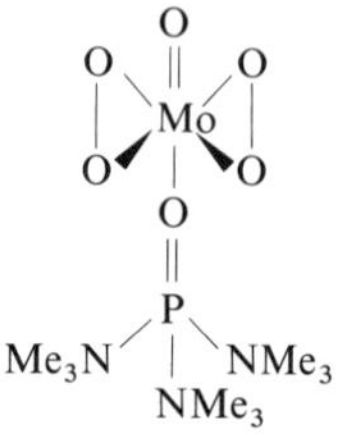

VII[29] 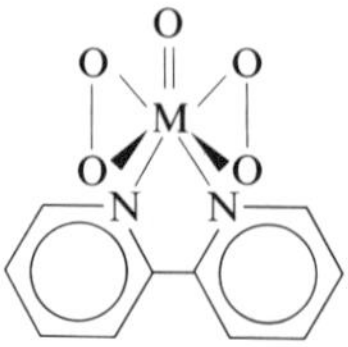 **VIII** (M = Mo, W[29], M = Cr[74]) 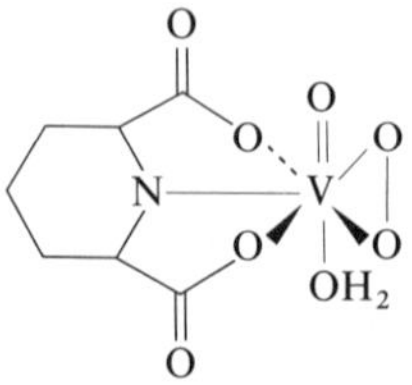 **IX**[73]

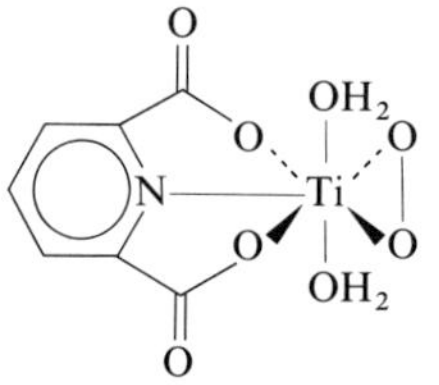

X[72] 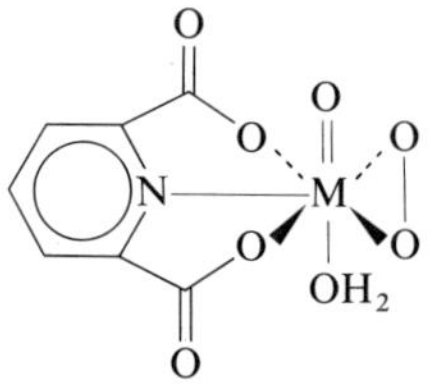 **XI** (M = Mo, W[75])

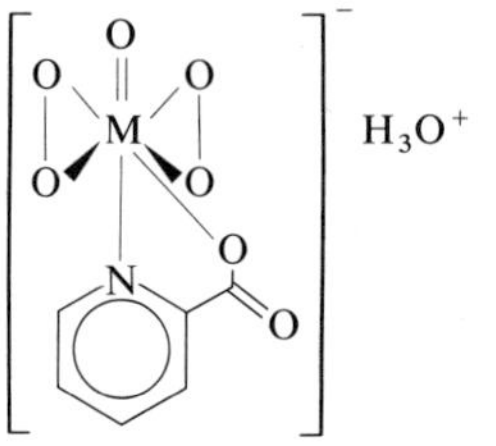

XII (M = Mo, W[76]) 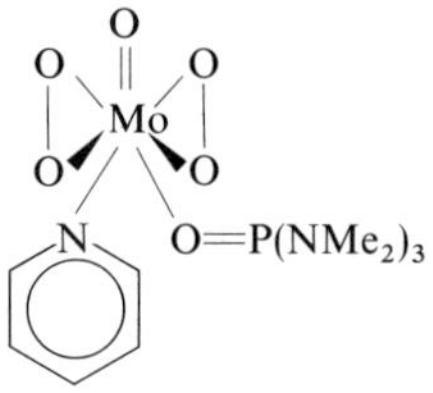 **XIII**[29]

Heterobimetallic μ-peroxo complexes containing both cobalt and molybdenum have been prepared,[78]

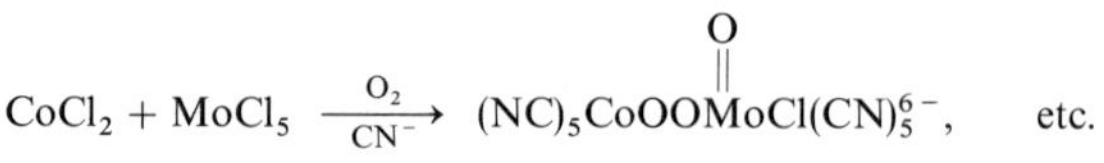

$$CoCl_2 + MoCl_5 \xrightarrow[CN^-]{O_2} (NC)_5CoOO\overset{O}{\overset{\|}{Mo}}Cl(CN)_5^{6-}, \quad \text{etc.}$$

which, upon loss of the chloro ligand, spontaneously rearrange to the peroxo(oxo)molybdenum complex.

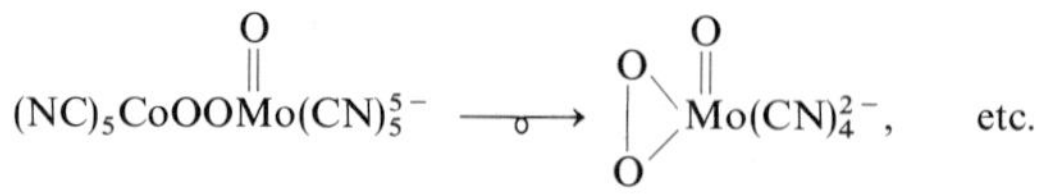

$$(NC)_5CoOO\overset{O}{\overset{\|}{Mo}}(CN)_5^{5-} \longrightarrow (O_2)\overset{O}{\overset{\|}{Mo}}(CN)_4^{2-}, \quad \text{etc.}$$

Since the latter can also be synthesized independently from a μ-oxodimolybdenum(V) complex by ligand displacement with sodium peroxide, it represents a unique example in which direct peroxidation of a molybdenum(V, VI) compound has been achieved with dioxygen and not peroxides.

Mimoun *et al.*[79] showed that complex **VII** and the analogous Mo(VI) and W(VI) complexes selectively epoxidize olefins in a stoichiometric reaction at room temperature.

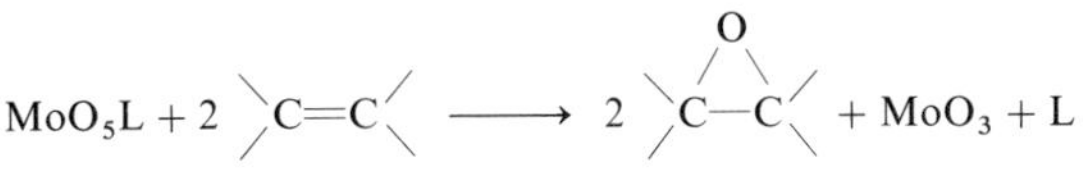

$$MoO_5L + 2\ \rangle C{=}C\langle \longrightarrow 2\ \rangle C\overset{O}{-}C\langle + MoO_3 + L \tag{47}$$

The increase in rates attendant upon alkyl substitution at the double bond is consistent with an electrophilic activation process for epoxidation. Such a mechanism is accommodated in the rate-limiting coordination of olefin in Scheme VII.[79]

Scheme VII:

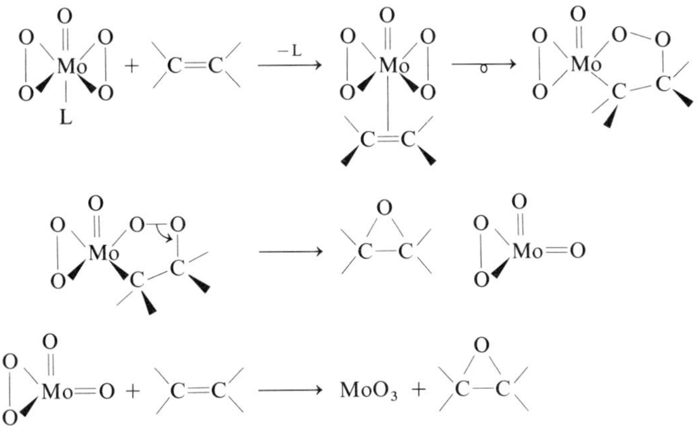

$$MoO(O_2)_2L + \rangle C{=}C\langle \xrightarrow{-L} MoO(O_2)_2(\rangle C{=}C\langle) \longrightarrow \text{peroxymetallocycle}$$

$$\text{peroxymetallocycle} \longrightarrow \rangle C\overset{O}{-}C\langle + MoO_2(O_2)$$

$$MoO_2(O_2) + \rangle C{=}C\langle \longrightarrow MoO_3 + \rangle C\overset{O}{-}C\langle$$

Indeed, epoxidation is severely inhibited by polar coordinating solvents, which block coordination sites and prevent the approach of the olefin. In this mechanism, the intramolecular attack by the peroxo ligand on the coordinated olefin to form the peroxymetallocycle must occur subsequent to the rate-limiting olefin coordination. However, an alternative mechanism involving a three-membered transition state,

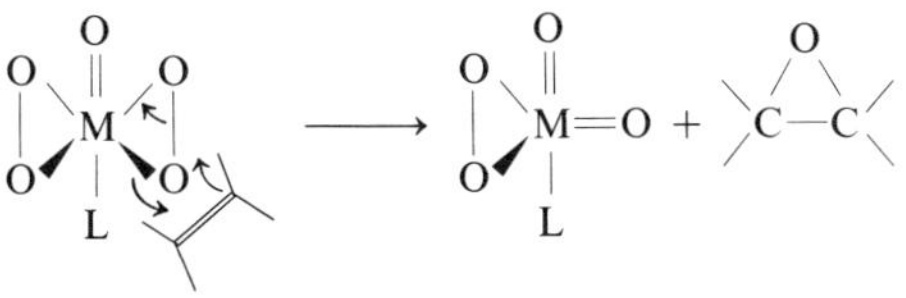

$$MO(O_2)_2L \cdots \rangle C{=}C\langle \longrightarrow MO_2(O_2)L + \rangle C\overset{O}{-}C\langle \tag{48}$$

proposed by Sharpless,[80] cannot be discounted. Optically active oxodiperoxomolybdenum(VI) complexes containing chiral ligands have been used as reagents for the asymmetric epoxidation of olefins.[81]

Epoxidation with complex **VII** can be made catalytic by using excess H_2O_2 to regenerate the peroxo complex. However, it does not afford any special advantages over other Mo(VI) compounds for the catalytic processes described in Chapter 3. The analogous complex **XIII** has been used for the selective, low-temperature hydroxylation of enolate anions, e.g.,[82]

$$\mathrm{RC(=O)CH_2R'} \xrightarrow{\text{base}} \mathrm{RC(O^-){=}CHR'} \xrightarrow{\text{XIII}} \mathrm{RC(=O){-}CH(O^-)R'} \qquad (49)$$

Similarly, arylmagnesium bromides react with complex **XIII** at $-78°C$ to afford the corresponding phenols.[83] Both of these reactions are related to the conversion of butyllithium to lithium butoxide by complex **XII** described earlier.[35]

Peroxomolybdenum(VI) species stabilized by polydentate ligands, such as complexes **XI** and **XII**, have been used by Mares and co-workers[84] as catalysts for the Baeyer–Villiger oxidation of cyclic ketones to lactones with H_2O_2 at 60°C. The oxidation could involve a peroxymetallocycle intermediate formed by addition of the 1,3-dipole to the carbonyl group.

$$\overset{+}{\mathrm{M}}{-}\mathrm{O}{-}\mathrm{O^-} + \mathrm{O{=}C}(\text{ring}) \longrightarrow [\mathrm{M{-}O{-}O{-}C(ring){-}O}] \longrightarrow \mathrm{M{=}O} + \text{lactone } \mathrm{(C(=O){-}O \text{ ring})} \qquad (50)$$

Peroxo complexes containing monodentate ligands, such as complexes **VII** and **XIII**, decompose readily at 60°C and are not suitable as catalysts, owing to a competing formation of oligomeric ketone peroxides.

Polydentate complexes **XII**, where M = Mo or W, stoichiometrically oxidize secondary alcohols to ketones.[85] The tungsten complex is significantly more active than the molybdenum analog, and it can function as a catalyst for the oxidation of secondary alcohols in the presence of excess hydrogen peroxide. The ketone presumably arises from the decomposition of an alkoxometal hydroperoxide) intermediate formed by alcoholysis of the peroxometal bond.

$$\mathrm{M}\langle\mathrm{O_2}\rangle + \mathrm{{>}C(H)OH} \longrightarrow \left[\mathrm{M(OOH)(O{-}C(H){<})}\right] \xrightarrow[-H_2O]{} \mathrm{M{=}O} + \mathrm{{>}C{=}O} \qquad (51)$$

$$\mathrm{M{=}O} + \mathrm{H_2O_2} \xrightarrow[-H_2O]{} \mathrm{M}\langle\mathrm{O_2}\rangle, \quad \text{etc.}$$

A related reaction is the oxidation of secondary alcohols and amines with molecular oxygen catalyzed by $RuCl_3$.[86] The dehydrogenation of 2-aminoalkanes to the imines may proceed by an alkylaminoruthenium(III) hydroperoxide intermediate,

$$\mathrm{Ru^{III}(-O-OH\cdots H-C-N(H)-)} \longrightarrow \mathrm{O{=}Ru^{III}} + \mathrm{{>}C{=}NH} + \mathrm{H_2O} \tag{52}$$

or by a direct route involving dioxygen as the terminal oxidant.

$$\mathrm{Ru^{III}{-}N(H){-}C(H){<}} \xrightarrow{-H^+} \mathrm{{>}C{=}NH} + \mathrm{Ru^{I}} \xrightarrow{O_2} \mathrm{Ru^{III}}, \qquad \text{etc.}$$

In the foregoing oxidations with peroxomolybdenum(VI) and peroxotungsten(VI) complexes, the product is an oxometal complex from which the peroxometal complex *cannot* be regenerated with dioxygen. However, there is a recent report[87] describing the formation of a mixed Mo(VI) metal–peroxo complex **XIV** with molecular oxygen.

$$\mathrm{O_2 + CoCl_2 + MoCl_4} \xrightarrow[H_2O]{KCN} \mathrm{K_6\left[(NC)_5Co^{III}{-}O{-}O{-}Mo^{VI}(Cl)(=O)(CN)_5\right]} \tag{53}$$

XIV

Reactions of this complex have not yet been reported.

Weiss and co-workers[88] have described the preparation of a *trans*-diperoxomolybdenum(VI) porphyrin, which affords a *cis*-dioxo complex upon photolysis.

$$\mathrm{(O_2)Mo(O_2)\ porphyrin} \xrightarrow{h\nu} \mathrm{(O)_2Mo\ porphyrin} + \mathrm{O_2} \tag{54}$$

The expected stoichiometry leads to a molecule of singlet oxygen by chelotropic [2 + 2]-elimination.

$$\mathrm{(O_2)Mo^{VI}(O_2)} \longrightarrow \mathrm{(O_2)Mo^{IV}} + \mathrm{O{=}O} \longrightarrow \mathrm{Mo(=O)_2} \tag{55}$$

In this context, it is interesting that Hayashi and co-workers[89] favored singlet oxygen in the formation of dienone hydroperoxides,

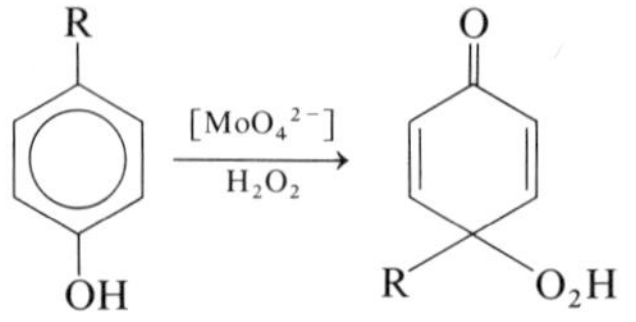

$$\text{R-C}_6\text{H}_4\text{-OH} \xrightarrow[H_2O_2]{[MoO_4^{2-}]} \text{4-R-4-O}_2\text{H-cyclohexa-2,5-dienone} \tag{56}$$

from *p*-alkylphenols and tetraperoxomolybdate $[Mo(O_2)_4]^{2-}$. The latter, formed from molybdate and excess hydrogen peroxide in neutral aqueous media, decomposes thermally to two molecules of oxygen.

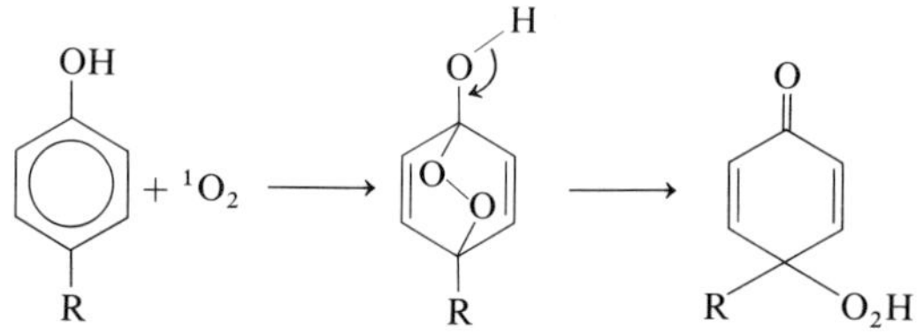

$$\text{HO-C}_6\text{H}_4\text{-R} + {}^1O_2 \longrightarrow \text{endoperoxide} \longrightarrow \text{4-R-4-O}_2\text{H-cyclohexa-2,5-dienone}$$

A peroxotitanium(IV)–octaethylporphyrin complex has also been described.[90]

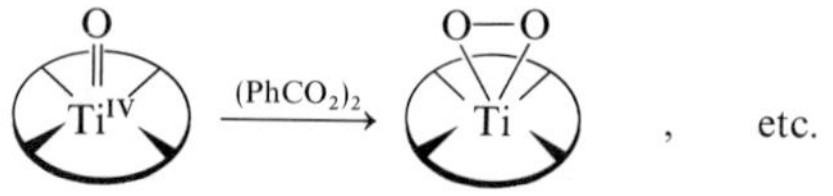

$$\text{O=Ti}^{IV}\text{(porphyrin)} \xrightarrow{(PhCO_2)_2} \text{(O-O)Ti(porphyrin)}, \quad \text{etc.}$$

Finally, a class of metallaoxaziridines, analogous to peroxometal complexes, has been prepared from N-substituted hydroxylamines and either dioxo complexes or peroxo complexes.[91,92]

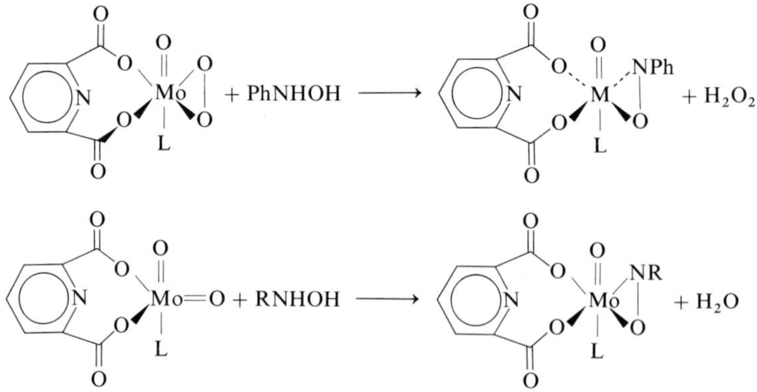

$$\text{(dipic)Mo(O)(O}_2\text{)L} + \text{PhNHOH} \longrightarrow \text{(dipic)M(O)(ONPh)L} + H_2O_2 \tag{57}$$

$$\text{(dipic)Mo(O)}_2\text{L} + \text{RNHOH} \longrightarrow \text{(dipic)Mo(O)(ONR)L} + H_2O$$

where M = Mo, W and L = H_2O, py, HMPA. These molybdaoxaziridines react with olefins to produce allylic amines, rather than the anticipated aziridines (by analogy with the formation of epoxides from the peroxo complexes).[91]

$$\text{alkene} + (\text{O}{=})\text{Mo}(\text{O})(\text{NAr}) \longrightarrow \text{allylic amine (NHAr)} + (\text{O}{=})\text{M}{=}\text{O} \qquad (58)$$

With cyclohexanone, the 2-arylaminocyclohexanone resulted.[92]

$$\text{cyclohexanone} + (\text{O}{=})\text{Mo}(\text{O})(\text{NPh}) \longrightarrow \text{2-(NHPh)cyclohexanone} + (\text{O}{=})\text{Mo}{=}\text{O} \qquad (59)$$

V. SUPEROXO- AND μ-PEROXOMETAL COMPLEXES

The metalloenzymes that are responsible for the utilization of molecular oxygen in biological systems usually contain one or more of the redox couples from the group Cu(I)/Cu(II), Fe(II)/Fe(III), Mn(II)/Mn(III), Co(II)/Co(III).[15–19,93] Thus, the oxygen transport proteins utilize iron (hemoglobin, myoglobin) or copper (hemocyanin). The oxidases, which reduce peroxide to water, and the oxygenases, which effect the oxygenation of organic substrates, generally contain iron or copper. Cobalt is thought to utilize oxygen in certain reactions mediated by vitamin B_{12},[94] and manganese has been implicated[95] as an oxygen transport agent in photosynthesis. Finally, Cu-, Mn-, and Fe-containing superoxide dismutases, which catalyze the reaction,

$$2\,O_2^{\overline{\cdot}} + 2\,H^+ \longrightarrow H_2O_2 + O_2 \qquad (60)$$

in biological systems, have been identified.[95,96]

In contrast to the metals described in the preceding two sections, this group of metals has a marked tendency to form superoxometal or μ-peroxometal complexes (see Scheme II). In the following sections, the chemistry of the peroxocobalt, peroxomanganese, peroxocopper, and peroxoiron complexes will be described individually.

A. Cobalt

Since cobalt is the most studied metal in this group, numerous Co(II) chelates are known to form superoxo and/or μ-peroxo complexes by reaction with dioxygen.[15–18] The best known of these is (Salen)Co(II), the progenitor of a series of Co(II)–Schiff base complexes studied extensively by Calvin and Martell.[97] Similarly, Nishinaga and co-workers[98] studied the reactions (*vide infra*) of organic substrates with molecular oxygen in the presence of

(Salen)Co(II) and the related complex (Salpr)Co(II), which reversibly binds O_2.

$$(\text{Salpr})\text{Co}^{II} + O_2 \rightleftharpoons (\text{Salpr})\text{Co}{-}O{-}O\cdot \tag{61}$$

(Salpr)Co

(Salen)Co(II) generally forms a μ-peroxo complex, although a superoxo complex is formed in pyridine or in the presence of imidazole derivatives. Moreover, it is possible that these complexes react via the superoxo complex, which is in equilibrium with the μ-peroxo complex in solution.

Before considering some of the reactions catalyzed by superoxometal complexes, it is worthwhile to consider briefly the properties of the superoxide ion.[25,99] Electrochemical studies[100] show that superoxide ion is a moderate reducing agent and a very weak oxidant. Furthermore, solutions of superoxide behave as bases,[100] although the pK_a of $HO_2^{\cdot}$ in water is 4.88. Thus, the first step in the reactions of superoxide with dipolar, protic substrates is probably proton removal to produce the hydroperoxy radical, as in the oxidation of 3,5-*tert*-butylcatechol.[100]

$$\text{ArOH(OH)} + O_2^{\cdot-} \rightleftharpoons [\text{ArO}^-(\text{OH}) + \cdot O_2H] \longrightarrow \text{ArO}\cdot(\text{OH}) + HO_2^- \longrightarrow \text{products} \tag{62}$$

Superoxometal complexes also appear to react first as bases (*vide infra*), although one might, *a priori*, expect superoxo complexes to abstract a hydrogen atom from hydrocarbons, by analogy with alkylperoxy (R—O—O·) radicals.

$$M{-}O{-}O\cdot + RH \longrightarrow M{-}O{-}OH + R\cdot \tag{63}$$

Indeed, such an "oxygen activation" mechanism has often been proposed in the catalytic effect of certain transition metal compounds during hydrocarbon autoxidations. In all those cases examined critically, however, it was shown later that the effect arose from the metal-catalyzed homolysis of alkyl hydroperoxides.[53,54,101,102] Thus, the enhancement with metal phthalocyanines has been ascribed by several authors[103] to oxygen activation, since a number of transition metal phthalocyanines readily form superoxo complexes with molecular oxygen.[104–106] These compounds bear close structural similarity to metal porphyrins, which constitute the prosthetic group of several oxidative enzymes. (See Chapter 8.)

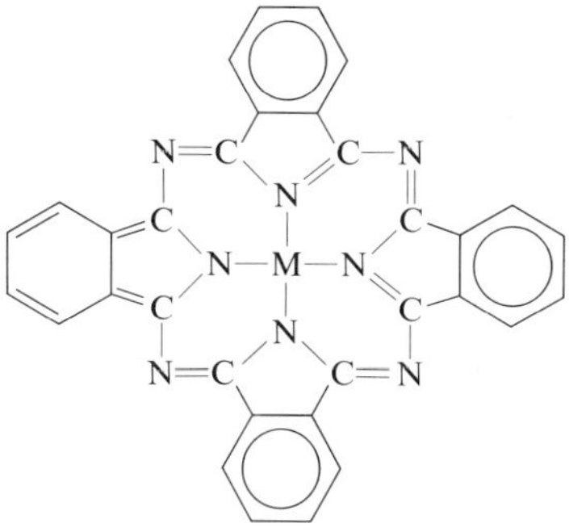

Metal phthalocyanines 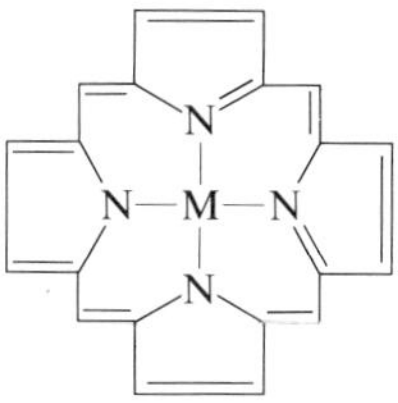 Metal porphyrins

Van Tilborg and Vreugdenhil[102] carried out a detailed kinetic investigation of the autoxidation of cumene in the presence of a variety of metal phthalocyanine catalysts (Cu, V, Ni, Co, Zn, Mg) and demonstrated unequivocally that initiation arose from the metal-catalyzed homolysis of adventitious hydroperoxides. At present there is no evidence that reaction 63 is feasible with hydrocarbons, certainly at rates that can compete with other initiation pathways. Although the superoxo complex is too stable to react with hydrocarbons, it may be rendered more reactive if both the dioxygen and the substrate were coordinated to the same metal. Thus, James and Ochiai[107] suggested reaction 64 as the key step in the stoichiometric oxidation of a cyclooctene–Rh(I) complex with dioxygen.

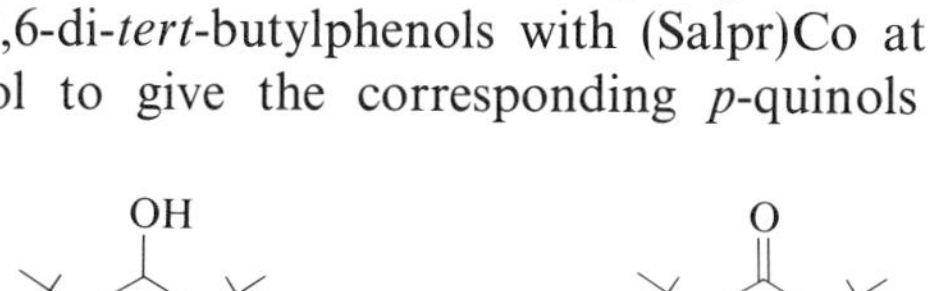

$$\cdot O{-}O{-}Rh^{II}(Cl)(L)(\text{cyclooctene}) \longrightarrow HO{-}O{-}Rh^{II}(Cl)(L)(\text{cyclooctenyl}\cdot) \tag{64}$$

Reaction 63 should also be more favorable with substrates, such as phenols, which afford stable radicals. Nishinaga reported the catalytic oxidation of 4-alkyl-2,6-di-*tert*-butylphenols with (Salpr)Co at room temperature in methanol to give the corresponding *p*-quinols in quantitative yield.[98,108–110]

$$\text{ArOH} + O_2 \xrightarrow[\text{MeOH}]{[(\text{Salpr})\text{Co}]} p\text{-quinol (R, OH)} \tag{65}$$

If the reaction is carried out in CH_2Cl_2, or in MeOH at 0°C, the reaction proceeds stoichiometrically to give an alkylperoxycobalt(III) complex, which is the putative intermediate.

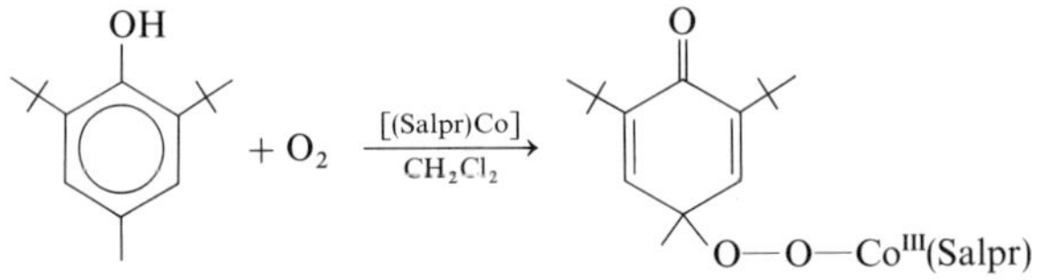

$$\text{ArOH} + O_2 \xrightarrow[CH_2Cl_2]{[(\text{Salpr})\text{Co}]} \text{O=C}_6\text{(dienone)}\text{–O–O–Co}^{III}(\text{Salpr}) \quad (66)$$

Nishinaga[98,110] favors a mechanism involving an initial hydrogen abstraction from the phenol by the superoxo complex,

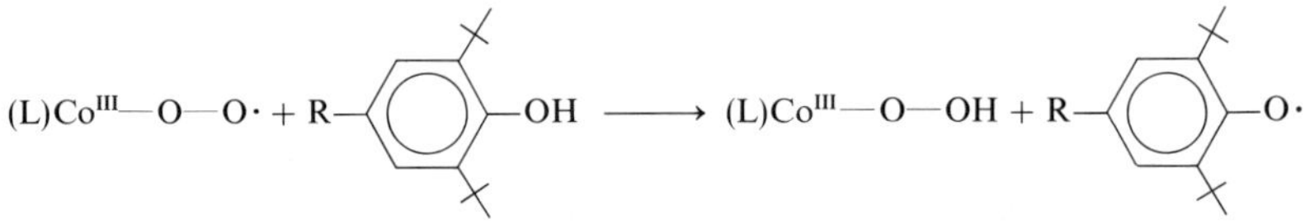

$$(\text{L})\text{Co}^{III}\text{—O—O}\cdot + \text{R–Ar–OH} \longrightarrow (\text{L})\text{Co}^{III}\text{—O—OH} + \text{R–Ar–O}\cdot \quad (67)$$

followed by (electron transfer) reduction of the phenoxy radical to give an organocobalt complex that inserts oxygen.

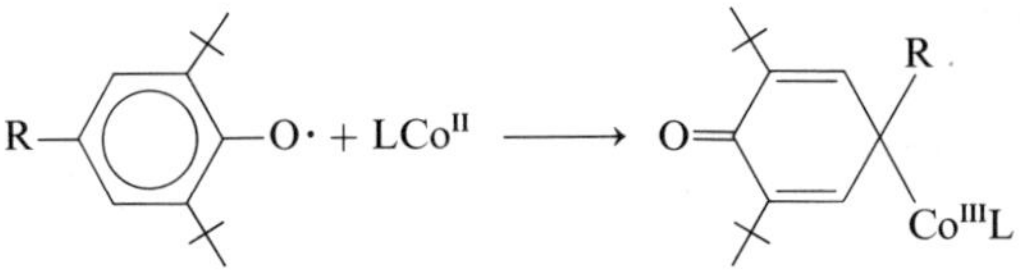

$$\text{R–Ar–O}\cdot + \text{LCo}^{II} \longrightarrow \text{O=(dienone)(R)Co}^{III}\text{L} \quad (68)$$

$$\text{O=(dienone)(R)Co}^{III}\text{L} + O_2 \longrightarrow \text{O=(dienone)(R)O—O—Co}^{III}\text{L} \quad (69)$$

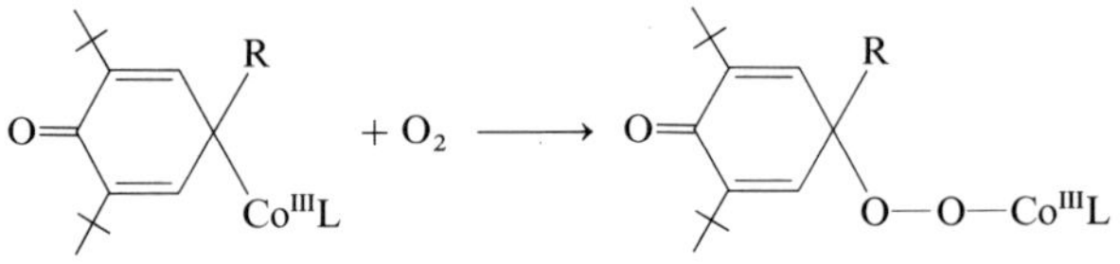

Bearing in mind the basic properties of superoxide described earlier, an alternative mechanism can be formulated in which the superoxo complex initially removes a proton,

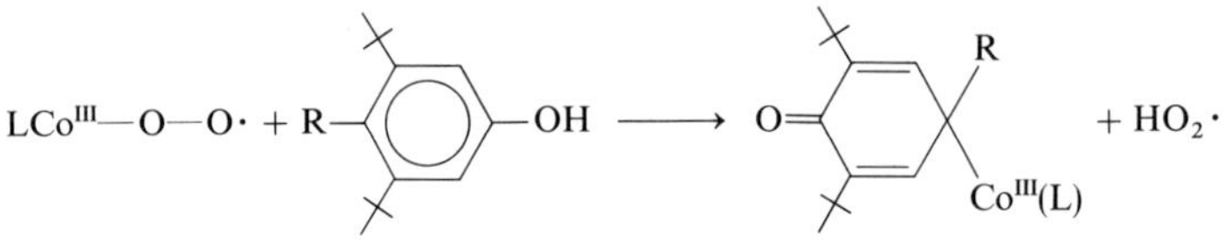

$$\text{LCo}^{III}\text{—O—O}\cdot + \text{R–Ar–OH} \longrightarrow \text{O=(dienone)(R)Co}^{III}(\text{L}) + HO_2\cdot \quad (70)$$

followed by oxygen insertion which probably involves a radical chain mechanism (compare eqs 19–21).

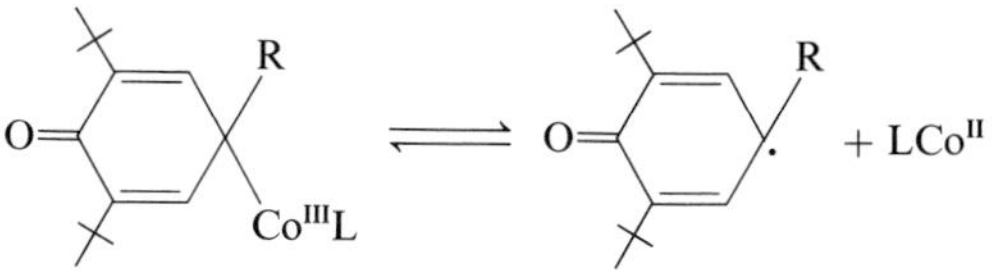

R, $Co^{III}L$ ⇌ R + LCo^{II} (71)

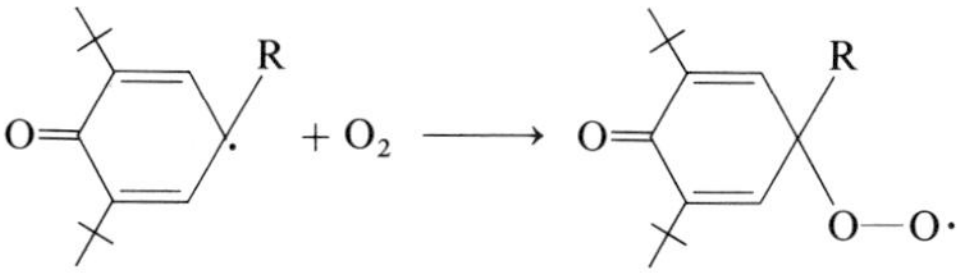

R + O_2 ⟶ R, O—O· (72)

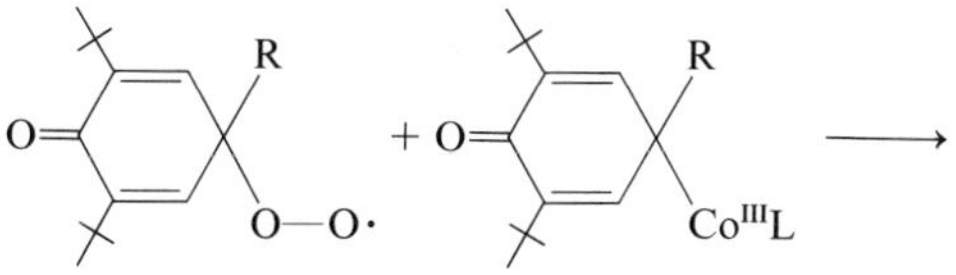

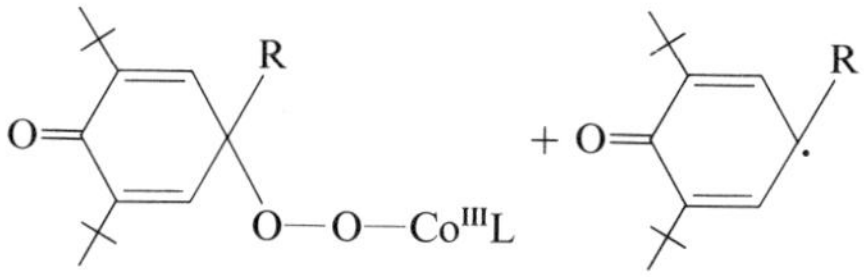

R, O—O· + R, $Co^{III}L$ ⟶

R, O—O—$Co^{III}L$ + R (73a)

or:

R, O—O· + LCo^{II} ⟶ R, O—O—$Co^{III}L$ (73b)

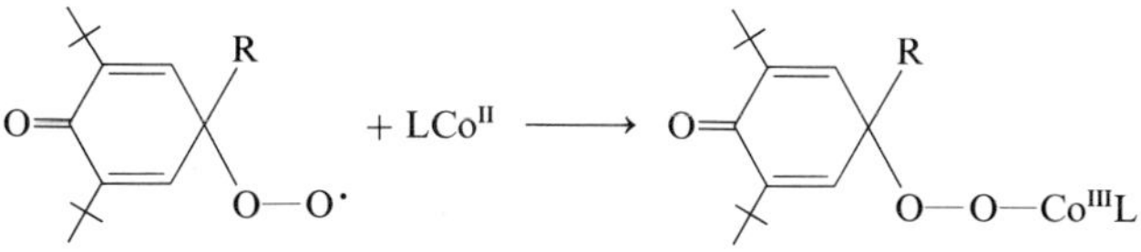

Reaction of the alkylperoxocobalt(III) complex with the phenol would regenerate the organocobalt complex and the hydroperoxide, which yields quinol by subsequent cobalt catalysis.

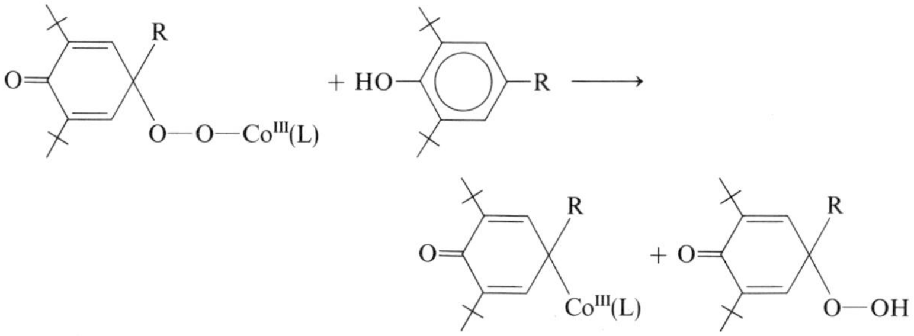

R, O—O—$Co^{III}(L)$ + HO—R ⟶

R, $Co^{III}(L)$ + R, O—OH (74)

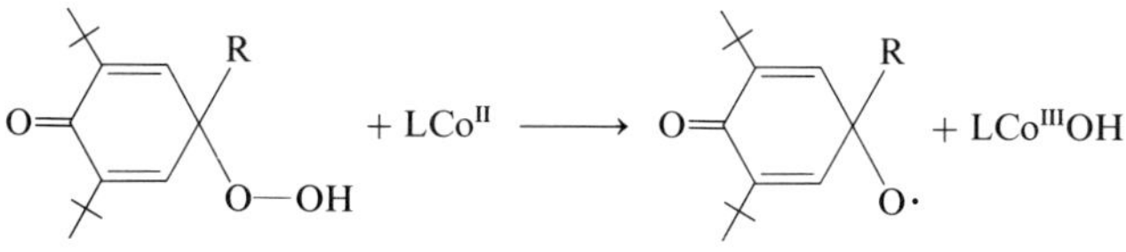

R, O—OH + LCo^{II} ⟶ R, O· + $LCo^{III}OH$ (75)

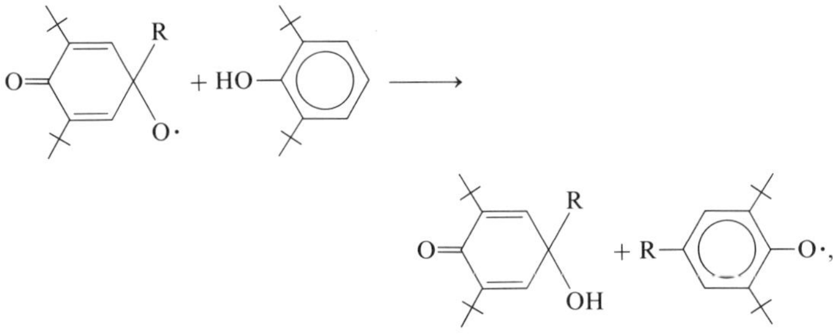

$$\text{(76)}$$

etc. (76)

Alternatively, quinol formation may occur via the reaction of the organoperoxycobalt(III) intermediate with the alcohol solvent, i.e.,

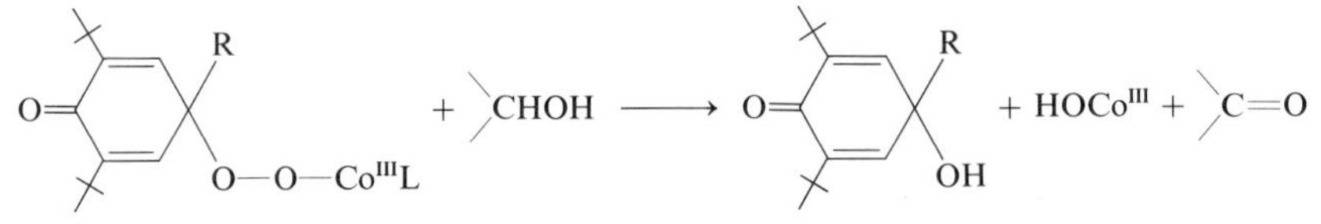

$$+ \;\rangle\text{CHOH} \longrightarrow \; + \text{HOCo}^{\text{III}} + \;\rangle\text{C}{=}\text{O}$$

Indeed, Nishinaga has shown that stoichiometric amounts of aldehyde are formed when primary alcohols are used as solvents.[109] The overall reaction constitutes a chain process initiated by the solvolysis of the superoxocobalt(III) complex, suggesting that a cobalt(III) complex containing an easily displaced anion should, in principle, initiate the reaction.

A related reaction is the (Salen)Co(II)-catalyzed oxidation of phenols to the corresponding quinones, first studied by van Dort and Geursen[111] and later by others.[112–115] The best yields are obtained in DMF as solvent.[113–114]

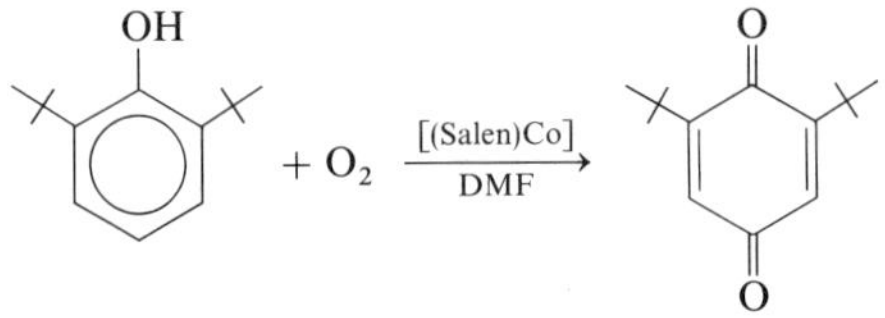

$$+ \text{O}_2 \xrightarrow[\text{DMF}]{[(\text{Salen})\text{Co}]} \qquad (77)$$

A mechanism has been proposed[114] in which hydrogen abstraction by the superoxocobalt(III) complex is the initial step. However, nucleophilic displacement by the phenol on either the peroxo- or μ-peroxocobalt(III) complex is an equally attractive possibility.

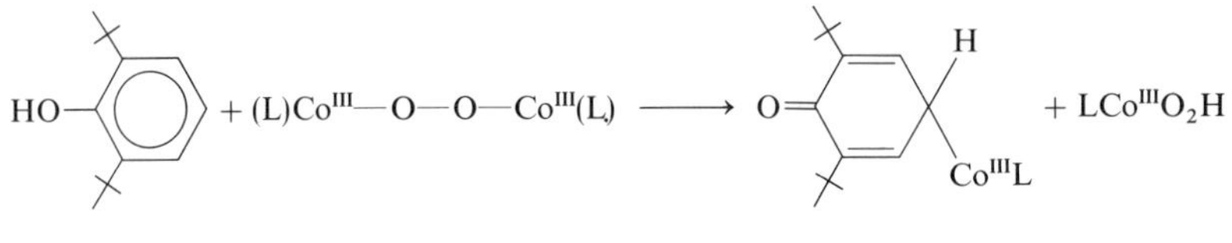

$$\text{HO—Ar} + (\text{L})\text{Co}^{\text{III}}\text{—O—O—Co}^{\text{III}}(\text{L}) \longrightarrow \text{O}{=}\text{C}_6\text{H}_2(t\text{-Bu})_2(\text{H})\text{Co}^{\text{III}}\text{L} + \text{LCo}^{\text{III}}\text{O}_2\text{H} \qquad (78)$$

A subsequent insertion of oxygen into the Co(III) complex (compare eqs 71–73), followed by a second nucleophilic displacement would regenerate the organocobalt(III) complex:

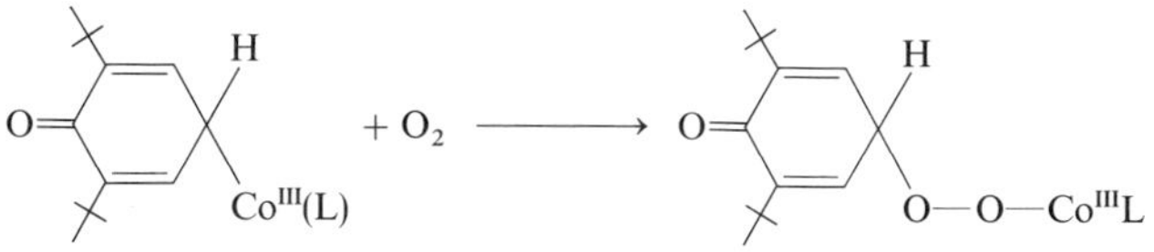

(79)

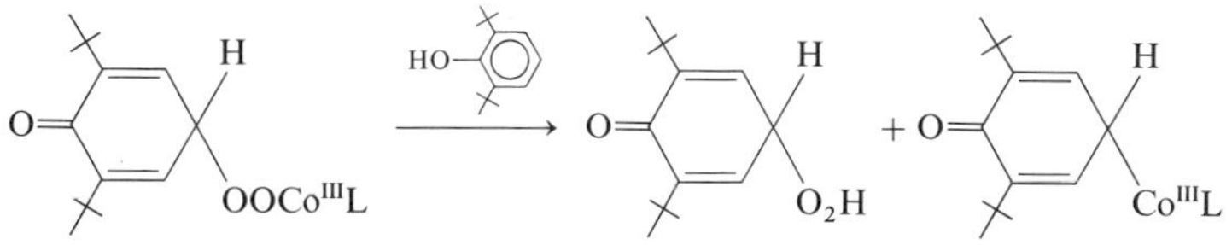

(80)

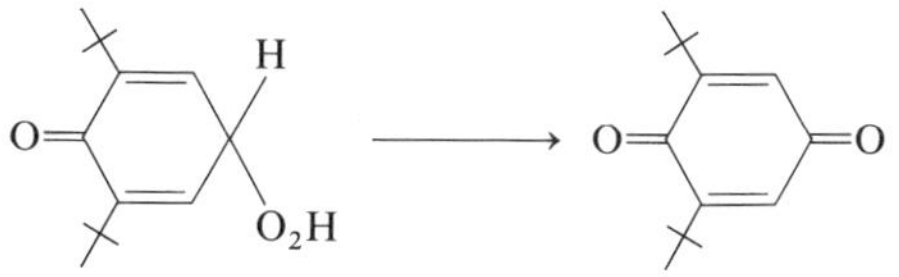

(81)

A γ-hydrogen elimination from the alkylperoxocobalt(III) complex is, however, more likely.

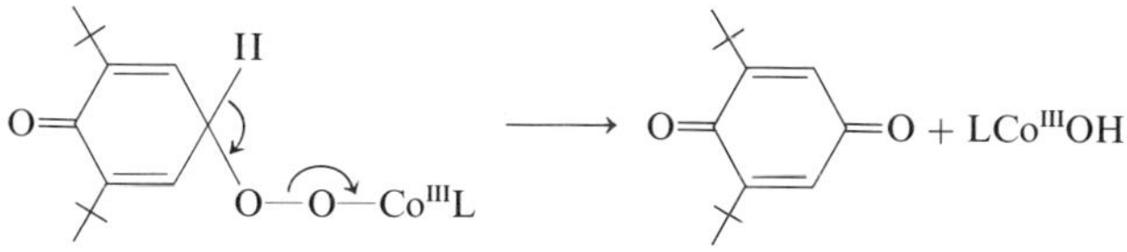

(82)

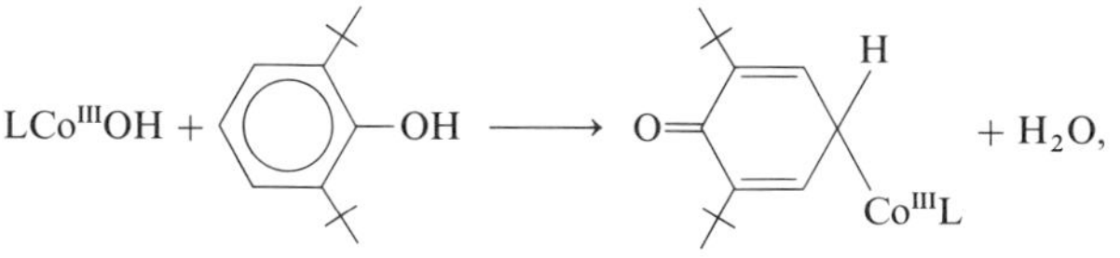

(83)

Insertion of molecular oxygen into the cobalt–carbon bond is the key step in these processes, similar to the autoxidation of vitamin B_{12} and related alkylcobalt(III) complexes.[116] Although the organocobalt(III) intermediate has not been completely identified, the problem is not severe since their interconversion should be facile, i.e.,

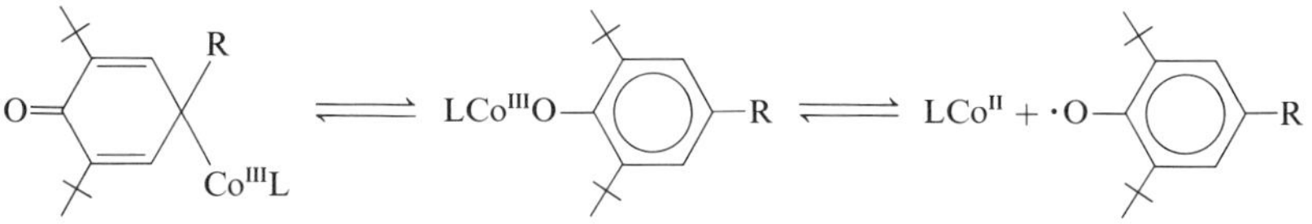

The oxidative cleavage of 3-substituted indoles[117–119] in eq 84 is a model reaction for the conversion of tryptophan to formylkynurenin by the enzyme tryptophan 2,3-dioxygenase in eq 85.

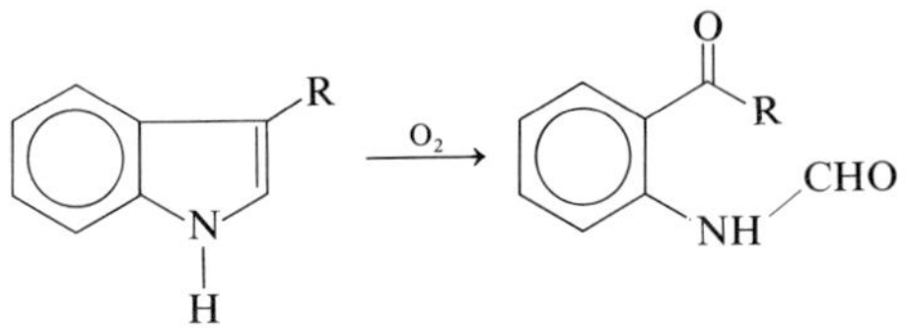

(84)

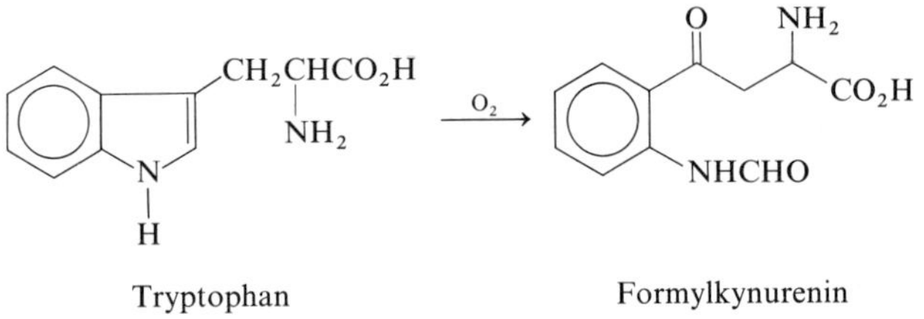

(85)

Tryptophan Formylkynurenin

Reaction 84 proceeds at ambient temperatures in the presence of (Salen)Co(II)[117] and cobalt- and copper-porphyrin complexes in CH_2Cl_2,[118] as well as manganese(II) phthalocyanine in DMF.[119] Interestingly, the reaction with (Salen)Co is faster in CH_2Cl_2 than in MeOH and is markedly retarded by strongly coordinating solvents such as DMF and pyridine. Moreover, the reactivities of substituted indoles can be correlated with their electron donor properties,[117] suggesting prior coordination of the substrate. A pathway involving nucleophilic displacement and oxygen insertion is analogous to reactions 79–81 discussed above.

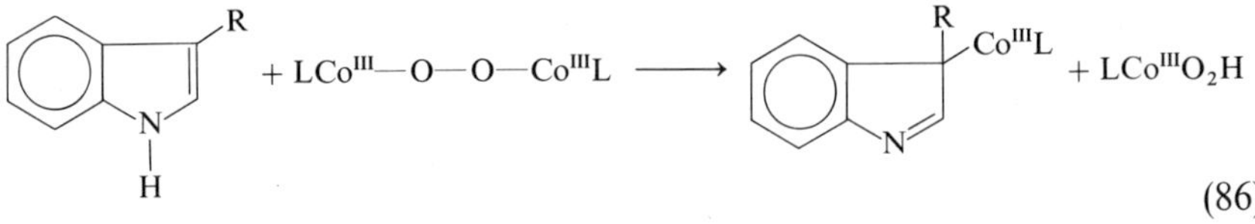

$+ LCo^{III}—O—O—Co^{III}L \longrightarrow$ $+ LCo^{III}O_2H$ (86)

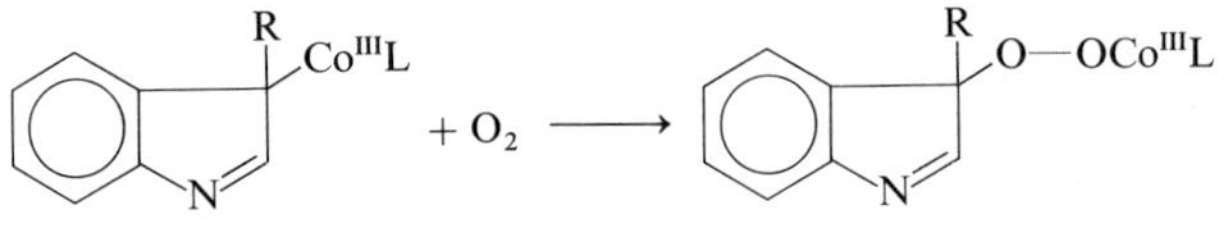

$+ O_2 \longrightarrow$ (87)

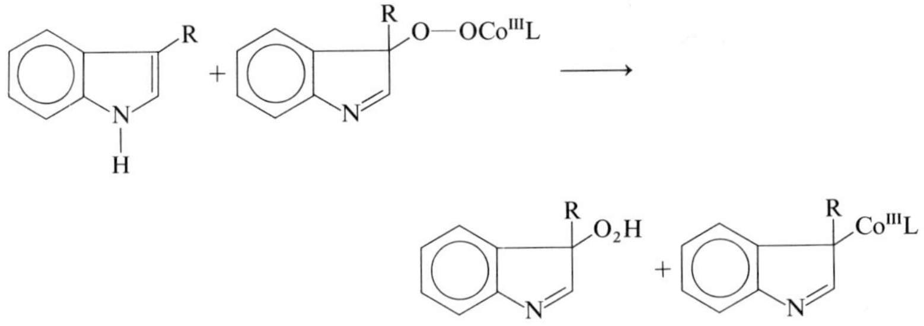

(88)

(89)

B. Manganese

(Salen)Mn(II) is irreversibly oxidized by air to a μ-oxo complex,[120] but in the presence of triethylamine it catalyzes the oxidation of β-isophorone (**XV**) to the diketone **XVI** in high yield[121] when carried out in acetonitrile, chloroform, or ethereal solvents at ambient temperatures. (Salen)Co is active but less selective than (Salen)Mn.

$$\textbf{XV} + O_2 \xrightarrow[Et_3N,\ 25^\circ C]{[(Salen)Mn]} \textbf{XVI} + H_2O \qquad (90)$$

Triethylamine is capable of generating a carbanion from **XV**, which subsequently affords an alkylperoxomanganese(III) intermediate via the sequence of nucleophilic displacement and oxygen insertion proposed for (Salen)Co above.

$$+\ LMn^{III}OOMn^{III}L \xrightarrow{-LMn^{III}O_2} (H,\ Mn^{III}L) \xrightarrow{O_2} (H,\ OOMn^{III}L) \qquad (91)$$

$$(H,\ O{-}OMn^{III}L) \longrightarrow \quad + LMn^{III}OH, \quad \text{etc.} \qquad (92)$$

Manganese(II) tetraphenylporphyrin reversibly complexes dioxygen at low temperatures ($< -78^\circ C$), but no reactions of the complex have yet been described.[122] An interesting recent development is the preparation of reversible oxygen complexes at ambient temperatures from simple phosphine–Mn(II) complexes,[123]

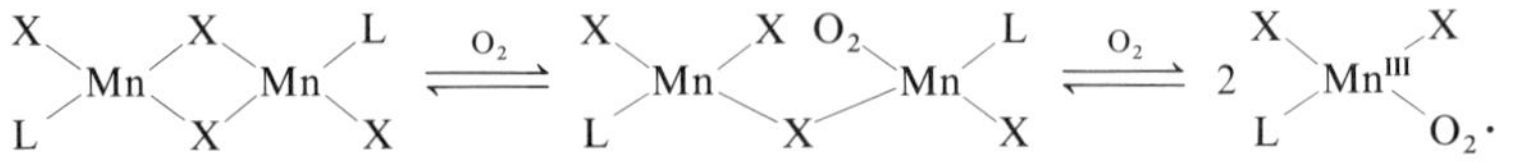

$$\mathrm{L(X)Mn(\mu\text{-}X)_2Mn(L)X} \underset{}{\overset{O_2}{\rightleftharpoons}} \mathrm{L(X)Mn(\mu\text{-}X)(\mu\text{-}O_2)Mn(L)X} \underset{}{\overset{O_2}{\rightleftharpoons}} 2\ \mathrm{L(X)_2Mn^{III}O_2\cdot} \qquad (93)$$

where L = R_3P and X = Cl, Br.

C. Copper

Although copper is present in a variety of metalloenzymes that utilize molecular oxygen in biological oxidations,[96c] there are very few synthetic copper complexes that form well-characterized adducts with dioxygen reversibly. Simmons and Wilson[124] have described a reversible adduct with the Cu(I) macrocycle **XVII**, a model for the metalloprotein, hemocyanin.

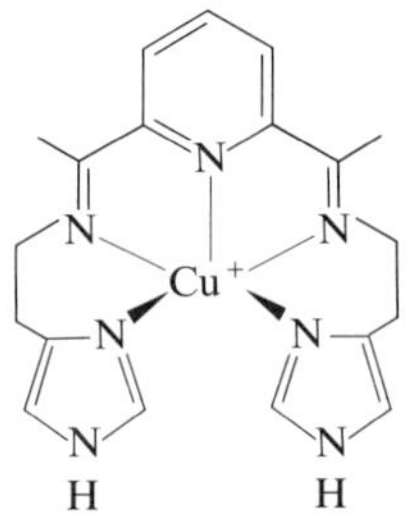

XVII

Reaction of cuprous chloride with oxygen in pyridine affords an active catalyst in the oxidative coupling of acetylenes, aromatic amines, and phenols. The structure of this complex has not been elucidated. It may be a coordinated oxocopper(II), $[(py)_n\mathrm{Cu(II)O}]_n$, rather than a peroxo complex.[125] Tsuji and co-workers[126] have used this complex as a catalyst for the oxidative cleavage of catechol with molecular oxygen in a mixture of pyridine and methanol to produce *cis,cis*-muconic acid monomethyl ester **XVIII**, which is analogous to the oxidative cleavage of catechol to *cis,cis*-muconic acid by the enzyme pyrocatechase. (See Chapter 8.)

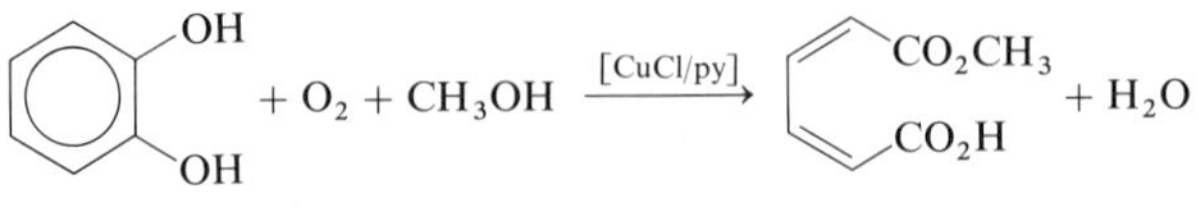

$$\mathrm{C_6H_4(OH)_2} + O_2 + CH_3OH \xrightarrow{[\mathrm{CuCl/py}]} \mathrm{HO_2C{-}CH{=}CH{-}CH{=}CH{-}CO_2CH_3} + H_2O$$

XVIII

The same system is capable of the oxidative cleavage of 3-methylindole, similar to the enzyme-catalyzed transformation in eq 84.[126]

A recent study of the CuCl/py-catalyzed cleavage of catechol has shown that the active oxidant is a dimeric cupric methoxy hydroxide complexed with pyridine prepared by the addition of an equivalent of water to cupric methoxy chloride, Cu(OMe)Cl, in pyridine.[127,127a] This complex oxidizes catechol to **XVIII** under anaerobic conditions. Since there is no direct reaction with the substrate, dioxygen only serves to reoxidize the Cu(I). The various Cu(II) species in solution are outlined below.[127,127a]

$$4\,CuCl + O_2 \xrightarrow{py} (py)_2Cu(\mu\text{-}O)_2Cu(py)_2 + 2\,py_2CuCl_2 \quad (94)$$

$$2\,py_2CuCl_2 \underset{}{\overset{MeOH}{\rightleftharpoons}} 2\,py_2Cu(OMe)(OH) \overset{-py}{\rightleftharpoons} [(MeO)(py)Cu(\mu\text{-}OH)_2Cu(py)(OMe)]$$

XIX

The active species is likely to be the monomer **XIX**, which corresponds to the methanol adduct (hemialcoholate) of an oxocopper(II) complex. These reactions are examples of oxidations with oxometal species in which the reagent is regenerated with molecular oxygen (see Chapter 6). Rogic and Demmin further demonstrated that this reagent converts phenol to catechol and *o*-benzoquinone, followed by cleavage to **XVIII**.[127,127a]

$$o\text{-benzoquinone} + MeOH \rightleftharpoons \text{(OMe)(OH) adduct} \xrightarrow[-MeOH]{XIX} \text{O–Cu(py)}_2\text{–OH adduct} + H_2O \longrightarrow \textbf{XVIII} \quad (95)$$

The reaction of ammonia with the same copper(II)–oxygen species generated as in eq 94 produces a new copper reagent designated as "CuO/NH_3," which reacts with *o*-benzoquinones, catechols, and phenols in the presence of dioxygen to afford the corresponding mononitriles of muconic acid, e.g.,[127b]

$$t\text{-Bu}\text{–}C_6H_4\text{–}OH + \tfrac{3}{2}\,O_2 \xrightarrow{[CuO/NH_3]} t\text{-Bu}\text{–}C(\text{CH=CH–}CO_2H)(\text{CH=CH–}CN)$$

By analogy with eq 94, the active component may be viewed as a dimeric copper(II) amide hydroxide in equilibrium with the corresponding monomeric and oligomeric species complexed with pyridine or ammonia, e.g.,

$$2\,Py_2Cu(NH_2)(OH) \rightleftharpoons (Py)(H_2N)Cu(\mu\text{-}OH)_2Cu(NH_2)(Py) + 2\,Py$$

In contrast, cuprous chloride in acetonitrile solutions catalyzes the selective oxidation of phenol to *p*-benzoquinone with dioxygen under mild conditions.[127c]

$$C_6H_5OH + O_2 \xrightarrow[CH_3CN]{[CuCl]} O{=}C_6H_4{=}O + H_2O \qquad (96)$$

The reaction with oxygen at 1000 psi and 40°C for 1 hr afforded *p*-benzoquinone in 80% selectivity at 93% phenol conversion. This copper catalysis compares with analogous oxidations catalyzed by (Salen)Co, in which smooth reaction can be observed only with phenols activated by at least one, and preferably two or more, alkyl groups. By analogy with the (Salen)Co-catalyzed oxidations of phenols to benzoquinones, one might expect the catalytic phenol oxidations mediated by copper to proceed via an analogous series of reactions, i.e.,

$$ArOCu^{II} \longrightarrow Cu^{I} + ArO\cdot$$

$$ArO\cdot + O_2 \longrightarrow O{=}C_6H_4(H)(O{-}O\cdot)$$

$$O{=}C_6H_4(H)(O{-}O\cdot) + ArOCu^{II} \longrightarrow O{=}C_6H_4(H)(O{-}O{-}Cu^{II}) + ArO\cdot$$

or

$$O{=}C_6H_4(H)(O{-}O\cdot) + Cu^{I} \longrightarrow O{=}C_6H_4(H)(O{-}O{-}Cu^{II}), \quad \text{etc.}$$

D. Iron

The reaction of molecular oxygen with simple iron(II) complexes is rapid and irreversible. Initial formation of a superoxo–Fe(III) complex is followed by the rapid reaction with a second Fe(II) to produce a μ-peroxo complex. The latter is unstable and decomposes to give μ-oxo–Fe(III) complexes.

$$
\begin{aligned}
\mathrm{Fe^{II}} + \mathrm{O_2} &\longrightarrow \mathrm{Fe^{III}OO\cdot} \xrightarrow{\mathrm{Fe^{II}}} \mathrm{Fe^{III}OOFe^{III}} \\
\mathrm{Fe^{III}OOFe^{III}} &\longrightarrow 2\,\mathrm{O{=}Fe^{IV}} \xrightarrow{\mathrm{Fe^{II}}} \mathrm{Fe^{III}OFe^{III}}
\end{aligned}
\qquad (97)
$$

A third type of dioxygen–iron complex can be prepared by the treatment of chloro(tetraphenylporphyrinato)iron(III) with potassium superoxide in DMF at -50°C. It has been formulated as a mononuclear, high-spin ferric peroxo complex, $(\mathrm{TPP})\mathrm{Fe}{<}\begin{matrix}\mathrm{O}\\|\\\mathrm{O}\end{matrix}$.[128]

The formation of a μ-peroxo complex is supported by the isolation of a μ-peroxo complex after exposure of a toluene solution of an iron(II) porphyrin at -50°C to O_2.[129] Iron(II) complexes react more readily than the corresponding cobalt(II) complexes with dioxygen, but the resultant superoxo and μ-peroxo complexes are much less stable than their cobalt(II) analogs. The difference has been attributed to the relative stabilities of the O=Co(IV) and O=Fe(IV) species formed by cleavage of the O—O bond in the μ-peroxo complex.[129a]

In the oxygen-binding hemoproteins, hemoglobin (Hb) and myoglobin (Mb), the active site is an iron(II) porphyrin (heme, **XX**),

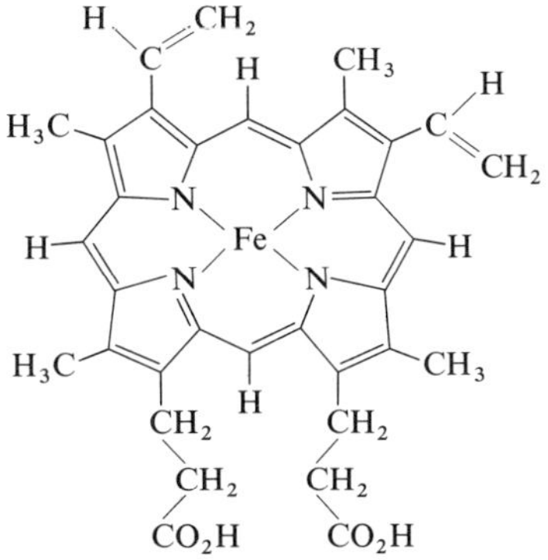

XX

which is tightly bound to a protein (globin) through numerous hydrophobic interactions and a single coordinate bond to the imidazole group of a proximal histidine residue. In the deoxy form, the Fe(II) center is five-coordinate, with the high-spin Fe(II) projecting out of the mean plane of the four porphyrin nitrogen atoms toward the coordinated imidazole moiety. A stable 1:1 superoxoiron(III) complex is formed with molecular oxygen, owing to the hindered approach of a second heme–Fe(II) by the steric encumbrance of the surrounding ligands. In recent years much effort has been devoted to designing synthetic models that mimic the oxygen-binding properties of Hb and Mb.[16,19] Wang [130] was the first to demonstrate, in 1958, that simple iron(II) porphyrins can be reversibly oxygenated when immobilized on a solid polymer matrix. Recently, reversible oxygenation of simple iron(II) porphyrins and other macrocyclic complexes has been observed in solution at low temperatures ($-50°C$) and high dilution ($\sim 10^{-4}$ *M*).[131] The problem encountered in the design of synthetic models using simple iron(II) porphyrins and strongly coordinating N-donor ligands is the strong preference of iron for six coordination,

$$\text{Fe}^{II} \overset{B}{\rightleftharpoons} \text{B-Fe}^{II} \overset{B}{\rightleftharpoons} \text{B-Fe}^{II}\text{-B} \qquad (98)$$

where B = py, piperidine, and *N*-methylimidazole. The hexacoordinated complex is inert to oxygen, but the four-coordinate complex undergoes rapid, irreversible oxidation to *μ*-oxoiron(III) species. The use of sterically hindered porphyrin ligands such as Baldwin's "capped" porphyrins[132] and Collman's "picket-fence" porphyrins,[133] in which steric hindrance inhibits the bonding of an axial base to one side of the porphyrin ring, has led to the synthesis of iron(II) porphyrins that reversibly bind dioxygen in nonaqueous solution at ambient temperatures. For example, crystalline superoxoiron(III)[133] and superoxocobalt(III)[134] complexes with structure **XXI** have been isolated and fully characterized. They are kinetically stable for prolonged periods in the solid state and for shorter periods in solution at ambient temperatures.

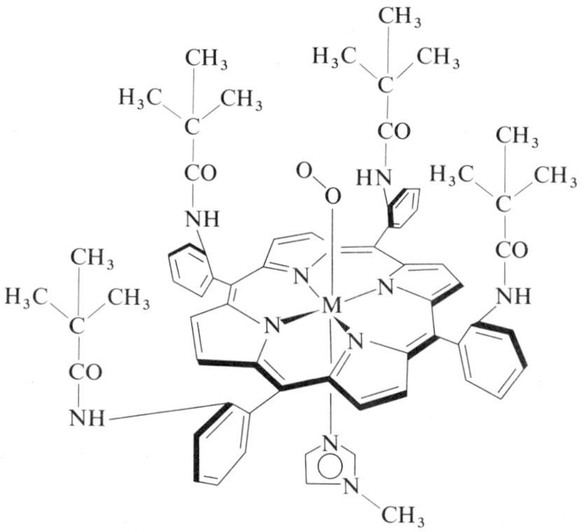

XXI (M = Fe, Co)

Although the "picket-fence" **XXII** the capped **XXIII**, and the basket-handle porphyrin **XXIV**[135] combine reversibly with oxygen, the "strapped" porphyrin structure **XXV** apparently does not provide sufficient steric shielding, and these complexes are oxidized irreversibly.[136]

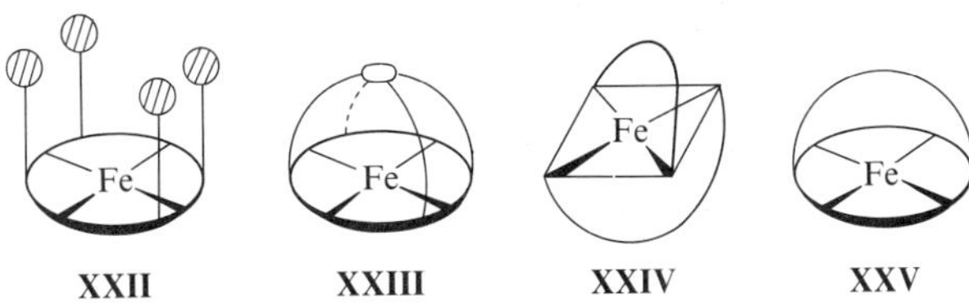

In these systems, the stability of the superoxo complex is largely dependent on the nature and concentration of the ligating base B, as illustrated schematically below.[132]

$$Fe^{II} \underset{}{\overset{B}{\rightleftharpoons}} B\text{—}Fe^{II} \overset{O_2}{\rightleftharpoons} B\text{—}Fe^{III}\text{—}O_2$$

$$2\ Fe^{II} \xrightarrow{O_2} \longrightarrow Fe^{III}\text{—}O\text{—}Fe^{III}$$

In the presence of B, the superoxo ligand is situated in the protected environment of the porphyrin cavity. In the absence of B, coordination occurs from the unshielded side of the porphyrin ring to afford an unprotected superoxo complex, which is rapidly converted to the μ-oxo complex.

These elegant synthetic contributions have made it possible to construct model Fe(II) complexes that mimic the oxygen transport enzymes Hb and Mb. Another important group of iron-containing enzymes are the various oxygenases that activate molecular oxygen for reaction with organic substrates (see Chapter 8), as in the ubiquitous cytochrome *P*-450 oxygenases, which catalyze the hydroxylation of a variety of organic substrates. These enzymes also contain at their active site an oxygen-binding hemoprotein. The nature of the axial ligands is a matter of conjecture, but the thought that a cysteinyl thiol group is covalently bonded to an iron(III) porphyrin unit has stimulated the synthesis of model iron–porphyrin complexes with the $[N_4Fe(III)SR]$ structural unit.[137]

The active center in these enzymes is believed to be an oxoiron(V) species initiated by a one-electron reduction of a superoxoiron(III) complex.

$$\mathrm{Fe^{III}{-}O{-}O\cdot \xrightarrow[H^+]{e^-} Fe^{III}{-}O{-}OH \longrightarrow Fe^{V}{=}O(OH)} \qquad (99)$$

Model hydroxylation reactions have been reported which in many cases appear to involve oxygen transfer from oxoiron species. These oxidations are discussed in Chapter 6 together with analogous reactions with oxometal compounds.

Related to the iron–porphyrin complexes are the ruthenium(II) complexes of *meso*-tetraphenylporphyrin and octaethylporphyrin,[138] which reversibly bind molecular oxygen at ambient temperatures in the presence of labile axial ligands such as acetonitrile. Enhanced oxygenation in polar aprotic solvents, as in the iron systems, was rationalized in terms of stabilization of a superoxoruthenium(III) complex.

The study of the properties and reactions of these superoxometal complexes will undoubtedly continue to be an area of intense activity in the context of their relevance to enzymatic oxidation mechanisms. The design of model systems capable of effecting mild selective oxidations of organic substrates, particularly hydrocarbons, is an important goal.

VI. OXYGEN TRANSFER FROM COORDINATED LIGANDS

A new approach to oxygen activation has recently emerged, in which oxygen is transferred from a ligand such as a nitro group to a substrate.[139] This step is followed by the reoxidation of the reduced form of the ligand (in this case NO) by molecular oxygen. Both steps are mediated by a transition metal, as illustrated in the general principle below,

$$\mathrm{M{-}XO_n + S \longrightarrow M{-}XO_{n-1} + SO} \qquad (100)$$

$$\mathrm{M{-}XO_{n-1} + \tfrac{1}{2}\,O_2 \longrightarrow M{-}XO_n} \qquad (101)$$

where X may be a main group element and S represents the substrate. In this system, the formal oxidation state of the metal does not change. Such studies emerge from Clarkson and Basolo,[140] who showed that five-coordinate cobalt–nitrosyl complexes react with dioxygen at ambient temperature in the presence of a base to form the six-coordinate nitro complexes,

$$\mathrm{LCo{-}NO + B + \tfrac{1}{2}\,O_2 \longrightarrow (B)LCo{-}NO_2} \qquad (102)$$

where L = tetradentate ligand. Tovrog and co-workers[139] have recently shown that Co(Saloph)py(NO_2) (where Saloph = *N*,*N'*-bisalicylidene-*o*-phenylenediamino) oxidizes Ph_3P stoichiometrically to Ph_3PO via an oxygen transfer mechanism.

$$\mathrm{Co^{III}(Saloph)py(NO_2) + Ph_3P \longrightarrow Co^{III}(Saloph)NO + Ph_3PO} \qquad (103)$$

Since the electronic structure of the nitrosyl complex is consistent[140] with the formulation as a nitrosylcobalt(III) species, reaction 103 occurs without a change in the formal oxidation state of cobalt. In the presence of molecular oxygen, a catalytic oxidation of Ph_3P results.

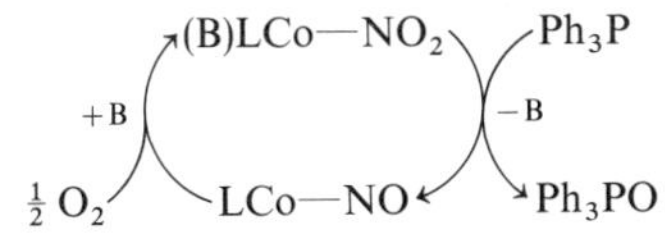

$$\text{(B)LCo—NO}_2 \xrightarrow[-\text{B}]{\text{Ph}_3\text{P} \to \text{Ph}_3\text{PO}} \text{LCo—NO} \xrightarrow[+\text{B}]{\frac{1}{2}\text{O}_2} \text{(B)LCo—NO}_2$$

The nitrocobalt(III) complex is a stronger oxidant than the oxomolybdenum(VI) complex $Mo(VI)O_2(S_2CNR_2)_2$, which has also been shown to transfer oxygen to Ph_3P (see Chapter 6). Thus, the following reaction proceeded quantitatively in dichloroethane at 60°C under an argon atmosphere.

$$\text{Co}^{\text{III}}\text{(Saloph)py(NO}_2\text{)} + \text{Mo}^{\text{IV}}\text{O(S}_2\text{CNR}_2\text{)}_2 \xrightarrow{-\text{py}} \text{Co}^{\text{III}}\text{(Saloph)NO} + \text{Mo}^{\text{VI}}\text{O}_2\text{(S}_2\text{CNR}_2\text{)}_2 \quad (104)$$

The addition of Lewis acids such as a transition metal complex (e.g., Mo^{6+}) or a main group Lewis acid (e.g., BF_3) enhances the capacity of these nitrocobalt(III) complexes to oxidize secondary alcohols and dienes. Oxygen atom transfer from nitro ligands on palladium(II)–π-olefin complexes has also been carried out in an oxygen-free anhydrous medium.[141] Thus, ethylene and propylene have been oxidized quantitatively to acetaldehyde and acetone, respectively, with the concomitant reduction to the nitrosylcobalt analog and palladium(II).

$$\text{LCoNO}_2 + (\text{RCH=CH}_2)\text{Pd}^{\text{II}} \longrightarrow \text{LCoNO} + \text{Pd}^{\text{II}} + \text{RC(=O)CH}_3 \quad (105)$$

When the transformation in eq 105 is coupled with the reoxidation of the nitrosocobalt complex by dioxygen in eq 102, a catalytic cycle shown in Scheme VIII has been proposed.

Scheme VIII:

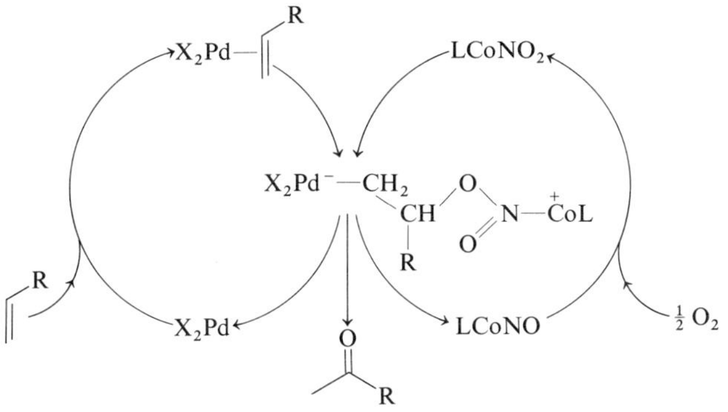

X_2Pd—(RCH=CH$_2$); $LCoNO_2$; X_2Pd^-—CH_2—CH(R)—O—N(=O)—$\overset{+}{Co}L$; X_2Pd; RC(=O)CH$_3$; LCoNO; RCH=CH$_2$; $\frac{1}{2}O_2$

Turnover numbers as high as 12 have been achieved in diglyme. In acetic acid, the cooxidation of ethylene catalyzed by a combined palladium acetate/nitrocobalt catalyst affords vinyl acetate as the sole product.

There are other examples of oxygen atom transfer from coordinated nitro ligands to various substrates.[142] Oxygen transfer from alkyl- and aryl-nitro compounds can also be mediated by metals.[143]

REFERENCES

1. P. A. Kilty and W. M. H. Sachtler, *Catal. Rev.* **10**, 1 (1974).
2. L. Vaska, *Accts. Chem. Rsch.* **9**, 175 (1976).
3. J. S. Valentine, *Chem. Rev.* **73**, 235 (1973).
4. V. J. Choy and C. J. O'Connor, *Coord. Chem. Rev.* **9**, 145 (1972–1973).
5. G. Henrici-Olivé and S. Olivé, *Angew. Chem.* **86**, 1 (1974).
6. A. V. Savitskii and V. I. Nelyubin, *Russ. Chem. Rev.* **44**, 110 (1975).
7. (a) L. Klevan, J. Poene, and S. K. Madan, *J. Chem. Educ.* **50**, 670 (1973).
   (b) T. Yoshida, K. Tatsumi, and S. Otsuka, *Pure Appl. Chem.* **52**, 713 (1980).
8. C. Bocard and I. Sérée de Roch, *Rev. Inst. Fr. Pet.* **28**, 891 (1973).
9. H. Mimoun, *Rev. Inst. Fr. Pet.* **33**, 259 (1978).
10. M. M. Taqui Khan and A. E. Martell, "Homogeneous Catalysis by Metal Complexes," Vol. 1, Ch. 2. Academic Press, New York, 1974.
11. J. E. Lyons, *in* "Aspects Homogeneous Catalysis," Vol. 3, Ch. 1, (R. Ugo, ed.), D. Reidel, Dordrecht, 1977.
12. J. E. Lyons, *in* "Fundamental Research in Homogeneous Catalysis" (M. Tsutsui and R. Ugo, eds.), Vol. 1, p. 1. Plenum, New York, 1971.
13. J. P. Franck, C. Bocard, I. Sérée de Roch, and L. Sajus, *Rev. Inst. Fr. Pet.* **24**, 710 (1969).
14. L. H. Vogt, H. M. Faigenbaum, and S. E. Wiberley, *Chem. Rev.* **63**, 269 (1963).
15. C. McLendon and A. E. Martell, *Coord. Chem. Rev.* **19**, 1 (1976).
16. F. Basolo, B. M. Hoffmann, and J. A. Ibers, *Accts. Chem. Rsch.* **8**, 384 (1975); R. D. Jones, D. A. Summerville, and F. Basolo, *Chem. Rev.* **79**, 139 (1979).
17. (a) A. G. Sykes and J. A. Weil, *Prog. Inorg. Radiochem.* **13**, 1 (1970).
    (b) R. G. Wilkins, *Adv. Chem. Series* **100**, 111 (1971).
18. (a) S. Nemeth, Z. Szeverenyi, and L. I. Simandi, *Inorg. Chim. Acta Lett.* **44**, L107 (1980).
    (b) L. P. Velyutin, V. M. Potekhim, and V. I. Ovchinnikov, *J. Gen. Chem. USSR* **49**, 2028 (1980).
    (c) R. S. Drago, J. Gaul, A. Zombeck, and D. K. Straub, *J. Am. Chem. Soc.* **102**, 1033 (1980).
19. J. P. Collman, *Accts. Chem. Rsch.* **10**, 265 (1977); T. G. Traylor, *ibid.*, **14**, 102 (181).
20. E. Fremy, *Justus Liebigs Ann. Chem.* **83**, 227, 289 (1852).
21. T. Tsumaki, *Bul. Chem. Soc. Japan* **13**, 252 (1938).
22. C. Floriani and F. Calderazzo, *J. Chem. Soc. A* p. 946 (1969).
23. A. L. Crumbliss and F. Basolo, *Science* **164**, 1168 (1969); *J. Am. Chem. Soc.* **92**, 55 (1970), see also W. M. Coleman, and L. T. Taylor, *Coord. Chem. Rev.* **32**, 1 (1980).
24. L. Vaska, *Science* **140**, 809 (1963).
25. J. Wilshire and D. T. Sawyer, *Accts. Chem. Rsch.* **12**, 105 (1979); see also D. T. Sawyer, M. J. Gibian, M. M. Morrison, and E. T. Seo, *J. Am. Chem. Soc.* **100**, 627 (1978). I. B. Afanasev, *Russ. Chem. Rev.* **48**, 527 (1979).

25. (a) See also C. L. Wong, J. A. Switzer, K. P. Balakrishnan, and J. F. Endicott, *J. Am. Chem. Soc.* **102**, 5511 (1980); R. S. Drago, J. P. Cannady, and K. A. Leslie, *ibid.* p. 6014; compare also J. P. Collman, I. Denisevich, Y. Konai, M. Marrocco, C. Koval, and F. C. Anson, *ibid.* p. 6027.

25. (b) W. J. Luow, T. I. A. Gerber and D. J. A. deWaal, *J. Chem. Soc., Chem. Commun.* 760 (1980). See also L. K. Hanson and B. M. Hoffman, *J. Am. Chem. Soc.* **102**, 4602 (1980).

26. (a) S. Muto, H. Ogata, and Y. Kamiya, Chem. Lett. p. 809 (1975).
(b) M. Pizzotti, S. Cenini, and G. LaMonica, *Inorg. Chim. Acta* **33**, 161 (1978).

27. B. S. Tovrog, S. E. Diamond, and F. Mares, *J. Am. Chem. Soc.* **101**, 5067 (1979).

27a. J. Billecke, W. Kokisch, and J. W. Buchler, *J. Am. Chem. Soc.* **102**, 3622 (1980).

28. J. A. Connor and E. A. V. Ebsworth, *Adv. Inorg. Chem. Radiochem.* **6**, 279 (1964).

29. H. Mimoun, I. Sérée de Roch, and L. Sajus, *Bul. Soc. Chim. France* p. 1481 (1969).

30. B. Bosnich, W. G. Jackson, J. T. D. Lo, and J. W. McLaren, *Inorg. Chem.* **13**, 2605 (1974).

31. (a) J. H. Bayston, N. K. King, F. D. Looney, and M. E. Winfield, *J. Am. Chem. Soc.* **91**, 2775 (1969).
(b) J. Ellis, J. M. Pratt, and M. Green, *J. Chem. Soc., Chem. Commun.* p. 781 (1973); see also E. W. Abel, J. M. Pratt, R. Whelan, and P. J. Wilkinson, *J. Am. Chem. Soc.* **96**, 7119 (1974).

32. W. J. Wallace, J. C. Maxwell, and W. C. Caughey, *Biochem. Biophys. Res. Commun.* **57**, 1104 (1974).

33. V. K. Kol'tover, O. I. Koifman, A. M. Khenkin, and A. A. Shteinman, *Bul. Acad. Sci. USSR, Div. Chem. Sci.* p. 1479 (1978); see also R. Catton, P. Premovic, L. Stavdal, and and P. West, *J. Chem. Soc., Chem. Commun.* p. 863 (1980).

33a. R. Guiland, J. M. Latour, C. Lecomte, J. C. Marchon, J. Protus, and D. Ripoll, *Inorg. Chem.* **17**, 1228 (1978); J. M. Latour, B. Galland, and J. C. Marchon, *J. Chem. Soc., Chem. Commun.* p. 570 (1979); M. Nakajima, J. M. Latour, and J. C. Marchon, *ibid.* p. 763 (1977); J. M. Latour, J. C. Marchon, and M. Nakajima, *J. Am. Chem. Soc.* **101**, 3974 (1979); J. C. Marchon, M. M. Latour, and C. J. Boreham, *J. Mol. Catal.* 7, 227 (1980).

34. (a) P. J. Chung, H. Suzuki, Y. Moro-oka, and T. Ikawa, *Chem. Lett.* p. 63 (1980).
(b) F. Sakurai, H. Suzuki, Y. Moro-oka, and T. Ikawa, *J. Am. Chem. Soc.* **102**, 1749 (1980).

35. S. L. Regen and G. M. Whitesides, *J. Organometal. Chem.* **59**, 293 (1973).

36. J. H. Bayston and M. E. Winfield, *J. Catal.* **3**, 123 (1964).

37. G. R. Hughes and D. M. P. Mingos, *Transition Met. Chem.* **3**, 381 (1978); S. Bhaduri, P. R. Raithby, C. I. Zuccaro, M. B. Hursthouse, L. Casella, and R. Ugo, *J. Chem. Soc., Chem. Commun.* p. 991 (1978).

38. G. W. Hooper, Br. Patent 1,056,121 (1967) to ICI; P. N. Dyer and F. Mosely, U.S. Patent 4,128,627 (1978) to Air Products; L. Kim and G. W. Schoenthal, Ger. Patent 2,615,625 (1976) to Shell.

39. J. K. Kochi, "Organometallic Mechanisms and Catalysis," pp. 517ff. Academic Press, New York, 1978.

40. D. A. Ryan and J. H. Espenson, *J. Am. Chem. Soc.* **101**, 2488 (1979).

41. R. Ugo, F. Conti, S. Cenini, R. Mason, and G. B. Robertson, *J. Chem. Soc., Chem. Commun.* p. 1498 (1968).

41a. R. Ugo, G. M. Zanderighi, A. Fusi, and D. Carreri, *J. Am. Chem. Soc.* **102**, 3745 (1980).

42. R. A. Sheldon and J. A. van Doorn, *J. Organometal. Chem.* **94**, 115 (1975).

43. M. J. Y. Chen and J. K. Kochi, *J. Chem. Soc., Chem. Commun.* p. 204 (1977).

44. H. Mimoun, R. Charpentier, A. Mitschler, J. Fischer, and R. Weiss, *J. Am. Chem. Soc.* **102**, 1047 (1980).

45. G. Wilke, H. Shott, and P. Heimbach, *Angew. Chem. Int. Ed.* **6**, 92 (1967).

46. A. Sen and J. Halpern, *J. Am. Chem. Soc.* **99**, 8337 (1977).
47. D.-H. Chin, G. N. La Mar, and A. L. Balch, *J. Am. Chem. Soc.* **102**, 5945 (1980).
48. J. P. Collman, M. Kubota and J. W. Hosking, *J. Am. Chem. Soc.* **89**, 4809 (1967).
49. (a) J. Blum, H. Rosenman, and E. D. Bergmann, *Tetrahedron Lett.* p. 3665 (1967).
(b) J. Blum, J. Y. Becker, H. Rosenman, and E. D. Bergmann, *J. Chem. Soc. B* p. 1000 (1969).
50. (a) A. J. Birch and G. S. Subba Rao, *Tetrahedron Lett.* p. 2917 (1968).
(b) E. W. Stern, *J. Chem. Soc., Chem. Commun.* p. 736 (1970).
51. V. P. Kurkov, J. Z. Pasky, and J. B. Lavigne, *J. Am. Chem. Soc.* **90**, 4743 (1968).
52. L. W. Fine, M. Grayson, and V. H. Suggs, *J. Organometal. Chem.* **8**, 219 (1970).
53. R. A. Sheldon, *J. Chem. Soc., Chem. Commun.* p. 788 (1971).
54. A. Fusi, R. Ugo, F. Fox, A. Pasini, and S. Cenini, *J. Organometal. Chem.* **26**, 417 (1970).
55. J. E. Lyons and J. O. Turner, *J. Org. Chem.* **37**, 2881 (1972).
56. J. E. Lyons and J. O. Turner, *Tetrahedron Lett.* p. 2903 (1972).
57. H. Arzoumanian, A. A. Blanc, J. Metzger, and J. E. Vincent, *J. Organometal. Chem.* **82**, 261 (1974).
58. W. Strohmeier and E. Eder, *J. Organometal. Chem.* **94**, C14 (1975).
59. J. E. Lyons, *Adv. Chem. Series* **132**, 64 (1974).
60. B. L. Booth, R. N. Haszeldine, and G. R. H. Neuss, *J. Chem. Soc., Chem. Commun.* p. 1074 (1972).
61. (a) C. Dudley and G. Read, *Tetrahedron Lett.* p. 5273 (1972).
(b) C. W. Dudley, G. Read, and P. J. C. Walker, *J. Chem. Soc., Dalton*, p. 1926 (1974).
62. R. Tang, F. Mares, N. Neary, and D. E. Smith, *J. Chem. Soc., Chem. Commun.* 274 (1979).
63. F. Igersheim and H. Mimoun, *J. Chem. Soc., Chem. Commun.* p. 559 (1978); *Nouv. J. Chim.* **4**, 161 (1980).
63a. H. Mimoun, M. M. Perez Machirant, and I. Sérée de Roch, *J. Am. Chem. Soc.* **100**, 5437 (1978).
64. G. Read and P. J. C. Walker, *J. Chem. Soc., Dalton* p. 883 (1977).
65. J. K. Stille, L. F. Hines, R. W. Fries, P. K. Wong, D. E. James, and K. Lau, *Adv. Chem. Series* **132**, 90 (1974); J. E. Backvall, B. Akermark, and S. O. Ljunggren, *J. Am. Chem. Soc.* **101**, 2411 (1979).
66. (a) J. Farrar, D. Holland, and D. J. Milner, *J. Chem. Soc., Dalton* p. 815 (1975).
(b) D. Holland and D. J. Milner, *J. Chem. Soc., Dalton* p. 2440 (1975).
67. K. Takao, M. Wayaku, Y. Fujiwara, T. Imanaka, and S. Teranishi, *Bul. Chem. Soc. Japan* **43**, 3898 (1970).
68. K. Takao, H. Azuma, Y. Fujiwara, T. Imanaka, and S. Teranishi, *Bul. Chem. Soc. Japan* **45**, 2003 (1972).
69. G. Read, *J. Mol. Catal.* **4**, 83 (1978).
70. F. Mares and R. Tang, *J. Org. Chem.* **43**, 4631 (1978).
71. H. S. Ryang and C. S. Foote, *J. Am. Chem. Soc.* **102**, 2129 (1980).
72. (a) D. Schwarzenbach, *Helv. Chim. Acta* **55**, 2990 (1972).
(b) D. Schwarzenbach, *Inorg. Chem.* **9**, 2391 (1970).
73. R. Drew and F. Einstein, *Inorg. Chem.* **12**, 829 (1973); F. Offner and J. Dehand, *C. R. Hebd. Seances Acad. Sci., Series C* **273**, 50 (1971).
74. (a) R. Stomberg and I. Ainalem, *Acta Chem. Scand.* **22**, 1439 (1968).
(b) R. Stomberg, *Acta Chem. Scand.* **23**, 2755 (1969).
75. D. Westlake, R. Kergoat, and J. E. Guerchais, *C. R. Hebd. Seances Acad. Sci., Series C* **280**, 113 (1975).
76. S. Jacobson, R. Tang, and F. Mares, *Inorg. Chem.* **17**, 3055 (1978).
77. J. Y. Calves, J. E. Guerchais, R. Kergoat, and N. Kheddar, *Inorg. Chim. Acta* **33**, 95 (1979).

78. H. Arzoumanian, R. Lai, R. L. Alvarez, J. F. Petrignani, and J. Metzger, *J. Am. Chem. Soc.* **102**, 845 (1980); see also H. Arzoumanian, H. Bitar, and J. Metzger, *J. Mol. Catal.* **7**, 373 (1980).
79. (a) H. Mimoun, I. Sérée de Roch, and L. Sajus, *Tetrahedron* **26**, 37 (1970).
    (b) H. Arakawa, Y. Moro-oka, and A. Ozaki, *Bul. Chem. Soc. Japan* **47**, 2958 (1974).
80. K. B. Sharpless, J. M. Townsend, and D. R. Williams, *J. Am. Chem. Soc.* **94**, 295 (1972); see also A. A. Achrem, T. A. Timoschtschuk, and D. I. Metelitza, *Tetrahedron* **30**, 3165 (1974).
81. (a) H. B. Kagan, H. Mimoun, C. Mark, and V. Schurig, *Angew. Chem. Int. Ed.* **18**, 485, (1979).
    (b) See also W. Winter, C. Mark, and V. Schurig, *Inorg. Chem.* **19**, 2045 (1980).
82. E. Vedejs, D. A. Engler, and J. E. Telschow, *J. Org. Chem.* **43**, 188 (1978); E. Vedejs, *J. Am. Chem. Soc.* **96**, 5944 (1974).
83. N. J. Lewis, S. Y. Gabhe, and M. R. de la Mater, *J. Org. Chem.* **42**, 1479 (1977).
84. S. E. Jacobson, R. Tang, and F. Mares, *J. Chem. Soc., Chem. Commun.* p. 888 (1978).
85. S. E. Jacobson, D. A. Muccigrosso, and F. Mares, *J. Org. Chem.* **44**, 921 (1979).
86. R. Tang, S. E. Diamond, N. Neary, and F. Mares, *J. Chem. Soc., Chem. Commun.* p. 562 (1978).
87. H. Arzoumanian, R. Lopez Alvarez, A. D. Kowalak, and J. Metzger, *J. Am. Chem. Soc.* **99**, 5175 (1977).
88. (a) R. Chevrier, T. Diebald, and R. Weiss, *Inorg. Chim. Acta* **19**, L57 (1976).
    (b) H. Ledon and M. Bonnet, *J. Mol. Catal.* **7**, 309 (1980).
89. Y. Hayashi, S. Shioi, M. Toyami, and T. Sakan, *Chem. Lett.* p. 651 (1973).
90. R. Guilard, M. Fontesse, P. Fournari, C. Lecomte, and J. Protus, *J. Chem. Soc., Chem. Commun.* p. 161 (1976).
91. L. S. Liebeskind, K. B. Sharpless, R. D. Wilson, and J. A. Ibers, *J. Am. Chem. Soc.* **100**, 7061 (1978).
92. D. A. Muccigrosso, S. E. Jacobson, P. A. Apgar, and F. Mares, *J. Am. Chem. Soc.* **100**, 7063 (1978).
93. (a) T. P. Singer and R. N. Ondarza, eds., "Mechanisms of Oxidizing Enzymes," Dev. Biochem. Vol. 1, Elsevier, Amsterdam, 1978.
    (b) O. Hayaishi, ed., "Mechanisms of Oxygen Activation." Academic Press, New York, 1974.
94. J. H. Bayston, N. K. King, F. D. Looney, and M. E. Winfield, *J. Am. Chem. Soc.* **91**, 2775 (1969); see also G. N. Schrauzer and L. P. Lee, *ibid.* **92**, 1551 (1970).
95. G. D. Lawrence and D. T. Sawyer, *Coord. Chem. Rev.* **27**, 173 (1978).
96. (a) I. Fridovich, *Accts. Chem. Rsch.* **5**, 321 (1972).
    (b) I. Fridovich, *Adv. Enzymol.* **41**, 35 (1974); *Am. Sci.* **63**, 54 (1975).
    (c) J. A. Fee and G. J. McClune, *in* "Mechanisms of Oxidizing Enzymes" (T. P. Singer and R. N. Ondarza, eds.), p. 273 Elsevier, Amsterdam, 1978.
97. M. Calvin and A. E. Martell, "Chemistry of the Metal Chelate Compounds," p. 337. Prentice-Hall, Englewood Cliffs, New Jersey, 1952.
98. A. Nishinaga, H. Tomita, T. Shimizu, and T. Matsuura, *in* "Fundamental Research in Homogeneous Catalysis" (Y. Ishii and M. Tsutsui, eds.), Vol. 2, p. 241. Plenum, New York, 1978. See also *Tetrahedron Lett.* p. 4849, p. 4853 (1980).
99. (a) E. Lee-Ruff, *Chem. Soc. Rev.* **6**, 195 (1977).
    (b) D. T. Sawyer and M. J. Gibian, *Tetrahedron* **35**, 1471 (1979).
100. D. T. Sawyer, M. J. Gibian, M. M. Morrison, and E. T. Seo, *J. Am. Chem. Soc.* **100**, 627 (1978).
101. A. D. Vreugdenhil, *J. Catal.* **28**, 493 (1973).

102. W. J. M. van Tilborg and A. D. Vreugdenhil, *Tetrahedron* **31**, 2825 (1975); see also R. V. Norton, *Prepr., Div. Pet. Chem. Am. Chem. Soc.* **15**, B18 (1970).
103. See, for example, H. Kropf and K. Knaak, *Tetrahedron* **28**, 1143 (1972); Y. Kamiya, *Tetrahedron Lett.* p. 4965 (1968); E. Ochiai, *Tetrahedron* **20**, 1819 (1964).
104. A. B. P. Lever, *Adv. Inorg. Chem. Radiochem.* **7**, 28 (1965).
105. C. J. Pedersen, *J. Org. Chem.* **22**, 127 (1957).
106. J. H. Fuhrhop and D. Mauzerall, *J. Am. Chem. Soc.* **91**, 4174 (1969).
107. B. R. James and E. Ochiai, *Canad. J. Chem.* **49**, 976 (1971); see also B. R. James and F. T. T. Ng, *J. Chem. Soc., Chem. Commun.* p. 908 (1970); B. R. James, F. T. T. Ng, and E. Ochiai, *Canad. J. Chem.* **50**, 590 (1972).
108. (a) A. Nishinaga, K. Watanabe, and T. Matsuura, *Tetrahedron Lett.* p. 1291 (1974).
(b) A. Nishinaga, K. Nishizawa, H. Tomita, and T. Matsuura, *J. Am. Chem. Soc.* **99**, 1287 (1977).
109. (a) A. Nishinaga, H. Tomita, and T. Matsuura, *Tetrahedron Lett.* **21**, 1261, 2833 (1980).
(b) A. Nishinaga, T. Shimizu, and T. Matsuura, *Tetrahedron Lett.* **21**, 1265 (1980).
(c) A. Nishinaga, H. Tomita, T. Shimizu, and T. Matsuura, *in* "Fundamental Research in Homogeneous Catalysis" (Y. Ishii and M. Tsutsui, eds.), Vol. 2, p. 241. Plenum. New York, 1978.
110. A. Nishinaga and H. Tomita, *J. Mol. Catal.* **7**, 179 (1980).
111. H. M. van Dort and H. J. Geursen, *Recl. Trav. Chim. Pays-Bas* **86**, 520 (1967).
112. L. H. Vogt, J. G. Wirth, and H. L. Finkbeiner, *J. Org. Chem.* **34**, 273 (1969).
113. British Patent 1,268,653 (1972) to BASF.
114. V. M. Kothari and J. J. Tazuma, *J. Catal.* **41**, 180 (1976).
115. T. J. Fullerton and S. P. Ahern, *Tetrahedron Lett.* p. 139 (1976).
116. F. R. Jensen and R. C. Kiskis, *J. Am. Chem. Soc.* **97**, 5825 (1975).
117. A. Nishinaga, *Chem. Lett.* p. 273 (1975).
118. M. N. Dufour-Ricroch and A. Gaudemar, *Tetrahedron Lett.* p. 4079 (1976).
119. K. Uchida, M. Soma, S. Naito, T. Onishi, and K. Tamara, *Chem. Lett.* p. 471 (1978).
120. T. Yarino, T. Matsushita, I. Masuda, and K. Shinra, *J. Chem. Soc., Chem. Commun.* p. 1317 (1970); C. C. Patel and C. P. Prabhakoran, *J. Inorg. Nucl. Chem.* **31**, 3316 (1969).
121. M. Constantini, A. Dromard, M. Joffret, B. Brossard, and J. Varagnat, *J. Mol. Catal.* **7**, 89 (1980).
122. (a) B. Gonzalez, J. Kouba, S. Yee, C. A. Reed, J. Kirner, and W. R. Scheidt, *J. Am. Chem. Soc.* **97**, 3247 (1975).
(b) C. J. Weschler, B. M. Hoffman, and F. Basolo, *J. Am. Chem. Soc.* **97**, 5278 (1975).
(c) B. M. Hoffman, C. J. Weschler, and F. Basolo, *J. Am. Chem. Soc.* **98**, 5473 (1976).
(d) See also A. B. P. Lever, J. P. Wilshire, and S. K. Quan, *J. Am. Chem. Soc.* **101**, 3668 (1979).
123. D. A. O'Sullivan, *Chem. & Eng. News* **56** (Dec. 4), p. 24 (1978).
124. (a) M. G. Simmons and L. J. Wilson, *J. Chem. Soc., Chem. Commun.* p. 634 (1978).
(b) See also M. G. Burnett, V. McKee, S. M. Nelson, and M. G. B. Drew, *J. Chem. Soc., Chem. Commun.* p. 829 (1980).
125. I. Bodek and G. Davies, *Inorg. Chem.* **17**, 1814 (1978); G. E. Kramer, G. Davies, R. B. Davis, and R. W. Slaven, *J. Chem. Soc., Chem. Commun.* p. 606 (1975).
126. J. Tsuji, H. Takayanagi, and I. Sakai, *Tetrahedron Lett.* p. 1245 (1975); J. Tsuji and H. Takayanagi, *J. Am. Chem. Soc.* **96**, 7349 (1974); *Chem. Lett.* p. 65 (1980).
127. M. M. Rogic and T. R. Demmin, *J. Am. Chem. Soc.* **100**, 5472 (1978).
127a. M. M. Rogic, T. R. Demmin, and W. B. Hammond, *J. Am. Chem. Soc.* **98**, 7441 (1976).
127b. T. R. Demmin and M. M. Rogic, *J. Org. Chem.* **45**, 4210 (1980).
127c. E. L. Reilly, Br. Patent 1,511,813 (1977) to DuPont.

128. E. McCandlish, A. R. Miksztal, M. Nappa, A. Q. Sprenger, J. S. Valentine, J. D. Strong, and T. G. Spiro, *J. Am. Chem. Soc.* **102**, 4269 (1980).
129. D.-H. Chin, J. Del Gaudio, G. N. LaMar, and A. L. Balch, *J. Am. Chem. Soc.* **99**, 5486 (1977); D.-H. Chin, G. N. LaMar, and A. L. Balch, *ibid.* **102**, 4344, 5945 (1980); I. Collamati, *Inorg. Chim. Acta* **35**, L303 (1979).
129a. E. I. Ochiai, *Inorg. Nucl. Chem. Lett.* **10**, 453 (1974).
130. J. H. Wang, *J. Am. Chem. Soc.* **80**, 3168 (1958); see also H. Ledon and Y. Brigandat, *J. Organometal. Chem.* **165**, C25 (1979).
131. C. K. Chang and T. G. Traylor, *J. Am. Chem. Soc.* **95**, 5810, 8475, 8477 (1973); J. Almog, J. E. Baldwin, R. L. Dyer, J. Huff, and C. J. Wilkerson, *ibid.* p. 5600; D. L. Anderson, C. J. Weschler, and F. Basolo, *ibid.* **96**, 5599 (1974).
132. J. Almog, J. E. Baldwin, and J. Huff, *J. Am. Chem. Soc.* **97**, 227 (1975).
133. J. P. Collman, R. Gagne, C. A. Reed, T. R. Halbert, G. Lang, and W. T. Robinson, *J. Am. Chem. Soc.* **97**, 1427 (1975); J. P. Collman and K. S. Suslick, *Pure Appl. Chem.* **50**, 951 (1978).
134. J. P. Collman, J. I. Brauman, K. M. Doxsee, T. R. Halbert, S. E. Hayes, and K. S. Suslick, *J. Am. Chem. Soc.* **100**, 2761 (1978); J. P. Collman, R. R. Gagne, J. Kouba, and H. Ljusberg-Wahren, *ibid* **96**, 6800 (1974).
135. M. Momenteau, B. Loock, J. Mispelter, and E. Bisagni, *Nouv. J. Chim.* **3**, 77 (1979).
136. A. R. Battersby, D. G. Buckley, S. G. Hartley, and M. D. Turnbull, *J. Chem. Soc., Chem. Commun.* p. 879 (1976); J. E. Baldwin, T. Klose, and M. Peters, *ibid.* p. 881.
137. J. P. Collman, T. N. Sorrell, and B. M. Hoffman, *J. Am. Chem. Soc.* **97**, 913 (1975); S. Koch, S. C. Tang, R. H. Holm, R. B. Frankel, and J. A. Ibers, *ibid.* p. 916; S. Koch, S. C. Tang, R. H. Holm, and R. B. Frankel, *ibid.* p. 914.
138. N. Farrel, D. H. Dolphin, and B. R. James, *J. Am. Chem. Soc.* **100**, 324 (1978).
139. B. S. Tovrog, S. E. Diamond, and F. Mares, *J. Am. Chem. Soc.* **101**, 270 (1979).
140. S. G. Clarkson and F. Basolo, *Inorg. Chem.* **12**, 1528 (1973).
141. B. S. Tovrog, F. Mares and S. E. Diamond, *J. Am. Chem. Soc.* **102**, 6617 (1980).
142. (a) D. T. Doughty, G. Gordon, and R. P. Stewart, *J. Am. Chem. Soc.* **101**, 2645 (1979).
    (b) R. D. Feltham and J. C. Kriege, *J. Am. Chem. Soc.* **101**, 5064 (1979).
    (c) H. D. Abruña, J. L. Walsh, T. J. Meyer, and R. W. Murray, *J. Am. Chem. Soc.* **102**, 3764 (1980).
143. R. S. Berman and J. K. Kochi, *Inorg. Chem.* **19**, 248 (1980).
144. F. Igersheim and H. Mimoun, *Nouv. J. Chim.* **4**, 711 (1980).
145. M. Roussel and H. Mimoun, *J. Org. Chem.* **45**, 5381 (1980).

ADDITIONAL READING

B. R. James, A. W. Addison, M. Cairns, D. Dolphin, N. P. Farrell, D. R. Paulson, and S. Walker, Progress in dioxygen complexes of metalloporphyrin. In "Fundamental Research in Homogeneous Catalysis" (Y. Ishii and M. Tsutsui, eds.), Vol. 3, p. 751. Plenum, New York, 1979.

T. G. Spiro, ed., "Metal Ion Activation of Dioxygen." Wiley, New York, 1980.

Chapter 5

Direct Homolytic Oxidation by Metal Complexes

Catalytic oxidations as described in Chapters 2 and 3 involve the interaction of the metal catalyst with either dioxygen or alkyl hydroperoxide, the latter being formed *in situ* or in a separate step. There is another class of catalytic oxidations that involve the direct attack on the substrate by the metal complex, the catalyst being regenerated with dioxygen or peroxidic intermediates. The industrially important oxidation of *p*-xylene to terephthalic acid with a cobalt catalyst is a relevant example of the latter.

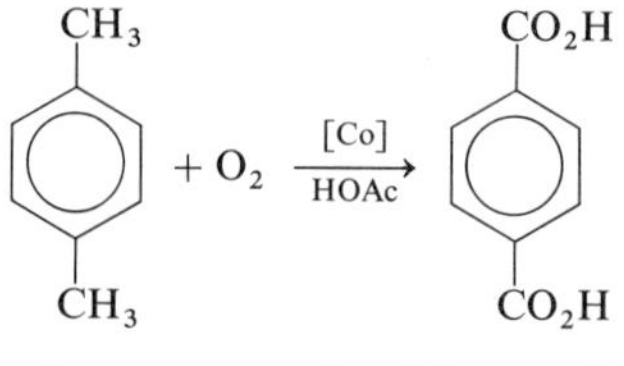

$$p\text{-}CH_3C_6H_4CH_3 + O_2 \xrightarrow[\text{HOAc}]{[\text{Co}]} p\text{-}HO_2CC_6H_4CO_2H$$

The direct interaction of strong metal oxidants with organic substrates can lead to the production of radical intermediates in two ways. Both processes are depicted below for the reaction of a metal triacetate with a hydrocarbon RH.

Electron transfer

$$M(OAc)_3 + RH \rightleftharpoons RH^{+\cdot} + M(OAc)_3^- \quad (1)$$

$$RH^{+\cdot} \longrightarrow R\cdot + H^+ \quad (2)$$

Electrophilic substitution

$$RH + M(OAc)_3 \longrightarrow RM(OAc)_2 + HOAc \quad (3)$$

$$RM(OAc)_2 \longrightarrow R\cdot + M(OAc)_2 \quad (4)$$

The net result in both cases is a one-electron reduction of the metal oxidant with concomitant formation of the substrate radical (R·). The ease of electron transfer oxidation of hydrocarbons to produce the cation-radical ($RH^{+\cdot}$) is related to the ionization potential listed in Table I. Electrophilic

TABLE I Ionization Potentials of Organic Substrates[a]

Substrate	Ionization potential (eV)	Substrate	Ionization potential (eV)
Alkanes		Aromatic compounds	
Methane	12.8	Nitrobenzene	10.0
Ethane	11.5	Benzene	9.25
Propane	11.1	Fluorobenzene	9.2
n-Butane	10.7	Chlorobenzene	9.1
Isobutane	10.6	Bromobenzene	9.0
n-Pentane	10.5	Toluene	8.8
n-Hexane	10.4	Ethylbenzene	8.75
Cyclohexane	9.9	Cumene	8.7
		o-Xylene	8.55
Alkenes		Mesitylene	8.4
Ethylene	10.45	Phenol	8.5
Propylene	9.75	Anisole	8.2
1-Butene	9.6	Aniline	7.7
2-Butene	9.15	Naphthalene	8.1
Isobutene	9.2	Biphenyl	8.3
1-Hexene	9.45	Anthracene	7.3
Cyclohexene	8.95	Methylnaphthalene	7.9
Butadiene	9.1	Methylanisole	7.9
		Thioanisole	7.9
		Acetophenone	9.4
		Benzyl chloride	9.2
		Pyridine	9.25
		Methylpyridine	9.0
		t-Butylbenzene	8.35
		Benzaldehyde	9.5

[a] Ionization potentials, appearance potentials and heats of formation of gaseous positive ions. *Natl. Stand. Ref. Data Series* (*U.S., Natl. Bur. Stand.*) **NSRDS–NBS** (1969). (Values for thioanisole and *t*-butylbenzene are probably incorrect.)

substitution is also expected to parallel electron transfer, and the distinction between electron transfer and electrophilic substitution processes based on structure–reactivity relationships alone is often difficult to make. Moreover, they are probably competing processes in a number of instances, and the dominant pathway with a particular oxidant may vary with a change in substrate. These two processes are used as the mechanistic basis for the discussion of redox reactions of various organic substrates with metal complexes, as discussed individually below.

I. AROMATIC COMPOUNDS

The oxidation of alkylbenzenes at relatively low temperatures (90–110°C) and oxygen pressures (0.2–1 bar) can be achieved in acetic acid in the presence of relatively high concentrations ($\sim 0.1\ M$) of cobaltic acetate.[1] Methyl ethyl ketone or ozone can be used as an activator to reconvert Co(II) in the presence of dioxygen to Co(III), since long induction periods are otherwise observed.[1,2] The results of numerous studies of the reaction of cobaltic acetate in acetic acid with alkylbenzenes are compatible with the following generally accepted mechanism.[3–14]

Scheme I:

$$\mathrm{ArCH_3 + Co^{III} \rightleftharpoons [ArCH_3]^{+\cdot} + Co^{II}} \tag{5}$$

$$\mathrm{[ArCH_3]^{+\cdot} \longrightarrow ArCH_2\cdot + H^+} \tag{6}$$

Benzyl acetate is derived from the subsequent reaction of the benzyl radical with cobaltic acetate.

Scheme II:

$$\mathrm{ArCH_2\cdot + Co^{III} \longrightarrow ArCH_2{}^+ + Co^{II}} \tag{7}$$

$$\mathrm{ArCH_2{}^+ + HOAc \longrightarrow ArCH_2OAc + H^+} \tag{8}$$

However, under autoxidizing conditions, the benzyl radical is trapped by dioxygen, and aromatic aldehydes are the primary products, formed by reaction of benzylperoxy radicals with Co(II) [which simultaneously regenerates the Co(III) oxidant].

Scheme III:

$$\mathrm{ArCH_2\cdot + O_2 \longrightarrow ArCH_2O_2\cdot} \tag{9}$$

$$\mathrm{ArCH_2O_2\cdot + Co^{II} \longrightarrow ArCH_2O_2Co^{III}} \tag{10}$$

$$\mathrm{ArCH(H){-}O{-}O{-}Co^{III} \longrightarrow ArCHO + HOCo^{III}} \tag{11}$$

$$\mathrm{HOCo^{III} + HOAc \longrightarrow AcOCo^{III} + H_2O} \tag{12}$$

The usual reaction of alkylperoxy radicals, namely,

$$ArCH_2O_2\cdot + ArCH_3 \longrightarrow ArCH_2OOH + ArCH_2\cdot \qquad (13)$$

is largely circumvented by the efficient trapping of the benzylperoxy radical with the relatively high concentration of Co(II) present. In the absence of oxygen, benzyl acetates are formed in a stoichiometric reaction.

The aldehyde is susceptible to further autoxidation to the corresponding carboxylic acid,[15,16] although it is possible to isolate it in moderate yields in some cases.[17] Since the oxidation of benzaldehyde by Co(III) is less favorable than that of the alkylbenzene substrate,

$$C_6H_5\text{—CHO} + Co^{III} \rightleftharpoons [C_6H_5]^{+\cdot}\text{—CHO} + Co^{II} \qquad (14)$$

a relatively high Co(III) concentration and low oxygen pressure favor aldehyde production.

The reversibility of the cation-radical formation in eq 5 is consistent with the retardation of the rate by added Co(II). The kinetics are further complicated by the dimer–monomer equilibrium[18] of cobalt(III) acetate in acetic acid (the monomer is the more active catalyst). Onopchenko and coworkers[11–13] found the relative rates of oxidation of alkylbenzenes by Co(III) acetate in acetic acid to be the reverse of that expected from a classical free radical mechanism. For example, toluene is oxidized approximately 10 times faster than cumene. Similarly, the oxidation of *p*-cymene with stoichiometric amounts of Co(III) acetate occurs virtually exclusively at the methyl group,

$$p\text{-}CH_3C_6H_4CH(CH_3)_2 \xrightarrow[\text{HOAc}]{Co(OAc)_3} \underset{(81\%)}{p\text{-}AcOCH_2C_6H_4CH(CH_3)_2} + \underset{(15\%)}{p\text{-}HO_2CC_6H_4COCH_3} \qquad (15)$$

and under autoxidizing conditions, *p*-isopropylbenzoic acid is the major product.

$$p\text{-}CH_3C_6H_4CH(CH_3)_2 + O_2 \xrightarrow[\text{HOAc}]{[Co(OAc)_2]} \underset{(90\%)}{p\text{-}HO_2CC_6H_4CH(CH_3)_2} + \underset{(10\%)}{p\text{-}OHCC_6H_4CH(CH_3)_2} \qquad (16)$$

Similarly, *p*-ethyltoluene, *sec*-butyltoluene, and 1,1-di-*p*-tolylethane afforded products mainly derived from the oxidation of the methyl groups, which can be accounted for by proton loss from the intermediate cation-radical in eq 6, as governed by stereoelectronic considerations and not by the thermodynamic stability of the radical formed.[12] Indeed, the preferred conformation of the tertiary hydrogen in the isopropyl group of the *p*-cymene cation-radical is located in the nodal plane of the benzene ring, where its interaction with the π orbital is minimized in the transition state. On the other hand, the loss of any one of the hydrogens on the methyl group is not conformationally restricted, owing to the low barrier to rotation.

Although the observed kinetics have generally been reconciled with an electron transfer mechanism via reactions 5 and 6, some authors[19] prefer a mechanism involving the direct, reversible formation of a benzyl radical without the intermediacy of a cation-radical. The difference can be reconciled if the lifetime of the cation-radical, as a discrete intermediate, varies with reactivity. Thus, the rapid proton transfer from reactive cation-radicals such as those from monoalkylbenzenes may occur predominantly within the solvent cage, and free radical-cations as such are not formed.

$$C_6H_5CH_3 + AcOCo^{III} \rightleftharpoons \left[C_6H_5CH_3^{+\cdot} \; AcOCo^{II} \right] \xrightarrow{\text{rapid}} C_6H_5CH_2\cdot + Co^{II} + HOAc \qquad (17)$$

Heiba and co-workers[20] observed the esr spectra of the cation-radicals of arenes upon treatment with Co(III) acetate in trifluoroacetic acid. (Cation-radicals are generally more stable in trifluoracetic acid because of its low nucleophilicity, but stable cation-radicals from electron-rich aromatic compounds can be observed even in acetic acid.[10]) Further evidence for cation-radicals is provided by the following reaction, which is difficult to explain by a conventional free radical mechanism.[10]

$$p\text{-}CH_3OC_6H_4CH_2SPh \xrightarrow{Co^{III}} [p\text{-}CH_3OC_6H_4CH_2SPh]^{+\cdot} \xrightarrow{-PhS\cdot} p\text{-}CH_3OC_6H_4CH_2^{+} \xrightarrow{HOAc} p\text{-}CH_3OC_6H_4CH_2OAc \qquad (18)$$

Manganic acetate is also capable of oxidizing alkylaromatic compounds in a direct reaction. However, it is a weaker oxidant than Co(III), and an electron transfer mechanism is observed only with aromatic compounds such as *p*-anisole and methylnaphthalene, having ionization potentials less than about 8 eV.[21–24] With the less reactive benzene and toluene, the products result from the reaction of the arene and carboxymethyl radicals generated from the thermolysis of Mn(III) acetate.[24–27] The complete reaction scheme is illustrated in Scheme IV for toluene.

Scheme IV:

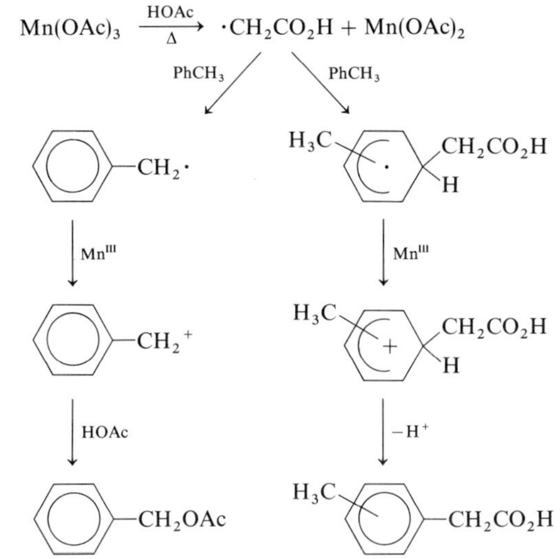

$$Mn(OAc)_3 \xrightarrow[\Delta]{HOAc} \cdot CH_2CO_2H + Mn(OAc)_2$$

PhCH$_3$ (left): $C_6H_5CH_2\cdot \xrightarrow{Mn^{III}} C_6H_5CH_2^+ \xrightarrow{HOAc} C_6H_5CH_2OAc$

PhCH$_3$ (right): H_3C–cyclohexadienyl radical(H)(CH_2CO_2H) $\xrightarrow{Mn^{III}}$ H_3C–cyclohexadienyl cation(H)(CH_2CO_2H) $\xrightarrow{-H^+}$ H_3C–C_6H_4–CH_2CO_2H

The arylacetic acids are subsequently decarboxylated by Mn(III) acetate to a mixture of xylyl acetates.

$$ArCH_2CO_2H \xrightarrow{Mn^{III}} ArCH_2\cdot + CO_2 + H^+ \quad (19)$$

$$ArCH_2\cdot \xrightarrow[HOAc]{Mn^{III}} ArCH_2OAc + H^+ \quad (20)$$

Electron transfer can be suppressed by potassium acetate or Mn(II) acetate or by working under completely anhydrous conditions. Similar processes are observed in oxidations of aromatic compounds with Ce(IV) acetate.[28,29]

The difference between cobaltic and manganic oxidations is illustrated in the autoxidation of *p*-cymene. Manganese catalysis affords mainly *p*-toluic acid and *p*-methylacetophenone by degradation of the isopropyl substituent,[13]

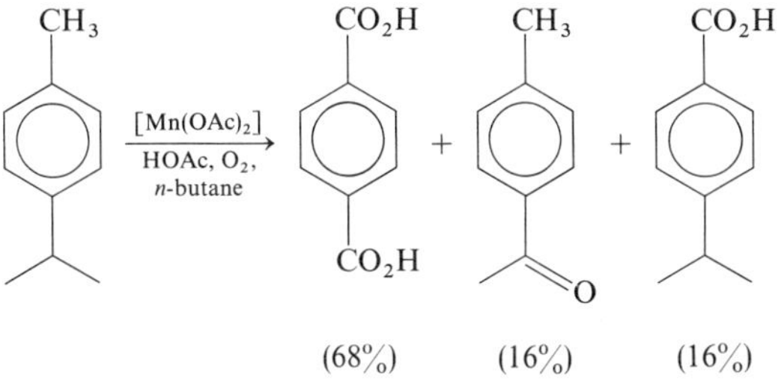

$$p\text{-}CH_3C_6H_4CH(CH_3)_2 \xrightarrow[HOAc,\ O_2,\ n\text{-butane}]{[Mn(OAc)_2]} p\text{-}HO_2CC_6H_4CO_2H + p\text{-}CH_3C_6H_4COCH_3 + p\text{-}HO_2CC_6H_4CH(CH_3)_2 \quad (21)$$

(68%) (16%) (16%)

whereas cobalt catalysis leads to the oxidation of the methyl group in eq 16. Furthermore, Mn(III) acetate is usually a less efficient catalyst than Co(III) acetate in the autoxidations of alkylaromatic compounds involving the direct oxidation of the substrate.[1]

A. Effect of Additives

Schemes I, II, and III represent the basic steps involved in the oxidation of aralkanes by metal acetates in acetic acid media. In addition, halide salts, other metal additives, and strong acids also affect the efficiency of the oxidation, as described individually below.

1. *Synergism by Halide*

Bromide, as hydrogen bromide, sodium bromide, or organic bromide, has a pronounced synergistic effect on the cobalt- and manganese-catalyzed autoxidations of alkylaromatic hydrocarbons.[30–38] Thus, *p*-xylene is readily oxidized to terephthalic acid in excellent yields, at atmospheric pressure and temperatures as low as 60°C, with a catalyst consisting of a mixture of cobaltic acetate and sodium bromide. Generally, high concentrations of catalyst ($\sim 0.1\ M$) are required, and none of the other halogens approaches bromide in activity. The active catalyst is considered to be acetatobromo-cobalt(II), formed in the equilibrium:

$$\mathrm{Co(OAc)_2 + NaBr \longrightarrow Co(OAc)Br + NaOAc} \tag{22}$$

A small amount of *tert*-butyl hydroperoxide is usually required to initiate the reaction, in which cobalt(II) is oxidized to cobalt(III) This is followed by the reactions shown in Scheme V, in which bromine atoms are the chain transfer agents.[38a]

Scheme V:

$$\mathrm{AcOCo^{III}Br \longrightarrow AcOCo^{II} + Br\cdot} \tag{23}$$

$$\mathrm{Br\cdot + ArCH_3 \longrightarrow HBr + ArCH_2\cdot} \tag{24}$$

$$\mathrm{ArCH_2\cdot + O_2 \longrightarrow ArCH_2O_2\cdot} \tag{25}$$

$$\mathrm{ArCH_2O_2\cdot + AcOCo^{II} \longrightarrow ArCHO + AcOCo^{III}OH} \tag{26}$$

$$\mathrm{AcOCo^{III}OH + HBr \longrightarrow AcOCo^{III}Br + H_2O} \tag{27}$$

In the presence of bromide, there is apparently no direct reaction of Co(III) with the substrate. A similar mechanism can be written for oxidations carried out with a mixture of $Mn(OAc)_2$ and NaBr.[37] The operation of different mechanisms is illustrated in Table II by a comparison of the relative rates of oxidation of alkylbenzenes catalyzed by $Co(OAc)_3$ alone (column 2)

TABLE II Relative Reactivities of Hydrocarbons toward Cobalt Oxidation

	Relative reactivity (per active hydrogen)			
Hydrocarbon	$Co(OAc)_3$ (65°C)[a]	$Co(OAc)_2$–NaBr (60°C)[b]	$RO_2\cdot$ (30°C)[c]	Br· (40°C)[d]
Toluene	1.0	1.0	1.0	1.0
Ethylbenzene	1.3	8.3	9.3	17
Cumene	0.3	16.8	15.9	37
p-Methoxytoluene	71[e]	3.4	—	—
Durene	275[f]	3.8	—	—
p-Xylene	10.3	1.5	—	—

[a] Heiba *et al.*[10]
[b] Kamiya.[34]
[c] Cumylperoxy radical [J. A. Howard and K. U. Ingold, *Canad. J. Chem.* **46**, 1017 (1968)].
[d] G. A. Russell and C. J. de Boer, *J. Am. Chem. Soc.* **85**, 3136 (1963).
[e] At 105°C (Onopchenko *et al.*[12]
[f] Co(III) + LiCl (Heiba *et al.*[10]).

with those observed in the presence of added bromide (column 3). The difference in selectivity between $Co(OAc)_2$–NaBr and radical bromination (column 5) is slight and may be attributed to differences in solvent, temperature, etc. under which they were measured.

The rate of oxidation of toluene by Co(III) acetate is also enhanced considerably in the presence of chloride.[10] The relative rates of various alkylbenzenes as well as the products are consistent with an electron transfer mechanism. The rate enhancement was attributed to the formation of a chlorocobalt(III) complex of higher oxidation potential.[10] [Halide ions may also induce the dissociation of the dimeric cobaltic acetate, to yield more reactive mononuclear Co(III) complexes.] A comparison of the synergistic effects between bromide and chloride at high temperatures (180°C) indicated the latter to be less selective.[39] For example, oxidation of *p*-*tert*-butyltoluene in the presence of $Co(OAc)_2$ and NaBr gave *p*-*tert*-butylbenzoic acid in 94% yield. On the other hand, $Co(OAc)_2$ and HCl afforded considerable amounts of products resulting from the degradation of the *tert*-butyl group. These results accord with bromine and chlorine atoms as the chain transfer agents. Bromine atoms are known to be capable of abstracting only benzylic hydrogens, whereas the less selective chlorine atoms can attack methyl groups. These results suggest that the oxidations of aromatic hydrocarbons carried out with high concentrations of cobalt catalysts involve two competing processes, namely, the direct oxidation of the substrate to the cation-radical and hydrogen abstraction by a ligand-derived radical X.

$$XCo^{III} \xrightarrow{ArH} XCo^{II} + ArH^{+\cdot}, \quad \text{etc.} \tag{28}$$

$$XCo^{III} \longrightarrow Co^{II} + X\cdot, \quad \text{etc.} \tag{29}$$

The relative rates of the two processes are dependent on several factors: (1) the ionization potential of ArH, (2) the oxidation potential of the ligand X, in which the relative ease of oxidation is in the order $X = Br^- > Cl^- \gg AcO^-$, and (3) the temperature. Electron transfer oxidation of ArH is generally observed with X = AcO, whereas ligand oxidation predominates with X = Br. Chloride occupies an intermediate position favoring both processes, but ligand oxidation is favored at higher temperatures. Intramolecular competition between electron transfer from the aromatic ring (i.e., cation-radical formation) and from the carboxylate group (i.e., decarboxylation) has recently been shown in the oxidation of 4-phenylbutyric acid by cobalt(III).[40]

Hronec and Vesely[41] reported another ligand effect in these systems, namely, enhancement by amine ligands. Thus, the yield of trimesic acid from the $CoBr_2$-catalyzed oxidation of mesitylene in acetic acid at 140°C,

$$\text{1,3,5-}(CH_3)_3C_6H_3 \xrightarrow[\text{[CoBr}_2\text{], HOAc}]{O_2} \text{1,3,5-}(CO_2H)_3C_6H_3 \tag{30}$$

dramatically increased in the presence of amines such as benzylamine, *N,N*-diethylaniline, and triethanolamine. The origin of the salutary effect of amines is not clear but could possibly lie in the promotion of the dissociation of dimeric Co(III) complexes into reactive monomeric species. Alternatively, amines may remove traces of strong acids such as HBr, which catalyze the heterolytic decomposition of hydroperoxides into phenolic inhibitors.

2. *Doped Cobalt Catalysts*

The cobalt-catalyzed autoxidation of *p*-xylene to terephthalic acid is limited by the oxidation of the intermediate.

$$p\text{-}CH_3C_6H_4CH_3 \xrightarrow[\text{[Co]}]{O_2} p\text{-}HO_2CC_6H_4CH_3 \xrightarrow[\text{[MEK, Co]}]{O_2} p\text{-}HO_2CC_6H_4CO_2H \tag{31}$$

The ionization potential of the *p*-toluic acid is significantly higher than that of *p*-xylene (see Table I), making oxidation via electron transfer more difficult.

Consequently, additives such as methyl ethyl ketone (MEK) and bromide are used to promote complete oxidation to terephthalic acid[1,42,43] by maintaining high concentrations of active cobalt(III) species. Chester and co-workers[44] have recently reported that the rate of the $Co(OAc)_2$-catalyzed oxidation of *p*-xylene to terephthalic acid can be enhanced dramatically by small amounts of zirconium(IV) and hafnium(IV) acetates [$ZrO(OAc)_2$ and $HfO(OAc)_2$]. The addition of $ZrO(OAc)_2$ (corresponding to 15% of the cobalt level) led to a rate which was higher than that observed with a tenfold increase in the cobalt concentration. The relative rates of oxidation of substituted toluenes were not affected by the additives, suggesting no change in the basic mechanism. The cocatalytic effect was attributed to two effects: (1) the ability of the metal to attain greater than six coordination in solution and (2) the high stability toward reduction. Indeed, the availability of more than six coordination sites is common to zirconium(IV) and hafnium(IV), which are also quite resistant to reduction in solution. It was further suggested that the Zr(IV) or Hf(IV) redistribute the dimer–monomer equilibrium of cobaltic acetate by forming a weak complex with the active monomeric Co(III) species.

3. *Effect of Strong Acids*

The rates of oxidation of aromatic hydrocarbons by metal acetate oxidants are also dramatically affected by strong acids. Thus, Mn(III) and Co(III) acetates in the presence of sulfuric or trifluoroacetic acid (TFA) rapidly and selectively oxidize aromatic side chains at 25°C.[45]

$$ArCH_3 + Co(OAc)_3 \xrightarrow{TFA-HOAc} \begin{cases} \xrightarrow{N_2} ArCH_2OAc & (32) \\ \xrightarrow{O_2} ArCO_2H & (33) \end{cases}$$

$$ArCH_3 + Mn(OAc)_3 \xrightarrow[O_2]{HOAc-H_2SO_4} ArCHO, \quad \text{etc.} \qquad (34)$$

Enhancement of the oxidizing (electrophilic) properties by strong acids such as TFA is a general feature of oxidations with metal acetates. It has been observed with cobalt(III),[45–48] manganese(III),[45,49] lead(IV),[50–54] thallium(III),[55–59] cerium(IV),[60,61] copper(II),[62] and palladium(II).[63,64] Similarly, the electrophilic properties of copper(I)[65] and mercury(II)[66] acetates are strongly promoted by the replacement of acetate by trifluoroacetate, owing to the increased dissociation into cationic species.

$$(CF_3CO_2)_nM \rightleftharpoons (CF_3CO_2)_{n-1}M^+ + CF_3CO_2^- \qquad (35)$$

In the presence of strong acids, even Mn(III) is capable of oxidizing relatively unreactive alkylbenzenes such as toluene via an electron transfer process.[45] Furthermore, electron-poor arenes, such as benzene and halobenzenes, are readily oxidized at room temperature by Co(III) in TFA to give the corresponding aryl trifluoroacetates in high yield.[48] By contrast, these arenes are inert to Co(III) acetate in acetic acid, even at high temperatures.

Kinetic and esr studies are consistent with a mechanism involving two successive one-electron transfers, as shown in Scheme VI for benzene.[48]

Scheme VI:

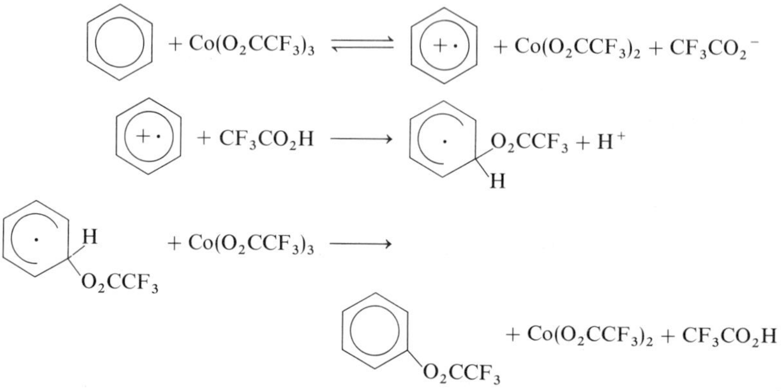

$$C_6H_6 + Co(O_2CCF_3)_3 \rightleftharpoons [C_6H_6]^{+\cdot} + Co(O_2CCF_3)_2 + CF_3CO_2^- \quad (36)$$

$$[C_6H_6]^{+\cdot} + CF_3CO_2H \longrightarrow [C_6H_6(H)(O_2CCF_3)]^{\cdot} + H^+ \quad (37)$$

$$[C_6H_6(H)(O_2CCF_3)]^{\cdot} + Co(O_2CCF_3)_3 \longrightarrow C_6H_5O_2CCF_3 + Co(O_2CCF_3)_2 + CF_3CO_2H \quad (38)$$

Cation-radical formation in eq 36 is probably preceded by a charge transfer complex between the arene and cobalt(III).[67]

B. Oxidative Nucleophilic Substitution

The electron transfer mechanism for the oxidative substitution of arenes by Co(III) in TFA contrasts with the analogous oxidation of the same arenes with Pb(IV) trifluoroacetate in TFA, in which the isolation of aryllead(IV) intermediates supports an electrophilic substitution mechanism.[51–53]

Scheme VII:

$$ArH + Pb(O_2CCF_3)_4 \longrightarrow ArPb(O_2CCF_3)_3 + CF_3CO_2H \quad (39)$$

$$ArPb(O_2CCF_3)_3 \longrightarrow ArO_2CCF_3 + Pb(O_2CCF_3)_2 \quad (40)$$

Similarly, unactivated arenes readily react with thallium(III) trifluoroacetate in TFA to give $ArTl(O_2CCF_3)_2$. These arylthallium(III) trifluoroacetates[55–58] are stable and do not readily undergo reductive elimination to aryl trifluoroacetates and Tl(I). Similarly, the rate of electrophilic mercuration of arenes is increased by a factor of 7×10^5 in TFA relative to acetic acid as solvent.[66]

Electron transfer and electrophilic substitution mechanisms both depend on π-electron availability, since the orbital from which the electron is removed by charge transfer has the same symmetry as the orbital that participates in electrophilic attack.[68] In other words, mechanistic distinctions cannot easily be made on the basis of substituent effects—electron releasing substituents facilitate and electron-attracting substituents hinder both processes. The difference between Co(III) and Pb(IV) oxidation is, however, well illustrated by the reaction with toluene. Lead(IV) affords a mixture of tolyl trifluoroacetates in high yield, whereas Co(III) under the same condition produces oligomeric products resulting from further reaction of the toluene cation-radical with toluene:

$$H_3C\text{—}C_6H_5^{+\cdot} + C_6H_5\text{—}CH_3 \longrightarrow \text{(H}_3\text{C—C}_6\text{H}_4^{+}\text{(H)—(H)C}_6\text{H}_4^{\cdot}\text{—CH}_3\text{)}, \quad \text{etc.} \tag{41}$$

The competition between oxidative oligomerization and oxidative substitution is dependent on the reactivity of the arene, as well as the molar ratio of Co(III) and arene.

Electrophilic aromatic substitutions may involve initial charge transfer, but the role of cation-radicals as discrete intermediates would depend on the facility of the subsequent bond formation.

$$C_6H_6 + E^+ \underset{\text{transfer}}{\overset{\text{electron}}{\rightleftharpoons}} [C_6H_6^{+\cdot} \quad E] \xrightarrow[\text{formation}]{\text{bond}} C_6H_6E^{+} \tag{42}$$

In arene oxidations with "hard" metal ions such as Co(III), Mn(III), and Ce(IV), free cation-radicals are formed owing to the relatively low stability of the aryl–metal bond. With "soft" metals such as Pb(IV), Tl(III), and Pd(II) (see Chapter 7), the existence of a more stable aryl–metal bond can lead preferentially to arylmetal intermediates. For the latter group of oxidants, free radical-cations are observed with electron-rich arenes which form stable cation-radicals. Thus, the oxidation of several electron-rich arenes with lead(IV)[54] and thallium(III)[59] involves discrete cation-radical intermediates in TFA. These intermediates are trapped by external nucleophiles to afford substitution products from what is formally an oxidative nucleophilic substitution.[55,69]

$$ArH + 2\,M(O_2CCF_3)_3 + X^- \longrightarrow ArX + 2\,M(O_2CCF_3)_2 + CF_3CO_2H + CF_3CO_2^- \tag{43}$$

Products arise from nucleophilic substitution on an arylmetal intermediate, e.g., $ArTl(O_2CCF_3)_2$,[55] or by reaction of the intermediate cation-radical with the nucleophile.[69]

Arenes, particularly those with electron-releasing substituents, undergo oxidative dehydrodimerization with thallium(III) trifluoroacetate in trifluoroacetic acid solution or in carbon tetrachloride or acetonitrile containing boron trifluoride etherate.[56]

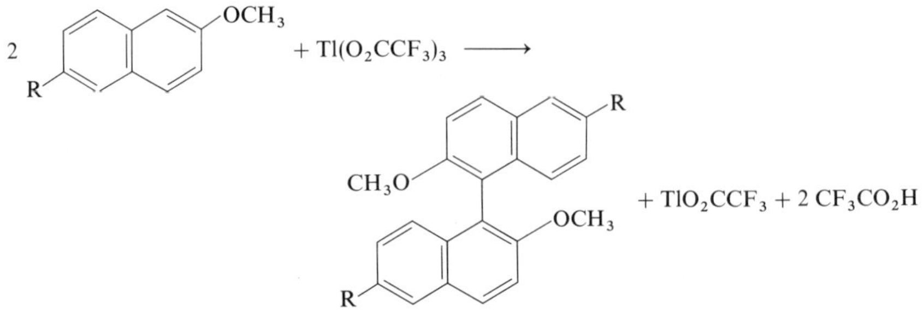

$$2\ \text{R-C}_{10}\text{H}_6\text{-OCH}_3 + \text{Tl(O}_2\text{CCF}_3)_3 \longrightarrow (\text{R-C}_{10}\text{H}_5\text{-OCH}_3)_2 + \text{TlO}_2\text{CCF}_3 + 2\ \text{CF}_3\text{CO}_2\text{H}$$

Aromatic compounds that contain powerful electron-withdrawing substituents such as NO_2, CO_2R, or CN do not react under these conditions. The coupling reaction is postulated to proceed via an arene cation-radical formed by the electron transfer process depicted in the scheme below.

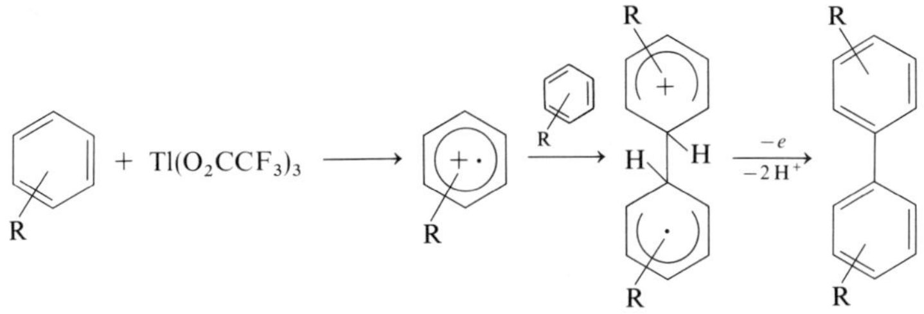

$$\text{RC}_6\text{H}_5 + \text{Tl(O}_2\text{CCF}_3)_3 \longrightarrow [\text{RC}_6\text{H}_5]^{+\cdot} \xrightarrow{\text{RC}_6\text{H}_5} [\text{R(H)C}_6\text{H}_5\text{-C}_6\text{H}_5\text{(H)R}]^{+\cdot} \xrightarrow[-2\text{H}^+]{-e} \text{RC}_6\text{H}_4\text{-C}_6\text{H}_4\text{R}$$

This formulation accords well with the results of an earlier esr study in which the direct observation of the arene cation-radicals was related to aromatic thallation by the ion-pair intermediate in eq 44.[59]

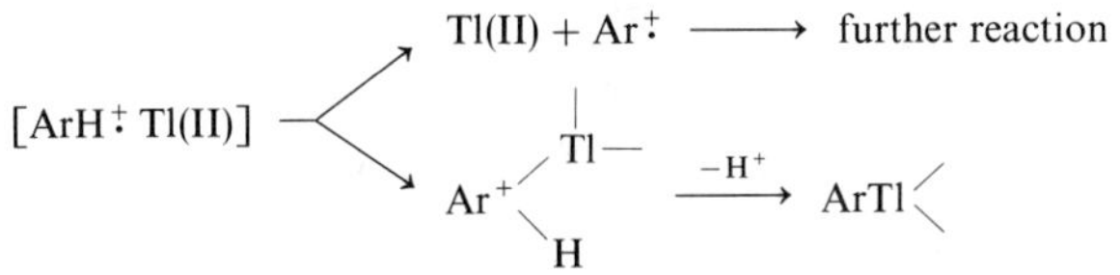

$$[\text{ArH}^{+\cdot}\ \text{Tl(II)}] \begin{cases} \text{Tl(II)} + \text{Ar}^{+\cdot} \longrightarrow \text{further reaction} \\ \text{Ar}^{+}(\text{H})\text{-Tl} \xrightarrow{-\text{H}^+} \text{ArTl}< \end{cases} \quad (44)$$

The stoichiometric coupling of phenols to biphenols by oxidation with the electron transfer oxidant manganese(III) acetylacetonate has been reported,[70] e.g.,

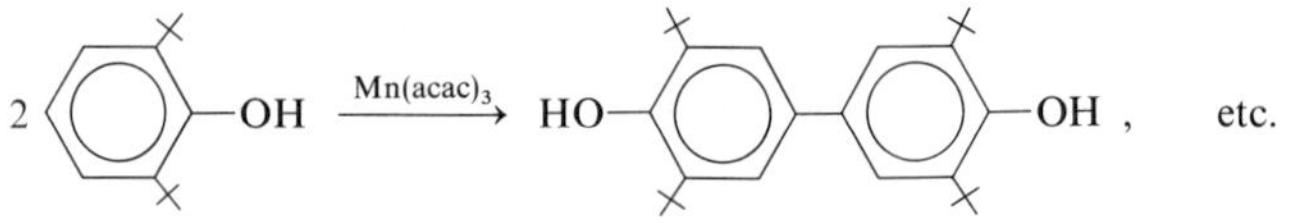

$$2\ \text{(2,6-di-}t\text{-Bu-C}_6\text{H}_3)\text{-OH} \xrightarrow{\text{Mn(acac)}_3} \text{HO-(}t\text{-Bu}_2\text{C}_6\text{H}_2)\text{-(C}_6\text{H}_2 t\text{-Bu}_2)\text{-OH}, \quad \text{etc.}$$

The biphenol results from the dimerization of intermediate phenoxy radicals, formed by either the loss of a proton from the cation-radical,

$$C_6H_5\text{—OH} + Mn^{III} \xrightarrow[-Mn^{II}]{} (C_6H_5)^{+\cdot}\text{—OH} \longrightarrow C_6H_5\text{—O}\cdot + H^+ \quad (45)$$

or the homolytic cleavage of an aryloxymanganese(III) intermediate.

$$C_6H_5\text{—OH} + Mn^{III}(acac) \xrightarrow[-H(acac)]{} C_6H_5\text{—OMn}^{III} \longrightarrow C_6H_5\text{—O}\cdot + Mn^{II} \quad (46)$$

Oxidations of phenols with (Salen)Co(II) and related complexes, in combination with molecular oxygen as noted in Chapter 4, probably involve aryloxy radicals from Co(III) and the phenol.

II. ALKENES

The direct oxidation of alkenes by metal acetate may proceed by either cation-radicals or organometallic intermediates. For example, the oxidation of ethylene by cobaltic trifluoroacetate in TFA affords ethylene glycol di(trifluoroacetate).[46] The formation of the cation-radical by one-electron transfer in reaction 47 is supported by the observation of the esr spectra of cation-radicals during reaction of $Co(O_2CCF_3)_3$ with several alkenes in TFA solutions.[71]

Scheme VIII:

$$CH_2{=}CH_2 + Co^+(O_2CCF_3)_2 \longrightarrow \overset{+}{C}H_2\text{—}\overset{\bullet}{C}H_2 + Co(O_2CCF_3)_2 \quad (47)$$

$$\overset{+}{C}H_2\text{—}\overset{\bullet}{C}H_2 + CF_3CO_2H \longrightarrow CF_3CO_2CH_2CH_2\cdot + H^+ \quad (48)$$

$$CF_3CO_2CH_2CH_2\cdot + Co^+(O_2CCF_3)_2 \longrightarrow CF_3CO_2CH_2CH_2{}^+ + Co(O_2CCF_3)_2 \quad (49)$$

$$CF_3CO_2CH_2CH_2{}^+ + CF_3CO_2H \longrightarrow CF_3CO_2CH_2CH_2O_2CCF_3 + H^+ \quad (50)$$

Cation-radical formation was also considered to be the rate-determining step in the reaction of alkenes with Co(III) in aqueous medium.[72] Under these conditions the cation-radical reacts further with water to form a complex mixture of aldehydes, ketones, carboxylic acids, etc. Surprisingly few studies[73,74] have been made of the autoxidation of olefins in the presence of high concentrations of Co(III) acetate in acetic acid, compared to the attention paid to the oxidation of alkylaromatic compounds described in Section I. One might expect competition between 1,2-addition and allylic substitution to depend on the rate of proton loss from the intermediate cation-radical, as shown below.

$$[RCH_2CH{=}CH_2]^{\dot{+}} \xrightarrow{AcO^-} RCH_2\dot{C}HCH_2OAc \longrightarrow \text{products} \quad (51)$$

$$[RCH_2CH{=}CH_2]^{\dot{+}} \xrightarrow{-H^+} RCH\overset{\cdot}{\underset{CH}{\frown}}CH_2 \longrightarrow \text{products} \quad (52)$$

The stoichiometric oxidation of olefins by manganese(III) acetate in acetic acid affords γ-lactones[75–79] by the addition of carboxymethyl radicals to the double bond, followed by the oxidation of the resulting radical.[75]

$$Mn(OAc)_3 \xrightarrow{\Delta} Mn(OAc)_2 + \cdot CH_2CO_2H \quad (53)$$

$$\cdot CH_2CO_2H + RCH{=}CH_2 \longrightarrow R\dot{C}H{-}CH_2CH_2CO_2H \quad (54)$$

$$R\dot{C}HCH_2CH_2CO_2H + Mn^{III} \longrightarrow R\overset{+}{C}HCH_2CH_2CO_2H + Mn^{II} \quad (55)$$

$$R\overset{+}{C}HCH_2CH_2CO_2H \longrightarrow \underset{O-C(=O)-CH_2}{RCH{-}CH_2} + H^+ \quad (56)$$

The oxidation of olefins in the presence of Mn(III) and Cu(II) acetates in a mixture of acetic anhydride and acetic acid leads to unsaturated acids.[80] For example, 1-octene affords mainly 4-decenoic acid, together with some 3-decenoic acid. Under these conditions, the intermediate alkyl radical in eq 54 is efficiently trapped by Cu(II) to produce olefins via oxidative elimination.

$$Cu^{II} + RCH_2\dot{C}HCH_2CH_2CO_2H \longrightarrow \left.\begin{matrix} RCH{=}CHCH_2CH_2CO_2H \\ + \\ RCH_2CH{=}CHCH_2CO_2H \end{matrix}\right\} + Cu^{I} + H^+ \quad (57)$$

In the presence of bromide, cyclohexene is rapidly oxidized by Mn(III) acetate in acetic acid at 70°C to cyclohexenyl acetate in 83% yield.[81] The catalytic effect of bromide and the observed relative reactivities of various olefins accord with a mechanism involving hydrogen abstraction by bromine atom as the chain transfer agent.

$$Mn^{III} + Br^- \longrightarrow Mn^{II} + Br\cdot \quad (58)$$

$$Br\cdot + RCH_2CH{=}CH_2 \longrightarrow HBr + RCH\overset{CH}{\overset{\cdot}{\frown}}CH_2 \quad (59)$$

$$RCH\overset{CH}{\overset{\cdot}{\frown}}CH_2 + Mn^{III} \xrightarrow[HOAc]{} \left[\underset{OAc}{RCH}CH{=}CH_2 + RCH{=}CH\underset{OAc}{CH_2}\right] + Mn^{II} \quad (60)$$

A process for the manufacture of ethylene glycol by the air oxidation of ethylene to ethylene 1,2-diacetate with a combination of metal catalyst and halide has been described.[82,83]

$$CH_2{=}CH_2 + O_2 \xrightarrow[\text{HOAc}]{[\text{Te}^{\text{IV}}]/\text{Br}^-} AcOCH_2CH_2OAc + H_2O$$

A catalyst consisting of tellurium dioxide and bromide salts leads to high selectivities (>95%) to ethylene diacetate. The details of the mechanism are unknown, but it may involve oxidative bromination in one form or another, e.g.,

Scheme IX:

$$Te^{IV} + 2\,Br^- \longrightarrow Te^{II} + Br_2 \quad (61)$$

$$Br_2 + CH_2{=}CH_2 \xrightarrow[\text{HOAc}]{} BrCH_2CH_2OAc + HBr \quad (62)$$

$$BrCH_2CH_2OAc + HOAc \longrightarrow AcOCH_2CH_2OAc + HBr \quad (63)$$

$$Te^{II} + 2\,HBr + \tfrac{1}{2}\,O_2 \longrightarrow Te^{IV} + H_2O + 2\,Br^- \quad (64)$$

(See Scheme X for an alternative mechanism.)

In contrast to reactions with Co(III) and Mn(III) described above, the oxidation of olefins by soft metal oxidants such as lead(IV),[84] thallium(III),[85–91] and mercury(II)[92] involves organometallic intermediates formed by the electrophilic addition to the double bond, which is generally referred to as *oxymetallation.*[91] For example, the oxidation of olefins with aqueous thallium(III) produces glycols and carbonyl compounds via β-hydroxyalkylthallium(III) intermediates.[87,88]

$$RCH{=}CH_2 + X_2Tl^+ \xrightarrow[H_2O]{} \underset{\displaystyle OH}{RCH}CH_2TlX_2 + H^+ \quad (65)$$

Carbonyl compounds result from β-hydrogen elimination via the enol,

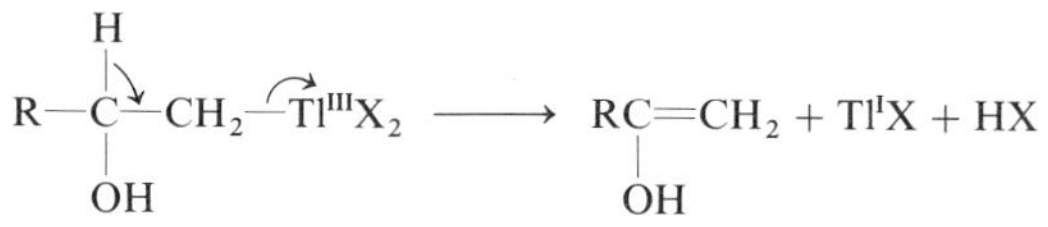

$$R{-}\overset{\displaystyle H}{\underset{\displaystyle OH}{C}}{-}CH_2{-}Tl^{III}X_2 \longrightarrow \underset{\displaystyle OH}{RC}{=}CH_2 + Tl^{I}X + HX \quad (66)$$

and glycols can be formed by either intermolecular or intramolecular reductive elimination of Tl(I), e.g.,

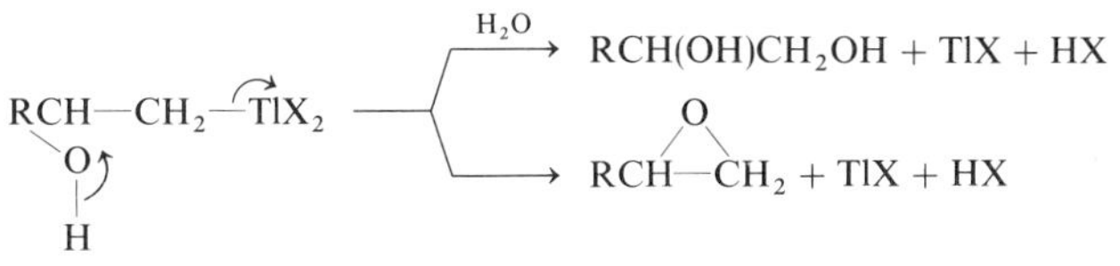

$$\underset{\displaystyle O{-}H}{RCH}{-}CH_2{-}TlX_2 \begin{cases} \xrightarrow{H_2O} RCH(OH)CH_2OH + TlX + HX \\ \longrightarrow RCH\overset{O}{—}CH_2 + TlX + HX \end{cases} \quad (67)$$

When the reaction is performed with Tl(III) acetate in aqueous acetic acid, epoxides are obtained in good yields from propylene and isobutylene.[89,90] The intermediate β-hydroxyalkylthallium(III) adduct has been isolated at low temperatures.[89]

$$Tl(OAc)_3 + (CH_3)_2C{=}CH_2 \xrightarrow[-HOAc]{H_2O} HO{-}\overset{\displaystyle CH_3}{\underset{\displaystyle CH_3}{C}}{-}CH_2Tl(OAc)_2$$

Processes have been described[90] for the *stoichiometric* epoxidation of olefins using the Tl(III) acetate in aqueous acetic acid. It is possible for this system to be made catalytic, since Tl(I) can be reconverted to Tl(III) with dioxygen in the presence of a supported platinum catalyst and tetraalkylammonium salts as promoters.[92] The Tl(III) may also be regenerated by electrochemical oxidation of Tl(I) (*vide infra*).

A direct catalytic process, analogous to an oxychlorination, has been developed for the conversion of ethylene to ethylene glycol by using copper(II) in conjunction with thallation.[93] The catalytic sequence can be described by the reactions,

$$CH_2{=}CH_2 + TlCl_3 \xrightarrow[H_2O]{} ClCH_2CH_2OH + TlCl + HCl$$

$$ClCH_2CH_2OH \xrightarrow[H_2O]{} HOCH_2CH_2OH + HCl$$

$$TlCl + 2\,CuCl_2 \longrightarrow TlCl_3 + 2\,CuCl$$

$$2\,CuCl + 2\,HCl + \tfrac{1}{2}\,O_2 \longrightarrow 2\,CuCl_2 + H_2O$$

leading to the net reaction,

$$CH_2{=}CH_2 + H_2O + \tfrac{1}{2}\,O_2 \rightarrow HOCH_2CH_2OH$$

The process is more direct than the two-step acetoxylation route developed by the Halcon Company (*vide supra*), but it suffers from a lower yield of ethylene glycol (75%).

Alternatively, there may be other metal complexes that have the desired properties for both oxymetallation and regeneration by dioxygen. Indeed, the glycolization of ethylene by tellurium(IV) in Scheme IX may be such an example.

Scheme X:

$$CH_2{=}CH_2 + Te^{IV} \xrightarrow{HOAc} AcOCH_2CH_2Te^{IV} + H^+$$

$$AcOCH_2CH_2Te^{IV} \xrightarrow[HOAc]{} AcOCH_2CH_2OAc + Te^{II} + H^+ \qquad (68)$$

$$Te^{II} + 2\,HBr + \tfrac{1}{2}\,O_2 \longrightarrow Te^{IV} + H_2O + 2\,Br^-$$

The subsequent reductive elimination in eq 68 is analogous to that of the organothallium(III) intermediate in eq 67. The mechanisms in Schemes IX and X should be readily distinguishable.

III. ALKANES

Owing to their low reactivity, relatively high temperatures are generally required to effect autoxidation of alkanes by the metal-catalyzed homolytic processes described in Chapter 3. Extensive degradation of labile intermediates into smaller fragments usually results from the extreme conditions required. Furthermore, the slight differences in reactivity of various C—H bonds to free radicals lead to indiscriminate attack along the alkane chain. Improvements in selectivity in the conversion of alkanes under relatively mild conditions continues to be an important goal of petrochemical research.

Simple alkanes can be selectively oxidized by molecular oxygen in the presence of relatively high concentrations of cobaltic acetate in acetic acid.[94,94a] For example, the autoxidation of *n*-butane employing small amounts of metal catalysts requires temperatures in excess of 170°C. Acetic acid is formed with roughly 40% selectivity as the major component of a complex mixture of oxygenated products. By contrast, the oxidation of *n*-butane in the presence of high concentrations of Co(II) acetate in acetic acid (with methyl ethyl ketone as promoter) proceeded readily at 100°–125°C to afford acetic acid in 83% selectivity at 80% conversion of *n*-butane.[94] The oxidation of Co(II) to Co(III) preceded the attainment of the maximum rate, in accord with a mechanism involving the direct interaction of butane with Co(III). Manganic acetate was ineffective as a catalyst under these conditions, as expected from the lower oxidation potential of the Mn(III, II) couple.

Cyclohexane is also readily oxidized by Co(III) acetate in acetic acid at 80°C to give cyclohexyl acetate and 2-acetoxycyclohexanone as the major products.[94a] (Cyclohexane is only half as reactive as toluene under these conditions.) In the presence of oxygen, adipic acid is the major product and is formed in approximately 75% selectivity at 80% conversion of cyclohexane.[95]

$$\text{cyclohexane} \xrightarrow[O_2]{[Co^{II}(OAc)_2]} HO_2C(CH_2)_4CO_2H \tag{69}$$

In contrast to free radical autoxidations, alkanes containing tertiary C—H bonds such as isobutane were less reactive than *n*-butane, and methylcyclohexane was less reactive than cyclohexane. These results are reminiscent of the lower reactivity of cumene compared to toluene observed under similar conditions (see eqs 15 and 16). In contrast to the oxidation of alkylaromatic compounds, no deuterium kinetic isotope effect was observed, which was reconciled[94,94a] with a mechanism involving reversible one-electron transfer to produce the cation-radical followed by proton loss to the alkyl radical.

Scheme XI:

$$RH + Co^{III} \underset{k_{-1}}{\overset{k_1}{\rightleftharpoons}} RH^{+\cdot} + Co^{II}$$

$$RH^{+\cdot} \xrightarrow{k_2} R\cdot + H^+ \tag{71}$$

$$R\cdot \xrightarrow{Co^{III}} R^+ \xrightarrow{HOAc} ROAc \tag{72}$$

$$R\cdot \xrightarrow{O_2} RO_2\cdot \longrightarrow \text{products} \tag{73}$$

For alkylaromatic substrates, reaction 70 is fast and k_2 is rate determining, which would explain the kinetic isotope effect. In contrast, electron transfer from alkanes is less favorable by virtue of their lower ionization potentials, and the rate would be governed by the ratio of rate constants k_{-1}/k_2. On the basis of ionization potentials alone, isobutane is expected to be more reactive than *n*-butane, which suggests that steric effects play a significant role in determining the rate of electron transfer. Another apparent anomaly lies in the observation that benzene, which has a significantly lower ionization potential than cyclohexane in Table I, is inert to Co(III) acetate. These anomalies can be partly resolved if the electron transfer step is considered to be reversible. Thus, for substrates such as benzene, which cannot form a stable radical by proton loss, k_{-1} is larger than k_2, whereas the reverse applies to alkanes and alkylbenzenes. The relative rates of oxidation of various cycloalkanes by Co(III) acetate were measured in acetic acid at 90°C.[95] From the pattern of reactivities, it was concluded that complex formation between Co(III) and the alkane was rate determining and strongly influenced by steric factors.

The rate of oxidation of cyclohexane by Co(III) acetate is enhanced in the presence of bromide ions in acetic acid.[34] By analogy with alkylaromatic oxidations discussed in Section I, these reactions probably involve homolytic chain transfer by bromine atoms [compare eqs 23–27]. The rates of oxidation of alkanes are also dramatically accelerated by strong acids such as TFA, which is another feature in common with arene oxidations. For example, *n*-heptane is readily oxidized at 25°C by Co(III) in a mixture of TFA and HOAc to afford 2-heptyl acetate as the major product (81% selectivity). In the presence of dioxygen, 2-heptanone is formed with 83% selectivity.[96,97]

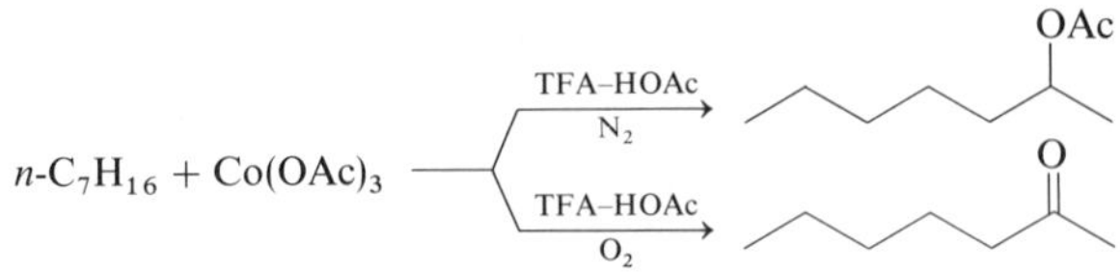

$$n\text{-}C_7H_{16} + Co(OAc)_3 \xrightarrow[N_2]{\text{TFA-HOAc}} \text{2-heptyl acetate (OAc)} \tag{74}$$

$$n\text{-}C_7H_{16} + Co(OAc)_3 \xrightarrow[O_2]{\text{TFA-HOAc}} \text{2-heptanone} \tag{75}$$

Similarly, 2-heptyl chloride is formed when chlorine atom donors such as CCl_4 are present, or when trichloro acetic acid is used as the strong acid activator to intercept the 2-heptyl radicals by chlorine atom transfer.

$$2\text{-}C_7H_{15}\cdot + R'Cl \rightarrow 2\text{-}C_7H_{15}Cl + R'\cdot \tag{76}$$

Unusual selectivities analogous to those observed in the absence of strong acid were observed in the oxidation of 2-methylpentane.[96,97]

$$CH_3CH(CH_3)CH_2CH_2CH_3 + Co(OAc)_3 \xrightarrow[\text{HOAc}]{Cl_3CCO_2H} RCl \left\{ \begin{array}{l} \text{Selectivity (\%)} \\ \begin{array}{ccccc} & C & & & \\ & | & & & \\ C- & C- & C- & C- & C \end{array} \\ \begin{array}{ccccc} 3 & 5 & 2 & 74 & 13 \end{array} \end{array} \right. \tag{77}$$

Hanotier and co-workers[96] concluded that the interaction of Co(III) and alkanes leads to the reversible formation of alkyl radicals by a direct process, the precise nature of which was not further elucidated.

The electrochemical oxidation of bridgehead hydrocarbons was compared to the oxidations carried out with Co(III), Mn(III), and Pb(IV) acetates in TFA.[98] Anodic oxidations of various alkyladamantanes afforded products derived from fragmentation, but the metal oxidants afforded high yields of the adamantyl trifluoroacetates by loss of a bridgehead hydrogen.

(78)

1-*t*-Bu-adamantane —(Anodic)→ 1-O_2CCF_3-adamantane + [*t*-Bu^+]

1-*t*-Bu-adamantane —(Co^{III}, TFA)→ 1-O_2CCF_3-3-*t*-Bu-adamantane

Jones and Mellor concluded, on the basis of product distributions, that oxidations with the metal reagents do not involve cation-radical intermediates, but rather electrophilic attack at the C—H bond. These reactions may constitute examples of a general class of electrophilic substitutions at saturated carbon centers, in which attack at a σ bond occurs via a trigonal (three-center) transition state,[99,100] e.g.,

$$-\overset{|}{\underset{|}{C}}-H + CoX_2^+ \longrightarrow \left[-\overset{|}{\underset{|}{C}}\cdots\begin{matrix} H \\ CoX_2 \end{matrix} \right]^+ \xrightarrow{-H^+} -\overset{|}{\underset{|}{C}}-CoX_2$$

$$-\overset{|}{\underset{|}{C}}-CoX_2 \longrightarrow -\overset{|}{\underset{|}{C}}\cdot + CoX_2 \tag{79}$$

The reactions of alkanes with other large electrophiles, such as PCl_5, are also subject to pronounced steric effects and exhibit unusually high reactivity ratios for the cleavage of secondary and tertiary C—H bonds. The differences, however, may stem from the association of the cation-radicals with the inorganic counterion, which could significantly influence their behavior.[98]

Rates of oxidation of benzene by Co(III) trifluoroacetate in TFA that are 10 times faster than those of cyclohexane[101] must also be explained, since they conflict with the analogous oxidation with Co(III) acetate in acetic acid in which cyclohexane undergoes facile oxidation (*vide supra*). The dichotomy between electron transfer and electrophilic processes in these systems is clearly not resolved.

For additional examples of the homolytic activation of alkanes, see Chapter 7, Section VII.

IV. CARBONYL COMPOUNDS

Autoxidation provides a simple means for converting linear aldehydes, obtained by hydroformylation of terminal olefins, to linear carboxylic acids. These reactions are readily carried out under relatively mild conditions, even without metal catalysts, by the conventional radical chain processes described in Chapter 2.[101] In the presence of metal catalysts such as Mn(II), Co(II), and Fe(II), chain initiation involves the direct, rate-determining reaction of the metal oxidant with the aldehydic substrate,[102–113] e.g.,

$$\mathrm{RCHO} + \mathrm{Co^{III}} \longrightarrow \mathrm{R\dot{C}O} + \mathrm{Co^{II}} + \mathrm{H^+} \qquad (80)$$

The resultant acyl radicals are converted to the peracid, which is involved in the regeneration of Co(III), as discussed in Chapter 3.

$$\mathrm{R\dot{C}O} + \mathrm{O_2} \longrightarrow \mathrm{RCO_3\cdot} \qquad (81)$$

$$\mathrm{RCO_3\cdot} + \mathrm{RCHO} \longrightarrow \mathrm{RCO_3H} + \mathrm{R\dot{C}O} \qquad (82)$$

$$\mathrm{RCO_3H} + \mathrm{Co^{II}} \longrightarrow \begin{cases} \mathrm{HOCo^{III}} + \mathrm{RCO_2\cdot} \\ \mathrm{RCO_2Co^{III}} + \mathrm{HO\cdot} \end{cases} \qquad (83)$$

Depending on the reaction conditions, the metal-catalyzed autoxidation of acetaldehyde can be used for the production of acetic acid or peracetic acid.[114] Furthermore, autoxidation in the presence of both Co(II) and Cu(II) acetates produces acetic anhydride,[115] in which the key step involves electron transfer oxidation of intermediate acetyl radicals by Cu(II), in competition with their reaction with oxygen.

$$\mathrm{CH_3\dot{C}O + Cu^{II} \longrightarrow CH_3\overset{+}{C}O + Cu^{I}} \tag{84}$$

$$\mathrm{CH_3\overset{+}{C}O + CH_3CO_2H \longrightarrow (CH_3CO)_2O + H^+} \tag{85}$$

Copper(II) is more effective than other metal oxidants for the oxidation of radicals to the corresponding cations.[116] With branched aldehydes, decarbonylation can occur by reaction 86, which becomes predominant with aldehydes that afford *tert*-alkyl radicals,[117] e.g.,

$$\mathrm{(CH_3)_3C\dot{C}O \longrightarrow (CH_3)_3C\cdot + CO} \tag{86}$$

Nikishin and co-workers[118–124] have examined the reactions of aliphatic aldehydes with olefins in the presence of cobaltic and manganic acetates in acetic acid to producc a variety of interesting products. For example, the reaction of olefins with aldehydes, in the presence of Co(II) acetate and a limited supply of oxygen, generates ketones via the following chain transfer sequence:[118,119]

$$\mathrm{R\dot{C}O + R'CH{=}CH_2 \longrightarrow R'\dot{C}HCH_2\overset{O}{\overset{\|}{C}}R} \tag{87}$$

$$\mathrm{R'\dot{C}HCH_2\overset{O}{\overset{\|}{C}}R + RCHO \longrightarrow R'CH_2CH_2\overset{O}{\overset{\|}{C}}R + R\dot{C}O} \tag{88}$$

In the presence of stoichiometric amounts of Mn(III) acetate in acetic acid, α-formylalkyl radicals are produced which add to olefins:[120–124]

$$\mathrm{RCH_2CHO + Mn^{III} \longrightarrow R\dot{C}HCHO + Mn^{II} + H^+} \tag{89}$$

$$\mathrm{R\dot{C}HCHO + R'CH{=}CH_2 \longrightarrow R'\dot{C}HCH_2CH(R)CHO} \tag{90}$$

The resulting alkyl radical can undergo hydrogen transfer with the aldehyde or electron transfer oxidation by Mn(III).

$$\mathrm{R'\dot{C}HCH_2CH(R)CHO + RCH_2CHO \longrightarrow R'CH_2CH_2CH(R)CHO + RCH_2\dot{C}O} \tag{91}$$

$$\mathrm{R'\dot{C}HCH_2CH(R)CHO + Mn^{III}OAc \longrightarrow R'\underset{\underset{OAc}{|}}{C}HCH_2CH(R)CHO + Mn^{II}} \tag{92}$$

Unsaturated aldehydes are formed in the presence of catalytic amounts of Cu(II) acetate.[124]

$$\mathrm{R'\dot{C}HCH_2CH(R)CHO + Cu^{II} \longrightarrow R'CH{=}CHCH(R)CHO + Cu^{I} + H^+} \tag{93}$$

These reactions show that oxidations by metal oxidants in high concentrations proceed by different pathways compared to those carried out with catalytic amounts of metal complexes.

The initial formation of α-formylalkyl radicals in these systems most likely proceeds by homolytic cleavage of the metal enolate,[125] e.g.,

$$\mathrm{RCH{=}CH{-}OMn^{III} \longrightarrow R\dot{C}HCHO + Mn^{II}} \tag{94}$$

Similarly, ketones and esters are oxidized by Mn(III) and Ce(IV) acetates to α-oxoalkyl radicals, which can add to olefins.[126] For example, acetone produces the acetonyl radical which affords adducts with olefins.

$$CH_3\overset{O}{\overset{\|}{C}}CH_2\cdot + RCH{=}CH_2 \longrightarrow R\dot{C}HCH_2CH_2\overset{O}{\overset{\|}{C}}CH_3 \qquad (95)$$

$$R\dot{C}HCH_2CH_2\overset{O}{\overset{\|}{C}}CH_3 \xrightarrow{AcOMn^{III}} \begin{cases} RCH(OAc)CH_2CH_2\overset{O}{\overset{\|}{C}}CH_3 \\ \quad + \\ RCH{=}CHCH_2\overset{O}{\overset{\|}{C}}CH_3 \end{cases} \qquad (96)$$

The radical derived from diethyl malonate similarly adds to olefins and provides a convenient route to γ,δ-unsaturated acids and γ-lactones.[127]

$$\cdot CH(CO_2Et)_2 + RCH_2CH{=}CH_2 \longrightarrow RCH_2\dot{C}HCH_2CH(CO_2Et)_2 \qquad (97)$$

$$RCH_2\dot{C}HCH_2CH(CO_2Et)_2 \xrightarrow{Cu(OAc)_2} RCH{=}CHCH_2CH(CO_2Et)_2 \qquad (98)$$

$$RCH{=}CHCH_2CH(CO_2Et)_2 \xrightarrow[H_2O]{NaOH} RCH{=}CHCH_2CH_2CO_2H \qquad (99)$$

$$RCH{=}CHCH_2CH(CO_2Et)_2 \xrightarrow[H_2O]{H^+} RCH_2\text{-(γ-butyrolactone)}{=}O \qquad (100)$$

A novel oxidative addition of ketones to enol acetates with manganic acetate has been reported, e.g.[127a]

$$CH_3(CH_2)_5\overset{O}{\overset{\|}{C}}CH_3 + CH_2{=}CHOAc \xrightarrow[AcOH/Ac_2O]{Mn(OAc)_3} CH_2\begin{cases} CH{=}CHOAc \\ C(=O)(CH_2)_5CH_3 \end{cases}$$

Mild hydrolysis followed by cyclization of the ketoaldehyde affords a cyclopentenone related to dihydrojasmone.

Ketones are readily oxidized by dioxygen to carboxylic acids with manganic acetate as a catalyst in acetic acid.[128] For example, acetophenones yield benzoic acids.

$$ArCOCH_3 \xrightarrow[O_2]{[Mn(OAc)_2]} ArCO_2H \qquad (101)$$

Carboxylic acids, the secondary products of many hydrocarbon autoxidations, undergo oxidative decarboxylation with Mn(III), Co(III), and Ce(IV),[99,129]

$$RCO_2H + M^{n+} \longrightarrow M^{(n-1)+} + R\cdot + CO_2 + H^+ \qquad (102)$$

in competition with the formation of α-carboxylalkyl radicals (compare Scheme IV). Under autoxidizing conditions, the alkyl radicals react further with dioxygen. Except for carboxylic acids yielding stable alkyl radicals (e.g., from pivalic acid), this reaction is relatively slow under the conditions usually employed in metal-catalyzed autoxidations. The rates are, however, markedly enhanced by strong acids such as TFA.

Finally, thallium(III) complexes, such as Tl(III) nitrate, effect a variety of stoichiometric oxidative transformations of carbonyl compounds recently detailed by McKillop and Taylor.[130]

V. ALCOHOLS AND GLYCOLS

The autoxidation of primary and secondary alcohols produces carbonyl compounds and hydrogen peroxide as the primary products via the following sequence of steps.[131]

Scheme XII:

$$R_2CHOH \xrightarrow{\text{initiation}} R_2\dot{C}OH \tag{103}$$

$$R_2\dot{C}OH + O_2 \longrightarrow R_2C\begin{smallmatrix}OH \\ O_2\cdot\end{smallmatrix} \tag{104}$$

$$R_2C\begin{smallmatrix}OH \\ O_2\cdot\end{smallmatrix} + R_2CHOH \longrightarrow R_2C\begin{smallmatrix}OH \\ O_2H\end{smallmatrix} + R_2\dot{C}OH \tag{105}$$

$$R_2C\begin{smallmatrix}OH \\ O_2H\end{smallmatrix} \longrightarrow R_2C{=}O + H_2O_2 \tag{106}$$

Alcohols are generally not autoxidized as readily as olefins or aldehydes, mainly owing to the high rate of termination of the α-hydroxyperoxy radicals. In the presence of cobalt and manganese salts, initiation commonly involves a rate-limiting homolytic cleavage of metal alkoxides.[125,132] For example, in the stoichiometric oxidation of *tert*-amyl alcohol by Co(III) in aqueous acid, acetone is formed via the following sequence:

$$EtMe_2COH + XCo^{III} \rightleftharpoons EtMe_2COCo^{III} + HX \tag{107}$$

$$EtMe_2COCo^{III} \longrightarrow EtMe_2CO\cdot + Co^{II} \tag{108}$$

$$EtMe_2CO\cdot \longrightarrow Me_2CO + Et\cdot \tag{109}$$

The selectivity in the C—C bond cleavage of alkoxy radicals follows the stability of the liberated alkyl radical.[133] The *extent* of C—C cleavage of intermediate alkoxy radicals is often influenced by metal ions such as cobalt, suggesting the concerted cleavage of the metal alkoxide intermediate.

$$\mathrm{R{-}\overset{|}{\underset{|}{C}}{-}O{-}Co^{III}} \longrightarrow \mathrm{{>}C{=}O + R\cdot + Co^{II}} \quad (110)$$

Similarly, the observation of an α-deuterium kinetic isotope effect in the oxidation of secondary alcohols by Co(III) suggests rate-limiting removal of hydrogen.[125]

The oxidative cleavage of 1,2-glycols to carbonyl compounds is an important reaction in organic synthesis. It is usually achieved by the use of stoichiometric quantities of rather expensive reagents such as periodic acid.[134] The mechanistic distinctions between one-electron and two-electron oxidations of glycols by heavy-metal acetates have been made.[135] The former involve cleavage of intermediate alkoxy radicals:

$$\mathrm{R\overset{O\cdot}{\overset{|}{C}}H{-}\overset{OH}{\overset{|}{C}}HR'} \longrightarrow \mathrm{RCHO + R'\dot{C}HOH} \quad (111)$$

The cobalt-catalyzed cleavage of 1,2-glycols by molecular oxygen in aprotic polar solvents at 100°C has been reported.[136] Depending on the reaction time, aldehydes or carboxylic acids can be isolated as the major products. The reaction presumably involves initiation by a one-electron oxidation,

$$\mathrm{R\overset{OH}{\overset{|}{C}}H{-}\overset{OH}{\overset{|}{C}}HR' + Co^{III}} \longrightarrow \mathrm{R\overset{O\cdot}{\overset{|}{C}}H{-}\overset{OH}{\overset{|}{C}}HR' + Co^{II} + H^+} \quad (112)$$

followed by the recycling of the Co(II) by reoxidation with peracids formed by further oxidation of the aldehydes.

Alternatively, carboxylic acids could be formed together with Co(III) (compare reactions 10 and 11).

$$\mathrm{R\underset{\underset{OH}{|}}{C}H\cdot + O_2} \longrightarrow \mathrm{R\underset{\underset{OH}{|}}{C}HO_2\cdot} \quad (113)$$

$$\mathrm{R\underset{\underset{OH}{|}}{C}HO_2\cdot + Co^{II}} \longrightarrow \mathrm{R{-}\overset{H}{\overset{|}{\underset{\underset{OH}{|}}{C}}}{-}O{-}O{-}Co^{III}} \longrightarrow \mathrm{RCO_2H + HOCo^{III}} \quad (114)$$

VI. ELECTROCHEMICAL GENERATION OF METAL OXIDANTS

The similarity between the chemical oxidation of alkenes, arenes, and alkanes by electron transfer oxidants and electrochemical oxidations[137–143] of the same substrates is noteworthy, since the anodic processes are known to involve successive one-electron transfers and cation-radicals are intermediates.

A variety of alkylbenzenes undergo anodic acetoxylation, in which the loss of an α proton and solvation of the cation-radical form the basis of side-chain and nuclear acetoxylation, respectively.[140,141] Cation-radicals are similarly implicated in the anodic oxidation of alkenes and alkanes.[144–146] Since anodic oxidations are often not as selective as the stoichiometric oxidations with metal salts described in this chapter, the oxidation of substrates with electrochemically generated metal oxidants represents an alternative approach. Such a process would constitute a metal-catalyzed oxidation in which electricity replaces dioxygen. An example of this technique is the oxidation of substituted toluenes to the corresponding benzaldehydes with electrogenerated Mn(III) salts.[147] Electrochemically generated MnO_2 has also been used for the selective oxidation of alkylbenzenes to the corresponding benzaldehydes, such as the conversion of *p*-methoxytoluene to *p*-anisaldehyde.[148] The electrochemical process has an additional advantage in consistently producing an active form of MnO_2, which is a notoriously capricious oxidant, performing satisfactorily only when it is prepared in an active, finely divided state.

The electrochemical method may also apply to the activation and functionalization of saturated hydrocarbons by rhodium(0) species.[149] Thus, the rhodium(I) complex $(\text{diphos})_2\text{Rh}^+$ is readily reduced electrochemically at -2.1 V versus $Ag/AgNO_3$ in benzonitrile solutions.

$$(\text{diphos})_2\text{Rh}^+ \underset{}{\overset{\pm e}{\rightleftharpoons}} (\text{diphos})_2\text{Rh}^0 \tag{115}$$

In the presence of cyclohexane, the rhodium(0) species is oxidized to the hydride, $(\text{diphos})_2\text{RhH}$, via a hydrogen atom transfer:

$$(\text{diphos})_2\text{Rh}^0 + \text{C}_6\text{H}_{12} \longrightarrow (\text{diphos})_2\text{Rh}^\text{I}\text{H} + \text{C}_6\text{H}_{11}^{\bullet} \tag{116}$$

The regeneration of the rhodium(I) cation with acids,

$$(\text{diphos})_2\text{RhH} + \text{HX} \longrightarrow (\text{diphos})_2\text{Rh}^+\text{X}^- + \text{H}_2 \tag{117}$$

thus constitutes an electrochemical cycle leading to the net reaction:

$$\text{C}_6\text{H}_{12} \longrightarrow \text{C}_6\text{H}_{11}^{\bullet} + \tfrac{1}{2}\text{H}_2 \tag{118}$$

Anodic oxidation of the acylchromium(0) anion $RCOCr(CO)_5^-$ affords the corresponding paramagnetic chromium(I) species, which are transient on the electrochemical time scale of a millisecond.[150] The reactive benzoylchromium(I) radical attacks the tetraethylammonium countercation to afford the hydroxycarbene complex $Ph(HO)C{=}Cr^0(CO)_5$. The sequential transformation of acylchromium species,

$$PhC(=O)Cr(CO)_5^- \xrightarrow{-e^-} PhC(=O)Cr(CO)_5 \xrightarrow{RH} PhC(OH){=}Cr(CO)_5 + R\cdot \quad (119)$$

where $RH = Et_4N^+$, formally represents the catalytic cycle (Scheme XIII),

Scheme XIII:

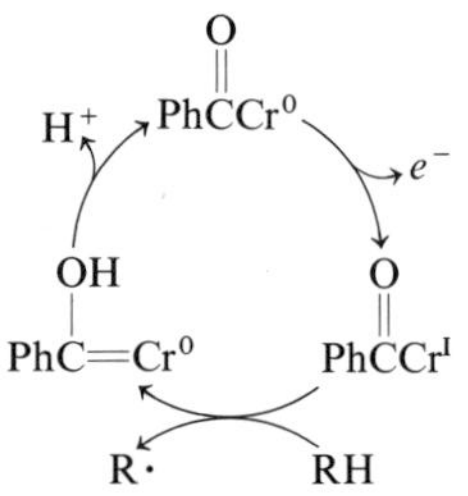

$$PhC(=O)Cr^0 \xrightarrow{-e^-} PhC(=O)Cr^I \xrightarrow{RH \to R\cdot} PhC(OH){=}Cr^0 \xrightarrow{-H^+} PhC(=O)Cr^0$$

in which the overall transformation is tantamount to the (electrochemical)-oxidation of the hydrogen-donor substrate, i.e.,

$$RH \longrightarrow R\cdot + H^+ + e^- \quad (120)$$

A two-electron electrocatalytic system is represented in the ready reversibility of the Ru(II)/Ru(IV) complex.[151] The net two-electron oxidation of *cis*$(bipy)_2pyRu(OH_2)^{2+}$ in aqueous solution at 0.6–0.8 V versus a standard calomel electrode, depending on the pH, leads to the oxoruthenium(IV) species $(bipy)_2pyRuO^+$, which is capable of oxidizing isopropyl alcohol to acetone, *p*-toluic acid to terephthalic acid, and cyclohexene to cyclohexenone.

$$(CH_3)_2CHOH \longrightarrow (CH_3)_2C{=}O + 2\,H^+ + 2\,e^- \quad (121)$$

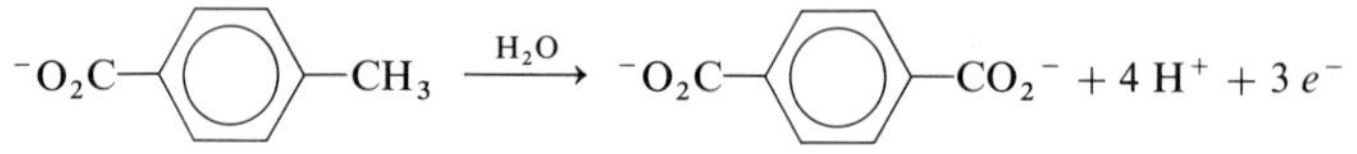

$$^-O_2C{-}C_6H_4{-}CH_3 \xrightarrow{H_2O} {}^-O_2C{-}C_6H_4{-}CO_2^- + 4\,H^+ + 3\,e^- \quad (122)$$

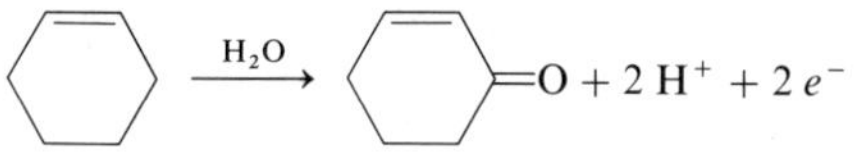

$$\text{cyclohexene} \xrightarrow{H_2O} \text{cyclohexenone}{=}O + 2\,H^+ + 2\,e^- \quad (123)$$

The overall electrocatalytic scheme would be represented by a net oxidative dehydrogenation of a substrate.

REFERENCES

1. W. F. Brill, *Ind. Eng. Chem.* **52**, 837 (1960).
2. A. S. Hay, J. W. Eustance, and H. S. Blanchard, *J. Org. Chem.* **25**, 616 (1960).
3. C. F. Hendriks, H. C. A. van Beek, and P. M. Heertjes, *Ind. Eng. Chem. Prod. Rsch. Dev.* **17**, 256 (1978).

4. E. J. Y. Scott and A. W. Chester, *J. Phys. Chem.* **76**, 1520 (1972).
5. V. N. Sapunov and L. Abdenur, *Kinet. Katal.* (Eng. Trans.) **15**, 20 (1974).
6. K. Sakota, Y. Kamiya, and N. Ohta, *Canad. J. Chem.* **47**, 387 (1969).
7. M. Kashima and Y. Kamiya, *Bul. Chem. Soc. Japan* **47**, 481 (1974).
8. Y. Kamiya and M. Kashima, *J. Catal.* **25**, 326 (1972).
9. Y. Kamiya and M. Kashima, *Bul. Chem. Soc. Japan* **46**, 905 (1973).
10. E. I. Heiba, R. M. Dessau, and W. J. Koehl, *J. Am. Chem. Soc.* **91**, 6830 (1969); *Prepr., Div. Pet. Chem., Am. Chem. Soc.* p. A44 (1969).
11. A. Onopchenko, J. G. D. Schulz, and R. Seekircher, *J. Chem. Soc., Chem. Commun.* p. 939 (1971).
12. A. Onopchenko, J. G. D. Schulz, and R. Seekircher, *J. Org. Chem.* **37**, 1414 (1972). See also E. Baciocchi, L. Mandolini, and C. Rol, *Ibid.*, **45**, 3906 (1980).
13. A. Onopchenko and J. G. D. Schulz, *J. Org. Chem.* **37**, 2564 (1972).
14. T. Morimoto and Y. Ogata, *J. Chem. Soc. B* **62**, 1353 (1967).
15. C. F. Hendriks, H. C. A. van Beek, and P. M. Heertjes, *Ind. Eng. Chem. Prod. Rsch. Dev.* **17**, 260 (1978); **16**, 270 (1977).
16. A. N. Nemecek, C. F. Hendriks, H. C. A. van Beek, M. A. de Bruyn, and E. J. H. Kerckhoffs, *Ind. Eng. Chem. Prod. Rsch. Dev.* **17**, 133 (1978).
17. J. Imamura, M. Takehara, K. Chigasaki, and K. Kizawa, Ger. Patent 2,605,678 (1975) to Agency of Industrial Science and Technology; Sanko Chemical Co.
18. (a) S. S. Lande, C. D. Falk, and J. K. Kochi, *J. Inorg. Nucl. Chem.* **33**, 4101 (1971). (b) C. F. Hendricks, H. C. A. van Beek, and P. M. Heertjes, *Ind. Eng. Chem. Prod. Rsch. Dev.* **18**, 43 (1979).
19. J. Hanotier and H. Hanotier-Bridoux, *J. Chem. Soc., Perkin II* p. 1036 (1973).
20. R. M. Dessau, S. Shih, and E. I. Heiba, *J. Am. Chem. Soc.* **92**, 412 (1970).
21. P. J. Andrulis, M. J. S. Dewar, R. Dietz, and R. L. Hunt, *J. Am. Chem. Soc.* **88**, 5473 (1966).
22. T. Aratani and M. J. S. Dewar, *J. Am. Chem. Soc.* **88**, 5479 (1966).
23. P. J. Andrulis and M. J. S. Dewar, *J. Am. Chem. Soc.* **88**, 5483 (1966); see also J. R. Gilmore and J. M. Mellor, *Tetrahedron Lett.* p. 3977 (1971).
24. E. I. Heiba, R. M. Dessau, and W. J. Koehl, *J. Am. Chem. Soc.* **91**, 138 (1969).
25. R. E. van der Ploeg, R. W. de Korte, and E. C. Kooyman, *J. Catal.* **10**, 52 (1968).
26. H. J. den Hertog and E. C. Kooyman, *J. Catal.* **6**, 347, 357 (1966).
27. R. van Helden and E. C. Kooyman, *Recl. Trav. Chim. Pays-Bas* **80**, 57 (1961).
28. E. I. Heiba and R. M. Dessau, *J. Am. Chem. Soc.* **93**, 995 (1971).
29. R. O. C. Norman, C. B. Thomas, and P. J. Ward, *J. Chem. Soc., Perkin I* p. 2914 (1973). See also E. Baciocchi, C. Rol, and L. Mandolini, *J. Am. Chem. Soc.* **102**, 7597 (1980).
30. D. A. S. Ravens, *Trans. Faraday Soc.* **55**, 1768 (1959).
31. A. S. Hay and H. S. Blanchard, *Canad. J. Chem.* **43**, 1306 (1965).
32. M. Hronec, V. Vesely, and J. Herain, *Col. Czech. Chem. Commun.* **44**, 3362 (1979).
33. (a) Y. Kamiya, *Tetrahedron* **22**, 2029 (1966);
    (b) Y. Kamiya, *Adv. Chem. Series* **76**, 193 (1968).
34. Y. Kamiya, *J. Catal.* **33**, 480 (1974).
35. I. V. Zakharov and V. M. Muratov, *Dokl. Akad. Nauk SSSR* **196**, 156 (1971).
36. F. F. Shcherbina and T. V. Lysukho, *Kinet. Catal.* (Eng. Trans.) **19**, 872 (1978).
37. J. R. Gilmore and J. M. Mellor, *J. Chem. Soc., Chem. Commun.* p. 507 (1970).
38. C. E. H. Bawn and T. K. Knight, *Discuss. Faraday Soc.* **46**, 164 (1968).
38a. See also T. Okada and Y. Kamiya, *Bul. Chem. Soc.* Japan. **52**, 3321 (1979) for the effect of copper(II).
39. H. D. Holtz, *J. Org. Chem.* **37**, 2069 (1972); *J. Chem. Soc., Chem. Commun.* p. 1166 (1971).
40. C. Giordano, A. Belli, and A. Citterio, *J. Org. Chem.* **45**, 345 (1980).

41. M. Hronec and V. Vesely, *Col. Czech. Chem. Commun.* **40**, 2165 (1975); **42**, 1851 (1977); **41**, 350 (1976); **44**, 3362 (1979).
42. H. S. Bryant, C. A. Duval, L. E. McMakin, and J. I. Savoca, *Chem. Eng. Prog.* **67**, 69 (1971).
43. (a) Y. Ichikawa, G. Yamashita, M. Tokashiki, and T. Yamaji, *Ind. Eng. Chem.* **62**, 38 (1970). (b) Y. Ichikawa and Y. Takeuchi, *Hydrocarbon Process.* **51**, 103 (1972).
44. A. W. Chester, P. S. Landis, and E. J. Y. Scott, *Chemtech.* p. 366 (1978).
45. J. Hanotier, H. Hanotier-Bridoux, and P. de Radzitzky, *J. Chem. Soc., Perkin II* pp. 381, 1035 (1973).
46. R. T. Tang and J. K. Kochi, *J. Inorg. Nucl. Chem.* **35**, 3845 (1973).
47. S. S. Lande and J. K. Kochi, *J. Am. Chem. Soc.* **90**, 5196 (1968).
48. J. K. Kochi, R. T. Tang, and T. Bernath, *J. Am. Chem. Soc.* **95**, 7114 (1973).
49. J. M. Anderson and J. K. Kochi, *J. Am. Chem. Soc.* **92**, 2450 (1970).
50. R. E. Partch, *J. Am. Chem. Soc.* **89**, 3662 (1967).
51. J. R. Kalman, J. T. Pinhey, and S. Sternhell, *Tetrahedron Lett.* p. 5369 (1972).
52. J. R. Campbell, J. R. Kalman, J. T. Pinhey, and S. Sternhell, *Tetrahedron Lett.* p. 1763 (1972).
53. H. C. Bell, J. R. Kalman, J. T. Pinhey, and S. Sternhell, *Tetrahedron Lett.* pp. 853, 857 (1974).
54. R. O. C. Norman, C. B. Thomas, and J. S. Willson, *J. Chem. Soc. B* p. 518 (1971); *J. Chem. Soc., Perkin II* p. 325 (1973).
55. (a) A. McKillop and E. C. Taylor, *Chem. Br.* **8**, 4 (1972). (b) E. C. Taylor and A. McKillop, *Accts. Chem. Rsch.* **3**, 338 (1970).
56. A. McKillop, A. G. Turrell, D. W. Young, and E. C. Taylor, *J. Am. Chem. Soc.* **102**, 6504 (1980).
57. K. Ichikawa, S. Uemura, T. Nakano, and E. Uegaki, *Bul. Chem. Soc. Japan* **44**, 545 (1971).
58. J. M. Broidy and K. A. Moore, *J. Chem. Soc., Perkin I* p. 2917 (1974).
59. I. H. Elson and J. K. Kochi, *J. Am. Chem. Soc.* **95**, 5061 (1973).
60. R. A. Sheldon and J. K. Kochi, *J. Am. Chem. Soc.* **90**, 6686 (1968).
61. R. O. C. Norman, C. B. Thomas, and P. J. Ward, *J. Chem. Soc., Perkin I* p. 2917 (1974).
62. C. L. Jenkins and J. K. Kochi, *J. Am. Chem. Soc.* **94**, 843 (1972).
63. F. R. S. Clark, R. O. C. Norman, C. B. Thomas, and J. S. Willson, *J. Chem. Soc., Perkin I* p. 1289 (1974).
64. G. G. Arzoumanidis and F. C. Rauch, *Chemtech.*, **3**, 700 (1973).
65. R. G. Salomon and J. K. Kochi, *J. Chem. Soc., Chem. Commun.* p. 559 (1972).
66. H. C. Brown and R. A. Wirkkala, *J. Am. Chem. Soc.* **88**, 1447, 1453 (1966).
67. T. Szymańska-Buzar and J. J. Ziółkowski, *J. Mol. Catal.* **5**, 341 (1979).
68. E. B. Pedersen, T. E. Pedersen, K. Torssell, and S. O. Lawesson, *Tetrahedron* **29**, 579 (1973).
69. M. E. Kurz and G. W. Hage, *J. Org. Chem.* **42**, 4080 (1977).
70. M. J. S. Dewar and T. Nakaya, *J. Am. Chem. Soc.* **90**, 7134 (1968).
71. R. M. Dessau, *J. Am. Chem. Soc.* **92**, 6356 (1970).
72. C. E. H. Bawn and J. A. Sharp, *J. Chem. Soc.* pp. 1854, 1866 (1957).
73. (a) M. Hirano and T. Morimoto, *J. Chem. Res., Synop.* p. 104 (1979). (b) See also M. Hirano, E. Kitamura, and T. Morimoto, *J. Chem. Soc., Perkin II* p. 569 (1980).
74. L. Verstraelen, M. M. Lalmand, A. J. Hubert, and P. Teyssie, *J. Chem. Soc., Perkin II* p. 1285 (1976).
75. E. I. Heiba, R. M. Dessau, and W. J. Koehl, *J. Am. Chem. Soc.* **90**, 5905 (1968).
76. E. I. Heiba, R. M. Dessau, and P. G. Rodewald, *J. Am. Chem. Soc.* **96**, 7977 (1974).
77. J. B. Bush and H. J. Finkbeiner, *J. Am. Chem. Soc.* **91**, 5903 (1969).
78. M. Okano, *Chem. Ind. (London)* p. 423 (1972).

79. E. I. Heiba, R. M. Dessau, and W. J. Koehl, *J. Am. Chem. Soc.* **90**, 2706 (1968).
80. W. J. de Klein, *Recl. Trav. Chim. Pays-Bas* **94**, 151 (1975).
81. J. R. Gilmore and J. M. Mellor, *J. Chem. Soc. C* p. 2355 (1971).
82. J. Kollar, U.S. Patents 3,689,535; 3,668,239 (1972); 3,778,468 (1973) to Halcon International.
83. J. R. Valbert, U.S. Patent 3,715,388 (1973).
84. R. M. Moriarty, *in* "Selective Organic Transformations" (B. S. Thyagarajan, ed.), p. 125. Wiley (Interscience), New York, 1972.
85. P. M. Henry, *Adv. Chem. Series* **70**, 126 (1968).
86. A. Lethbridge, R. O. C. Norman, and C. B. Thomas, *J. Chem. Soc., Perkin I* p. 2763 (1973).
87. R. R. Grinstead, *J. Org. Chem.* **26**, 238 (1961).
88. P. M. Henry, *J. Am. Chem. Soc.* **87**, 990 (1965).
89. W. Kruse and T. M. Bednarski, *J. Org. Chem.* **36**, 1154 (1971).
90. (a) Belgian Patent 853, 864 (1976) to Halcon International.
    (b) U.S. Patent 3,641,067 (1969) to Hercules.
91. W. Kitching, *Organometal. React.* **3**, 319 (1972).
92. R. A. Johnson, U.S. Patent 4,192,814 (1980) and W. F. Brill, U.S. Patent 4,115,420 (1978) to Halcon Research and Development Corporation.
93. Br. Patent 1,182,273 (1970) to Teijin.
94. A. Onopchenko and J. G. D. Schulz, *J. Org. Chem.* **38**, 909 (1978).
94a. A Onopchenko and J. G. D. Schulz, *J. Org. Chem.* **38**, 3729 (1978).
95. K. Tanaka, *Chemtech.*, p. 555 (1974); *Hydrocarbon Process*, **53**, 114 (1974); *Prepr., Pet. Chem. Div., Am. Chem. Soc.* **19**, 103 (1974). See also J. G. D. Schulz and A. Onopchenko, *J. Org. Chem.* **45**, 3716 (1980).
96. J. Hanotier, P. Camerman, H. Hanotier-Bridoux, and P. de Radzitsky, *J. Chem. Soc., Perkin II* p. 2247 (1972).
97. J. Vaerman, P. de Radzitzky, and J. Hanotier, Br. Patent 1,209,140 (1970) to Labofina.
98. S. R. Jones and J. M. Mellor, *J. Chem. Soc., Perkin II* p. 511 (1977).
99. R. A. Sheldon and J. K. Kochi, *Adv. Catal.* **25**, 272 (1976).
100. G. A. Olah, *Chem. Br.* **8**, 281 (1972); *Angew. Chem. Int. Ed.* **12**, 173 (1973).
101. R. A. Sheldon, unpublished results.
102. C. E. H. Bawn and J. B. Williamson, *Trans. Faraday Soc.* **47**, 721, 735 (1951).
103. C. E. H. Bawn, T. P. Hobin, and L. Raphael, *Proc. R. Soc. London, Ser. A* **237**, 313 (1956).
104. C. E. H. Bawn and J. E. Jolley, *Proc. R. Soc. London, Ser. A* **237**, 297 (1956).
105. G. C. Allen and A. Aguilo, *Adv. Chem. Series* **76**, 363 (1968).
106. F. Marta, E. Boga, and M. Matok, *Discuss. Faraday Soc.* **46**, 173 (1968).
107. W. F. Brill and F. Lister, *J. Org. Chem.* **26**, 565 (1961).
108. M. Zawadzki and J. J. Ziolkowski, *React. Kinet. Catal. Lett.* **10**. 119 (1973).
109. Y. Ohkatsu, M. Takeda, T. Hara, and A. Misono, *Bul. Chem. Soc. Japan* **40**, 1413 (1967).
110. Y. Ohkatsu, T. Osa, and A. Misono, *Bul. Chem. Soc. Japan* **40**, 2111 (1967).
111. M. E. Ladhabhoy and M. M. Sharma, *J. Appl. Chem.* **20**, 274 (1970).
112. A. W. Schwab, E. N. Frankel, E. J. Dufek, and J. C. Cowan, *J. Am. Oil Chem. Soc.* **49**, 75 (1972).
113. A. Schwab, *J. Am. Oil Chem. Soc.* **50**, 74 (1973).
114. J. A. John and F. J. Weymouth, *Chem. Ind.* (*London*) p. 62 (1962).
115. G. Twigg, *Chem. Ind.* (*London*) p. 476 (1966).
116. J. K. Kochi, *Record. Chem. Prog.* **27**, 207 (1966); *Science* **155**, 415 (1967); *Pure Appl. Chem.* **4**, 307 (1971); *Accts. Chem. Rsch.* **7**, 351 (1974).

117. G. I. Nikishin, M. G. Vinogradov, and S. P. Verenchikov, *Bul. Acad. Sci. USSR, Div. Chem. Sci.* p. 1698 (1969).
118. G. I. Nikishin, M. G. Vinogradov, and R. V. Kereselidze, *Bul. Acad. Sci. USSR, Div. Chem. Sci.* p. 1570 (1967); p. 1083 (1966).
119. M. G. Vinogradov, R. V. Kereselidze, G. G. Gachechiladze, and G. I. Nikishin, *Bul. Acad. Sci. USSR, Div. Chem. Sci.* p. 276 (1969); see also H. Inoue, Y. Kimura, and E. Imoto, *Bul. Chem. Soc. Japan* **46**, 3303 (1973).
120. M. G. Vinogradov, S. P. Verenchikov, and G. I. Nikishin, *Bul. Acad. Sci. USSR, Div. Chem. Sci.* p. 947 (1972).
121. G. I. Nikishin, M. G. Vinogradov, S. P. Verenchikov, I. N. Kostyukov, and R. V. Kereselidze, *J. Org. Chem. USSR* **8**, 544 (1972).
122. G. I. Nikishin, M. G. Vinogradov, and G. P. Ilina, *J. Org. Chem. USSR* **8**, 1422 (1972).
123. M. G. Vinogradov, G. P. Ilina, A. V. Ignatenko, and G. I. Nikishin, *J. Org. Chem. USSR* **8**, 1425 (1972).
124. G. I. Nikishin, M. G. Vinogradov, and G. P. Ilina, *Synthesis* p. 376 (1972).
125. W. A. Waters and J. S. Littler, *in* "Oxidation in Organic Chemistry" (K. B. Wiberg, ed.), Part A, p. 186. Academic Press, New York, 1969.
126. E. I. Heiba and R. M. Dessau, *J. Am. Chem. Soc.* **93**, 524, 995 (1971); **94**, 2888 (1972).
127. G. I. Nikishin, M. G. Vinogradov, and T. M. Fedorova, *J. Chem. Soc., Chem. Commun.* p. 693 (1973).
127a. N. Fukumiya, M. Okano and T. Aratani, *Chem. Ind.* (*London*) p. 86 (1980)
128. R. van Helden and E. C. Kooyman, *Recl. Trav. Chim. Pays-Bas* **80**, 57 (1961); H. J. den Hertog and E. C. Kooyman, *J. Catal.* **6**, 357 (1966).
129. R. A. Sheldon and J. K. Kochi, *Org. React.* (*N.Y.*) **19**, 279 (1972).
130. See, for example, A. McKillop and E. C. Taylor, *Adv. Organometal. Chem.* **11**, 147 (1973); A. McKillop, *Pure Appl. Chem.* **43**, 463 (1975).
131. E. T. Denisov, N. I. Mitskevich, and V. E. Agabekov, "Liquid Phase Oxidation of Oxygen-Containing Compounds" (D. A. Paterson, Engl. Transl.), p. 23. Consultants Bureau, New York, 1977.
132. D. C. Nonhebel and J. C. Walton, "Free Radical Chemistry," p. 317. Cambridge Univ. Press, London and New York, 1974.
133. J. D. Bacha and J. K. Kochi, *J. Org. Chem.* **30**, 3272 (1965); J. K. Kochi, *J. Am. Chem. Soc.* **84**, 1193 (1962); C. Walling and A. Padwa, *ibid.* **85**, 1593 (1963).
134. C. A. Bunton, *in* "Oxidation in Organic Chemistry" (K. B. Wiberg, ed.), Part A, p. 367. Academic Press, New York, 1969.
135. W. S. Trahanovsky, *Methods Free-Radical Chem.* **4**, 133 (1973); W. S. Trahanovsky, L. H. Young, and M. H. Bierman, *J. Org. Chem.* **34**, 869 (1969).
136. G. de Vries and A. Schors, *Tetrahedron Lett.* p. 5869 (1968).
137. A. P. Tomilov, S. G. Mairanovsky, M. Y. Fioshin, and V. A. Smirnov, "Electrochemistry of Organic Compounds." Wiley, New York, 1972.
138. M. M. Baizer, ed., "Organic Electrochemistry." Dekker, New York, 1973.
139. A. J. Fry, "Synthetic Organic Electrochemistry." Harper, New York, 1973.
140. L. Eberson and H. Schäfer, *Fortschr. Chem. Forsch.* **21**, 5 (1971).
141. L. Eberson and K. Nyberg, *Accts. Chem. Rsch.* **6**, 106 (1973).
142. N. L. Weinberg and H. R. Weinberg, *Chem. Rev.* **68**, 449 (1968).
143. J. H. P. Utley, *Chem. Ind.* (*London*) p. 230 (1972).
144. M. Fleischmann and D. Pletcher, *Tetrahedron Lett.* p. 6255 (1969).
145. V. R. Koch and L. L. Miller, *J. Am. Chem. Soc.* **95**, 8631 (1973).
146. T. M. Siegel, L. L. Miller, and J. Y. Becker, *J. Chem. Soc., Chem. Commun.* p. 341 (1974).
147. R. Ramaswamy, M. S. V. Pathy, and H. V. K. Udupa, *J. Electrochem. Soc.* **110**, 202 (1963).

148. D. H. Becking, U.S. Patent 3,985,809 (1976) to Oxy Metal Industries Corp.
149. J. A. Sofranko, R. Eisenberg, and J. A. Kampmeier, *J. Am. Chem. Soc.* **102**, 1165 (1980).
150. R. J. Klingler and J. K. Kochi, *Inorg. Chem.* **20**, 34 (1981).
151. B. A. Moyer, M. S. Thompson, and T. J. Meyer, *J. Am. Chem. Soc.* **102**, 2310 (1980).

ADDITIONAL READING

P. D. McDonald and G. A. Hamilton, Mechanisms of phenolic oxidative coupling. *In* "Oxidation in Organic Chemistry" (W. Trahanovsky ed.), Part B, p. 97. Academic Press, New York, 1973.

Chapter 6

Direct Oxidation by Oxometal (M=O) Reagents

I. OXOMETAL REAGENTS: MECHANISTIC FORMULATIONS

The stoichiometric oxidations of organic substrates by oxometal (M=O) reagents, such as permanganate,[1] chromic acid and chromyl compounds,[2] SeO_2,[3–6] OsO_4,[7] RuO_4,[8,9] and MnO_2,[10] are well known to chemists.

These reagents have traditionally played an important role in organic synthesis owing to their capacity for selective oxygen transfer to a wide variety of substrates under mild conditions. Participation by one or more M=O groups is a key mechanistic feature common to virtually all of these reactions (*vide infra*).

A second group of oxidation reactions involving oxometal reagents is the heterogeneous, gas-phase oxidation of hydrocarbons over metal oxide or mixed metal oxide catalysts.[11–26] These reactions are performed at elevated temperatures (300°–600°C) and form the basis of a number of important petrochemical processes.[27,28] Perhaps the best known of these are the vapor-phase oxidation and ammoxidation of propylene to acrolein and acrylonitrile, respectively, over bismuth molybdate catalysts.[25,26]

$$CH_3CH{=}CH_2 + O_2 \xrightarrow{[Bi_2Mo_2O_9]} CH_2{=}CHCHO + H_2O \qquad (1)$$

$$CH_3CH{=}CH_2 + O_2/NH_3 \xrightarrow{[Bi_2Mo_2O_9]} CH_2{=}CHCN + 2\,H_2O \qquad (2)$$

It is now generally accepted[11–13] that these and analogous reactions involve initial oxidation of the substrate, followed by the regeneration of the oxometal catalyst by reaction of the reduced form with molecular oxygen, i.e.,

$$\text{substrate} + M_{ox} \longrightarrow \text{product} + M_{red} \qquad (3)$$

$$M_{red} + O_2 \longrightarrow M_{ox}, \quad \text{etc.} \qquad (4)$$

This model was first suggested by Mars and van Krevelen[29] in 1954 for the oxidation of aromatic hydrocarbons with V_2O_5 and subsequently by Batist and co-workers[30] for bismuth molybdate oxidations. In most cases, the reaction of the oxidant with the substrate in eq 3 is rate limiting.

In recent years there has been considerable speculation concerning the nature of the active oxidant in the iron-containing enzymes (oxygenases), which mediate the incorporation of molecular oxygen into a variety of organic substrates. Current opinion appears to favor an oxoiron species, probably O=Fe(V) (*vide infra*). Sharpless and Flood[31] first called attention to the striking similarities between reactions catalyzed by oxygenases (such as the stereoselective hydroxylation of alkanes, epoxidation of alkenes, and hydroxylation of arenes) and those effected by oxometal reagents.

Oxometal groups are capable of effecting the following types of oxidative transformations,

$$O{=}M^{n+} \xrightarrow{RH} M^{(n-2)+} + ROH \qquad (5)$$

$$O{=}M^{n+} \xrightarrow{>C=C<} M^{(n-2)+} + >C\overset{O}{\frown}C< \qquad (6)$$

$$O{=}M^{n+} \xrightarrow{L} M^{(n-2)+} + LO \qquad (7)$$

where L = trialkylphosphines and trialkylamines as well as dialkyl sulfides.

Similarly, dioxometal reagents can be involved at two sites, such as

$$O_2M^{n+} + \rangle C{=}C\langle \longrightarrow M^{(n-4)+} + \rangle C{=}O + O{=}C\langle \qquad (8)$$

$$O_2M^{n+} + \rangle CH{-}\overset{|}{C}H{-}\overset{|}{C}{=}C\langle \longrightarrow (HO)_2M^{(n-2)+} + \rangle C{=}\overset{|}{C}{-}\overset{|}{C}{=}C\langle \qquad (9)$$

$$O_2M^{n+} + \rangle CHOH \longrightarrow (HO)_2M^{(n-2)+} + \rangle C{=}O \qquad (10)$$

followed by

$$(HO)_2M^{(n-2)+} \longrightarrow O{=}M^{(n-2)+} + H_2O$$

Oxometal functions can be represented as the resonance hybrid of the covalent and the dipolar canonical forms.

$$O{=}M \longleftrightarrow {}^{-}O{-}M^{+}$$

$$O_2M \longleftrightarrow ({}^{-}O)(O{=})M^{+}$$

As Sharpless and co-workers have pointed out,[32] the $O{=}M^{n+}$ functional group constitutes one example of a general class of high-valent XM^{n+} species comprising various metal ylides, e.g.,

$$O{=}M^{n+} \qquad RN{=}M^{n+} \qquad R_2C{=}M^{n+} \qquad Cl_2M^{n+}$$

which are capable of transferring the group X to various substrates, especially olefins. The mechanisms by which such transformations occur are subject to the same ambiguities noted in Chapter 5 for the oxidations of organic substrates with high-valent metals. Thus, electron transfer processes (involving radical intermediates) as well as electrophilic mechanisms (involving organometallic intermediates) are possible. For example, oxygen insertion into the C—H bond of a hydrocarbon RH can be distinguished by the following mechanistic alternatives.

Electron transfer

$$O{=}\overset{\overset{O}{\|}}{M}{}^{n+} + RH \longrightarrow HO{-}\overset{\overset{O}{\|}}{M}{}^{(n-1)+} + R\cdot$$

$$HO{-}\overset{\overset{O}{\|}}{M}{}^{(n-1)+} + R\cdot \longrightarrow O{=}M^{(n-2)+} + ROH$$

Electrophilic substitution[32]

$$O{=}\overset{\overset{O}{\|}}{M}{}^{n+}\;(H{-}R) \longrightarrow HO{-}\overset{\overset{O}{\|}}{\underset{R}{M}}{}^{n+} \longrightarrow HO{-}\underset{OR}{M}{}^{(n-2)+}$$

$$HO{-}\underset{OR}{M}{}^{(n-2)+} \longrightarrow O{=}M^{(n-2)+} + ROH$$

Any reasonable mechanism must account for two important experimental observations; namely, many of these reactions exhibit substantial deuterium kinetic isotope effects and proceed with partial retention of configuration.[1,2,33,34] In an electron transfer mechanism, this would require the C—H bond-breaking step to be rate limiting. Furthermore, the alkyl radical so formed is not free but is rapidly oxidized further within the solvent cage. Retention of configuration in the cage reactions of radical pairs has been demonstrated.[35,36] Such a step could also involve an initial reaction at the metal to form an organometallic intermediate, i.e.,

$$HOM^{(n-1)+} + R\cdot \longrightarrow \begin{matrix} HO \\ R \end{matrix}\!\!>M^{n+} \longrightarrow ROH + M^{(n-2)+} \qquad (11)$$

With regard to electrophilic attack, the reactions of hydrocarbons with oxometal reagents were generally considered to proceed by attack of the nucleophilic substrate on oxygen until Sharpless[32] pointed out that attack of the nucleophile on the positive metal center is mechanistically more reasonable (reasoning by analogy with the nucleophilic additions to carbonyl groups in organic chemistry). Organometallic species are a direct consequence of this hypothesis and hitherto had not generally been considered to be viable intermediates in the reactions of oxometal reagents. We shall now consider the stoichiometric and catalytic oxidations of various organic substrates by oxometal reagents within this mechanistic framework.

II. ALLYLIC OXIDATIONS

A. Stoichiometric Oxidations

The interaction of olefins with oxometal reagents, as with other oxidants, can involve reaction either at the double bond or at the allylic C—H bond. This competition is reflected in the reaction of olefins with the simplest "oxo" species, namely, singlet oxygen.[37]

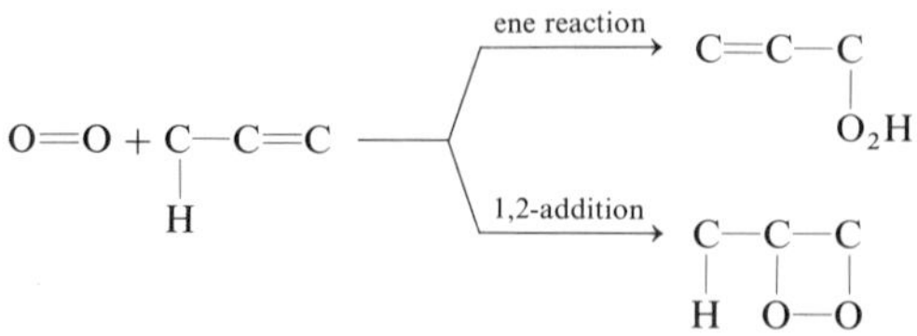

$$O{=}O + \underset{\mathrm{H}}{\underset{|}{C}}{-}C{=}C \longrightarrow \begin{cases} \xrightarrow{\text{ene reaction}} C{=}C{-}\underset{O_2H}{\underset{|}{C}} & (12) \\ \xrightarrow{\text{1,2-addition}} \underset{H}{\underset{|}{C}}{-}\underset{O}{\underset{|}{C}}{-}\underset{O}{\underset{|}{C}} \;(\text{O{-}O}) & (13) \end{cases}$$

Even in this conceptually straightforward example, the mechanistic details are not completely resolved as to whether these reactions are concerted or proceed by perepoxide or other intermediates.

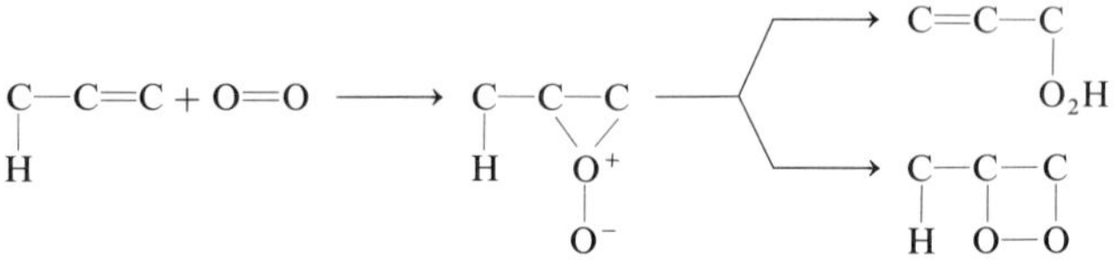

$$\underset{\mathrm{H}}{\underset{|}{C}}{-}C{=}C + O{=}O \longrightarrow \underset{\mathrm{H}}{\underset{|}{C}}{-}C{-}C \;(\text{with } O^{+}{-}O^{-}) \longrightarrow \begin{cases} C{=}C{-}\underset{O_2H}{\underset{|}{C}} & (14) \\ \underset{H}{\underset{|}{C}}{-}\underset{O}{\underset{|}{C}}{-}\underset{O}{\underset{|}{C}} \;(\text{O{-}O}) & (15) \end{cases}$$

Selenium dioxide is the most commonly used reagent for the allylic oxidation of olefins.[3–6] The mechanism of this reaction has been shown by Sharpless and co-workers[38–40] to involve an initial ene addition of an Se=O ($Se^{+}{-}O^{-}$) moiety to produce an organoselenium intermediate (i.e., an allylseleninic acid), first suggested by Wiberg.[41]

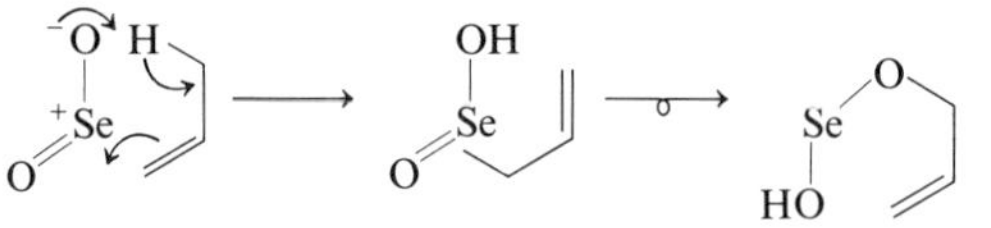

(16)

This insertion is followed by oxidative elimination to carbonyl compounds

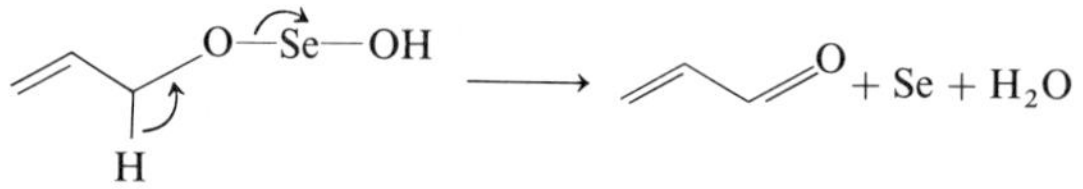

$$\text{CH}_2{=}\text{CH}{-}\text{CH(H)}{-}\text{O}{-}\text{Se}{-}\text{OH} \longrightarrow \text{CH}_2{=}\text{CH}{-}\text{CH}{=}\text{O} + \text{Se} + H_2O \qquad (17)$$

or hydrolysis to alcohols:

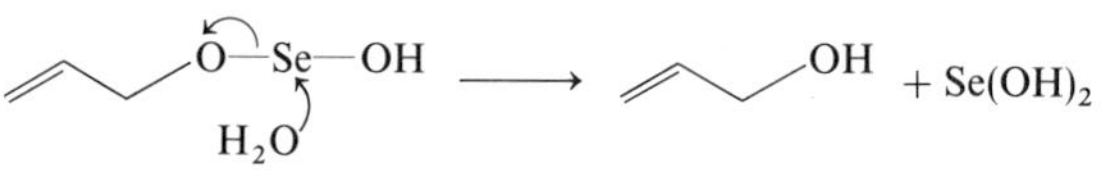

$$\text{CH}_2{=}\text{CH}{-}\text{CH}_2{-}\text{O}{-}\text{Se}{-}\text{OH} \;(+\,H_2O) \longrightarrow \text{CH}_2{=}\text{CH}{-}\text{CH}_2{-}\text{OH} + Se(OH)_2 \qquad (18)$$

Convincing evidence for an allylseleninic acid intermediate was provided by its trapping as a selenolactone in reactions with suitable substrates.[40] The initial step is analogous to the Prins reaction of formaldehyde with alkenes.[42]

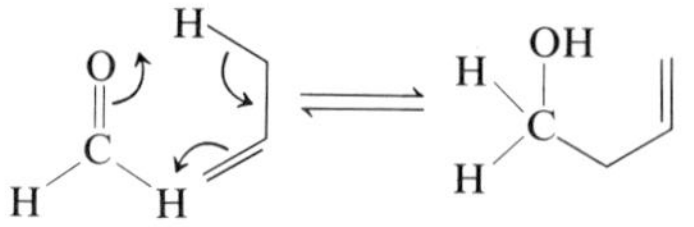

(19)

The isomerization of olefins in liquid sulfur dioxide also involves an ene reaction.[43]

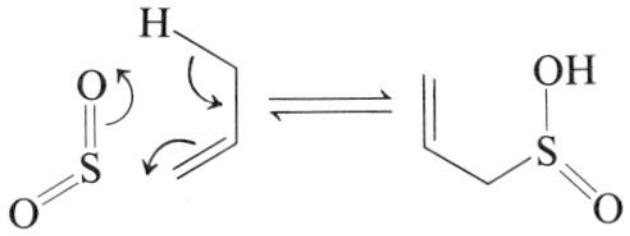

(20)

A careful study of the stereochemistry of the hydroxylation of deuterium-labeled alkenes with selenium dioxide has revealed the presence of at least two pathways.[44] In *tert*-butyl alcohol solutions containing pyridine, the major path for the allylic hydroxylation is operationally concerted in accord with the mechanism proposed by Sharpless and co-workers. In *tert*-butyl alcohol alone, however, a stepwise process, involving carbonium ion intermediates, provides a competing, stereorandom path.

In the presence of excess *tert*-butyl hydroperoxide and water, the SeO_2-catalyzed oxidations of olefins afford allylic alcohols and α,β-unsaturated carbonyl compounds.[45] Since catalytic amounts (2–50%) of SeO_2 are employed, the reduced forms of selenium are avoided, being reoxidized by *tert*-butyl hydroperoxide. Acetylenes show a strong tendency to undergo α,α'-dioxydihydroxylation with the SeO_2–TBHP reagent, e.g.,[46]

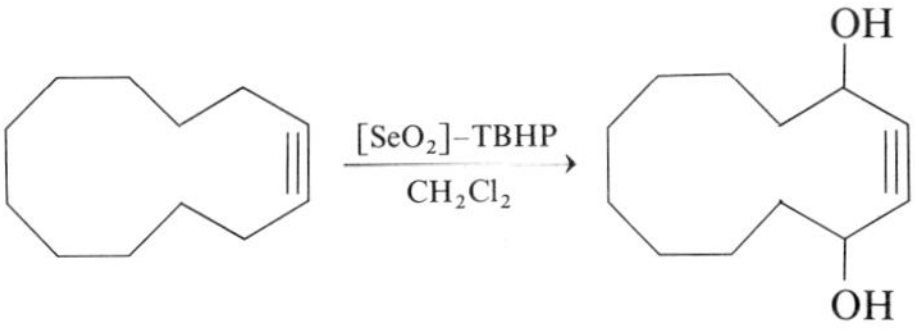

$$\xrightarrow[CH_2Cl_2]{[SeO_2]\text{–TBHP}}$$

(55%)

Allylic aminations using imidoselenium(IV) reagents have also been described,[47] e.g.,

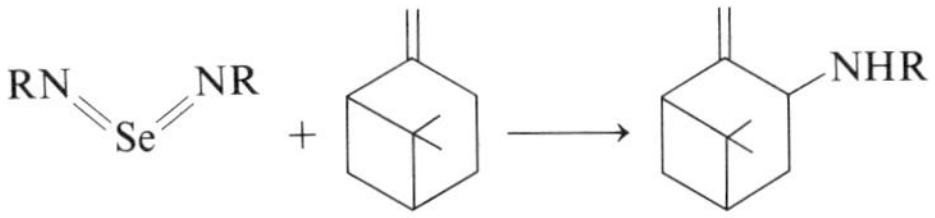

RN=Se=NR + ⟶ NHR, etc. (21)

Another well-known reaction of SeO_2 is the oxidation of carbonyl compounds to 1,2-dicarbonyl compounds. Evidence has been presented[48] in favor of a mechanism analogous to that discussed above for olefin oxidations, i.e.,

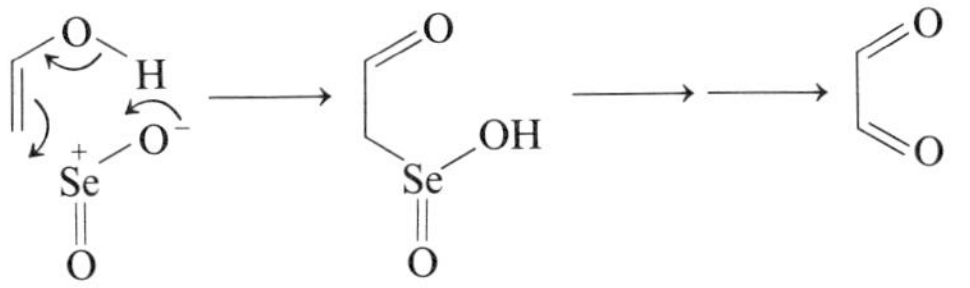

etc. (22)

Aldehydes can be readily converted to nitriles using hydroxylamine and selenium dioxide.[49]

$$\mathrm{RCHO + H_2NOH \longrightarrow RCH{=}NOH + H_2O}$$

$$\mathrm{RCH{=}NOH + SeO_2 \longrightarrow [R(H)C{=}N{-}O{-}Se({=}O)OH] \longrightarrow RC{\equiv}N + SeO(OH)_2}$$

B. Catalytic Oxidations

The conversion of propylene selectively to acrolein over a cuprous oxide catalyst in the gas phase at 300°–400°C was reported by Hearne and Adams[50] in 1948. This event marked the beginning of an era of research on oxide catalysis which gathered impetus with the discovery of the bismuth molybdate catalyst for the selective oxidation and ammoxidation of propylene[51] (see reactions 1 and 2), as well as the dehydrogenation of butene to butadiene.[52] Shortly after the introduction of the bismuth molybdate catalysts, an even more selective uranium antimonate (USb_3O_{10}) catalyst[53] and afterward a series of third-generation, multicomponent oxide catalyst were developed.[54] The selectivity to acrolein in these heterogeneous processes is generally high, in some cases exceeding 90%. In recent years, there have been many publications concerned with the mechanisms of these commercially important processes.[11–28,55] The salient mechanistic features of these reactions are the rate-limiting direct interaction of the oxometal reagent with the substrate (see the Mars–van Krevelen model above[29]). Furthermore, it has been unequivocally established by the use of isotopic carbon tracer experiments that the oxidation of propylene over bismuth molybdate proceeds via the initial formation of a symmetric allylic intermediate.[56] With regard to the activity–selectivity properties of various oxometal catalysts, Sachtler and de Boer[57] proposed as a working hypothesis that easily reducible catalysts (i.e., strong oxidants) should be active but nonselective in producing only extensive degradation to CO_2. At the other extreme, metal oxides that are difficult to reduce were deemed inactive. Only those metal oxides of intermediate reducibility (i.e., moderate oxidants) were considered to be capable of producing selective oxidation. However, the nature of the active site in the bismuth molybdate catalyst, as well as the precise origin of the superior selectivity–activity exhibited by certain combinations of oxides, has yet to be completely resolved.[12,14] In other words, there is no unequivocal theoretical basis for the prediction of catalytic activity and selectivity. At this juncture, the new developments in this field are largely derived on an empirical basis.

Trifiro[58] appears to have been the first to call attention to the many similarities between the catalytic oxidations of olefins over oxide catalysts and the stoichiometric oxidations with oxometal reagents. Parallels were noted for both allylic oxidations and those involving overall reaction at the double bond (*vide infra*). The basic, underlying feature of all these reactions is the participation of the oxometal groups.[58–61] In general, it appears that allylic oxidation stems from moderately strong oxidants (electrophiles) such as Se(IV) and Mo(VI), whereas reaction at the double bond occurs with the very strong oxidants, e.g., Cr(VI), Mn(VII), Ru(VIII), and V(V) (*vide infra*). By analogy with the currently accepted mechanism for allylic oxidation with SeO_2, the catalytic oxidations could also involve organometallic intermediates.

$$\text{HO-Mo}^{\text{VI}}(=\text{O})\text{-CH}_2\text{CH=CH}_2 \longrightarrow \text{CH}_2\text{=CHCH}_2\text{O-Mo}^{\text{IV}}\text{-OH}, \quad \text{etc.} \tag{23}$$

Hydrolysis of the Mo(IV) intermediate (these reactions are usually carried out in the presence of water vapor) would afford allyl alcohol. Acrolein may be formed by the subsequent rapid oxidation of allyl alcohol by oxomolybdenum(VI). Alternatively, β-hydrogen elimination from the intermediate could afford acrolein and Mo(II) directly.

$$\text{CH}_2\text{=CHCH}_2\text{O-Mo}^{\text{IV}}\text{-OH} \longrightarrow \text{CH}_2\text{=CHCHO} + \text{Mo}^{\text{II}} + \text{H}_2\text{O} \tag{24}$$

The symmetric allylic intermediate could arise by equilibration resulting from the reversibility of the initial step. Furthermore, the stability of organometallic intermediates at the elevated temperatures employed (>300°C) is likely to be quite limited, and homolytic cleavage to allylic radicals would be facile.

$$\text{Mo}^{\text{VI}}\text{—CH}_2\text{—CH=CH}_2 \rightleftharpoons \text{Mo}^{\text{V}} + \text{CH}_2\overset{\cdot}{\text{CH}}\text{CH}_2 \tag{25}$$

Allyl radicals may also be produced directly from the one-electron oxidation of propylene.[62]

$$\text{OMo}^{\text{VI}} + \text{CH}_2\text{=CHCH}_3 \longrightarrow \text{HOMo}^{\text{V}} + \text{CH}_2\overset{\cdot}{\text{CH}}\text{CH}_2 \tag{26}$$

Further oxidation to allyl alcohol may proceed directly or by a two-step process involving an organometallic intermediate.

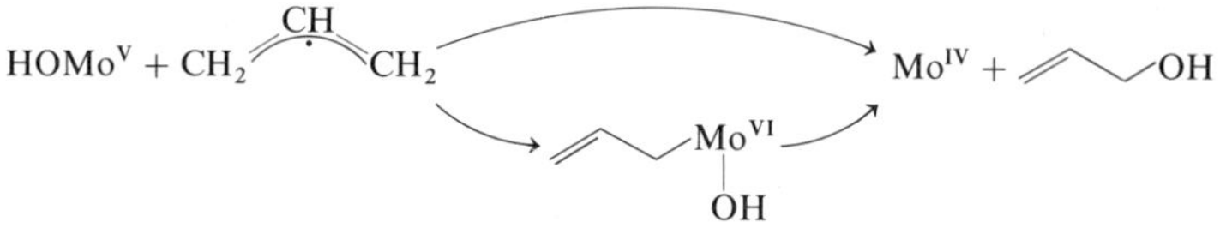

$$HOMo^{V} + CH_2\overset{\dot{C}H}{\cdots}CH_2 \longrightarrow Mo^{IV} + CH_2{=}CHCH_2OH \quad (27)$$

$$\longrightarrow CH_2{=}CHCH_2{-}Mo^{VI}(OH) \longrightarrow \quad (28)$$

In either case, the formation of *free* allyl radicals, as such, does not take place to any significant extent, since this route would lead to nonselective autoxidation via the usual free radical chain mechanism described in Chapter 2.

Dehydrogenation observed with higher olefins results from elimination either of the initially formed organometallic intermediate or of the rearrangement product, as illustrated below for the oxidation of 1-butene to butadiene:

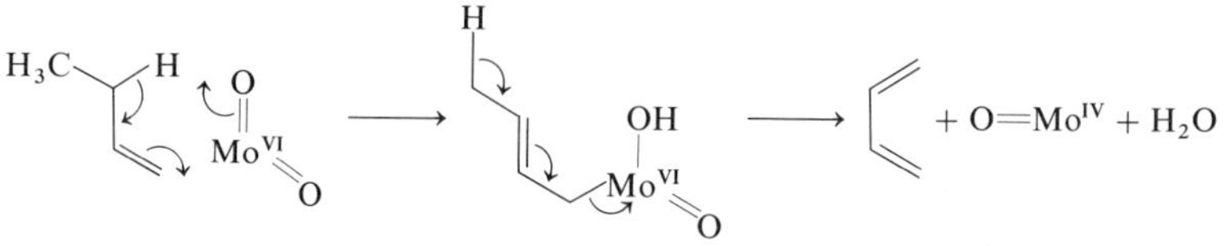

$$H_3C{-}CH(H){-}CH{=}CH_2 + O{=}Mo^{VI}{=}O \longrightarrow CH_2{=}CH{-}CH(CH_2H){-}Mo^{VI}(OH){=}O \longrightarrow CH_2{=}CH{-}CH{=}CH_2 + O{=}Mo^{IV} + H_2O \quad (29)$$

$$CH_3CH{=}CHCH_2{-}Mo^{VI}(=O)OH \longrightarrow CH_2{=}CH{-}CH(CH_2H){-}O{-}Mo^{IV}{-}OH \longrightarrow CH_2{=}CH{-}CH{=}CH_2 + (HO)_2Mo^{IV} \quad (30)$$

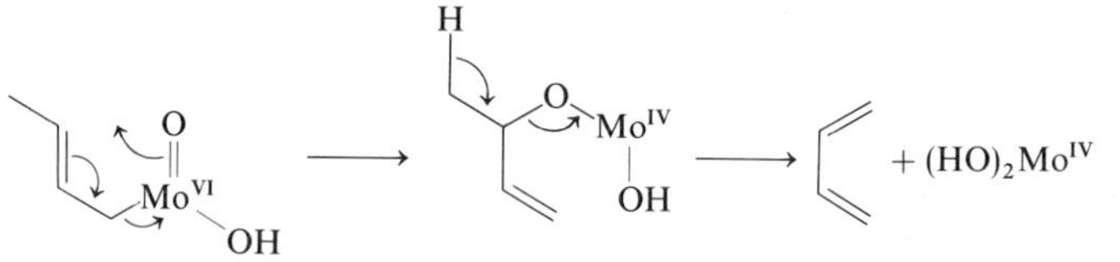

Alternatively, the diene may be formed via dehydration of an allylic alcohol intermediate, the latter being formed via electron transfer and/or organometallic pathways described above. Bismuth molybdate is known to be both an excellent oxidation and dehydration catalyst for the conversion of allyl alcohol and methyl vinyl carbinol to acrolein and butadiene, respectively.[63]

Since the activation energy for the oxidation of propylene in eq 1 is approximately the same as that for ammoxidation to acrylonitrile in eq 2, the same rate-limiting formation of a common allylic intermediate is suggested.[25] Kinetic evidence further suggests that acrolein is not an intermediate in the formation of acrylonitrile.[25] Reasoning by analogy with the imidoselenium(IV) aminating agents (*vide infra*), acrylonitrile could be formed by a 2,3-sigmatropic shift of an organometallic intermediate formed from an imidomolybdenum(VI) reagent.

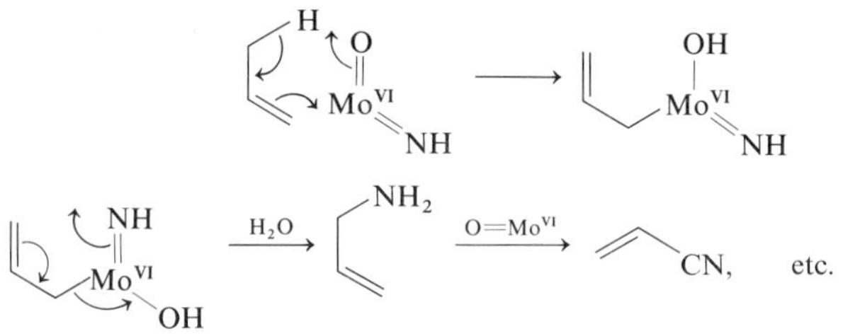

$$CH_2{=}CHCH_2H + O{=}Mo^{VI}{=}NH \longrightarrow CH_2{=}CHCH_2{-}Mo^{VI}(OH){=}NH$$

$$CH_2{=}CHCH_2{-}Mo^{VI}(=NH)OH \xrightarrow{H_2O} CH_2{=}CHCH_2NH_2 \xrightarrow{O=Mo^{VI}} CH_2{=}CHCN, \quad \text{etc.} \quad (31)$$

The detection and isolation of organometallic intermediates under the conditions employed in these reactions are highly unlikely. Thus, the experimental verification of the mechanisms postulated above would be possible only if oxomolybdenum(VI) complexes could be found which were sufficiently electrophilic to react with olefins at moderate temperatures so that the organometallic intermediates could be isolated. In this context, the report that dioxomolybdenum(VI) dialkyldithiocarbamate complexes $(R_2NCS_2)_2MoO_2$ are able to transfer oxygen to phosphines under mild conditions is noteworthy.[64] Oxygen transfer presumably proceeds by prior attack of the nucleophilic substrate at the metal center:

$$Ph_3P + \overset{\overset{O}{\|}}{\underset{\underset{O}{\|}}{Mo^{VI}}} \longrightarrow Ph_3\overset{+}{P}-\overset{\overset{O^-}{|}}{\underset{\underset{O}{\|}}{Mo^{VI}}} \longrightarrow Ph_3\overset{+}{P}-O-\overset{-}{Mo}=O$$

$$Ph_3POMoO \longrightarrow Ph_3PO + O{=}Mo^{IV} \tag{32}$$

However, these complexes are only very weak oxidants which do not react directly with olefins.[65] (In fact, the reduced forms are strong reducing agents capable of transferring oxygen from sulfoxides.[66]) These and analogous oxomolybdenum(VI) complexes[66–69] are of additional interest in relationship to analogous oxidation reactions mediated by molybdoenzymes.[70]

Some model studies for ammoxidation were recently carried out with bisimidotungsten(VI) and bisimidomolybdenum(VI) complexes, $(t\text{-BuN})_2W(O\text{-}t\text{-Bu})_2$ and $(t\text{-BuN})_2Mo(O\text{-}t\text{-Bu})_2$, which upon metathesis with dimethylzinc afforded the bisimidodimethyl derivatives as binuclear complexes, representing the first examples of organotungsten(VI) and organomolybdenum(VI) imides of the type presented in eq 31.[71] Furthermore, indirect evidence for the intramolecular rearrangement of a carbon ligand to an imido functionality,

$$M\begin{matrix}{=}NR\\ \diagdown Ph\end{matrix} \longrightarrow M-N\begin{matrix}\diagup R\\ \diagdown Ph\end{matrix}$$

was obtained in the chromium(VI) analog.

Regeneration of the oxomolybdenum(VI) species in these reactions, and in the catalytic allylic oxidations, probably involves the following sequence of steps:[71a]

$$Mo^{IV} + O_2 \longrightarrow Mo^{V}-O-O\cdot \xrightarrow{Mo^{IV}} Mo^{V}-O-O-Mo^{V}$$

$$Mo^{V}-O-O-Mo^{V} \xrightarrow{\Delta} 2\ O{=}Mo^{VI} \tag{33}$$

The function of the second component in bismuth molybdate catalysts is not clear. For example, bismuth oxide(Bi_2O_3) itself has a low activity and is nonselective for propylene oxidation; MoO_3 is even less reactive but is fairly selective. Remarkable selectivity and activity obtain only upon the combination of the two. This situation is somewhat reminiscent of the catalyst combination consisting of titanium dioxide on silica for epoxidation described in Chapter 3. Perhaps the bismuthyl ligands simply provide the oxomolybdenum(VI) center with the correct stereochemical and electronic environment by increasing its electrophilicity and stabilizing the reactive *cis*-dioxomolybdenum(VI) species.

Methylbenzenes undergo ammoxidation to the corresponding benzonitriles with dioxygen and ammonia over oxometal catalysts.[72]

These processes differ from the selective, more mild oxidation of ketones to oximes, i.e., ammoximation;

$$>C{=}O + NH_3 + \tfrac{1}{2}\,O_2 \longrightarrow\ >C{=}NOH + H_2O$$

which is effectively carried out with porous, amorphous silica and alumina catalysts.[73]

III. OXIDATION AT THE CARBON–CARBON DOUBLE BOND

Attack on the carbon–carbon double bond is typical of strong oxidants (electrophiles) such as osmium and ruthenium tetroxides, permanganate, and chromium(VI). The first three reagents are considered to effect oxidation by way of the cyclic esters, e.g.

$$>C{=}C< \; + \; O_2MO_2 \longrightarrow \begin{matrix} >C-O \\ | \\ >C-O \end{matrix}\!\!>M(=O)_2 \qquad (34)$$

(where M = Os, Ru, Mn), which are either hydrolyzed to glycols (in the case of MnO_4^- and OsO_4) or undergo C—C bond cleavage to carbonyl compounds, e.g.,

$$\begin{matrix} >C-O \\ | \\ >C-O \end{matrix}\!\!>M^{VI}(=O)_2 \xrightarrow{2\,H_2O} >\underset{OH}{C}-\underset{OH}{C}< + (HO)_2M^{VI}(=O)_2 \qquad (35)$$

$$\begin{matrix} >C-O \\ | \\ >C-O \end{matrix}\!\!>M^{VI}(=O)_2 \longrightarrow 2 >C{=}O + M^{IV}(=O)_2 \qquad (36)$$

A. Stoichiometric Oxidations

The interaction of olefins with chromyl compounds [$Cr(VI)O_2X_2$] is notorious for producing complex mixtures of products. However, recent studies by Sharpless and co-workers[32,74] showed that oxidation of olefins with chromyl chloride CrO_2Cl_2 at low temperatures basically affords three primary products: epoxide, chlorohydrin, and vicinal dichloride. It was further established[32] that all three of these products arise from cis-addition processes. The results were explained by a novel mechanism involving attack of the substrate on the chromium center to produce an organometallic intermediate, in contrast to previous mechanisms that invariably invoked attack of the substrate on oxygen. This approach represents a turning point in the understanding of these often complex reactions. The mechanism is depicted in Scheme I.

Scheme I:

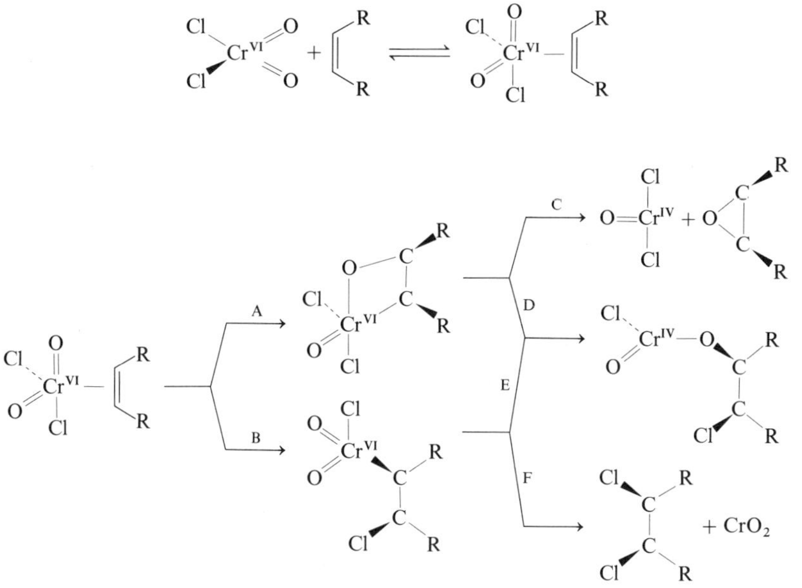

Cl, Cl, Cr^{VI}, O, O + R, R ⇌ Cl, O, Cr^{VI}, O, Cl — R, R

Cl, O, Cr^{VI}, O, Cl — R, R → A: O—C(R), C(R), Cl, Cr^{VI}, O, Cl; B: O, Cl, Cr^{VI}, O, C(R), C(R), Cl

C: O=Cr^{IV}(Cl)(Cl) + O, C(R), C(R)

D: Cl, Cr^{IV}—O, O, C(R), C(R), Cl

E

F: Cl, C(R), C(R), Cl + CrO_2

Pathways A and B, for decomposition of the initially formed π complex, both lead to alkylchromium(VI) intermediates by the insertion of the olefin into a O=Cr(VI) or a ClCr(VI) group, respectively. The former constitutes a [2 + 2]-cycloaddition of an olefin to an oxometal function and has a precedent in the analogous stereospecific cycloaddition of sulfur trioxide to olefins to afford cyclic sultones.[75]

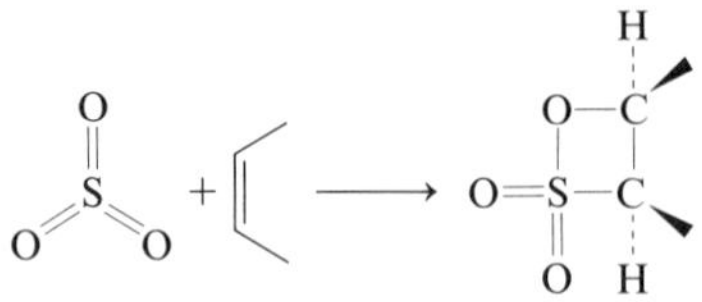

$$O_2S{=}O + \text{olefin} \longrightarrow \text{O=S(=O)–O–C(H)–C(H) (four-membered ring)} \quad (37)$$

It is essentially the microscopic reverse of the olefin-forming step in the well-known Wittig reaction.[76]

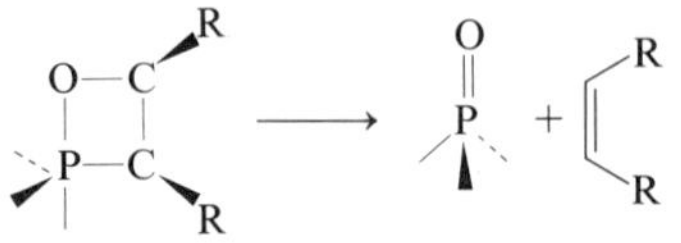

$$\text{P–C–C–O (four-membered ring, R, R)} \longrightarrow O{=}P + RCH{=}CHR \quad (38)$$

The intermediate metallacycle can decompose to give the epoxide (pathway C) or a Cr(IV) complex of the chlorohydrin (pathway D). The former is essentially the reverse of the stereoselective deoxygenation of epoxides by tungsten(IV) reagents.[77]

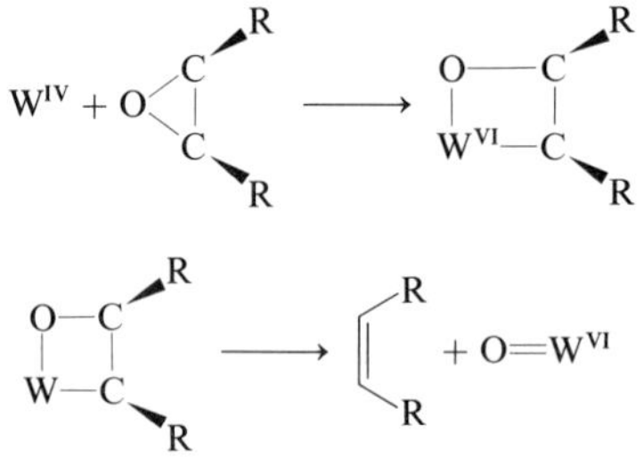

$$W^{IV} + \text{epoxide (R, R)} \longrightarrow \text{O–C–C–}W^{VI}\text{ (four-membered ring)}$$

$$\text{O–C–C–W (four-membered ring)} \longrightarrow RCH{=}CHR + O{=}W^{VI} \quad (39)$$

Chromyl acetate has also been reported[78] to form epoxides from olefins with retention of configuration. Pathway D constitutes reductive elimination of carbon and chlorine ligands, which is the reverse of oxidative addition, a process well known in organometallic chemistry and often proceeding with retention of configuration at carbon. Similarly, pathway F also involves reductive elimination to form the vicinal dichloride. Pathway E in Scheme I represents a 1,2-carbon to oxygen migration of the metal with concomitant two-electron reduction of the metal. (The microscopic reverse process is well known in phosphorus chemistry.) A transition metal analog of this reaction is the rearrangement of a vanadyl [O=V(V)] organometallic derivative,[79] e.g.,

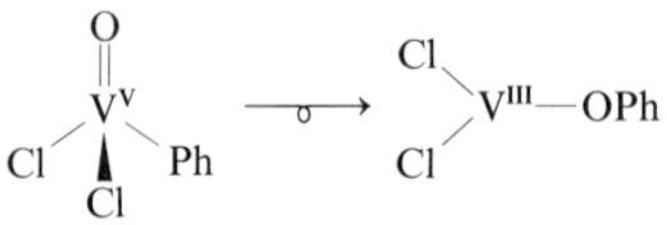

$$O{=}V^{V}(Cl)_2(Cl)Ph \xrightarrow{\sigma} Cl_2V^{III}{-}OPh \quad (40)$$

Sharpless[32] also reported that perrhenyl chloride (ReO_3Cl) reacts with olefins to effect cis-addition of HOCl, whereas the monooxo reagents $CrOCl_3$ and

$MnOCl_3$ afforded only chlorinated products. Cis-dichlorination of olefins with $MoCl_5$[80] was also explained by a mechanism involving successive insertion and reductive elimination.[32]

A few years ago, the proposal of high-valent organometallic intermediates would probably have been rejected out of hand. However, the literature now abounds with examples of stable, high-valent organometallic derivatives of chromium,[81] niobium,[82] tantalum,[82] tungsten,[83] and rhenium.[84] The concept of [2 + 2]-addition to oxometal functions can also be extended to the reactions of other useful reagents such as OsO_4 with olefins. (There is recent spectroscopic evidence for the prior formation of a olefin complex with OsO_4 using diadamantilydene as a nonreacting alkene.[85]) This well-known reaction has long been considered to proceed via a thermally allowed [4 + 2]-cycloaddition process involving attack on oxygen.

(41)

Sharpless[32] proposed an alternative [2 + 2]-cycloaddition to produce an organoosmium(VIII) intermediate, followed by reductive insertion of the Os—C bond into an Os=O bond.

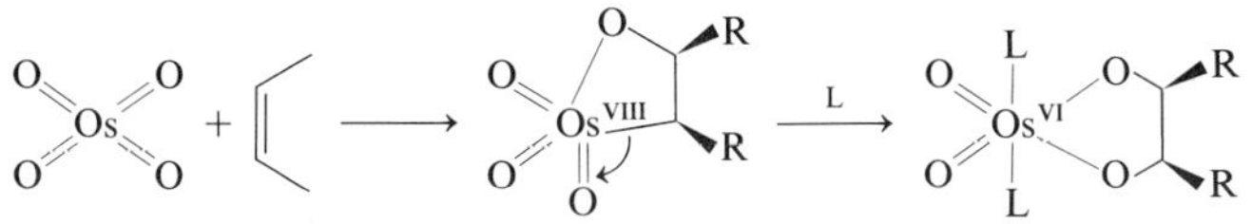

(42)

The latter is facilitated by the coordination of extra ligands (e.g., L = pyridine is often used in conjunction with OsO_4). Indeed, Hengtes and Sharpless have exploited this catalysis in order to induce chirality in moderate optical yields (40–70% ee) during the hydroxylation of various alkenes with OsO_4 using chiral pyridines such as (−)-2-(2-menthyl)pyridine.[86]

Imido analogs of OsO_4 react with olefins to produce β-amino alcohols by a cis-addition.[87,88]

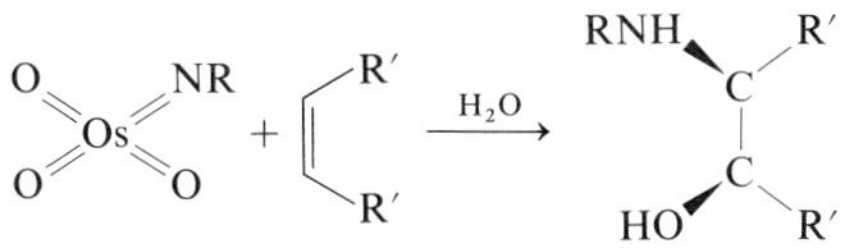

(43)

The oxyamination reaction can be made catalytic by employing chloramine salts of arylsulfonamides ($ArSO_2NClNa$)[87] or carbamates (ROCONClAg)[88] in conjunction with catalytic amounts of OsO_4. A reaction path involving a four-membered organometallacycle intermediate is favored over a [4 + 2]-cycloaddition.[32]

$$OsO_3(=NR) + R'CH=CHR' \longrightarrow \text{[RN–C(R')–C(R')–OsO}_3\text{ metallacycle]} \longrightarrow \text{product} \qquad (44)$$

That these reagents exhibit a strong preference for cycloaddition through the imido group, rather than the oxo group, is further illustrated in the diimidoosmium analog which reacts with olefins to afford 1,2-diamines.[89,89a] An alternative pathway for product formation in the reaction of oxometal groups with double bonds has been mentioned in Chapter 4. Thus, the formation of products arising from the cleavage of the double bond during the rhodium- and silver-catalyzed oxidations with molecular oxygen may result from the following pathways.[90]

$$M{=}O + {>}C{=}C{<} \longrightarrow \text{[M–O / C–C four-membered ring]} \longrightarrow M{=}C{<} + O{=}C{<} \qquad (45)$$

$$M{=}C{<} + O_2 \longrightarrow \text{[M–O / C–O four-membered ring]} \longrightarrow M{=}O + {>}C{=}O \qquad (46)$$

where M = Rh(III), Ag(II), etc.

In the vapor-phase allylic oxidations over heterogeneous oxometal catalysts, such as bismuth molybdate discussed above, a certain fraction of the olefin is completely degraded. The carbon dioxide originates from the competing reaction at the double bond, which probably involves [2 + 2]-cycloaddition as the initial step, followed by the subsequent cleavage depicted in eq 45. This formulation accords with the postulation of metal ylides in the catalyst activation step during the olefin metathesis performed with heterogeneous catalysts such as WO_3, MoO_3, and Re_2O_7 (usually on inorganic supports such as SiO_2 or Al_2O_3), e.g.,[90a]

$$W{=}O + R^1R^2C{=}CR^3R^4 \longrightarrow \text{[W–O / C–C ring]} \longrightarrow W{=}CR^1R^2 + O{=}CR^3R^4 \qquad (47)$$

$$W{=}CR^1R^2 + R^3R^4C{=}CR^1R^2 \longrightarrow \text{[W–C(R}^1\text{)(R}^2\text{) / R}^3\text{(R}^4\text{)C–C(R}^1\text{)(R}^2\text{) ring]} \longrightarrow W{=}CR^3R^4 + R^1R^2C{=}CR^1R^2 \qquad (48)$$

It is now generally accepted that olefin metathesis proceeds via metal ylides and metallacyclobutane intermediates, as in eq 48, a mechanism originally proposed by Herrison and Chauvin.[91] The microscopic reverse process of eqs 45 and 47 is presumably involved in olefin formation from a carbonyl compound and a tantalum(V) ylide,[92]

$$R_3Ta{=}C(H)C(CH_3)_3 + R^1R^2C{=}O \longrightarrow R_3Ta^+{-}CH(C(CH_3)_3){-}C(R^1)(R^2){-}O^-$$

$$R_3Ta{-}CH(C(CH_3)_3){-}C(R^1)(R^2){-}O \text{ (four-membered ring)} \longrightarrow R_3Ta{=}O + R^1R^2C{=}C(H)C(CH_3)_3 \qquad (49)$$

which is the structural analog of the Wittig reagent.

$$R_3P{=}C\langle + O{=}C\langle \longrightarrow R_3P{-}C\langle \,/\, O{-}C\langle \text{ (ring)} \longrightarrow R_3PO + \rangle C{=}C\langle \qquad (50)$$

Similarly, the formation of carbonyl compounds in the reaction of electrophilic chromium ylides with dioxygen[93] may also proceed by a [2 + 2]-cycloaddition analogous to eq 46, as suggested by Mimoun.[90] With ground-state dioxygen, however, such a pathway may proceed either stepwise by a superoxo intermediate,

$$\rangle C{=}M^{n+} + O_2 \longrightarrow \rangle C{=}M^{(n+1)+}{-}O{-}O\cdot \longrightarrow \rangle C{-}M^{n+} \text{ (with } O{-}O \text{ bridge)} \qquad (51)$$

or by a concerted addition, depending on whether the metal ylide is diamagnetic or paramagnetic.

Ab initio methods have been employed to calculate the thermochemistry of the cycloaddition of ethylene to chromyl chloride, shown in the schematic potential energy curve (Fig. 1). Although carbon–carbon bond cleavage is the higher-energy, less favored route with chromyl chloride, the calculations predict that the molybdenum and tungsten analogs undergo competitive reductive elimination (epoxide) and bond cleavage owing to increased σ-bond strengths. This explains why molybdenum and tungsten species, but not chromium, are effective metathesis catalysts.

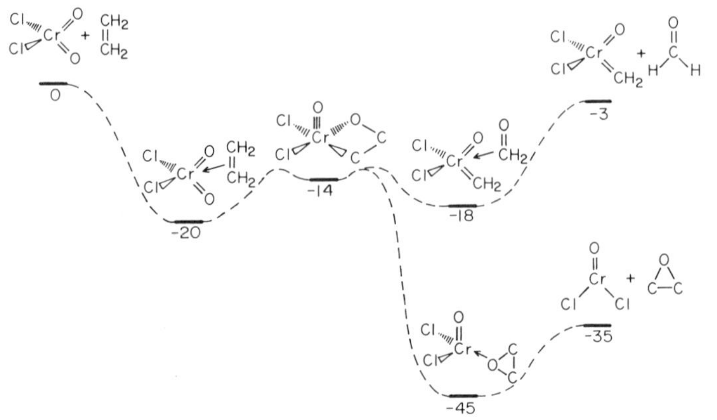

FIG. 1. Schematic potential energy curve for cycloaddition of ethylene to chromyl chloride (energies given in kilocalories per mole). From Rappe and Goddard.[94]

B. Catalytic Processes

With few exceptions, the vapor-phase oxidations of olefins with molecular oxygen over metal oxide catalysts at elevated temperatures lead to complete degradation to CO and CO_2 when the attack occurs at the double bond.[58] However, selective catalytic oxidations of double bonds can be effected when oxometal reagents are employed in conjunction with other oxidants such as peroxides as the active oxidizing agents. Thus, catalytic oxidations employing the combination of OsO_4 and H_2O_2, RO_2H, or amine oxides,[95] and RuO_4 together with sodium hypochlorite[8,9] are well established. Similarly, as noted earlier, selenium(IV) oxidations can be made to be catalytic by employing *tert*-butyl hydroperoxide or chloramine salts in stoichiometric amounts.

IV. ALLYLIC OXIDATION VERSUS DOUBLE BOND ATTACK

Although reagents such as SeO_2 and SO_2 react with olefins at the labile allylic C—H bonds (ene reaction), high-valent oxometal compounds such as O=Cr(VI), O=Os(VIII), and O=S(VI) effect reaction at the double bond (1,2-addition). It has been suggested[96,97] that the mode of attack depends on (1) the degree of covalency or polarization of the M=O bond and (2) the presence of nonbonding electrons on the metal. High-valent oxometal compounds having a polar $M^+—O^-$ bond, and no nonbonding electrons generally effect [2 + 2]-cycloadditions. On the other hand, lower-valent compounds preferentially undergo the ene addition, in which non-bonding electrons play a key role. This distinction is illustrated in the comparison of the oxo or imido complexes of osmium(VIII) with those of selenium(IV) or sulfur(IV) in their interaction with olefins, as discussed above, e.g.,[98]

$$(RN{=})_2M + CH_2{=}CHCH_3 \xrightarrow{M = S^{IV}, Se^{IV}} CH_2{=}CHCH_2NHR \quad (52)$$

$$(RN{=})_2M + CH_2{=}CHCH_3 \xrightarrow{M = Os^{VIII}} RNHCH_2CH(CH_3)NHR \quad (53)$$

Thc effect of nonbonding electrons may be related to the stabilization of a three-membered intermediate or transition state,

(54)

analogous to the perepoxide intermediate postulated in reactions of singlet oxygen with olefins discussed earlier.

(55)

This explanation is not entirely satisfactory since it does not adequately explain allylic oxidation with the high-valent oxometal reagents such as O=Mo(VI) in bismuth molybdate catalysts. Such a mechanistic dichotomy is reminiscent of the situation previously encountered in the reactions of metal salts with π systems in Chapter 5. A feature characteristic of both is the competition between electron transfer with hard oxidants such as cobalt(III) and electrophilic attack with soft electrophiles, giving organometallic intermediates with Tl(III) and Hg(II). If this concept is extrapolated to the oxometal reagents, the hard reagents such as O=Cr(VI), O=Os(VIII), and O=S(VI) are expected to effect electron transfer since their oxidation potentials are sufficient to accept an electron from olefins. For example, the initial electron transfer with chromyl chloride in eq 56 would be followed by the solvent cage combination of the cation-radical with the oxo anion in pathway B or with chloride in pathway A in Scheme II. All of the observed products can be derived from these intermediates, as shown in Scheme II.

Scheme II:

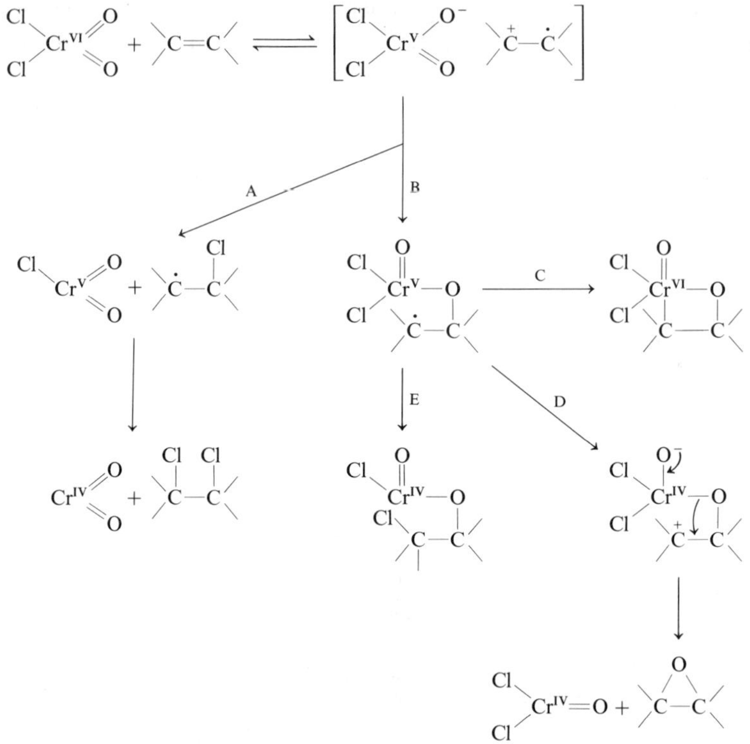

$$\mathrm{Cl_2Cr^{VI}O_2} + \mathrm{>C{=}C<} \rightleftharpoons [\mathrm{Cl_2Cr^{V}(O)O^-} \quad \mathrm{>\overset{+}{C}{-}\dot{C}<}] \tag{56}$$

Pathways: A → $\mathrm{Cl\,Cr^{V}O_2}$ + $\mathrm{>\dot{C}{-}C(Cl)<}$ → $\mathrm{Cr^{IV}O_2}$ + $\mathrm{>C(Cl){-}C(Cl)<}$; B → $\mathrm{Cl_2Cr^{V}(=O){-}O{-}C{<}{-}\dot{C}<}$; C → $\mathrm{Cl_2Cr^{VI}(=O)}$ metallaoxetane; E → $\mathrm{Cl\,Cr^{IV}(=O){-}O}$ with Cl transferred to C; D → $\mathrm{Cl_2Cr^{IV}(O^-){-}O{-}C{-}\overset{+}{C}}$ → $\mathrm{Cl_2Cr^{IV}{=}O}$ + epoxide.

Electron transfer reduction of the alkyl radical by Cr(V) in pathway C leads to the Sharpless intermediate (*vide supra*). However, an electron or ligand transfer oxidation by Cr(V), as in pathways D and E, respectively, would seem to be more likely. According to this formulation, the predominant cis stereochemistry derives from the intramolecular or cage processes stemming from the initially formed ion pair.

In contrast to chromyl chloride, the oxidation potentials of oxidants such as O=Se(IV), O=S(IV), and O=Mo(VI) are not sufficient to effect electron transfer from most olefins. In these cases, electrophilic addition of the metal to the double bond can be concerted with proton transfer from the allylic position to oxygen, i.e.,

$$\mathrm{M(=O)\cdots C{=}C{-}C{-}H} \longrightarrow \mathrm{M(OH){-}C{-}C{=}C} \tag{57}$$

The primary organometallic intermediate should be favored with those oxidants capable of forming carbon–metal bonds.

In summary, the novel concept of including organometallic intermediates such as alkylmetals, metal ylides, and metallacycles in the reactions of oxometal compounds encompasses both stoichiometric processes in solution and catalytic, heterogeneous oxidations.[32,90,96] Its application, mainly by Sharpless and co-workers,[32] has thrown a completely new light on this subject, which will no doubt lead to a flourish of new activity as experimental verification of these concepts are sought.

V. OXOIRON SPECIES IN ENZYMATIC HYDROXYLATION AND EPOXIDATION

Heme-containing monooxygenases, exemplified by the cytochrome *P*-450 group of enzymes, catalyze the hydroxylation and epoxidation of a variety of organic substrates by molecular oxygen in biological systems. Many of these reactions bear a marked resemblance to the stoichiometric oxidations with oxometal reagents such as chromyl compounds.[31] Both classes of oxidations show large deuterium kinetic isotope effects and retention of configuration during the oxidation of C—H bonds. Evidence has accumulated in recent years implicating oxoiron species as the active oxidants in the enzymatic systems.[99]

Fenton's reagent, consisting of a combination of iron(II) and hydrogen peroxide, effects the hydroxylation of organic substrates through the intermediacy of hydroxy radicals, as described in Chapter 3. These reactions have been generally carried out in aqueous solution with water-miscible substrates. Groves and van der Puy[99] have observed pronounced regio- and stereoselectivity in the hydroxylation of cyclohexanol to cyclohexanediols in 90% $CH_3CN–H_2O$ solutions. The major product was the *cis*-1,3-diol, which was formed in greater than 90% stereoselectivity. The active oxidant was considered to be an oxoiron(IV) species, formed either by the rearrangement of an iron(II) hydroperoxide,

$$\mathrm{HO{-}O{-}Fe^{II}} \longrightarrow \mathrm{O{=}Fe^{IV}(OH)} \tag{58}$$

or from the reaction of Fe(III) with hydroxy radical.[100]

$$\mathrm{Fe^{III}} + \mathrm{HO}\cdot \longrightarrow \mathrm{O{=}Fe^{IV}} + \mathrm{H^+} \tag{59}$$

The formation of the *cis*-1,3-diol was formulated as an intramolecular process, as shown below.

$$\text{HO(Fe}^{IV}\text{=O)-cyclohexane-H} \longrightarrow \text{HO(Fe}^{III}\text{—OH)-cyclohexyl} \longrightarrow \text{HO(Fe}^{II}\text{)-cyclohexane-OH} \tag{60}$$

The stereoselective formation of the *cis*-1,3-diol was explained[99] by a directed oxidation of the incipient alkyl radical by the proximate iron(III), involving either a ligand transfer or an electron transfer process. In the latter, competition for the incipient carbonium ion by water could account for the less than 100% stereoselectivity.

Oxoiron(IV) species have also been implicated in the analogous stereoselective hydroxylation of cyclohexanol with a combination of iron(II) and peracids in acetonitrile.[101]

$$\mathrm{RC(=O)O{-}O{-}Fe^{II}} \longrightarrow \mathrm{O{=}Fe^{IV}(O_2CR)} \tag{61}$$

It was shown that the peroxy oxygen was selectively incorporated, the diol showing less than 10% equilibration with water. Hydrogen abstraction, electron transfer, and nucleophilic capture of the incipient carbonium ion must be rapid with respect to molecular rotation to account for these results.

Complexes of iron(IV) are rare, and no well-characterized low-molecular-weight complexes containing the O=Fe(IV) functionality are available. Recently,[102] however, it was found that the peroxo-bridged $\mathrm{PFe^{III}OOFe^{III}P}$ (where P = tetraarylporphyrin), which is derived from the addition of dioxygen to $\mathrm{PFe^{II}}$, yields new complexes, described as ferryl macrocycles, on treatment with nitrogenous bases, such as B = *N*-methylimidizole.

$$\mathrm{2\,PFe^{II} + O_2 \longrightarrow PFe^{III}OOFe^{III}P}$$

$$\mathrm{PFe^{II}OOFe^{III}P + 2B \longrightarrow 2\,P(B)Fe^{IV}O}$$

The ferryl species formed in eq 61 is capable of direct oxygen atom transfer to triphenylphosphine even at −80°C.[102]

$$\mathrm{P(B)Fe^{IV}{=}O + Ph_3P \xrightarrow{B} P(B)_2Fe^{II} + Ph_3P{=}O}$$

In biochemical oxidations mediated by the cytochrome *P*-450 monooxygenases, the active oxidant is thought [103,103a] to be an oxoiron(V) species formed by the reductive activation of dioxygen. Two electrons are supplied by the cofactor (see Chapter 8), and the overall stoichiometry of oxygen activation is consistent with the following steps.

$$\mathrm{XFe^{III} \xrightarrow[H^+]{+e^-} Fe^{II} + HX} \tag{62}$$

$$\mathrm{Fe^{II} + O_2 \longrightarrow {\cdot}O{-}O{-}Fe^{III}} \tag{63}$$

$$\mathrm{{\cdot}O{-}O{-}Fe^{III} \xrightarrow[H^+]{+e^-} HO{-}O{-}Fe^{III}} \tag{64}$$

$$\mathrm{HO{-}O{-}Fe^{III} \xrightarrow{-OH^-} O{=}Fe^{V}(OH)} \tag{65}$$

The active oxidant can also be generated by directly treating the iron(III) enzyme with active oxygen donors such as hydroperoxides, peroxy acids, and iodosobenzene, which circumvent the steps in eqs 62–64,[104] i.e.,

$$XFe^{III} + RO_2H \longrightarrow RO{-}O{-}Fe^{III} + HX \tag{66}$$

$$RO{-}O{-}Fe^{III} \longrightarrow O{=}Fe^{V}(OR) \tag{67}$$

where R = alkyl or acyl. Alternatively,

$$Fe^{III} + PhIO \longrightarrow O{=}Fe^{V} + PhI \tag{68}$$

Groves[103,105] favors a homolytic mechanism for the subsequent hydroxylation of a C—H bond analogous to that outlined above for O=Fe(IV). In contrast, Sharpless[32] prefers an electrophilic substitution to produce an organoiron intermediate. This difference represents another example of the recurring dichotomy between electron transfer and electrophilic substitution for oxidative pathways.

Electron transfer[103,105]

$$\begin{array}{ccc} O{=}Fe^{V} & & \xrightarrow{} [O^{-}{-}Fe^{IV}\ RH^{+\cdot}] \longrightarrow \text{etc.} \\ \updownarrow & \xrightarrow{RH} & \downarrow \\ \cdot O{-}Fe^{IV} & & \xrightarrow{} [HO{-}Fe^{IV}\ R\cdot] \longrightarrow Fe^{III} + ROH \end{array} \tag{69}$$

Electrophilic substitution[32]

$$O{=}Fe^{V} + RH \longrightarrow HO{-}Fe^{V}(R) \longrightarrow ROH + Fe^{III} \tag{70}$$

or:

$$O^{-}{-}Fe^{V}({=}O) + RH \longrightarrow HO{-}Fe^{V}({=}O)(R) \longrightarrow HO{-}Fe^{III}(RO), \quad \text{etc.} \tag{71}$$

(The dioxo formulation in eq 71, although extant in many oxometal reagents, is unlikely to be important for the active oxoiron species in enzymatic systems owing to the insufficiency of coordination sites on iron bound to porphyrin ligands.) Groves and co-workers[103,105] examined the hydroxylation and epoxidation of hydrocarbons by iodosobenzene mediated by chloroiron(III)–porphyrin complexes as model reactions for cytochrome *P*-450 oxygenations. This system converted unactivated, saturated C—H bonds to the corresponding alcohols, and olefins were stereoselectively epoxidized. Although cyclohexane afforded cyclohexanol in only 8% yield, the lack of further oxidation to cyclohexanone was considered to be significant. In addition, a large deuterium kinetic isotope effect, comparable to those observed in permanganate oxidations, was detected. Cyclohexene afforded cyclohexene oxide in 55% yield, and 1,3-cyclohexadiene afforded the monoepoxide in 74% yield.

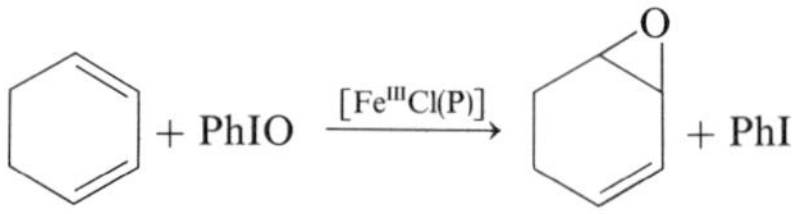

$$\text{1,3-cyclohexadiene} + \text{PhIO} \xrightarrow{[\text{Fe}^{\text{III}}\text{Cl(P)}]} \text{monoepoxide} + \text{PhI} \quad (72)$$

The complete retention of configuration observed in the epoxidation of *cis*- and *trans*-stilbenes is in contrast to the observation of only the *trans*-epoxide from either *cis*- or *trans*-stilbene with another model system, consisting of $Fe(acac)_3$ and H_2O_2.[106]

The stereoselective epoxidation of olefins with chromyl acetate was presented earlier.[78] By analogy with the mechanism in Scheme I, the oxoiron(V) porphyrin intermediate may participate in a 1,2-addition to the olefinic double bond.

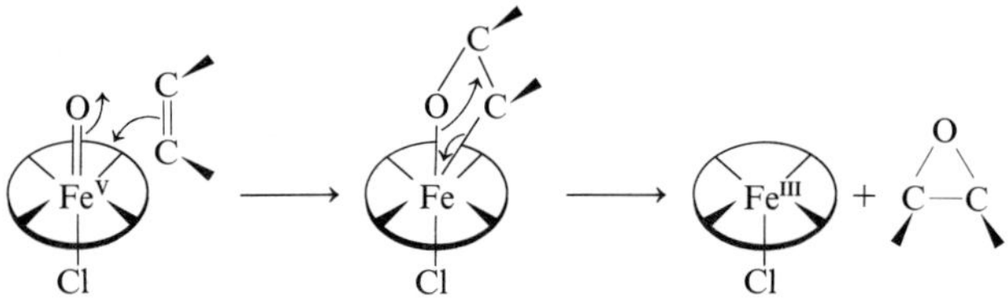

$$\text{Fe}^{\text{V}}(=\text{O})\text{Cl} + \text{C}=\text{C} \longrightarrow \text{Fe} \longrightarrow \text{Fe}^{\text{III}}\text{Cl} + \text{C}-\text{C (O)} \quad (73)$$

This process is essentially the microscopic reverse of the stereoselective deoxygenation of olefins by W(IV) complexes described in reaction 39.[77] [Addition can be facilitated by the dissociation of one of the porphyrin nitrogen ligands or chloride to reduce steric crowding in the seven-coordinate (17-electron) intermediate.]

The corresponding electron transfer mechanism is depicted below.

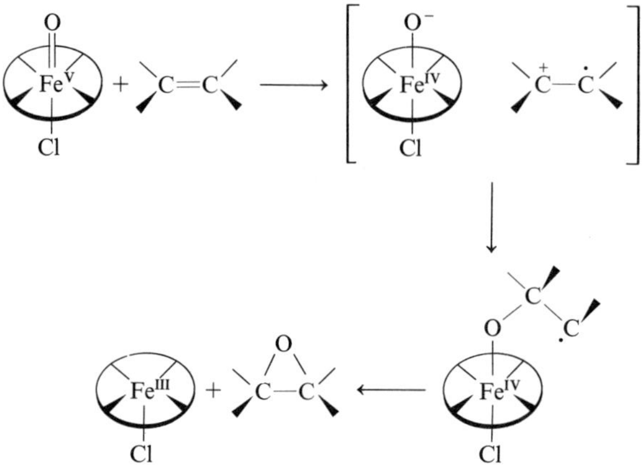

$$\text{Fe}^{\text{V}}(=\text{O})\text{Cl} + \text{C}=\text{C} \longrightarrow \left[\text{Fe}^{\text{IV}}(\text{O}^-)\text{Cl} \quad \overset{+}{\text{C}}-\dot{\text{C}} \right] \longrightarrow \text{Fe}^{\text{IV}}(\text{O}-\text{C}-\dot{\text{C}})\text{Cl} \longrightarrow \text{Fe}^{\text{III}}\text{Cl} + \text{C}-\text{C (O)} \quad (74)$$

Retention of configuration would arise if oxygen transfer were faster than molecular rotation. Oxoiron intermediates have also been implicated in the iron(III) porphyrin-catalyzed autoxidation of cyclohexene.[107]

An analogous oxochromium(V) complex has been generated by the treatment of chloro(tetraphenylporphyrinato)chromium(III), P(Cl)Cr^{III}, with either iodosylbenzene or *m*-chloroperbenzoic acid.[108] It is noteworthy that the oxo ligand in this complex is labile to ^{18}O exchange with water and transferable to various olefins to form epoxides in moderate to excellent yields, as outlined in Scheme III.

Scheme III:

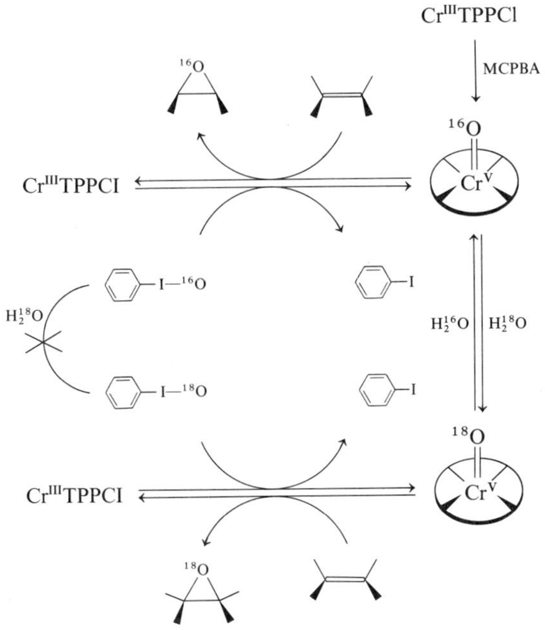

Cr^{III}TPPCl

MCPBA

^{16}O

Cr^{V}

Cr^{III}TPPCI

$H_2^{18}O$

PhI—^{16}O → PhI

PhI—^{18}O → PhI

$H_2^{16}O$ | $H_2^{18}O$

^{18}O

Cr^{V}

Cr^{III}TPPCI

Chlorination and hydroxylation of alkanes can also be effected with chloro(tetraphenylporphyrinato)manganese(III) in the presence of iodosobenzene,[109] e.g.,

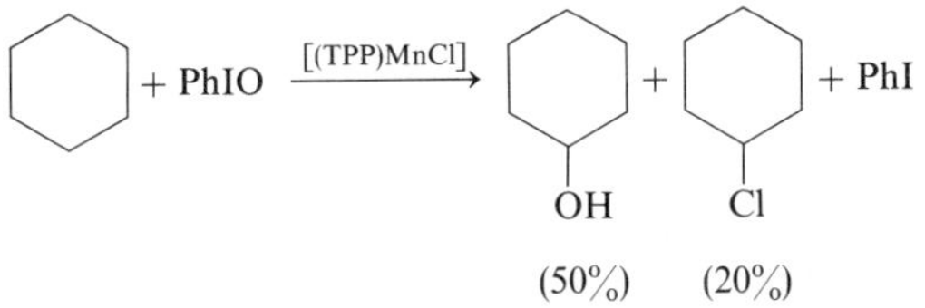

$$\text{cyclohexane} + \text{PhIO} \xrightarrow{[(\text{TPP})\text{MnCl}]} \text{cyclohexanol (OH)} + \text{chlorocyclohexane (Cl)} + \text{PhI}$$

(50%) (20%)

The active oxidant appears to be an oxo(TPP)Mn(V) complex with an occluded iodobenzene in which there is evidence for facile ^{18}O exchange with water similar to the chromium analog in Scheme III.

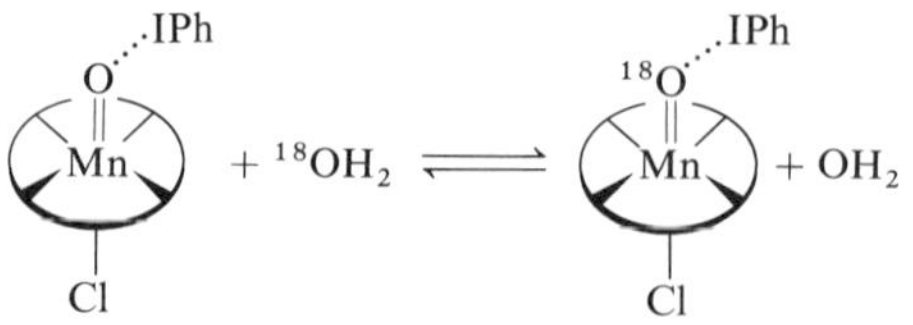

$$\text{O(}\cdots\text{IPh)=Mn–Cl} + {}^{18}OH_2 \rightleftharpoons {}^{18}\text{O(}\cdots\text{IPh)=Mn–Cl} + OH_2$$

It has been suggested that the hydroxylation and chlorination of alkanes by the oxomanganese(V) species proceeds by a stepwise, homolytic process,

$$O{=}Mn^{V}Cl + RH \longrightarrow [HOMn^{IV}ClR\cdot]$$

$$[HOMn^{IV}ClR\cdot] \longrightarrow HOMn^{III} + RCl$$

$$[HOMn^{IV}ClR\cdot] \longrightarrow ClMn^{III} + ROH$$

in which ligand transfer above may occur either by homolytic displacement or via an alkylmanganese(V) intermediate. The same oxomanganese(V) species is capable of epoxidizing olefins, but with loss of stereoselectivity. The lack of stereospecificity, which contrasts with the results of the closely related iron and chromium porphyrins discussed above, was accounted for by an addition of the oxomanganese(V) species to the double bond in a stepwise fashion to form a freely rotating free radical intermediate, as shown in Scheme IV.

Scheme IV:

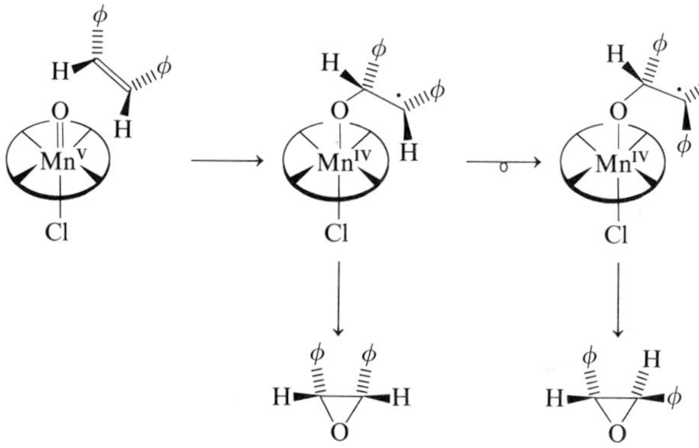

$$\text{O=Mn}^{V}\text{–Cl} + cis\text{-}\phi CH{=}CH\phi \longrightarrow \text{Mn}^{IV}\text{–O–CH}\phi\text{–}\dot{C}H\phi \xrightarrow{\circ} \text{Mn}^{IV}\text{–O–CH}\phi\text{–}\dot{C}H\phi\ (\text{rotated})$$

$$\downarrow \quad cis\text{-epoxide} \qquad \downarrow \quad trans\text{-epoxide}$$

Iron-containing monooxygenases are also involved in the hydroxylation of aromatic rings. An unusual characteristic of these reactions is the so-called

NIH shift, involving an intramolecular migration of the group displaced by hydroxyl to an adjacent nuclear position (see Chapter 8),

$$p\text{-}R\text{-}C_6H_4\text{-}X + O_2 \xrightarrow{[\text{enzyme}]} R\text{-}C_6H_3(X)\text{-}OH \qquad (75)$$

where X = D, Cl, alkyl, etc. Since Sharpless and Flood[31] showed that aromatic oxidations with chromyl reagents CrO_2X_2 also exhibit an NIH shift, an oxoiron species may play a role in these enzymatic systems. Lindsay-Smith and coworkers[110] have recently reported that the hydroxylation of deuterated aromatic hydrocarbons with Fenton's reagent in dipolar aprotic solvents produces NIH shifts of deuterium comparable to those observed with cytochrome *P*-450 enzymes. The magnitude of the NIH shift decreased significantly with increasing amounts of water in the solvent. The results were rationalized on the basis of competing intramolecular rearrangement and hydrolysis of an intermediate cyclohexadienyl cation.

$$X\text{-}C_6H_4\text{-}D + YO\cdot \xrightarrow{-e^-} [\text{cyclohexadienyl cation, X; YO, D}]^+ \xrightarrow{\text{NIH shift}} [\text{cyclohexadienone, X; H, D}] \longrightarrow X\text{-}C_6H_3(D)\text{-}OH \qquad (76)$$

$$[\text{cyclohexadienyl cation}]^+ \xrightarrow{H_2O} X\text{-}C_6H_3(H)\text{-}OH + DH_2O^+ \qquad (77)$$

The attacking species YO· may be either a hydroxyl radical or a ferryl species. No doubt, forthcoming studies in this exciting area will establish the precise role played by oxoiron species in hydroxylations and epoxidations with many iron-based oxidants.

VI. OXIDATIONS OF ALCOHOLS

Stoichiometric oxidations of alcohols to carbonyl compounds with oxochromium(VI),[2] manganese(VII),[1] and vanadium(V)[111] reagents are well known. These reactions probably involve the formation of an inner-sphere alkoxymetal intermediate that can decompose by either a homolytic,

$$\underset{\text{H}}{\overset{}{>\!\!\!\!\diagup\text{C}}}\text{—O—M}^{n+} \longrightarrow \underset{\text{H}}{>\!\!\!\!\diagup\text{C}}\text{—O}\cdot + \text{M}^{(n-1)+} \xrightarrow{-\text{H}^+} >\!\text{C=O} + \text{M}^{(n-2)+} \qquad (78)$$

or a heterolytic pathway.

$$\underset{\text{H}}{>\!\text{C}}\text{—O—M}^{n+} \xrightarrow{-\text{H}^+} >\!\text{C=O} + \text{M}^{(n-2)+} \qquad (79)$$

A deuterium kinetic isotope effect is expected for a concerted heterolytic process.

High selectivities to propionaldehyde have been observed[112] in the V_2O_5-catalyzed, vapor-phase oxidation of *n*-propanol with molecular oxygen in the temperature range 170°–250°C. Selectivities ranging from 90 to 97% were claimed with V_2O_5 modified with alkaline earth metal oxides. Oxidation of the alcohol by O=V(V) is followed by the regeneration of the oxidant by molecular oxygen,

$$\text{O=}\overset{\text{O}}{\overset{\|}{\text{V}^{\text{V}}}} + >\!\text{CHOH} \longrightarrow >\!\text{CHO—}\underset{\text{OH}}{\overset{\text{O}}{\overset{\|}{\text{V}^{\text{V}}}}} \longrightarrow >\!\text{C=O} + (\text{HO})_2\overset{\text{O}}{\overset{\|}{\text{V}^{\text{III}}}}$$

$$(\text{HO})_2\overset{\text{O}}{\overset{\|}{\text{V}^{\text{III}}}} + \tfrac{1}{2}\text{O}_2 \longrightarrow \text{O=}\overset{\text{O}}{\overset{\|}{\text{V}^{\text{V}}}} + \text{H}_2\text{O} \qquad (80)$$

The catalytic oxidation of secondary alcohols to ketones was also observed with the reagent consisting of $VO(acac)_2$ and TBHP in a study of the epoxidation of cyclic allylic alcohols.[113]

The catalytic oxidation of alcohols by oxometal reagents is implicated in the vapor-phase oxidation of olefins over a mixture of SnO_2 and MoO_3. Oxidation of propylene over this catalyst afforded acetone, in contrast to the formation of acrolein with bismuth molybdate, as discussed above. Isotopic labeling studies[114] demonstrated that the oxygen is derived from water by what is referred to as *oxyhydration* of the olefin.

$$\text{CH}_2\text{=CHCH}_3 + \text{H}_2\text{O} \rightleftharpoons (\text{CH}_3)_2\text{CHOH} \xrightarrow[\text{[SnO}_2\text{–MoO}_3\text{]}]{\text{O}_2} (\text{CH}_3)_2\text{C=O} + \text{H}_2\text{O} \qquad (81)$$

VII. PHASE TRANSFER CATALYSIS IN OXIDATIONS WITH OXOMETAL REAGENTS

Oxidations of hydrocarbons with oxometal reagents such as permanganate and dichromate often present problems owing to (1) the insolubility of the substrate in water and (2) the difficulty of finding a solvent that is not attacked by the oxidant. These reactions are often characterized by low yields and/or requirements for significantly greater than stoichiometric amounts of oxidant. In recent years phase transfer catalysis techniques[115,116] have been increasingly used to effect solubilization of these oxidants in relatively nonpolar solvents such as benzene, methylene chloride, and chloroform. In this way, considerable success has been achieved in promoting selective oxidations of a variety of organic substrates under mild conditions. Both stoichiometric and catalytic amounts of phase transfer agents, such as tetraalkylammonium and tetraalkylphosphonium salts and crown ethers, have been used to effect solubilization of a variety of oxometal anion reagents, such as $KMnO_4$,[117–129] $K_2Cr_2O_7$,[130–132] and K_2CrO_4[133,134] (see Table I).[135–137] In the presence of the phase transfer agent, the oxometal anion is transported into the organic phase as the tetraalkylammonium or phosphonium salt or as the crown ether complex. (For the principles of phase transfer catalysis see Weber and Gokel[115] and Starks and Liotta[116].)

The first example of a phase transfer catalytic oxidation (i.e., catalytic in the phase transfer agent) was reported by Gibson and Hosking,[117] who found that 1-octene, 1-propanol and 2-propanol, and 4-heptanol were selectively oxidized in a two-phase system consisting of chloroform and water in the presence of either stoichiometric or catalytic amounts of methyltriphenylarsonium chloride. Although no reaction was observed at room temperature between aqueous, neutral $KMnO_4$ and 1-octene, Starks[118] found that the addition of a catalytic amount of a quaternary ammonium salt, such as Aliquat 336, led to a smooth conversion to heptanoic acid. Similarly, 1-decene afforded nonanoic acid in 91% yield.

$$RCH{=}CH_2 \xrightarrow[R_4'NCl]{KMnO_4} RCO_2H + HCO_2H, \quad \text{etc.} \tag{82}$$

Under alkaline conditions Weber and Shepherd[121] obtained a 50% yield of *cis*-1,2-cyclooctanediol from cyclooctene, compared to a 7% yield obtained by the conventional method. If the glycol is appreciably soluble in water, overoxidation to cleavage products results. Foglia and co-workers[125] studied the phase transfer oxidation of long-chain olefins with $KMnO_4$ in CH_2Cl_2 and found that either *cis*-1,2-diols or carboxylic acids were produced,

TABLE I Solubilization of Inorganic Oxidants in Organic Solvents by Phase Transfer Agents

Anion	Phase transfer agent	Amount	Solvent	Substrates oxidized	Ref.
MnO_4^-	$CH_3(Ph)_3AsCl$	Catalytic	$CHCl_3$	Alkenes, alcohols	117
	Aliquat 336[a]	Catalytic	Benzene	Alkenes, alcohols	118, 119
	$(C_{12}H_{25})_3N$	Catalytic	—	Alkenes	120
	$PhCH_2(Et)_3NCl$	Catalytic	CH_2Cl_2	Alkenes	121
	$(C_{16}H_{33})(PhCH_2)Me_2NCl$	Catalytic	Benzene–acetic acid	Alkenes	122
	Bu_4PBr, Bu_4NBr, etc.	Catalytic	CH_2Cl_2	Alkenes	123
	Bu_4NBr	Stoichiometric	Pyridine	Alkenes, alcohols, aldehydes, alkylbenzenes	124
	Bu_4NBr, 18-crown-6	Catalytic	CH_2Cl_2	Alkenes	125
	$PhCH_2(Et)_3NCl$	Stoichiometric	CH_2Cl_2	Alkanes, cycloalkanes, alkylbenzenes	126
	$PhCH_2(Et)_3NCl$	Stoichiometric	CH_2Cl_2	Ethers	127
	Dicyclohexyl-18-crown-6	Stoichiometric	Benzene	Alkenes, alcohols, aldehydes, alkylbenzenes	128
	Aliquat 336[a]	Catalytic	—	Alkene	129
$Cr_2O_7^{2-}$	Adogen 464[b]	—	Benzene	Primary and secondary alcohols	130
	Bu_4NHSO_4	Catalytic	CH_2Cl_2	Primary alcohols	131
	Bu_4NHSO_4	Stoichiometric	$CHCl_3$	Alkyl halides	132
	Bu_4NHSO_4	Catalytic	$CHCl_3$, CH_2Cl_2	Primary alcohols	133
	Crown ethers	Catalytic	HMPA[c]	Alkyl halides	134
ClO^-	Bu_4NHSO_4	Catalytic	CH_2Cl_2, $CHCl_3$, benzene, EtOAc	Alcohols, amines	135
	Bu_4NHSO_4	Catalytic	$CHCl_3$	Polycyclic aromatics	136
	R_4NX, R_4PX	Catalytic	Esters	Alcohols, amines, aldehydes, etc.	137
IO_4^-	Bu_4PCl, Bu_4NBr, etc.	Stoichiometric	CH_2Cl_2, benzene	—	123

[a] Aliquat 336, a mixture of trialkyl(C_8–C_{12})methylammonium chlorides.
[b] Adogen 464, a mixture of trialkyl(C_6–C_{10})methylammonium chlorides.
[c] HMPA, hexamethylphosphorus triamide.

depending on the pH of the medium (this is also true in the absence of phase transfer agents).

$$\text{RHC=CHR} \xrightarrow[\text{Bu}_4\text{NBr}]{\text{KMnO}_4\text{-NaOH}} \text{R(H)(HO)C—C(OH)(H)R} \quad (80\%) \qquad (83)$$

$$\text{RHC=CHR} \xrightarrow[\text{Bu}_4\text{NBr}]{\text{KMnO}_4} 2\,\text{R—CO}_2\text{H} \quad (80\%) \qquad (84)$$

where $R = CH_3(CH_2)_7$. Similarly, Okimoto and Swern[123] observed stereospecific hydroxylation of oleyl and elaidyl alcohols with cold, alkaline $KMnO_4$ under phase transfer conditions. The presence of acetic acid was shown to facilitate the production of carboxylic acids.[122] In acidic solutions, trialkylamines also function as phase transfer agents for permanganate,[120] presumably in the form $R_3NH^+MnO_4^-$. Potassium permanganate is readily solubilized in benzene by complexation with crown ethers[128] or with acyclic polyethers.[138] The $KMnO_4$ complex of dicyclohexyl-18-crown-6 constitutes a mild, efficient oxidant for a variety of substrates.[128] For example, α-pinene afforded *cis*-pinonic acid in 90% yield.[128]

$$\alpha\text{-pinene} \xrightarrow[\text{dicyclohexyl-18-crown-6}]{\text{KMnO}_4\text{-C}_6\text{H}_6} \textit{cis}\text{-pinonic acid (CO}_2\text{H)} \qquad (85)$$

It was suggested[128] that oxidative cleavage involves a thermally allowed [2 + 4]-cycloaddition, followed by electron transfer and thermally allowed [2 + 2 + 2]-chelotropic elimination.

$$\text{RCH=CHR} + \text{MnO}_4^- \longrightarrow \text{(RC—O)(RC—O)Mn(=O)O}^- \xrightarrow{\text{MnO}_4^-} \text{(RC—O)(RC—O)Mn(=O)}_2$$

$$\text{(RC—O)(RC—O)Mn(=O)}_2 \xrightarrow{-\text{MnO}_2} 2\,\text{RCHO} \xrightarrow{\text{MnO}_4^-} 2\,\text{RCO}_2^- \qquad (86)$$

Lee and Chang[138] compared Adogen 464, dicyclohexyl-18-crown-6, and an acylic polyether(dimethylpolyethylene glycol) as phase transfer agents in alkene oxidations with $KMnO_4$. Yields and rates obtained with the latter compared well with the former two.

The permanganate ion solubilized in CH_2Cl_2 with triethylbenzylammonium ion allowed the direct spectrophotometric detection of the stable organomanganese intermediate derived by addition to dicyclopentadiene (see eq 86).[139]

The selective oxidation of primary alcohols to aldehydes with acidic K_2CrO_4[133] and $K_2Cr_2O_7$[131] has been reported under phase transfer conditions. Bis(tetrabutylammonium) dichromate in chloroform was used for the selective conversion of alkyl halides to carbonyl compounds.[132]

$$\gt CHX + (Bu_4N)_2Cr_2O_7 \longrightarrow \gt C{=}O, \quad \text{etc.} \tag{87}$$

Solubilization of neutral $K_2Cr_2O_7$ was achieved only with Adogen 464 in stoichiometric amounts.[130] The reagent was used for the selective oxidation of primary alcohols to aldehydes.

Hypochlorite ion is an effective oxidant for a variety of substrates under phase transfer conditions.[135] For example, tetrabutylammonium hydrogen sulfate catalyzes the selective oxidation of alcohols to aldehydes or ketones. Primary and secondary amines are oxidized to nitriles and ketones, respectively.

$$RCH_2NH_2 \xrightarrow[Bu_4N\text{-}HSO_4]{NaOCl} [R{-}\overset{H}{\overset{|}{C}}{=}NCl] \xrightarrow{-HCl} RC{\equiv}N \tag{88}$$

$$R_2CHNH_2 \xrightarrow[Bu_4N\text{-}HSO_4]{NaOCl} R_2C{=}NCl \xrightarrow[\text{(separate step)}]{H_2O} R_2C{=}O \tag{89}$$

Although methylene chloride, chloroform, carbon tetrachloride, and benzene are suitable solvents, ethyl acetate is the solvent of choice for these oxidations. Hypochlorite ion under phase transfer conditions has also been employed for the unusual conversion of certain polycyclic aromatic hydrocarbons to the corresponding arene oxides.[136]

The use of catalytic amounts of expensive oxometal reagents, such as OsO_4[7] and RuO_4,[8,9] in conjunction with oxygen donors, such as H_2O_2 and NaOCl, as primary oxidants is well established. The cis-hydroxylation of olefins with OsO_4 is generally more selective than that effected by alkaline permanganate. Ruthenium tetroxide RuO_4, being a stronger oxidant, effects oxidative cleavage of the double bond.

$$\gt C{=}C \lt \xrightarrow[H_2O_2]{[OsO_4]} \gt \underset{}{\overset{OH}{C}}{-}\overset{OH}{C} \lt \tag{90}$$

$$\gt C{=}C \lt \xrightarrow[NaOCl]{[RuO_4]} \gt C{=}O + O{=}C \lt \tag{91}$$

These reactions are generally considered to proceed by a thermally allowed [4 + 2]-cycloaddition. (However, see the earlier discussion on p. 165).

$$\text{C=C} + O_2M^{VIII}O_2 \xrightarrow{M = Os, Ru} \text{C(O)C(O)}M^{VI}O_2 \xrightarrow{M = Ru} 2\ \rangle C{=}O + M^{IV}O_2 \qquad (92)$$

The use of catalytic amounts of either RuO_2 or $RuCl_3$ in conjunction with NaOCl under phase transfer conditions has been reported to give excellent yields of cleavage products,[125] e.g.,

$$CH_3(CH_2)_{12}CH{=}CH_2 \xrightarrow[\text{NaOH, Bu}_4\text{NBr, CH}_2\text{Cl}_2]{[\text{RuO}_2],\ \text{NaOCl}} \underset{(100\%)}{CH_3(CH_2)_{12}CO_2H} \qquad (93)$$

Starks and co-workers[116,120] used OsO_4 and RuO_4 in combination with periodic acid under phase transfer conditions for the oxidation of olefins. Selective bishydroxylation is observed with both reagents. However with RuO_4, the initially formed 1,2-diol is further oxidized by periodate to aldehydes, which under carefully controlled conditions can be isolated as the major products.

Since a phase transfer agent is not necessary with water-soluble olefins, Foglia and co-workers could effect selective hydroxylation and oxidative cleavage of the alkali metal salts of long chain, unsaturated carboxylic acids with the combinations OsO_4–NaOCl and RuO_4–NaOCl, respectively.[140] The reactions were carried out under alkaline conditions in order to circumvent the formation of chlorinated products.

Recently, a new catalytic route for the epoxidation of styrene with sodium hypochlorite under PTC conditions has been reported.[140a]

$$PhCH{-}CH_2 \xrightarrow[R_4N^+\,Cl^-]{\text{NaOCl, [MnTPP]}} PhHC\overset{O}{—}CH_2$$

An oxomanganese(V) species was proposed as the active oxidant, which is formed by hypochlorite oxidation, e.g.,

$$LMn^{III}OAc + NaOCl \longrightarrow L\overset{O}{\overset{\|}{Mn^V}}OAc + NaCl$$

The recent developments of phase-transfer-catalyzed oxidations with oxometal reagents, coupled with the new mechanistic developments discussed earlier, have spurred a revival of interest in the reactions of these traditional oxidants. Doubtlessly, phase-transfer catalysis will be extended to a variety of other inorganic oxidants in the future. In this context, the recent report[141] of selective oxidations with potassium ferrate (K_2FeO_4), easily prepared by reaction of ferric nitrate with sodium hypochlorite, is noteworthy, especially in connection with the current interest in ferryl oxidants.

REFERENCES

1. R. Stewart, *in* "Oxidation in Organic Chemistry" (K. B. Wiberg, ed.), Part A, p.2. Academic Press, New York, 1965.
2. K. B. Wiberg, *in* "Oxidation in Organic Chemistry" (K. B. Wiberg, ed.), Part A, p. 69. Academic Press, New York, 1965.
3. H. J. Reich, *in* "Oxidation in Organic Chemistry" (W. S. Trahanovsky, ed.), Part C, p. 1. Academic Press, New York, 1978.
4. E. N. Trachtenberg, *in* "Oxidation" (R. L. Augustine, ed.), Vol. 1, Ch. 3. Dekker, New York, 1969.
5. N. Rabjohn, *Org. React.* (*N.Y.*) **24**, 261 (1978).
6. R. A. Jerussi, *in* "Selective Organic Transformations" (B. S. Thyagarajan, ed.), p. 301. Wiley, New York, 1970.
7. M. Schroeder, *Chem. Rev.* **80**, 187 (1980).
8. D. G. Lee and M. van der Engh, *in* "Oxidation in Organic Chemistry" Part B, (W. S. Trahanovsky, ed.), Ch. 4. Academic Press, New York, 1978.
9. P. N. Rylander, *Engelhard Ind., Tech. Bul.* **9**, 135 (1969); see also M. Schroder and W. P. Griffith, *J. Chem. Soc., Chem. Commun.* p. 58 (1979).
10. A. J. Fatiadi, *Synthesis* p. 65 (1976); J. S. Pizey, "Synthetic Reagents," Vol. 2, pp. 143–174. Wiley, New York, 1974.
11. B. C. Gates, J. R. Katzer, and G. C. A. Schuit, "Chemistry of Catalytic Processes," Ch. 4. McGraw-Hill, New York, 1979.
12. (a) G. W. Keulks, L. D. Krenzke, and T. M. Notermann, *Adv. Catal.* **27**, 183 (1978). (b) D. B. Dabyburjov, S. S. Jewur, and E. Ruckenstein, *Catal. Rev.* **19**, 293 (1979); (c) J. Haber, *Pure Appl. Chem.* **50**, 923 (1978).
13. G. W. Keulks, *in* "Catalysis in Organic Synthesis" (G. V. Smith, ed.), p. 109. Academic Press, New York, 1977.
14. G. C. A. Schuit, *Chim. Ind.* (*Milan*) **51**, 1307 (1969).
15. N. Giordano, *Chim. Ind.* (*Milan*) 53, 366 (1971); M. M. Rogic and D. Masilamani, *J.* Am. *Chem. Soc.* **99**, 5219 (1977); **100**, 4634 (1979).
16. S. Carra, *Chim. Ind.* (*Milan*) **53**, 366 (1971).
17. S. Carra, R. Ugo, and L. Zanderighi, *Inorg. Chim. Acta, Rev.* **2**, 55 (1969).
18. R. J. Sampson and D. Shooter, *Oxid. Combust. Rev.* **1**, 225 (1965).
19. L. Y. Margolis, *Adv. Catal.* **14**, 429 (1963).
20. H. H. Voge and C. R. Adams, *Adv. Catal.* **17**, 151 (1967).
21. H. H. Voge, *Adv. Chem. Series* **76**, 242 (1968).
22. W. M. H. Sachtler, *Catal. Rev.* **4**, 27 (1970).
23. C. R. Adams, *Ind. Eng. Chem.* **61**, 30 (1969); G. Simon and J. E. Germain, *Bul. Soc. Chim. France* **1**, 149 (1980).
24. D. A. Dowden, *Chem. Eng. Prog., Symp. Series* **63**, 73 (1967).
25. J. L. Callahan, R. K. Grasselli, E. C. Milberger, and H. A. Strecker, *Ind. Eng. Chem. Prod. Rsch. Dev.* **9**, 134 (1970).

26. J. L. Callahan and R. K. Grasselli, *AIChE J.* **9,** 755 (1963).
27. D. J. Hucknall, "Selective Oxidation of Hydrocarbons." Academic Press, New York, 1974.
28. T. Dumas and W. Bulani, "Oxidation of Petrochemicals: Chemistry and Technology." Appl. Sci. Pub., London, 1974.
29. P. Mars and D. W. van Krevelen, *Chem. Eng. Sci., Spec. Suppl.* **3**, 41 (1954).
30. P. A. Batist, C. J. Kapteijns, B. C. Lippens, and G. C. A. Schuit, *J. Catal.* **7**, 33 (1967).
31. K. B. Sharpless and T. C. Flood, *J. Am. Chem. Soc.* **93**, 2316 (1971).
32. K. B. Sharpless, A. Y. Teranishi, and J. E. Bäckvall, *J. Am. Chem. Soc.* **99**, 3120 (1977).
33. K. B. Wiberg and G. Foster, *J. Am. Chem. Soc.* **83**, 423 (1961).
34. R. H. Eastman and R. A. Quinn, *J. Am. Chem. Soc.* **82**, 4249 (1960).
35. J. P. Engström and F. D. Greene, *J. Org. Chem.* **37**, 968 (1972).
36. T. Koenig and J. M. Owens, *J. Am. Chem. Soc.* **96**, 4054 (1974); **95**, 8485 (1973).
37. W. Adam, *Chem. Ztg.* **99**, 142 (1975); L. M. Stephenson, M. J. Grdina and M. Orfanopolous, *Acc. Chem. Res.* **13**, 419 (1980); L. M. Stephenson, *Tetrahedron Lett.* **21**, 1005 (1980).
38. K. B. Sharpless and R. F. Lauer, *J. Am. Chem. Soc.* **94**, 7154 (1972).
39. H. P. Jensen and K. B. Sharpless, *J. Org. Chem.* **40**, 264 (1975).
40. D. Arigoni, A. Vasella, K. B. Sharpless, and H. P. Jensen, *J. Am. Chem. Soc.* **95**, 7917 (1973).
41. K. B. Wiberg and S. D. Nielsen, *J. Org. Chem.* **29**, 3353 (1964).
42. D. R. Adams and S. P. Bhatnagar, *Synthesis* p. 661 (1977).
43. D. Masilamani and M. M. Rogic, *J. Am. Chem. Soc.* **100**, 4634 (1978).
44. L. M. Stephenson and D. R. Speth, *J. Org. Chem.* **44**, 4683 (1979); see however, W. D. Woggon, I. Ruther, and H. Egli, *J. Chem. Soc., Chem. Commun.* p. 706, (1980).
45. M. A. Umbreit and K. B. Sharpless, *J. Am. Chem. Soc.* **99**, 5526 (1977).
46. B. Chabaud and K. B. Sharpless, *J. Org. Chem.* **44**, 4202 (1979)
47. K. B. Sharpless, T. Hori, L. K. Truesdale, and C. O. Dietrich, *J. Am. Chem. Soc.* **98**, 269 (1976); see also K. B. Sharpless and S. P. Singer, *J. Org. Chem.* **41**, 2504 (1976).
48. K. B. Sharpless and K. M. Gordon, *J. Am. Chem. Soc.* **98**, 300 (1976).
49. G. Sosnovsky, J. A. Krogh, and S. G. Umhoefer, *Synthesis* p. 722 (1979).
50. G. W. Hearne and M. L. Adams, U.S. Patent 2,451,485 (1948) to Shell Development Co.
51. (a) J. D. Idol, U.S. Patent 2,904,580 (1959) to SOHIO; see also J. N. Cosby, U.S. Patent 2,481,826 (1949) to Allied Chemical Corp..
    (b) J. L. Callahan, R. W. Foreman, and F. Veatch, U.S. Patent 3,044,966 (1962) to SOHIO.
    (c) F. Veatch, *Chem. Eng. Process* **56**, 65 (1960).
52. G. W. Hearne and K. W. Furman, U.S. Patent 2,991,320 (1961) to Shell Development Co.
53. (a) J. L. Callahan and B. Gertisser, U.S. Patent 3,198,750 (1965) to SOHIO.
    (b) R. K. Grasselli and J. L. Callahan, *J. Catal.* **14**, 93 (1969).
54. Ger. Patent 2,203,710 (1972) to SOHIO.
55. J. M. Peacock, A. J. Parker, P. G. Ashmore, and J. A. Hockey, *J. Catal.* **15**, 398 (1969) and preceding papers in this series.
56. C. C. McCain, G. Gough, and G. W. Godin, *Nature* (*London*) **198**, 989 (1963); C. R. Adams and T. J. Jennings, *J. Catal.* **2**, 63 (1963); **3**, 549 (1964).
57. W. M. H. Sachtler and H. N. de Boer, *Proc. Int. Congr. Catal. 3rd, 1964* Vol. 1, p. 252 (1965).
58. F. Trifiro, *Chim. Ind.* (*Milan*) **56**, 835 (1974).
59. (a) F. Trifiro and I. Pasquon, *J. Catal.* **12**, 412 (1968).
    (b) F. Trifiro, L. Kubelkova, and I. Pasquon, *J. Catal.* **19**, 121 (1970)
60. (a) F. Trifiro, P. Centola, and I. Pasquon, *J. Catal.* **10**, 86 (1968).
    (b) F. Trifiro, P. Centola, I. Pasquon, and P. Jiru, *Proc. Int. Congr. Catal., 4th, 1968* Vol. 1, p. 252 (1971).

61. Compare also P. -S. E. Dai and J. H. Lunsford, *J. Catal.* **64**, 173, 184 (1980).
62. (a) J. D. Burrington and R. K. Grasselli, *J, Catal.* **59**, 79 (1979); J. D. Burrington, C. T. Kartisek, and R. K. Grasselli, *ibid.* **63**, 235 (1980);
    (b) See however, P. G. Menon, *J. Catal.* **59**, 314 (1979); J. E. Germain, G. Pajonk, and S. J. Teichner, *ibid.* p. 317.
63. C. R. Adams, *J. Catal.* **10**, 355 (1968).
64. R. Barral, C. Bocard, I. Sérée de Roch, and L. Sajus, *Tetrahedron Lett.* p. 1693 (1972); *Kinet. Katal.* **14**, 130 (1973).
65. R. A. Sheldon, unpublished observations.
66. P. C. H. Mitchell and R. D. Scarle, *J. Chem. Soc., Dalton* p. 2552 (1975).
67. A. Nakamura, M. Nakayama, K. Sugihashi, and S. Otsuka, *Inorg. Chem.* **18**, 394 (1979).
68. G. Speier, *Inorg. Chim. Acta* **32**, 139 (1979).
69. G. D. Garner, R. Durant, and F. E. Mabbs, *Inorg. Chim. Acta* **24**, L29 (1977); compare also R. A. Clement, U. Klabunde, and G. W. Parshall, *J. Catal.* **4**, 87 (1978); R. Breslow, R. Q. Kluttz, and P. L. Khanna, *Tetrahedron Lett.* p. 3273 (1979); W. T. Reichle and W. L. Carrick, *J. Organometal. Chem.* **24**, 419 (1970).
70. (a) R. C. Bray and J. C. Swan, *Struct. Bonding (Berlin)* **11**, 107 (1972).
    (b) J. T. Spence, *Coord. Chem. Rev.* **4**, 475 (1969).
    (c) R. C. Bray, *Proc. Climate Int. Conf. Chem. Uses Molybdenum, 1st*, p. 216 (1973).
71. W. A. Nugent and R. L. Harlow, *J. Am. Chem. Soc.* **102**, 1760 (1980).
71a. See, however, J. R. Budge, B. M. K. Gatehouse, M. C. Nesbit, B. O. West, *J. Chem. Soc., Chem. Commun.* p. 370 (1981).
72. M. C. Sze and A. P. Gelbein, *Hydrocarbon Process.* p. 103 (1976); see also F. Porter, M. Erchak and J. N. Cosby, U.S. Patent 2,510,605 (1950) to Allied Chemical Co.
73. J. N. Armor, *J. Am. Chem. Soc.* **102**, 1453 (1980).
74. K. B. Sharpless and A. Y. Teranishi, *J. Org. Chem.* **38**, 185 (1973); see also S. K. Chung, *Tetrahedron Lett.* p. 3211 (1978).
75. M. Nagayama, O. Okumura, S. Noda, and A. Mori, *J. Chem. Soc., Chem. Commun.* p. 841 (1973).
76. E. Vedejs and K. A. J. Snoble, *J. Am. Chem. Soc.* **95**, 5778 (1973).
77. K. B. Sharpless, M. A. Umbreit, M. T. Nieh, and T. C. Flood, *J. Am. Chem. Soc.* **94**, 6538 (1972); A. Sattar, J. Forrester, M. Moir, J. S. Roberts, and W. Parker, *Tetrahedron Lett.* p. 1403 (1978).
78. W. Kruse, *J. Chem. Soc., Chem. Commun.* p. 1610 (1968).
79. W. J. Reichle and W. L. Carrick, *J. Organometal. Chem.* **24**, 419 (1970).
80. S. Uemura, A. Onoe, and M. Okano, *Bul. Chem. Soc. Japan* **41**, 790 (1968); J. San Filippo, A. F. Sowinsky, and L. J. Romano, *J. Am. Chem. Soc.* **97**, 1599 (1975).
81. J. Muller and W. Holzinger, *Angew. Chem. Int. Ed.* **14**, 760 (1975); W. Mowat and G. Wilkinson, *J. Organometal. Chem.* **38**, C35 (1972).
82. (a) R. R. Schrock, *Accts. Chem. Rsch.* **12**, 98 (1979).
    (b) C. Santani-Scampussi and J. G. Riess, *J. Chem. Soc., Dalton* p. 607 (1975).
83. W. Mowat, A. Shortland, G. Yagupski, N. J. Hill, M. Yagupski, and G. Wilkinson, *J. Chem. Soc., Dalton* p. 533 (1972).
84. K. Mertis, D. H. Williamson, and G. Wilkinson, *J. Chem. Soc., Dalton* p. 607 (1975); L. Galyer, K. Mertis, and G. Wilkinson, *J. Organometal. Chem.* **85**, C38 (1975).
85. W. A. Nugent, *J. Org. Chem.* **45**, 4533 (1980).
86. S. G. Hentges and K. B. Sharpless, *J. Am. Chem. Soc.* **102**, 4263 (1980).
87. K. B. Sharpless, D. W. Patrick, L. K. Truesdale, and S. A. Biller, *J. Am. Chem. Soc.* **97**, 2305 (1975); D. W. Patrick, L. K. Truesdale, S. A. Biller, and K. B. Sharpless, *J. Org. Chem.* **43**, 2628 (1978); A. O. Chong, K. Oshima, and K. B. Sharpless, *J. Am. Chem. Soc.* **99**, 3420 (1977). Compare also J. E. Bäckvall and E. E. Bjorkman, *J. Org. Chem.* **45**, 2893 (1980).

88. (a) K. B. Sharpless, A. O. Chong, and K. Oshima, *J. Org. Chem.* **41**, 177 (1976); E. Herranz and K. B. Sharpless, *ibid.* **43**, 2544 (1978).
(b) E. Herranz, S. A. Biller, and K. B. Sharpless, *J. Am. Chem. Soc.* **100**, 3596 (1978).
89. A. O. Chong, K. Oshima, and K. B. Sharpless, *J. Am. Chem. Soc.* (in press).
89a. For an alternative method, see R. N. Becker, M. A. White, and R. G. Bergman, *J. Am. Chem. Soc.* **102**, 5676 (1980).
90. H. Mimoun, *Rev. Inst. Fr. Pet.* **33**, 259 (1978).
90a. R. J. Haines and G. J. Leigh, *Chem. Soc. Rev.* **4**, 155 (1975); N. Calderon, J. P. Lawrence, and E. A. Ofstead, *Adv. Organometal. Chem.* **17**, 449 (1979), and references cited therein.
91. J. L. Herrison and Y. Chauvin, *Makromol. Chem.* **141**, 161 (1970). See also J. H. Wengrovius, R. R. Schrock, M. R. Churchill, J. R. Meissert and W. G. Young, *J. Am. Chem. Soc.* **102**, 4515 (1980); T. R. Howard, J. B. Lee and R. H. Grubbs, *Ibid.* **102**, 6876 (1980).
92. R. R. Schrock, *J. Am. Chem. Soc.* **98**, 5399 (1976).
93. E. O. Fischer and S. Riedmuller, *Chem. Berichte* **107**, 915 (1974).
94. A. K. Rappé and W. A. Goddard, III, *J. Am. Chem. Soc.* **102**, 5114 (1980).
95. V. van Rheenen, D. Y. Cha, and W. M. Hartley, *Org. Synth.* **58**, 43 (1978).
96. R. Ugo, "Relations between Homogeneous and Hetergeneous Catalysis," reprint of a lecture presented at the C.N.R.S. Symposium, Lyon, 3–6 November, 1977.
97. K. B. Sharpless, *Pap., Int. Symp. Oxygen Activation Selective Oxidations Catalysed Transition Met. 1st 1979* (1979).
98. K. B. Sharpless and T. Hori, *J. Org. Chem.* **41**, 176 (1976).
99. J. T. Groves and M. van der Puy, *J. Am. Chem. Soc.* **96**, 5274 (1974).
100. J. T. Groves and W. W. Swanson, *Tetrahedron Lett.* p. 1953 (1975).
101. J. T. Groves and G. A. McClusky, *J. Am. Chem. Soc.* **98**, 859 (1976).
102. D.-H. Chin, A. L. Balch, and G. N. La Mar, *J. Am. Chem. Soc.* **102**, 1446, 5945 (1980).
103. J. T. Groves, T. E. Nemo, and R. S. Myers, *J. Am. Chem. Soc.* **101**, 1032 (1979).
103a. V. Ullrich, *J. Mol. Catal.* **7**, 159 (1980).
104. G. D. Nordblum, R. E. White, and M. J. Coon, *Arch. Biochem. Biophys.* **175**, 524 (1976); E. G. Hrcay, J. Gustafsson, M. Ingelman-Sundberg, and L. Ernster, *Biochem. Biophys. Res. Commun.* **66**, 209 (1975); F. Lichtenberger, W. Nastainczyk, and V. Ullrich, *ibid.* **70**, 939 (1976).
105. J. T. Groves, W. J. Kruper, T. E. Nemo, and R. S. Myers, *J. Mol. Catal.* **7**, 169 (1980).
106. T. Yamamoto and M. Kimura, *J. Chem. Soc., Chem. Commun.* p. 948 (1977); see also M. Tohma, T. Tomita, and M. Kimura, *Tetrahedron Lett.* p. 4359 (1973).
107. H. Ledon, *C. R. Hebd. Seances Acad. Sci., Series C* **288**, 29 (1979); see also D. R. Paulson, R. Ullman, R. B. Sloane, and G. L. Closs, *J. Chem. Soc., Chem. Commun.* p. 186 (1974); M. Baccouche, J. Ernst, J. H. Fuhrhop, R. Schlozer, and H. Arzoumanian, *ibid.* p. 821 (1977).
108. J. T. Groves and W. J. Kruper, Jr., *J. Am. Chem. Soc.* **102**, 7615 (1980).
109. J. T. Groves, W. J. Kruper, Jr. and R. C. Haushalter, *J. Am. Chem. Soc.* **102**, 6375 (1980). See also C. L. Hill and B. C. Schardt, *Ibid.* **102**, 6374 (1980) and I. Willner, J. W. Otvos and M. Calvin, *J. Chem. Soc., Chem. Commun.* p. 964 (1980).
110. L. Castle, J. R. Lindsay-Smith, and G. V. Buxton, *J. Mol. Catal.* **7**, 235 (1980).
111. W. A. Waters and J. S. Littler, *in* "Oxidation in Organic Chemistry" (K. B. Wiberg, ed.), Part A, p. 186. Academic Press, New York, 1965.
112. Kh. M. Minachev, G. V. Antoshin, D. G. Klissurski, N. K. Guin, and N. Ts. Abadzhijeva, *React. Kinet. Catal. Lett.* **10**, 163 (1979).
113. T. Itoh, K. Jitsukawa, K. Kaneda, and S. Teranishi, *J. Am. Chem. Soc.* **101**, 159 (1979).
114. Y. Moro-oka, Y. Takita, and A. Ozaki, *Bul. Chem. Soc. Japan* **44**, 293 (1971).
115. W. P. Weber and G. W. Gokel, "Phase Transfer Catalysis in Organic Synthesis" Springer-Verlag, Berlin and New York, 1977.

116. C. M. Starks and C. Liotta, "Phase Transfer Catalysis. Principles and Techniques." Academic Press, New York, 1978.
117. N. A. Gibson and J. W. Hosking, *Austral. J. Chem.* **18**, 123 (1965).
118. C. M. Starks, *J. Am. Chem. Soc.* **93**, 195 (1971).
119. A. W. Herriott and D. Picker, *Tetrahedron Lett.* p. 1511 (1974).
120. C. M. Starks and P. H. Wasecheck, U.S. Patent 3,547,962 (1970) to Continental Oil Company.
121. W. P. Weber and J. P. Shepherd, *Tetrahedron Lett.* p. 4907 (1972).
122. A. P. Krapcho, J. R. Larson, and J. M. Eldridge, *J. Org. Chem.* **42**, 3749 (1977).
123. T. Okimoto and D. Swern, *J. Am. Oil Chem. Soc.* **54**, 862A, 867A (1977). See also W. Reischl and E. Zbiral, *Tetrahedron* **35**, 1109 (1979) and E. Santaniello, A Manzocchi and C. Farachi, *Synthesis*, p. 563 (1980).
124. M. J. Atherton and J. H. Holloway, *J. Chem. Soc., Chem. Commun.* p. 253 (1978).
125. T. A. Foglia, P. A. Barr, and A. J. Malloy, *J. Am. Oil Chem. Soc.* **54**, 858A (1977).
126. H. J. Schmidt and H. J. Schäfer, *Angew. Chem. Int. Ed.* **18**, 68 (1979).
127. H. J. Schmidt and H. J. Schäfer, *Angew. Chem. Int. Ed.* **18**, 69 (1979).
128. D. Sam and H. E. Simmons, *J. Am. Chem. Soc.* **94**, 4024 (1972).
129. M. S. Newman, H. M. Dali, and W. M. Hung, *J. Org. Chem.* **40**, 262 (1975).
130. R. D. Hutchins, N. R. Natale, W. J. Cook, and J. Ohr, *Tetrahedron Lett.* p. 4167 (1977).
131. D. Pletcher and S. J. D. Tait, *Tetrahedron Lett.* p. 1601 (1978).
132. D. Landini and F. Rolla, *Chem. Ind.* (*London*) p. 213 (1979); E. Santaniello and P. Ferraboschi, *Synthesis* **10**, 75 (1980).
133. D. Landini, F. Montanari, and F. Rolla, *Synthesis* p. 134 (1979); see also S. Cacchi, F. La Torre, and D. Misiti, *ibid.* p. 356.
134. G. Cardillo, M. Orena, and S. Sandri, *J. Chem. Soc., Chem. Commun.* p. 190 (1976).
135. G. A. Lee and H. H. Freedman, *Tetrahedron Lett.* p. 1641 (1976); U.S. Patent 3,996,259 (1975).
136. S. Krishnan, D. G. Kuhn, and G. A. Hamilton, *J. Am. Chem. Soc.* **99**, 8121 (1977).
137. U.S. Patent 4,079,075 (1977) to Dow Chemical Company.
138. D. G. Lee and V. S. Chang, *J. Org. Chem.* **43**, 1532 (1978).
139. T. Ogino, *Tetrahedron Lett.* **21**, 177 (1980).
140. T. A. Foglia, P. A. Barr, A. J. Malloy, and M. J. Constanzo, *J. Am. Oil Chem. Soc.* **54**, 870A (1977).

140a. E. Guilmet and B. Meunier, *Tetrahedron Lett.* **21**, 4449 (1980).

141. Y. Tsuda and S. Nakajima, *Chem. Lett.* p. 1397 (1978).

ADDITIONAL READING

J. T. Groves, Mechanisms of metal-catalyzed oxygen insertion. *In* "Metal Ion Activation of Dioxygen" (T. G. Spiro, ed.), Ch. 3. Wiley, New York, 1980.

M. Schroeder, Osmium tetraoxide cis hydroxylation of unsaturated substrates. *Chem. Rev.* **80**, 187 (1980).

W. S. Trahanovsky, Metal-ion oxidative cleavage of alcohols. *in* "Methods in Free-Radical Chemistry," (E. S. Huyser, ed.), Vol. 4, p. 133. Marcel Dekker, New York, 1973. (Includes oxidative cleavages of alcohols and 1,2-glycols by metal ions.)

Chapter 7

Activation by Coordination to Transition Metal Complexes

The *direct* interaction of organic substrates with high-valent metal complexes and oxometal reagents plays a key role in the oxidations presented in Chapters 5 and 6. These metal oxidants are generally strong electrophiles, which react with organic substrates via electrophilic attack or one-electron transfer. In this chapter we shall discuss another group of oxidative processes with a metal such as Pd(II) in a low oxidation state (i.e., "soft" center), which is not a particularly strong electrophile. The substrate, in most cases an olefin, is activated toward nucleophilic attack through π-complex formation with the metal. The result is an overall *oxidative nucleophilic substitution* of hydrogen with concomitant two-electron reduction of the metal catalyst in a *heterolytic process* involving organometallic intermediates. For example, in the Pd(II)-catalyzed oxidations of olefins in aqueous media, the overall reaction is:

$$Pd^{II}X_2 + \rangle C{=}C\langle_{H} + HOH \longrightarrow Pd^0 + \left[\rangle C{=}C\langle_{OH}\right] + 2\,HX \qquad (1)$$

The oxidant is regenerated in subsequent steps by reoxidation of the reduced Pd(0). Although Pd(II)-catalyzed oxidations bear a formal resemblance to Pb(IV) and Tl(III) oxidations, which also involve organometallic intermediates (see Chapter 5), the mechanisms are fundamentally different and exhibit opposite substrate structure–reactivity relationships.

An important landmark in the development of organotransition metal chemistry was the discovery in 1959 of the Wacker process for the conversion of ethylene to acetaldehyde. The commercial success of the Wacker process provided an enormous stimulus for further studies of palladium and other noble metal complexes as homogeneous catalysts.[1–14] This development led to the discovery of a variety of important homogeneous, liquid-phase processes involving noble metal catalysts such as hydroformylation, carbonylation, hydrogenation, isomerization, and oligomerization. Palladium complexes are generally superior catalysts for oxidation reactions, whereas other noble metals are more active in other reactions, e.g., rhodium for hydroformylation and carbonylation.

I. PALLADIUM-CATALYZED OXIDATIONS OF OLEFINS IN AQUEOUS MEDIA

The palladium-catalyzed oxidation of ethylene to acetaldehyde, commonly referred to as the Wacker process, was reported by Smidt and co-workers in 1959.[15–19] The process combines the stoichiometric oxidation of ethylene by Pd(II) in aqueous solution with the reoxidation of Pd(0) *in situ* by molecular oxygen in the presence of copper salts. The overall reaction constitutes a palladium-catalyzed oxidation of ethylene to acetaldehyde, with molecular oxygen as terminal oxidant. The following sequence of reactions has been proposed.

Scheme I:

$$C_2H_4 + PdCl_2 + H_2O \longrightarrow CH_3CHO + Pd + 2\,HCl \qquad (2)$$

$$Pd + 2\,CuCl_2 \longrightarrow PdCl_2 + Cu_2Cl_2 \qquad (3)$$

$$Cu_2Cl_2 + 2\,HCl + \tfrac{1}{2}\,O_2 \longrightarrow 2\,CuCl_2 + H_2O \qquad (4)$$

Acetaldehyde is formed in approximately 95% yield, and the side products include acetic acid (~2%), carbon dioxide (~1%), and chlorinated products

(~1%).[20] Propylene is oxidized under similar conditions to acetone in 90% yield, and 1-butene affords methyl ethyl ketone in ~85% yield.[13,14,17,21] The yields vary considerably with the higher olefins,[14] mixtures often being obtained as a result of olefin isomerization. Aqueous solutions of other Group VIII metal salts, such as salts of Pt(II), Ir(III), Ru(III), and Rh(III), oxidize olefins in an analogous manner[14] but significantly less effectively than Pd(II).

The kinetics and mechanism of the Pd(II)-catalyzed oxidations of olefins have been extensively studied and are the subject of continuing debate.[11,12,21–27] However, a mechanism involving the following basic steps is consistent with the experimental data.[26]

Scheme II:

$$PdCl_4^{2-} + C_2H_4 \rightleftharpoons [PdCl_3C_2H_4]^- + Cl^- \quad (5)$$

$$[PdCl_3C_2H_4]^- + H_2O \rightleftharpoons [PdCl_2(H_2O)C_2H_4] + Cl^- \quad (6)$$

$$[PdCl_2(H_2O)C_2H_4] + H_2O \rightleftharpoons [HOCH_2CH_2PdCl_2(H_2O)]^- + H^+ \quad (7)$$

$$[HOCH_2CH_2PdCl_2(H_2O)]^- \longrightarrow HOCH_2CH_2PdCl(H_2O) + Cl^- \quad (8)$$

$$HOCH_2CH_2PdCl(H_2O) \longrightarrow CH_3CHO + Pd + HCl + H_2O \quad (9)$$

The key features of this mechanism are the π–σ rearrangement of the coordinated olefin in eq 7 and the β-hydride elimination of the hydroxyethylpalladium(II) intermediate in eq 9. The first step is the formation of the π complex in eq 5, which reduces the electron density at the double bond, rendering it susceptible to nucleophilic attack. In accord with such a nucleophilic process, the rate decreases with increasing alkyl substitution at the double bond. This trend is in contrast to electrophilic attack in the analogous oxidations with Pb(IV) and Tl(III) complexes.[21] Palladium(II) catalyzes the nucleophilic attack on olefins more readily than Pt(II) for the following reasons:[3] (1) Pd(II)–olefin complexes are formed more rapidly than their Pt(II) analogs; (2) the back-donation of charge from metal to olefin is less for palladium than for platinum, resulting in a lower electron density around the olefin in the former; (3) the Pd—olefin bond is weaker than the Pt—olefin bond, leading to a lower activation energy for π–σ rearrangement; and (4) Pd(II) can readily expand its coordination sphere to accept a fifth or sixth ligand. In agreement with the presence of equilibria in eqs 5 and 6, the reaction is inhibited by chloride ion.

Recent mechanistic studies[26,27] have concentrated on the π–σ rearrangement or hydroxypalladation step in eq 7, which has been shown to proceed with trans stereochemistry by external attack of nucleophile (water), and not by rearrangement of a coordinated ligand (water or hydroxide).

Scheme III:

$$H_2O \curvearrowright CH_2{=}CH_2 \cdot PdCl_2(OH_2) \rightleftharpoons \left[HO{-}CH_2CH_2{-}PdCl_2(OH_2) \right]^- + H^+ \qquad (10)$$

$$\downarrow$$

$$HO{-}CH_2CH_2{-}Pd(Cl)(OH_2) + Cl^- \qquad (11)$$

$$\leftarrow_{\sigma} \quad (HO{-}CH{=}CH_2)PdCl(OH_2)H$$

$$\downarrow (12)$$

$$H{-}O{-}CH(CH_3){-}Pd(Cl)(OH_2) \longrightarrow CH_3CHO + Pd + HCl \qquad (13)$$

Bäckvall and co-workers[26] favor a reversible hydroxypalladation (eq 10), followed by the rate-limiting loss of chloride ion from the σ complex. The resulting coordinatively unsaturated 14-electron intermediate is more prone to a rapid β-hydride elimination in eq 11. This formulation is consistent with the absence of a significant isotope effect in the oxidation of C_2D_4.[21,23] The postulation of rapid intramolecular repalladation of the vinyl alcohol intermediate in eq 12, to form an α-hydroxyethylpalladium(II) species, was deemed necessary to explain the absence of deuterium incorporation in the acetaldehyde formed in D_2O.

When the Wacker oxidation is carried out at high copper(II) and chloride ion concentrations, 2-chloroethanol becomes the main product,[28] which is formed from the same β-hydroxyethylpalladium(II) intermediate.[26]

$$HO{-}CH_2CH_2{-}PdCl_2(OH_2) \xrightarrow[LiCl]{CuCl_2} HO{-}CH_2CH_2{-}Cl, \quad \text{etc.} \qquad (14)$$

This process is analogous to the formation of glycol acetates and related products during the oxidation of ethylene in acetic acid (*vide infra*). The role(s) of copper as the cocatalyst in these reactions is still not completely resolved. For example, copper(II) chloride is known to cleave carbon–palladium bonds readily, although the stereochemistry appears to be dependent on the chloride ion concentration.[29,30] Bäckvall and co-workers[26] further proposed that external attack on the coordinated alkene, resulting in trans stereochemistry, is also favored by nucleophiles such as acetate, alcohols, and amines (*vide infra*). Other nucleophiles such as aryl, alkyl, and hydride, which have only a limited existence outside the coordination sphere of palladium, prefer an intramolecular route, resulting in cis-addition. Nucleophiles such as chloride occupy an intermediate position and may react by both intra- and intermolecular processes.[26]

For the industrial process, a mixture of ethylene and oxygen saturated with water vapor is passed over a fixed bed of palladium and copper salts supported on active charcoal.[15] A heterogeneous catalyst consisting of V_2O_5 and Pd(II) is also highly active for the oxidation of ethylene to acetaldehyde[31] and acetic acid.[32]

II. PALLADIUM-CATALYZED OXIDATIONS OF OLEFINS IN NONAQUEOUS MEDIA

The formation of vinyl acetate from ethylene by reaction with $PdCl_2$ and NaOAc in acetic acid was first reported by Moiseev *et al.*[33] A catalytic process obtains with copper salts and dioxygen to reoxidize the palladium.[33–37a]

$$CH_2{=}CH_2 + HOAc + \tfrac{1}{2}O_2 \xrightarrow{[Pd^{II},\, Cu^{II}]} CH_2{=}CHOAc + H_2O \qquad (15)$$

The commercial process is reported to give a 90% yield of vinyl acetate based on ethylene and in excess of 95% yield based on acetic acid.[34]

Although ethylene is readily absorbed by solutions of $PdCl_2$ in acetic acid (π-complex formation), the presence of acetate ions (as NaOAc or LiOAc) is essential for the oxidative reaction. Van Helden and co-workers[35] found that ethylene reacts with $Pd(OAc)_2$ at 1 atm and 70°C to form palladium metal and vinyl acetate (40–50%), together with acetaldehyde, ethylidene diacetate, and acetic acid. In the presence of added sodium acetate the rate increased considerably, and the yield of vinyl acetate was enhanced to 80–90%. The similarity of this reaction to that in aqueous solutions is depicted in Scheme IV.

Scheme IV:

$$AcO^- + CH_2{=}CH_2 \cdot Pd^{II}(OAc)_2(HOAc) \longrightarrow AcO{-}CH_2CH_2{-}Pd^{II}(OAc)(HOAc) + AcO^- \qquad (16)$$

$$AcO{-}CH_2{-}CH_2{-}Pd(OAc)(HOAc) \longrightarrow AcOCH{=}CH_2 + HPdOAc + HOAc \qquad (17)$$

Nucleophilic attack by external acetate occurs concomitantly with π–σ rearrangement, leading to acetoxypalladation in eq 16. The subsequent β-hydride elimination in eq 17 is a commonly observed pathway for the decomposition of transition metal alkyls.[38,39] It may be facilitated by an acetate ligand as shown in eq 17, although it is usually considered to require coordinative unsaturation on the metal center to provide for a hydridopalladium intermediate.[38] The formation of some acetaldehyde, even under rigorously anhydrous conditions,[35–37a] may be due to the following Pd(II)-catalyzed reaction:

$$CH_2{=}CHOAc + HOAc \xrightarrow[HOAc]{[Pd^{II}]} CH_3CHO + (AcO)_2O \qquad (18)$$

The oxidation of higher olefins by Pd(II) in acetic acid often leads to a complex mixture of products.[12,14] The reactions are complicated by olefin isomerization and oligomerization, and the formation of π-allyl complexes.[40] Terminal olefins afford enol acetates as the major product of oxidation, whereas internal olefins give mainly allylic acetates, as illustrated below with propylene and 2-butene.

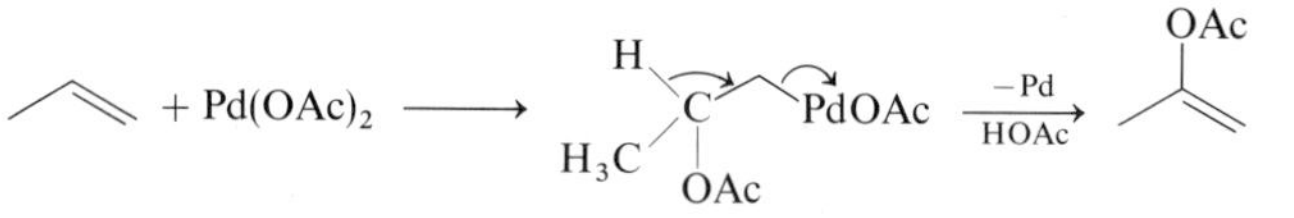

$$CH_3CH{=}CH_2 + Pd(OAc)_2 \longrightarrow (H)(H_3C)C(OAc)\text{–}CH_2PdOAc \xrightarrow[HOAc]{-Pd} CH_3C(OAc){=}CH_2 \qquad (19)$$

$$CH_3CH{=}CHCH_3 + Pd(OAc)_2 \longrightarrow H_3C\text{–}CH(OAc)\text{–}CH(PdOAc)\text{–}CH_2\text{–}H \xrightarrow[HOAc]{-Pd} CH_3CH(OAc)CH{=}CH_2 \qquad (20)$$

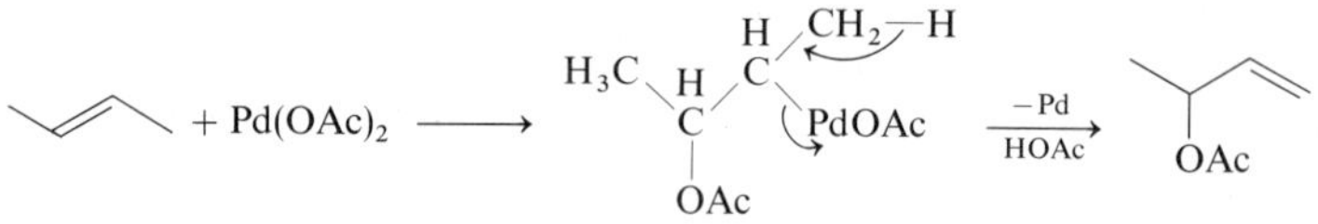

The regiospecificity indicates that β-hydride elimination occurs from the intermediate preferentially at the carbon center not containing the acetate group.

The reactions in eqs 15, 19, and 20 all constitute examples of a general class of Pd(II)-mediated nucleophilic substitutions at an sp^2 carbon center and described by the general equation:

$$X\text{—}Pd^{II}\text{—}Y + {>}C{=}C{<}_{Z} \longrightarrow {>}C{=}C{<}_{Y} + X\text{—}Pd^{II}\text{—}Z \qquad (21)$$

When Z = H, the result is *oxidative substitution* since the transient HPd(II)X species eliminates HX to form palladium metal. Oxidative substitutions are observed with a wide variety of nucleophiles[1] including alcohols, amines, and malonate ester:

$$\mathrm{>C{=}C<^{H}} + \mathrm{ROH} \xrightarrow{[\mathrm{Pd^{II}}]} \mathrm{>C{=}C<_{OR}}, \quad \text{etc.} \tag{22}$$

$$\mathrm{>C{=}C<^{H}} + \mathrm{R_2NH} \xrightarrow{[\mathrm{Pd^{II}}]} \mathrm{>C{=}C<_{NR_2}}, \quad \text{etc.} \tag{23}$$

$$\mathrm{>C{=}C<^{H}} + {}^{-}\mathrm{CH(CO_2Et)_2} \xrightarrow{[\mathrm{Pd^{II}}]} \mathrm{>C{=}C<_{CH(CO_2Et)_2}}, \quad \text{etc.} \tag{24}$$

When Z = halogen, a Pd(II)-catalyzed nucleophilic substitution results via successive reactions involving palladation and depalladation.[41]

$$\mathrm{ClCH{=}CH_2 + PdCl_2} \rightleftharpoons \mathrm{ClCH{=}CH_2 \cdot PdCl_2}$$

$$\mathrm{ClCH{=}CH_2 \cdot PdCl_2} \xrightleftharpoons{\mathrm{AcO^-}} \mathrm{(Cl)(AcO)CH{-}CH_2PdCl}$$

$$\mathrm{(Cl)(AcO)CH{-}CH_2PdCl} \longrightarrow \mathrm{AcOCH{=}CH_2 + PdCl_2} \tag{25}$$

The transvinylation reaction[11,42] between carboxylic acids and vinyl acetate follows a similar course.

$$\mathrm{CH_2{=}CHOAc + RCO_2H} \xrightarrow{[\mathrm{Pd^{II}}]} \mathrm{CH_2{=}CHO_2CR + HOAc} \tag{26}$$

The role of the copper cocatalyst in many of these oxidations is not always limited to catalysis of the reoxidation of Pd(0). Thus, in the presence of a high concentration of Cu(II), the reaction in acetic acid takes a different course and affords 1,2-disubstituted ethanes,[24,35–37a,43,44] described by the general equation 27.

$$\mathrm{CH_2{=}CH_2 + 2\,CuX_2 + AcO^-} \xrightarrow[\mathrm{HOAc}]{[\mathrm{Pd^{II}}]} \mathrm{XCH_2CH_2OAc + 2\,CuX + X^-} \tag{27}$$

1,2-Addition is also the major reaction in the presence of nitrate ion. Thus, ethylene and terminal olefins react with palladium(II) nitrate in acetic acid to produce glycol mono- and diacetates.[35–37] The reaction can be made catalytic by employing a catalyst consisting of a combination of Pd(II), Cu(II), Li^+, AcO^-, Cl^-, and $NO_3{}^-$, together with the olefin and molecular oxygen in acetic acid as solvent.[35] Ethylene and propylene glycol monoacetates can be manufactured in 90–95% yields using this process.[45] The most reasonable mechanism appears to be one involving the interception of the acetoxyalkylpalladium(II) intermediate by the oxidant, since it has been shown by Henry[46] that a variety of oxidants will effect formation of 1,2-addition products in the presence of Pd(II).

$$AcOCH_2CH_2PdX \longrightarrow CH_2{=}CHOAc + HPdX \quad (28)$$

$$AcOCH_2CH_2PdX \xrightarrow{CuX_2} AcOCH_2CH_2X + PdX_2 \quad (29)$$

The mode of interaction of the oxidant with the acetoxypalladation adduct is not completely understood. The competition between β-hydride elimination and oxidative substitution of the acetoxyalkylpalladium(II) intermediate bears many similarities to the competing oxidative elimination and oxidative substitution mechanisms observed in electron transfer reactions of alkyl radicals with Cu(II) complexes.[47] A mechanism has been suggested[30] involving alkyl transfer from Pd(II) to Cu(II), followed by spontaneous homolysis of a metastable alkylcopper(II) intermediate, i.e.,

$$AcOCH_2CH_2Pd^{II}X + Cu^{II}X_2 \longrightarrow AcOCH_2CH_2Cu^{II}X + PdX_2 \quad (30)$$

$$AcOCH_2CH_2Cu^{II}X \longrightarrow AcOCH_2CH_2\cdot + Cu^{I}X \quad (31)$$

$$AcOCH_2CH_2\cdot + Cu^{II}X_2 \longrightarrow AcOCH_2CH_2X + Cu^{I}X \quad (32)$$

The subsequent scavenging of the alkyl radical by copper(II) salts is known to be rapid.[47] The role of nitrate ion and other oxidants in promoting 1,2-addition is less clear, although several rationalizations have been proposed.[45,48] [Compare also the Pd-olefin oxidations induced by nitrocobalt(III) complexes, as discussed in Chapter 3, Section VI.]

Alkylpalladium(II) complexes have recently been isolated from the reaction of silyl enol ethers with $Cl_2(PhCN)_2Pd$, followed by rearrangement of the oxo-π-alkylpalladium(II) intermediate, e.g.,[49]

OSiMe$_3$ + $Cl_2(PhCN)_2Pd$ ⟶ [... PdCl]$_n$ + Me_3SiCl + 2 PhCN

PdCl $\xrightarrow[0]{MeCN}$ PdCl(MeCN)

Oxidative cleavages of these δ-oxoalkylpalladium(II) complexes by copper(II) chloride are accompanied by several interesting ring expansions,

PdCl(MeCN) + $CuCl_2$ ⟶ , etc

$CH_2PdCl(MeCN)$ + $CuCl_2$ ⟶ =O, etc.

The oxidative cyclization of *o*-crotylphenol, representing an intramolecular version of the Wacker reaction, is catalyzed by a chiral π-alkylpalladium(II) complex derived from β-pinene, i.e., (+)-[(3,2,10-η-pinene)PdOAc]$_2$.[50]

$$\text{o-crotylphenol (OH)} + O_2 \xrightarrow[\text{MeOH}]{[\mathbf{I}, Cu^{II}]} \text{2-vinyl-2,3-dihydrobenzofuran (*)} + \text{2-ethylbenzofuran}$$

Since the cyclized dihydrobenzofuran product is optically active, and the optical rotation remains invariant with the extent of reaction, the authors concluded that chiral ligand must have been associated with the palladium species throughout the catalytic cycle. To account for this conclusion, they suggested that the hydridopalladium species eliminated from the organopalladium intermediate (of eqs 17 and 28) retains the chiral π-alkyl ligand and furthermore that it remains intact in the subsequent oxidative regeneration of the active catalyst by dioxygen or copper(II).

III. OXIDATIVE CARBONYLATION OF OLEFINS

The conversion of cthylene to acrylic acid involves the Pd(II)-catalyzed oxidative carbonylation,[8,51] with copper salts used to reoxidize the palladium.

$$CH_2{=}CH_2 + CO + \tfrac{1}{2}\,O_2 \xrightarrow{[Pd^{II}, Cu^{II}]} CH_2{=}CHCO_2H \qquad (33)$$

A mechanism has been proposed,[51] in which the following steps are involved.

Scheme V:

$$PdCl_4^{2-} + CO + H_2O \longrightarrow \left[Cl_2(CO)Pd\overset{O}{\overset{\|}{C}}OH\right]^- + HCl + Cl^- \qquad (34)$$

$$\left[Cl_2(CO)Pd\overset{O}{\overset{\|}{C}}OH\right]^- + C_2H_4 \rightleftharpoons \left[Cl_2(C_2H_4)Pd\overset{O}{\overset{\|}{C}}OH\right]^- + CO \qquad (35)$$

$$\left[Cl_2(C_2H_4)Pd\overset{O}{\overset{\|}{C}}OH\right]^- \xrightarrow{CO} [Cl_2(CO)PdCH_2CH_2CO_2H]^- \qquad (36)$$

$$[Cl_2(CO)PdCH_2CH_2CO_2H]^- \longrightarrow CH_2{=}CHCO_2H + Pd + HCl + Cl^- + CO \qquad (37)$$

$$[Cl_2(CO)PdCH_2CH_2CO_2H]^- \xrightarrow{AcO^-} AcOCH_2CH_2CO_2H + Pd + 2\,Cl^- + CO \qquad (38)$$

The by-product, β-acetoxypropionic acid, formed in eq 38, may be thermally decomposed to yield additional acrylic acid.

$$AcOCH_2CH_2CO_2H \xrightarrow{\Delta} CH_2{=}CHCO_2H + AcOH \qquad (39)$$

Under anhydrous conditions, $PdCl_2$ and ethylene in the presence of carbon monoxide yield β-chloropropionyl chloride.[52]

$$PdCl_2 + CH_2{=}CH_2 + CO \longrightarrow ClCH_2CH_2COCl + Pd^0 \qquad (40)$$

Oxidative carbonylation of conjugated dienes such as 1,3-butadiene under anhydrous conditions affords unsaturated diesters.[53]

$$CH_2{=}CHCH{=}CH_2 + (\text{1,1-dimethoxycyclohexane}) + 2\,CO + \tfrac{1}{2}O_2 \xrightarrow[MeOH]{[Pd^{II}, Cu^{II}]} MeO_2CCH_2CH{=}CHCH_2CO_2Me + (\text{cyclohexanone})$$

Terminal acetylenes can also be oxidatively carbonylated by catalytic amounts of palladium(II) in the presence of cupric chloride.[54]

$$RC{\equiv}CH + CO + MeOH \xrightarrow[2\,CuCl_2]{[Pd^{II}]} RC{\equiv}CCO_2Me + 2\,HCl$$

IV. OXIDATION OF AROMATIC COMPOUNDS BY PALLADIUM(II) COMPLEXES

A. Oxidative Arylation

When benzene is heated with $PdCl_2$ and NaOAc at 90°C in acetic acid, biphenyl is formed together with palladium metal.[55] There is no reaction in the absence of NaOAc.

$$PhH + PdCl_2 + 2\,NaOAc \longrightarrow Ph_2 + Pd + 2\,NaCl + 2\,HOAc \qquad (41)$$

With monosubstituted benzenes, a mixture of isomeric biphenyls is obtained, the substitution pattern corresponding to that generally observed in electrophilic aromatic substitution. The latter is consistent with a mechanism involving a rate-determining electrophilic attack by $PdCl_2$. The rate is greatly accelerated by strong acids such as perchloric[56] and trifluoroacetic acids,[57] reminiscent of the enhancing effect of strong acids in electrophilic mercuration, thallation, and plumbation of arenes discussed in Chapter 5. [The σ-arylmercury(II) and σ-arylthallium(III) complexes are thermally stable and do not readily decompose to afford biaryls.] A mechanism was suggested[58] for biaryl formation, which involves a 1,2-addition of the arylpalladium(II) intermediate to the arene:

Scheme VI:

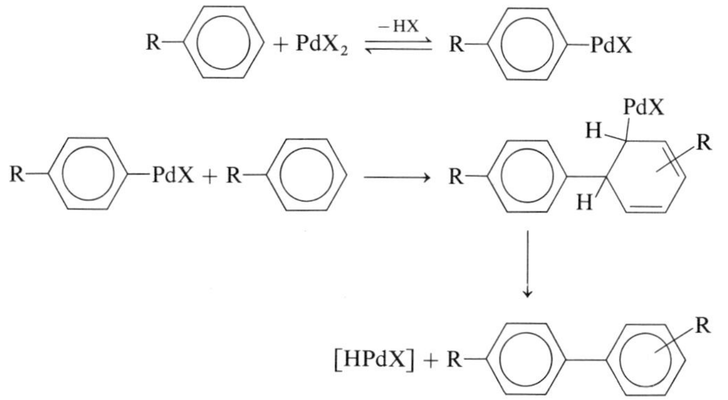

$$\text{R}-\text{C}_6\text{H}_5 + \text{PdX}_2 \xrightleftharpoons{-\text{HX}} \text{R}-\text{C}_6\text{H}_4-\text{PdX} \quad (42)$$

$$\text{R}-\text{C}_6\text{H}_4-\text{PdX} + \text{R}-\text{C}_6\text{H}_5 \longrightarrow \text{R}-\text{C}_6\text{H}_4-\text{C}_6\text{H}_5(\text{H})(\text{PdX})\text{R} \quad (43)$$

$$\downarrow$$

$$[\text{HPdX}] + \text{R}-\text{C}_6\text{H}_4-\text{C}_6\text{H}_4-\text{R} \quad (44)$$

An alternative mechanism involves the homolysis of the arylpalladium(II) intermediate, followed by addition of the aryl radical to the arene and electron transfer oxidation of the resulting cyclohexadienyl radical. It is difficult to predict substitution patterns for such a mechanism, since it involves both an electrophilic substitution and a homolytic addition.

Scheme VII:

$$\text{ArPd}^{\text{II}}\text{X} \xrightarrow{\Delta} \text{Ar}\cdot + \text{Pd}^{\text{I}}\text{X} \quad (45)$$

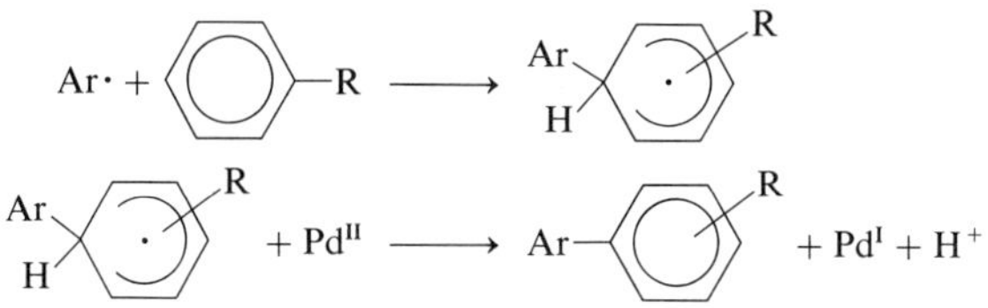

$$\text{Ar}\cdot + \text{C}_6\text{H}_5-\text{R} \longrightarrow \text{Ar}(\text{H})\text{C}_6\text{H}_5^{\cdot}-\text{R} \quad (46)$$

$$\text{Ar}(\text{H})\text{C}_6\text{H}_5^{\cdot}-\text{R} + \text{Pd}^{\text{II}} \longrightarrow \text{Ar}-\text{C}_6\text{H}_4-\text{R} + \text{Pd}^{\text{I}} + \text{H}^+ \quad (47)$$

The putative arylpalladium(II) intermediate is obtained more readily by ligand exchange between Pd(II) and an arylmercury(II) compound,

$$\text{ArHgX} + \text{PdX}_2 \longrightarrow \text{ArPdX} + \text{HgX}_2 \quad (48)$$

which then affords biaryls in a subsequent step. In the presence of oxygen and elevated temperatures (150°C), the oxidation of arenes to biaryls can be made catalytic in palladium.[59]

All the reactions described above appear to involve electrophilic substitution of arenes by Pd(II) and are therefore not fundamentally different from metallation with strong electrophiles such as Tl(III) and Pb(IV). Strictly speaking, these reactions do not truly represent activation of substrates by coordination. The novelty of the Pd(II) system lies in the reactions of the organopalladium(II) intermediate, which are not available for the arylthallium(III) and aryllead(IV) counterparts. For example, in the presence of olefins, *oxidative arylation*[60–63] of the olefin results.

$$\mathrm{ArPdX} + \mathrm{{>}C{=}C{<}_{H}} \longrightarrow \mathrm{{>}C{=}C{<}_{Ar}} + [\mathrm{HPdX}] \quad (49)$$

This synthetically useful reaction has been studied extensively by Heck,[62] as well as by Moritani and co-workers,[61,63] and is generally considered to proceed by an intramolecular nucleophilic addition of the aryl ligand to the coordinated olefin, followed by β-hydride elimination.[26,60–63]

$$\mathrm{C{=}C\cdots Pd(X)(Ar)} \longrightarrow \mathrm{(Pd{-}X)C{-}C(Ar)}, \quad \text{etc.} \quad (50)$$

$$\mathrm{Ar{-}C(H){-}C{-}Pd{-}X} \longrightarrow \mathrm{{}^{Ar}{>}C{=}C{<}} + [\mathrm{HPdX}] \quad (51)$$

Since palladium salts can be employed catalytically in the presence of silver or copper acetate and oxygen,[64,65] this process has been proposed as a viable route for the direct conversion of benzene to styrene.[64,65]

$$\mathrm{PhH} + \mathrm{C_2H_4} + \tfrac{1}{2}\,\mathrm{O_2} \xrightarrow[\mathrm{HOAc}]{[\mathrm{Pd^{II}},\ \mathrm{Cu^{II}}]} \mathrm{PhCH{=}CH_2} + \mathrm{H_2O} \quad (52)$$

A useful alternative to the oxidative arylation reaction is the Pd(0)-catalyzed vinylic substitution reaction studied extensively by Heck.[66]

$$\mathrm{RX} + \mathrm{{>}C{=}C{<}_{H}} + \mathrm{R_3N} \xrightarrow{[\mathrm{Pd^0}]} \mathrm{{>}C{=}C{<}_{R}} + \mathrm{R_3NH^+\,X^-} \quad (53)$$

It should be noted, however, that this reaction is catalytic in Pd(0) and does not involve an oxidation. In this case the putative aryl- or alkylpalladium(II) intermediate is obtained from oxidative addition of the organic halide to

Pd(0), i.e.,

$$RX + PdL_4 \longrightarrow R{-}\overset{\displaystyle L}{\underset{\displaystyle L}{Pd}}{-}X + 2\,L \tag{54}$$

$$R{-}\overset{\displaystyle L}{\underset{\displaystyle L}{Pd}}{-}X + R'CH{=}CH_2 \rightleftharpoons \begin{matrix} R'{-}CH \\ \| \\ H{-}CH \end{matrix} \quad \overset{\displaystyle X}{\underset{\displaystyle R}{Pd}}{-}L \tag{55}$$

$$\left[H{-}\overset{\displaystyle L}{\underset{\displaystyle L}{Pd}}{-}X\right] + R'CH{=}CHR \xleftarrow{L} R'CH(PdXL){-}CH(R)H \tag{56}$$

B. Oxidative Nucleophilic Substitution

Phenyl acetate is often formed as a by-product in the oxidation of benzene with $Pd(OAc)_2$.[56] This reaction is of practical interest, since the direct acetoxylation of arenes is not readily achieved. Thus, many workers have addressed the problem of increasing the yield of substitution products during the oxidation of arenes with Pd(II) compounds. Nucleophilic displacement at the α-carbon of σ-arylpalladium complexes should be feasible, by analogy with the corresponding σ-alkylpalladium(II) complexes considered above, i.e.,

$$Y^- \curvearrowright Ar{-}Pd^{II}{-}X \longrightarrow ArY + Pd^0 + X^- \tag{57}$$

Indeed, the following oxidative substitution of arenes has been reported by Henry.[67]

$$ArH + X^- + \text{oxidant} \xrightarrow{[Pd^{II}]} ArX, \qquad \text{etc.} \tag{58}$$

Oxidative substitution was observed with X = OAc, N_3, Cl, NO_2, CN, and SCN employing $K_2Cr_2O_7$, $Pb(OAc)_4$, $KMnO_4$, $NaClO_3$, $NaNO_3$, and $NaNO_2$ as the oxidants. The two pathways represented in eqs 59 and 60 were envisaged:

$$ArH + PdX_2 \xrightarrow{-HX} ArPdX \xrightarrow[X^-]{[\text{oxidant}]} ArX + PdX_2 \tag{59}$$

$$PdX_2 + [Ox] \xrightarrow{2\,X^-} Pd^{IV}X_4 \xrightarrow{ArH} ArX + PdX_2 + X^- \tag{60}$$

Eberson and co-workers[68–70] made extensive studies of the Pd(II)-mediated acetoxylation of arenes and showed that nuclear acetoxylation is promoted by dioxygen if excess of acetate ion is avoided. The addition of alkali metal acetates favors the formation of products derived from side-chain acetoxylation (*vide infra*). Lewis acids, on the other hand, promote the formation of biaryls (compare the action of perchloric and trifluoroacetic acids discussed earlier). Interestingly, a complete reversal of the usual pattern of isomer distribution for electrophilic aromatic substitution or anodic oxidation was observed in the Pd(II)-mediated acetoxylation of substituted arenes.[68] These results were rationalized on the basis of an addition–elimination sequence:

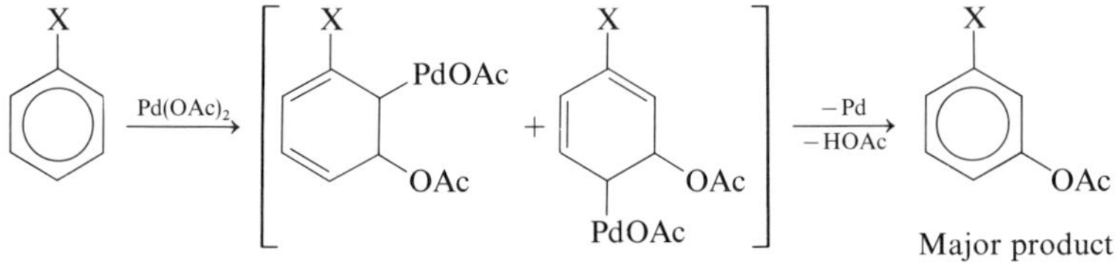
X
Pd(OAc)2
PdOAc
OAc
+
PdOAc
−Pd
−HOAc
Major product

(61)

where X = *t*-Bu, MeO, Cl, Br, etc. Oxidative acetoxylations employing $Pd(OAc)_2$ and dioxygen are generally slow, owing to the inefficient reoxidation of palladium(0). The rate can be increased significantly by the addition of stoichiometric amounts of nitrate, dichromate, or persulfate salts,[69–70] which then become the terminal oxidants (compare reaction 58). Alternatively, the reaction can be carried out at high temperatures with palladium dispersed on a solid support in order to promote reoxidation of the catalyst. Thus, Arpe and Hörnig[71] carried out a detailed investigation of the palladium-catalyzed acetoxylation of benzene in acetic acid. A selectivity of 78% was observed using a combination of palladium and gold on SiO_2 at 155°C.

Oxidative acetoxylation with alkylbenzenes as substrates can lead to substitution of both the aromatic nucleus and the side chain. For example, toluene reacts with $Pd(OAc)_2$ in acetic acid to afford a mixture of benzyl acetate and bitolyls.[56,72–74] The ratio of these products is dramatically influenced by the molar ratio of acetate relative to Pd(II). For example, at molar ratios of NaOAc and $PdCl_2$ equal to 5 and 20, the relative yields of bitolyl and benzyl acetate were 64 : 2 and 1 : 68, respectively.[72] A catalytic process was developed which employed catalyst consisting of $Pd(OAc)_2$ and $Sn(OAc)_2$ with dioxygen in acetic acid at 100°C.

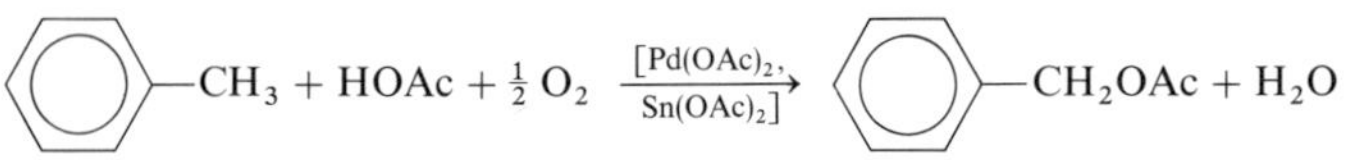
CH3 + HOAc + ½ O2
[Pd(OAc)2, Sn(OAc)2]
CH2OAc + H2O

(62)

The reaction is facilitated by electron-releasing groups; e.g., *p*-methoxytoluene afforded a 96% yield of *p*-methoxybenzyl acetate, but *p*-nitrotoluene gave only a 2% yield of *p*-nitrobenzyl acetate.[74] The mechanism of side-chain acetoxylation is still a matter of conjecture.[5,48,68] A scheme can be envisaged[48] involving several competing pathways for the initially formed σ complex.

CH_3–C_6H_5 $\xrightarrow{Pd(OAc)_2}$ σ complex (AcOPd, H; CH_3) $\xrightarrow{AcO^-}$ (AcOPd)(H)(OAc)-methylcyclohexadiene $\xrightarrow{-[HPdOAc]}$ $CH_3C_6H_4OAc$ (63)

σ complex $\xrightarrow{-HOAc}$ $CH_3C_6H_4PdOAc$ $\longrightarrow$ $C_6H_5CH_2PdOAc$ $\xrightarrow{-Pd}$ $C_6H_5CH_2OAc$ (64)

$CH_3C_6H_4PdOAc$ $\longrightarrow$ bitolyls

However, the promotion of side-chain acetoxylation by dioxygen and a high concentration of acetate is not readily explained.[48]

C. Oxidative Carbonylation of Arenes

By analogy with the Pd(II)-catalyzed oxidative carbonylation of olefins described earlier, oxidative carbonylation of arenes might be expected to be a feasible reaction, i.e.,

$$ArH + CO + \tfrac{1}{2}O_2 \xrightarrow{[Pd^{II}, Cu^{II}]} ArCO_2H, \quad \text{etc.} \tag{65}$$

An example of such a reaction has been reported, namely, the oxidative carbonylation of naphthalene to a mixture of naphthalenecarboxylic acids, with $Pd(OAc)_2$ as the catalyst in acetic acid at 100°C.[75]

$$C_{10}H_8 + CO + \tfrac{1}{2}O_2 \xrightarrow{[Pd(OAc)_2]} 1\text{-}C_{10}H_7CO_2H + 2\text{-}C_{10}H_7CO_2H, \quad \text{etc.} \tag{66}$$

V. OXIDATION OF ALCOHOLS TO CARBONYL COMPOUNDS

Group VIII metals such as Pt(II), Pd(II), Ru(III), and Rh(III) readily oxidize alcohols to carbonyl compounds by a process involving the β-hydride elimination of a metal alkoxide intermediate.[76]

$$\begin{array}{c} \diagdown \\ \diagup \end{array}\!\!\underset{\substack{| \\ H}}{C}-O-M^{n+} \longrightarrow \begin{array}{c} \diagdown \\ \diagup \end{array}\!\!C{=}O + HM^{n+} \tag{67}$$

Although this reaction is often employed for the synthesis of Group VIII metal hydrides, it has received little attention as a preparative procedure for the oxidation of alcohols.

When $PtCl_2(PEt_3)_2$ is refluxed in an ethanol solution of potassium hydroxide, the platinum hydride is obtained together with acetaldehyde.[77]

$$PtCl_2(PEt_3)_2 + KOH + EtOH \xrightarrow[-KCl]{KOH} HPtCl(PEt_3)_2 + CH_3CHO + H_2O \tag{68}$$

The reaction of alcohols with Pd(II) salts is reported to proceed catalytically when copper(II) is used as a cocatalyst at 70–120°C and 3 atm O_2 pressure.[78]

$$RCH_2OH + \tfrac{1}{2}\,O_2 \xrightarrow{[Pd^{II}, Cu^{II}]} RCHO + H_2O \tag{69}$$

A more recent study[79] found secondary alcohols to be readily oxidized at ambient temperature and 1 atm O_2 pressure in the presence of catalytic amounts of $PdCl_2$ and NaOAc. It is noteworthy that this reaction was retarded by Cu(II) salts.

In the presence of both carbon monoxide and dioxygen, alcohols undergo Pd(II)-catalyzed *oxidative carbonylation* to produce oxalate esters. The reaction is catalytic in Pd(II) when Cu(II) is used as a cocatalyst.[80]

$$2\,ROH + 2\,CO + \tfrac{1}{2}\,O_2 \xrightarrow[125^\circ C]{[Pd^{II}, Cu^{II}]} \begin{array}{c} CO_2R \\ | \\ CO_2R \end{array} + H_2O \tag{70}$$

A more recent development[81] is the use of nitrogen oxides, instead of copper salts, as the cocatalyst. (Compare the use of nitrate as cocatalyst in the oxidative addition to olefins.[35–37a]) Both liquid- and vapor-phase processes were described.[81] The reaction may proceed by the following sequence of reactions.

Scheme VIII:

$$ROH + N_2O_3 \longrightarrow 2\,RONO + H_2O \tag{71}$$

$$Pd^0 + 2\,RONO \longrightarrow Pd^{II}(OR)_2 + 2\,NO \tag{72}$$

$$Pd^{II}(OR)_2 + 2\,CO \longrightarrow Pd^{II}(CO_2R)_2 \tag{73}$$

$$Pd(CO_2R)_2 \longrightarrow Pd^0 + (CO_2R)_2 \tag{74}$$

$$2\,NO + \tfrac{1}{2}\,O_2 \longrightarrow N_2O_3 \tag{75}$$

VI. PALLADIUM(II)-CATALYZED OXIDATIONS WITH HYDROPEROXIDES

Mimoun[82] has recently reported the preparation of novel alkylperoxopalladium(II) carboxylates from *tert*-butyl hydroperoxide and a palladium(II) carboxylate.

$$Pd(O_2CR)_2 + BuO_2H \rightleftharpoons BuO_2PdO_2CR + RCO_2H \tag{76}$$

where R = $CH_3, CF_3, CCl_3, C_5F_{11}$. These compounds are soluble in organic solvents, and are capable of oxidizing terminal olefins to methyl ketones. The reaction is inhibited by pyridine, triphenylphosphine, and water, which is not characteristic of a Wacker process.

$$BuO_2PdO_2CR + R'CH{=}CH_2 \longrightarrow R'COCH_3 + BuOPdO_2CR \tag{77}$$

With $Pd(OAc)_2$ as the precursor in eq 76, the reaction is rather slow at ambient temperature. However, if Pd(II) trifluoroacetate (R = O_2CCF_3) is the precursor, a rapid reaction is observed with 1-octene in benzene at ambient temperature to afford 2-octanone in 98% yield. A mechanism was proposed[82] in which the alkylperoxopalladation of the olefin produces an open-chain analog of the peroxymetallocycles proposed as intermediates in the Rh(I)-catalyzed oxidations of terminal olefins with molecular oxygen discussed in Chapter 4.

Scheme IX:

$$\begin{array}{c} CH_2 \\ MeCO_2Pd-\| \\ |\quad CHR' \\ BuOO \end{array} \longrightarrow \begin{array}{c} H\quad H \\ C \quad H \\ MeCO_2Pd \quad C \\ O{-}O \quad R \\ Bu \end{array} \longrightarrow \begin{array}{c} MeCO_2PdOBu \\ + \\ O \\ \| \\ CH_3CR' \end{array} \tag{78a}$$

$$MeCO_2PdOBu + BuO_2H \longrightarrow MeCO_2PdO_2Bu + BuOH \tag{78b}$$

Ligand exchange in eq 78b allows the reaction to be catalytic in Pd(II) in the presence of excess TBHP. The reaction is not successful with cumene or ethylbenzene hydroperoxides, probably because the alkylperoxypalladium(II) intermediate spontaneously decomposes by aryl migration from carbon to oxygen, which is a common pathway for the acid-catalyzed decomposition of hydroperoxides.

$$\begin{array}{c} Ar \qquad H^+ \\ {-}C{-}O \\ \quad O{-}Pd \end{array} \longrightarrow HOPd^{II} + \overset{+}{C}{-}OAr, \quad \text{etc.} \tag{79}$$

The oxidation of α,β-unsaturated carbonyl compounds to 1,3-diketones with *tert*-alkyl hydroperoxides in the presence of catalytic amounts of Na_2PdCl_4

proceeds at 50°–80°C in aqueous acetic acid, isopropyl alcohol, or *N*-methylpyrrolidine.[83]

$$\mathrm{R{-}CH{=}CH{-}C({=}O){-}R} + t\text{-BuOOH} \xrightarrow{[\mathrm{Pd}]} \mathrm{R{-}C({=}O){-}CH_2{-}C({=}O){-}R} + t\text{-BuOH} \qquad (80)$$

Essentially the same mechanism as that presented in Scheme IX is undoubtedly involved in the palladium-catalyzed conversion of terminal olefins to methyl ketones with hydrogen peroxide,

$$\mathrm{RCH{=}CH_2 + H_2O_2} \xrightarrow{[\mathrm{Pd}]} \mathrm{RC({=}O)CH_3 + H_2O} \qquad (81)$$

as discussed in Chapter 4, Section III.

VII. ACTIVATION OF ALKANES BY METAL COMPLEXES

The interaction of saturated C—H and C—C bonds with supported noble metal catalysts (particularly platinum) at elevated temperatures forms the basis of several important petrochemical processes, such as the dehydrocyclization of alkanes to aromatic compounds. In recent years, much attention has been devoted to the selective activation of unactivated sp^3 hybridized C—H bonds by the use of soluble transition metal complexes under mild conditions.[84–88] Conceptually, the insertion of a low-valent Group VIII metal complex into a C—H bond should be feasible,

$$-\overset{|}{\underset{|}{\mathrm{C}}}-\mathrm{H} + \mathrm{M}^{n+} \rightleftharpoons -\overset{|}{\underset{|}{\mathrm{C}}}-\underset{\underset{\mathrm{H}}{|}}{\mathrm{M}}{}^{(n+2)+} \qquad (82)$$

by analogy with the oxidative addition of molecular hydrogen to these complexes.

$$\mathrm{H{-}H} + \mathrm{M}^{n+} \rightleftharpoons \mathrm{H}-\underset{\underset{\mathrm{H}}{|}}{\mathrm{M}}{}^{(n+2)+} \qquad (83)$$

Since the dissociation energy of the M—C bond is only slightly less than that of M—H bonds,[89] such an activation of alkanes is thermodynamically feasible. The C—H bond in an alkane can also be cleaved by strong eletrophiles, according to the formulation:

$$-\overset{|}{\underset{|}{\mathrm{C}}}-\mathrm{H} + \mathrm{M}^{n+} \rightleftharpoons -\overset{|}{\underset{|}{\mathrm{C}}}-\mathrm{M}^{n+} + \mathrm{H}^{+} \qquad (84)$$

This type of reaction has been discussed in Chapter 5 and is relevant to oxidations of alkanes with strong oxidants such as Co(III). We shall be concerned here with those reactions in which oxidative addition described by reaction 82 is the basic premise. This process has been compared[86] to the insertion of a singlet carbene, such as methylene, into a C—H bond. The unequivocal demonstration of the viability of oxidative addition of alkanes to metal centers is shown by the gas-phase ion–molecule reaction between isobutane and Co^+ (generated by the thermal decomposition of $CoCl_2$ and surface ionization of Co) in which both C—H and C—C insertions are observed.[90]

$$Co^+ + (CH_3)_3CH \longrightarrow \begin{cases} H\text{—}Co\text{—}C(CH_3)_3{}^+ \\ H\text{—}Co\text{—}CH_2CH(CH_3)_2{}^+ \\ CH_3\text{—}Co\text{—}CH(CH_3)_2{}^+ \end{cases}$$

Similar oxidative additions of alkanes occur with Fe^+ generated from $Fe(CO)_5$ by electron impact.[91] Even more relevant is the direct reaction of zirconium atoms (produced by the vaporization of zirconium metal on a tungsten filament) with isobutane and neopentane in condensed matrices at 77°K.[92] Evidence for both C—H and C—C insertions was based on the hydrolysis of the oxidative adducts, e.g.,

$$Zr + (CH_3)_4C \longrightarrow \begin{cases} HZrCH_2C(CH_3)_3 \xrightarrow{H_2O} (CH_3)_4C, \quad \text{etc.} \\ CH_3ZrC(CH_3)_3 \xrightarrow{H_2O} (CH_3)_3CH, \quad \text{etc.} \end{cases}$$

The reversible insertion of a metal into an aliphatic C—H bond is represented in the intramolecular example below.[93]

$$[(CH_3)_2PCH_2CH_2P(CH_3)_2]_2Ru^0 \rightleftharpoons [(CH_3)_2PCH_2CH_2P(CH_3)_2](CH_3)_2PCH_2CH_2P(CH_3)CH_2\text{—}Ru^{II}\text{—}H \qquad (85)$$

In the solid state, the equilibrium in eq 85 is shifted toward the Ru(II) adduct.[94]

Foley and Whitesides[95] have recently shown that the conversion of dineopentylbis(triethylphosphine)platinum(II) to bis(triethylphosphine)-3,3-dimethylplatinacyclobutane involves an initial dissociation of a phosphine ligand, followed by the insertion of the coordinatively unsaturated Pt(II) into an aliphatic C—H bond.

Scheme X:

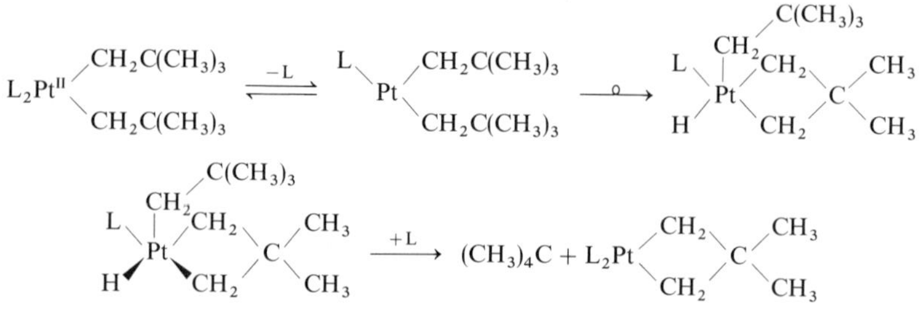

$$L_2Pt^{II}(CH_2C(CH_3)_3)_2 \underset{}{\overset{-L}{\rightleftharpoons}} L\,Pt(CH_2C(CH_3)_3)_2 \xrightarrow{\alpha} L(H)Pt(CH_2C(CH_3)_3)(CH_2)_2C(CH_3)_2$$

$$L(H)Pt(CH_2C(CH_3)_3)(CH_2)_2C(CH_3)_2 \xrightarrow{+L} (CH_3)_4C + L_2Pt(CH_2)_2C(CH_3)_2$$

where $L = Et_3P$. The dissociative mechanism in Scheme X suggests that coordinative unsaturation on the metal is a requirement for facile insertion, a situation which exists on metal surfaces.

Since the C—H bonds in alkanes are generally stronger than the corresponding C—M bonds, the equilibrium in eq 82 is likely to lie far to the left. However, the existence of such an equilibrium should lead to hydrogen–deuterium exchange under suitable conditions. Indeed, rapid deuteriation of alkanes is observed in mixtures of D_2O and CH_3CO_2D in the presence of chloroplatinum(II) complexes under comparatively mild conditions (100°C).[96,97] A mechanism was proposed involving the following reversible steps in a solvent (S).[86,87,98]

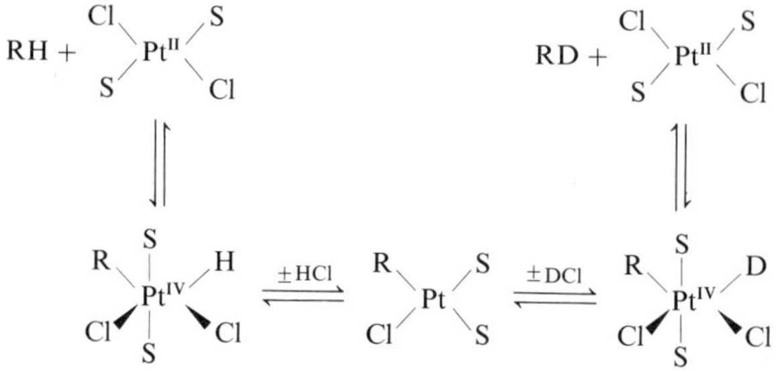

$$RH + Pt^{II}Cl_2S_2 \rightleftharpoons R(H)Pt^{IV}Cl_2S_2$$

$$R(H)Pt^{IV}Cl_2S_2 \underset{}{\overset{\pm HCl}{\rightleftharpoons}} R(Cl)PtS_2 \underset{}{\overset{\pm DCl}{\rightleftharpoons}} R(D)Pt^{IV}Cl_2S_2$$

$$R(D)Pt^{IV}Cl_2S_2 \rightleftharpoons RD + Pt^{II}Cl_2S_2$$

Indirect evidence for an alkylplatinum(II) intermediate was obtained from the formation of deuteriated alkanes during the reaction of Pt(II) complexes with alkylmercury(II) compounds in a mixture of D_2O and CH_3CO_2D.

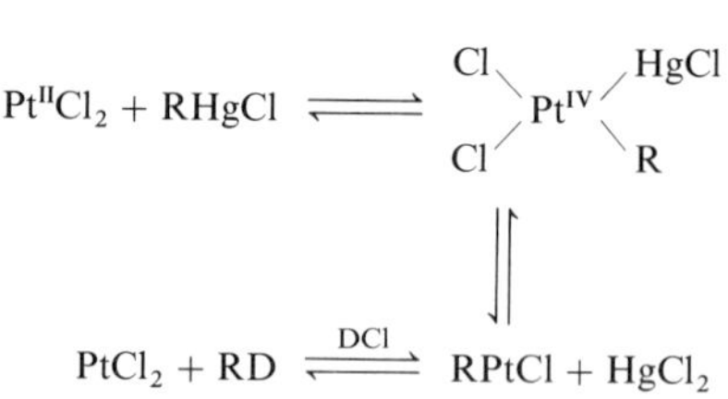

$$Pt^{II}Cl_2 + RHgCl \rightleftharpoons Cl_2Pt^{IV}(HgCl)(R)$$

$$Cl_2Pt^{IV}(HgCl)(R) \rightleftharpoons RPtCl + HgCl_2$$

$$PtCl_2 + RD \underset{}{\overset{DCl}{\rightleftharpoons}} RPtCl + HgCl_2$$

In principle, these results can also be explained with a mechanism involving electrophilic substitution, as in reaction 84. [Cyclometallation of nitrogen heterocycles with palladium(II) is an electrophilic process.[99]] However, there are several observations that militate against such a mechanism.[86] First, the rate of exchange is retarded in the more weakly solvating trifluoroacetic acid (TFA), which contrasts with the rate-enchancing effect of TFA on reactions of alkanes with strong electrophiles such as Co(III), as discussed in Chapter 5. Second, Pt(IV), which should be a stronger electrophile than Pt(II), is inactive. Masters reported that complexes of the type $Pt_2Cl_4L_2$, where $L = (n\text{-}Pr)_3P$ or $(n\text{-}Bu)_3P$, readily undergo selective H–D exchange at the C-3 position of the alkyl group in the phosphine ligand.[100] The following mechanism involving oxidative addition of a saturated C—H function to Pt(II) was proposed.

$$Pr_3P(Cl)Pt(\mu\text{-}Cl)_2Pt(Cl)PPr_3 \rightleftharpoons Pr_3P(Cl)PtCl \rightleftharpoons \text{(cyclometallated } Pr_2P\text{–}Pt(H)Cl_2)$$

In contrast to Pt(II), Pd(II) does not catalyze H–D exchange with alkanes.[101] Instead, alkanes reduce $Pd(O_2CCF_3)_2$ in TFA; e.g., cyclohexane is dehydrogenated to benzene.[102] (Compare the reactions of alkanes with Co(III) trifluoroacetate in Chapter 5.) These reactions exhibit all the characteristics of an electrophilic attack or electron transfer on the substrate (see Chapter 5). Rudakov[101] concluded that Pt(II) and Pd(II) interactions with alkanes involve different mechanisms—Pt(II) forming an alkylplatinum(II) intermediate via one of the following three transition states:

$$Pt\cdots(C\text{—}H) \qquad Pt\cdots C\text{—}H\cdots OH_2 \qquad \begin{matrix} Pt\cdots C\text{—} \\ | \quad\ \ | \\ L\cdots H \end{matrix}$$

whereas Pd(II) reacts by an outer-sphere electron transfer to give an olefin complex:

$$Pd^{2+} + RCH_2CH_3 \longrightarrow [Pd^{+}\ RCH_2CH_3^{\dot{+}}]$$

$$[Pd^{+}\ RCH_2CH_3^{\dot{+}}] \xrightarrow{-H^+} [Pd^{+}\ R\dot{C}HCH_3] \xrightarrow{-H^+} Pd^0\text{—}(RHC{=}CH_2)$$

However, the results could also be interpreted as an electrophilic substitution by Pd^{2+} as the initial step.

Alkane activation is not limited to Group VIII metals. The activation of methane toward H–D exchange by Ziegler–Natta type of catalysts such as $TiCl_4 - Me_2AlCl$ and $VCl_3 - Me_2AlCl$ in heptane solutions at 20° – 50°C was recently reported.[103] Reversible formation of metal ylides by α-elimination was postulated.

$$\rangle M(CH_3)_2 \rightleftharpoons \rangle M{=}CH_2 + CH_4 . \tag{86}$$

Further investigations of alkylidene–metal complexes[104] in this context is worthwhile.

Shilov and co-workers[105] have reported that C_1 to C_4 alkanes in the presence of stoichiometric amounts of $SnCl_2$ are oxidized to alcohols by molecular oxygen at room temperature in acetonitrile. The yield of alcohols varies from 7 to 15% with respect to tin. Since the relative reactivities of primary, secondary, and tertiary C—H bonds correspond closely with those observed in free radical reactions, it was concluded[84] that the key step involved hydrogen abstraction by a superoxotin(III) species.

$$RH + \cdot O{-}O{-}Sn^{III} \longrightarrow R\cdot + HO{-}O{-}Sn^{III} \tag{87}$$

Selective hydroxylation of alkanes under mild conditions has also been achieved with biomimetic reagents consisting of an iron(II) salt and a co-reductant. These reactions are discussed in Chapter 8.

It is also possible for the insertion in eq 82 to occur in stepwise fashion, involving hydrogen atom transfer as the initial step. Such a process has been clearly demonstrated[106] in the reaction of the electrochemically generated rhodium(0) complex $Rh(diphos)_2$ with unactivated alkanes and aralkanes in which alkyl radicals are generated e.g.,

$$(diphos)_2Rh + C_6H_{12} \longrightarrow (diphos)_2RhH + C_6H_{11}\cdot \tag{88}$$

Since the hydridorhodium(I) product reacts readily with acids to regenerate the reactant $(diphos)_2Rh^+$,

$$(diphos)_2RhH + HX \longrightarrow (diphos)_2Rh^+X^- + H_2 \tag{89}$$

an electrocatalytic process for the activation of C—H bonds may be possible.

Alkane activation is thus clearly feasible under mild conditions. However, the results obtained so far with soft metal catalysts have not led to the development of synthetically useful reactions. The putative alkylmetal intermediates have so far eluded interception by all reagents except protons and deuterons. The selective functionalization of alkanes under mild conditions remains an important challenge.

REFERENCES

1. P. M. Maitlis, "The Organic Chemistry of Palladium," Vols. 1 and 2. Academic Press, New York, 1971.
2. F. R. Hartley, "The Chemistry of Platinum and Palladium." Applied Science, London, 1973.
3. F. R. Hartley, *Chem. Rev.* **69**, 799 (1969); *J. Chem. Educ.* **50**, 263 (1973).
4. E. W. Stern, *Catal. Rev.* **1**, 73 (1967).
5. E. W. Stern, *in* "Transition Metals in Homogeneous Catalysis" (G. N. Schrauzer, ed.), p. 93. Dekker, New York, 1971.
6. M. Herberhold, "Metal π-Complexes," Vol. II, Part 1. Elsevier, Amsterdam, 1972; Part 2, 1974.
7. P. N. Rylander, "Organic Syntheses with Noble Metal Catalysts," Org. Chem. Monogr., Vol. 28. Academic Press, New York, 1973.
8. D. M. Fenton and K. L. Olivier, *Chemtech.* **2**, 220 (1972).
9. C. W. Bird, *Chem. Ind. (London)* p. 520 (1972); R. Huttel, *Synthesis* p. 225 (1970).
10. R. F. Heck, "Organotransition Metal Chemistry." Academic Press, New York, 1974.
11. P. M. Henry, *Accts. Chem. Rsch.* **6**, 16 (1973).
12. P. M. Henry, *Adv. Organometal. Chem.* **13**, 363 (1975).
13. B. C. Gates, J. R. Katzer, and G. A. Schuit, "Chemistry of Catalytic Processes," p. 129. McGraw-Hill, New York, 1979.
14. R. Jira and W. Freiesleben, *Organomet. React.* **3**, 1(1972).
15. J. Smidt, W. Hafner, R. Jira, J. Sedlmeier, R. Sieber, R. Ruttinger, and H. Kojer, *Angew. Chem.* **71**, 176 (1959).
16. J. Smidt, W. Hafner, R. Jira, J. Sedlmeier, and A. Sabel, *Angew. Chem. Int. Ed.* **1**, 80 (1962).
17. J. Smidt, *Chem. Ind. (London)* p. 54 (1962).
18. W. Hafner, R. Jira, J. Sedlmeier, J. Smidt, P. Fliegel, W. Friedrich, and A. Trommett, *Chem. Ber.* **95**, 1575 (1962).
19. R. Jira, J. Sedlmeier, and J. Smidt, *Justus Liebigs Ann. Chem.* **693**, 99 (1966).
20. G. Szonyi, *Adv. Chem. Series* **70**, 53 (1968).
21. P. M. Henry, *J. Am. Chem. Soc.* **88**, 1595 (1966).
22. A. Aguilo, *Adv. Organometal. Chem.* **5**, 321 (1967).
23. P. M. Henry, *J. Am. Chem. Soc.* **86**, 3246 (1964).
24. P. M. Henry, *Adv. Chem. Series* **70**, 126 (1968).
25. I. I. Moiseev, O. G. Levanda, and M. N. Vargaftik, *J. Am. Chem. Soc.* **96**, 1003 (1974).
26. J. E. Bäckvall, B. Åkermark, and S. O. Ljunggren, *J. Am. Chem. Soc.* **101**, 2411 (1979); *J. Chem. Soc., Chem. Commun.* p. 264 (1977).
27. J. K. Stille and R. Divakaruni, *J. Organometal. Chem.* **169**, 239 (1979); *J. Am. Chem. Soc.* **100**, 1303 (1978).
28. H. Stangl and R. Jira, *Tetrahedron Lett.* p. 3589 (1970).
29. J. E. Bäckvall, *Tetrahedron Lett.* p. 467 (1977).
30. R. A. Budnik, and J. K. Kochi, *J. Organometal. Chem.* **116**, C3 (1976).
31. A. B. Evnin, J. A. Rabo, and P. H. Kasai, *J. Catal.* **30**, 109 (1973).
32. J. L. Seoane, P. Boutry, and R. Montarnal, *J. Catal.* **63**, 182, 191 (1980).

33. I. I. Moiseev, M. N. Vargaftik, and Y. Syrkin, *Proc. Acad. Sci. USSR, Phys. Chem. Sect.* **133**, 801 (1960).
34. (a) S. A. Miller, *in* "Ethylene and Its Industrial Derivatives" (S. A. Miller, ed.), pp. 946-956. Benn, London, 1969.
    (b) H. Krekeler and W. Kronig, *Proc. 7th World Pet. Congr.* **5**, 41-48 (1967).
35. R. van Helden, C. F. Kohll, D. Medema, G. Verberg, and T. Jonkhoff, *Recl. Trav. Chim. Pays-Bas* **87**, 961 (1968).
36. D. R. Bryant and J. E. McKeon, *Prepr., Div. Pet. Chem., Am. Chem. Soc.* **14**, Bl (1969).
37. D. Clark, P. Hayden, and R. D. Smith, *Discuss. Faraday Soc.* **46**, 98 (1968).
37a. D. Clark and P. Hayden, *Prepr., Div. Pet. Chem., Am. Chem. Soc.* **11**, D5 (1966). R. G. Schultz and P. R. Rony, *ibid.* **12**, 139 (1967); P. M. Henry, *ibid.* **14**, B15 (1969); W. H. Clement and C. M. Selwitz, *Tetrahedron Lett.* p. 1081 (1962).
38. R. H. Grubbs, A. Miyashita, M. Liu, and P. L. Burk, *J. Am. Chem. Soc.* **100**, 2418 (1978); G. M. Whitesides, J. F. Gaasch, and E. R. Stedronsky, *ibid.* **94**, 5258 (1972); J. X. McDermott, J. F. White, and G. M. Whitesides, *ibid.* **98**, 6521 (1976).
39. G. Wilkinson, *Pure Appl. Chem.* **30**, 627 (1972); F. Calderazzo, *ibid.* **33**, 453 (1973); R. S. Braterman and R. J. Cross, *Chem. Soc. Rev.* **2**, 271 (1973).
40. H. C. Volger, *Ind. Eng. Chem. Prod. Res. Dev.* **9**, 311 (1970).
41. C. F. Kohll and R. van Helden, *Recl. Trav. Chim. Pays-Bas* **87**, 481 (1968); E. W. Stern, M. L. Spector, and H. P. Leftin, *J. Catal.* **6**, 152 (1966).
42. A. Sabel, J. Smidt, R. Jira, and H. Prigge, *Chem. Ber.* **102**, 2939 (1969).
43. P. M. Henry, *J. Org. Chem.* **32**, 2575 (1967); *J. Am. Chem. Soc.* **94**, 4437 (1972).
44. R. G. Schultz and D. E. Gross, *Adv. Chem. Series* **70**, 97 (1968).
45. A. Mitsutani, *Chem. Econ. Eng. Rev.* **5**, 32 (1973).
46. P. M. Henry, *J. Org. Chem.* **38**, 1681 (1973).
47. J. K. Kochi, *Pure Appl. Chem.* **4**, 377 (1971), and references cited therein.
48. R. A. Sheldon and J. K. Kochi, *Adv. Catal.* **25**, 272 (1976).
49. Y. Ito, H. Aoyama, and T. Saegusa, *J. Am. Chem. Soc.* **102**, 4519 (1980).
50. T. Hosokawa, T. Uno, and S. -I. Murahashi, *J. Am. Chem. Soc.* **103**, 2318 (1981).
51. D. M. Fenton, K. L. Olivier, and G. Biale, *Prepr., Div. Pet. Chem., Am. Chem. Soc.* **14**, C77 (1969).
52. J. Tsuji, M. Morikawa, and J. Kiji, *Tetrahedron Lett.* p. 1061 (1963).
53. H. S. Kesling and L. R. Zehner, U.S. Patent 4,171,450 (1979) to Atlantic Richfield Co.
54. J. Tsuji, M. Takahashi, and T. Takahashi, *Tetrahedron Lett.* **21**, 849 (1980).
55. R. van Helden and G. Verberg, *Recl. Trav. Chim. Pays-Bas* **84**, 1263 (1965).
56. J. M. Davidson and C. Triggs, *J, Chem. Soc. A* pp. 1324, 1331 (1968); *Chem. Ind.* (*London*) p. 457 (1966).
57. F. R. S. Clark, R. O. C. Norman, C. B. Thomas, and J. S. Willson, *J. Chem. Soc., Perkin I* p. 1289 (1974).
58. M. O. Unger and R. A. Fouty, *J. Org. Chem.* **34**, 18 (1969).
59. H. Itatani and H. Yoshimoto, *Chem. Ind.* (*London*) p. 674 (1971).
60. A. K. Yatsimirsky, A. D. Ryabov, and I. V. Berezin, *J. Mol. Catal.* **5**, 399 (1979).
61. I. Moritani and J. Fujiwara, *Synthesis* p. 524 (1973); *Tetrahedron Lett.* p. 1119 (1967).
62. R. F. Heck, *J. Am. Chem. Soc.* **90**, 5518, 5526, 5531, 5535, 5542 (1968); **91**, 6707 (1969).
63. M. Watanabe, M. Yamamura, I. Moritani, Y. Fujiwara, and A. Sonoda, *Bul. Chem. Soc. Japan* **47**, 1035 (1974), and previous publications in this series.
64. R. S. Shue, *J. Am. Chem. Soc.* **93**, 7116 (1971); *J. Chem. Soc., Chem. Commun.* p. 1510 (1971).
65. R. S. Shue, *J. Catal.* **26**, 112 (1972).
66. R. F. Heck, *Pure Appl. Chem.* **50**, 691 (1978), and references cited therein.

67. P. M. Henry, *J. Org. Chem.* **36**, 1886 (1971).
68. L. Eberson and L. Gomez-Gonzales, *Acta Chem. Scand.* **27**, 1162, 1249, 1255 (1973); *J. Chem. Soc., Chem. Commun.* p. 263 (1971).
69. L. Eberson and E. Jönsson, *Acta Chem. Scand., Series B* **28**, 771 (1974); *J. Chem. Soc., Chem. Commun.* p. 885 (1974).
70. L. Eberson and L. Jönsson, *Acta Chem. Scand., Series B* **28**, 597 (1974).
71. H. J. Arpe and L. Hörnig, *Erdoel Kohle, Erdgas, Petrochem. Brennst.-Chem.* **23**, 79 (1970).
72. D. R. Bryant, J. E. McKeon, and B. C. Ream, *Tetrahedron Lett.* p. 3371 (1968).
73. D. R. Bryant, J. E. McKeon, and B. C. Ream, *J. Org. Chem.* **33**, 4123 (1968).
74. C. H. Bushweller, *Tetrahedron Lett.* p. 2285 (1968).
75. German Patent 2,340,592 (1974) to Teijin Ltd.; *Angew. Chem.* **86**, 866 (1974).
76. H. D. Kaesz and R. B. Saillant, *Chem. Rev.* **72**, 231 (1972).
77. J. Chatt and B. L. Shaw, *Chem. Ind. (London)* 931 (1960); *J. Chem. Soc.* p. 5075 (1962).
78. W. G. Lloyd, *J. Org. Chem.* **32**, 2816 (1967).
79. T. F. Blackburn and J. Schwartz, *J. Chem. Soc., Chem. Commun.* p. 157 (1977).
80. D. M. Fenton and P. J. Steinwand, U.S. Patent 3,393,136 (1968).
81. K. Nishimura, S. Uchiumi, K. Fujii, K. Nishihira, and H. Itatani, *Prepr., Div. Pet. Chem., Am. Chem. Soc.* p. 355 (1979).
82. H. Mimoun, *J. Mol. Catal.* **7**, 1 (1980); see also H. Mimoun, R. Charpentier, A. Mitscher, J. Fischer, and R. Weiss, *J. Am. Chem. Soc.* **102**, 1047 (1980).
83. J. Tsuji, H. Nagashima, and K. Hori, *Chem. Lett.* p. 257 (1980).
84. (a) I. S. Kolomnikov, V. P. Kukolev, and M. E. Volpin, *Russ. Chem. Rev.* **43**, 399 (1974). (b) A. E. Shilov and A. A. Shteinman, *Kinet. Katal.* (Eng. Trans.) **14**, 117 (1973); **18**, 924 (1977).
85. A. E. Shilov, *in* "Some Theoretical Problems of Catalysis" (T. Kwan, G. K. Boreskov, and K. Tamaru, eds.), p, 165. Plenum, New York, 1973.
86. A. E. Shilov and A. A. Shteinman, *Coord. Chem. Rev.* **24**, 97 (1977).
87. D. E. Webster, *Adv. Organometal. Chem.* **15**, 147 (1977).
88. G. W. Parshall, *Accts. Chem. Rsch.* **8**, 113 (1975); *Chemtech.* **4**, 445 (1974).
89. P. J. Davidson, M. F. Lappert, and R. Pearce, *Accts. Chem. Rsch.* **7**, 209 (1974).
90. P. B. Armentrout and J. L. Beauchamp, *J. Am. Chem. Soc.* **102**, 1736 (1980).
91. J. Allison, R. B. Freas, and D. P. Ridge, *J. Am. Chem. Soc.* **101**, 1332 (1979).
92. R. J. Remick, T. A. Asunta, and P. S. Skell, *J. Am. Chem. Soc.* **101**, 1320 (1979).
93. J. Chatt and J. M. Davidson, *J. Chem. Soc.* p. 843 (1965).
94. F. A. Cotton, D. L. Hunter, and B. A. Frenz, *Inorg. Chim. Acta* **15**, 155 (1975).
95. P. Foley and G. M. Whitesides, *J. Am. Chem. Soc.* **101**, 2732 (1979).
96. N. F. Gol'dshleger, M. B. Tyabin, A. E. Shilov, and A. A. Shteinman, *Russ. J. Phys. Chem.* **43**, 1222 (1969).
97. R. J. Hodges, D. E. Webster, and P. B. Wells, *J. Chem. Soc. A* p. 3230 (1971); *J. Chem. Soc., Chem. Commun.* p. 462 (1971); *J. Chem. Soc., Dalton* p. 2571 (1972).
98. M. B. Tyabin, A. E. Shilov, and A. A. Shteinman, *Dokl. Akad. Nauk SSSR* **198**, 380 (1971).
99. A. J. Deeming and I. P. Rothwell, *Pure Appl. Chem.* **52**, 649 (1980).
100. C. Masters, *J. Chem. Soc., Chem. Commun.* p. 192 (1973); p. 1258 (1972).
101. E. S. Rudakov, *Dokl. Akad. Nauk SSSR* **238**, 396 (1978).
102. N. F. Goldsheger, M. L. Khidekel, A. E. Shilov, and A. A. Shteinman, *Kinet. Katal.* (Eng. Trans.) **15**, 235 (1974).
103. E. A. Grigorian, F. S. D'yachkovsky, and I. R. Mullagaliev, *Dokl. Akad. Nauk SSSR* **224,** 859 (1975).
104. R. R. Schrock, *Accts. Chem. Rsch.* **12**, 98 (1979).

105. N. Z. Muradov, A. E. Shilov, and A. A. Shteinman, *Kinet. Katal.* (Eng. Trans.) **13**, 1219 (1972).

106. J. A. Sofranko, R. Eisenberg, and J. A. Kampmeier, *J. Am. Chem. Soc.* **102**, 1163 (1980).

ADDITIONAL READING

P. M. Henry, "Palladium Catalyzed Oxidation of Hydrocarbons," Reidel Publ., Dordrecht, Netherlands, 1979.

R. Jira and W. Freiesleben, Olefin oxidation and related reactions with Group VIII noble metal compounds. *Organomet. React.* **3**, 1 (1972).

J. Tsuji, "Organic Synthesis by Means of Transition Metal Complexes." Springer-Verlag, Berlin and New York, 1975.

G. W. Parshall, "Homogeneous Catalysis: The Applications and Chemistry of Catalysis by Soluble Transition Metal Complexes." Wiley, New York, 1980.

D. E. Webster, Activation of alkanes by transition metal compounds. *Adv. Organometal. Chem.* **15**, 147 (1977).

T. H. Tulip and D. L. Thorn, Hydridometallacycloalkane Complexes of Iridium. Unassisted Intramolecular Distal C—H Activation. *J. Am. Chem. Soc.* **103**, 2448 (1981).

C. Masters, "Homogeneous Transition Metal Catalysis—A Gentle Art." Chapman and Hall, New York, 1981.

Chapter 8

Biochemical Oxidations

A constant supply of dioxygen is essential for the existence of most living organisms. Oxidation reactions are involved in many fundamental biological processes, such as energy transformation and storage, as well as the biosynthesis and metabolism of essential amino acids, vitamins, hormones, etc.

The discovery of the stereo- and regioselective hydroxylation of progesterone by *Rhizopus nigricans*, which provided a key step in the synthesis of cortisone,[1] stimulated interest in oxygenations with microorganisms.

(1)

A wide variety of microbial oxidations of simple hydrocarbons, terpenes, and steroids are now known.[2–6] In practice, two different types of processes, fermentation and enzymatic conversions, can be distinquished. In *fermentations*, whole microorganisms are used and the process can involve several enzymes in a complex, multistep process. Any required cofactors are synthesized by the organism from the substrate and added nutrients such as nitrogen, phosphorus, and essential ions. In contrast, *enzymatic processes* involve specific transformations catalyzed by a single, cell-free enzyme and require the stoichiometric presence of any necessary cofactors. Since enzymatic conversions are more amenable to mechanistic studies, we shall be concerned primarily with this type of process. It should be noted, however, that all commercial processes involving biochemical oxidations are fermentations. In the last two decades, enormous advances have been made in the understanding of enzymatic oxidation processes. Several enzymes have been isolated in a pure, crystalline form, and much structural information has been accumulated concerning the nature of prosthetic groups.

I. TERMINOLOGY

An *enzyme* is a protein having both catalytic activity and specificity for its substrate(s). Although many enzymes are composed of only a protein unit, others, including all *oxidoreductases*, require the presence of cofactors. A cofactor-requiring enzyme with all its components present is referred to as a *holoenzyme*. If the enzyme is lacking its cofactor(s), it is called an *apoenzyme*. *Cofactor*, or *coenzyme*, is a general term used to describe components other than the apoenzyme that are essential for the catalytic activity of the enzyme. When the coenzyme is tightly bound to the protein (often through covalent bonds, as in metalloenzymes and most flavoproteins), it is referred to as a *prosthetic group*. The prosthetic group is an essential component of the active site of the enzyme, to which it remains attached throughout the catalytic cycle. A second group of coenzymes are noncovalently bound additives or *cosubstrates*, which are consumed in stoichiometric amounts. They are usually low-molecular-weight organic compounds, which in oxidoreductases are electron and hydrogen donors (acceptors). Many biochemical oxidations involve a *cooxidation* of the substrate and the coenzyme, which is a hydrogen donor. The most commonly utilized cosubstrates are the *nicotine adenine nucleotides*, NAD and NADP, which are consumed in stoichiometric amounts in many reactions catalyzed by dehydrogenases and monooxygenases.

The International Union of Biochemistry has adopted a systematic classification of enzymes into six main categories according to the types of reactions they catalyze.[7,8] *Oxidoreductases* is the general name for enzymes that catalyze oxidations and reductions, and they are divided into subclasses on the basis of the type of substrate and the type of hydrogen or electron donor (acceptor) utilized. For example, the systematic name for alcohol dehydrogenase, which catalyzes the following reaction, is alcohol : NAD oxidoreductase (EC 1.1.1.1).

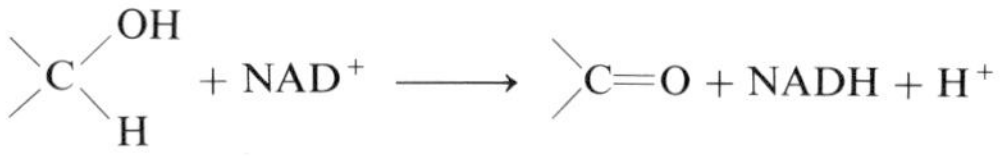

$$\text{>C(OH)(H)} + NAD^+ \longrightarrow \text{>C=O} + NADH + H^+ \quad (2)$$

II. COENZYMES AND PROSTHETIC GROUPS

A. Pyridine Nucleotides

The first coenzymes to be discovered were the pyridine nucleotides, nicotinamide adenine dinucleotide (NAD^+) and nicotinamide adenine dinucleotide phosphate ($NADP^+$), the structures of which are depicted below,

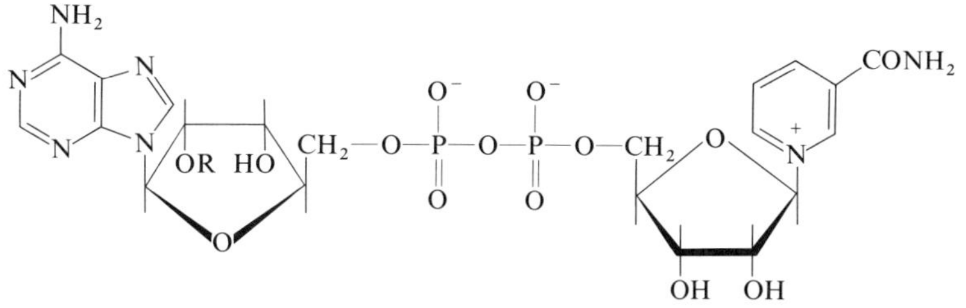

NH$_2$, N, N, N, N, OR, HO, O, CH$_2$—O—P(O$^-$)(=O)—O—P(O$^-$)(=O)—O—CH$_2$, O, N$^+$, CONH$_2$, OH, OH

where R is H in NAD^+ and $PO(OH)_2$ in $NADP^+$. $NADP^+$ has an additional phosphate group attached to the 2′ position of the adenosine moiety. These compounds are quaternary pyridinium salts, which can accept two hydrogens from suitable substrates. Under physiological conditions (pH ~ 7), this results in the release of a proton according to the equation:

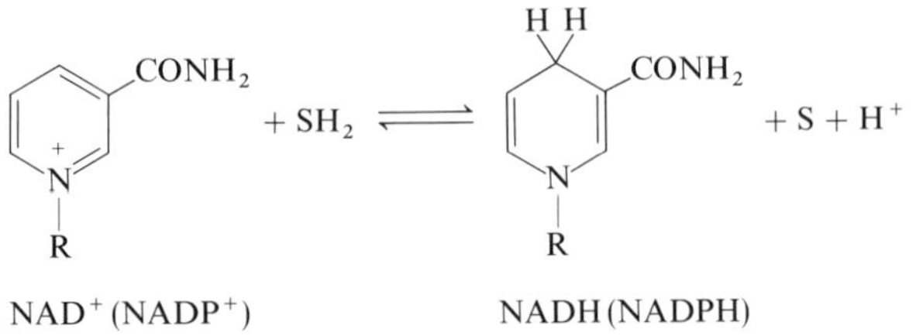

NAD$^+$ (NADP$^+$) + SH_2 ⇌ NADH (NADPH) + S + H^+ (3)

where SH_2 = reduced substrate and S = oxidized substrate. Many monooxygenases, which incorporate one atom of oxygen from molecular oxygen into organic substrates, utilize NADH or NADPH as a cosubstrate. The second oxygen atom is reduced to water, the hydrogens being provided by the cosubstrate according to eq 4,

$$S + O_2 + NADH + H^+ \longrightarrow SO + NAD^+ + H_2O \quad (4)$$

where S = reduced substrate and SO = oxidized substrate. A few monooxygenases (such as dopamine hydroxylase, a copper-containing protein) employ ascorbate as the cosubstrate, which is oxidized to dehydroascorbate.

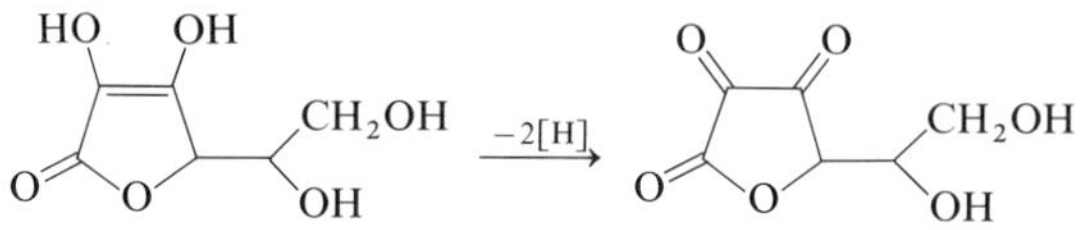

$$\text{ascorbate} \xrightarrow{-2[H]} \text{dehydroascorbate} \quad (5)$$

B. Reduced Pteridines

Phenylalanine hydroxylase, on the other hand, employs a reduced pteridine, *tetrahydrobiopterin*, as the cosubstrate:

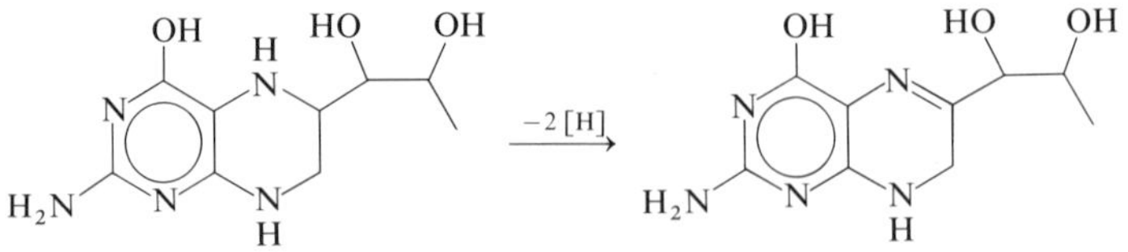

$$\text{tetrahydrobiopterin} \xrightarrow{-2[H]} \text{dihydrobiopterin} \quad (6)$$

C. Flavins

The two flavin coenzymes, flavin mononucleotide (FMN) and flavin adenine dinucleotide (FAD), are derivatives of riboflavin, vitamin B_2.

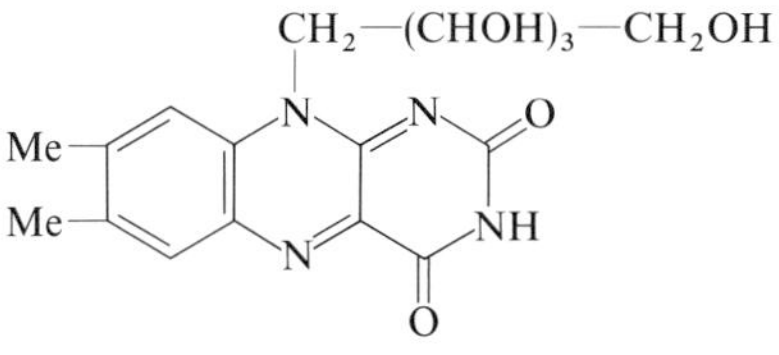

Riboflavin

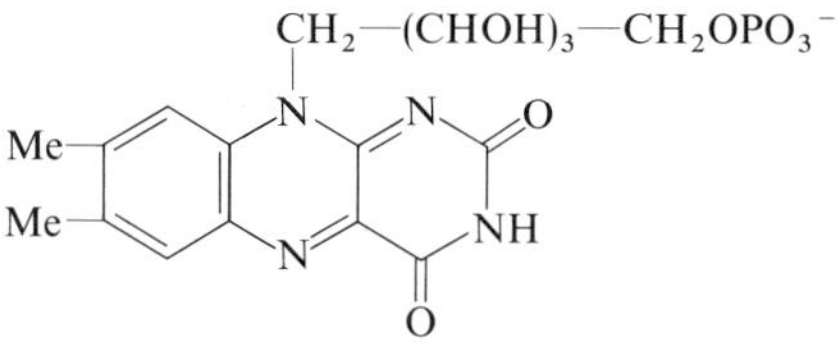

FMN

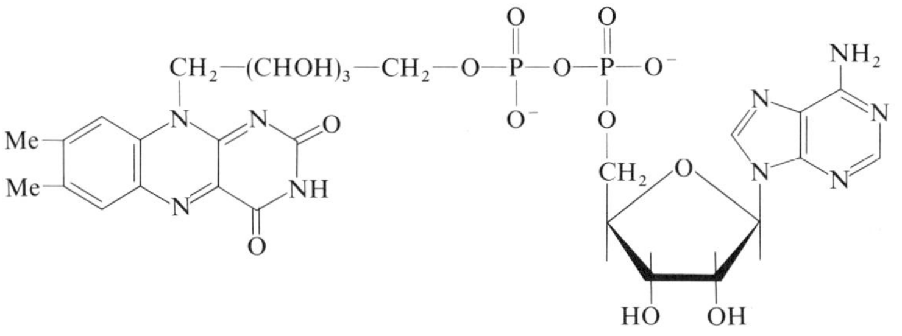

FAD

FMN and FAD function as coenyzmes to a large number of oxidoreductases. Their roles, however, differ from those of the pyridine nucleotides, which are freely dissociated from the enzymes and are consumed as co-substrates. The flavin coenzymes, on the other hand, are the prosthetic groups of *flavoproteins* and are covalently bound to the apoenzyme. They function as redox intermediates between substrates (electron donors) and their oxidants (electron acceptors). In doing so, they fulfill essentially the same role as metal ions in metalloenzymes. The chemistry of flavins has been reviewed.[9–11] They can function as redox intermediates by virtue of the facile, reversible reduction of the flavoquinone moiety via two successive one-electron changes.

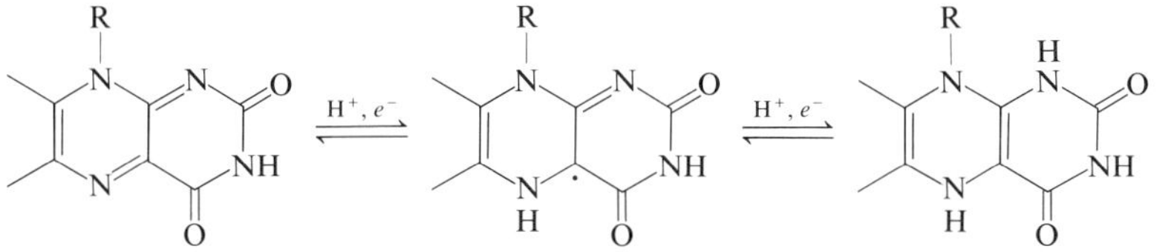

Flavoquinone (FMN or FAD) $\underset{}{\overset{H^+, e^-}{\rightleftharpoons}}$ Flavosemiquinone (FMNH· or FADH·) $\underset{}{\overset{H^+, e^-}{\rightleftharpoons}}$ Flavohydroquinone ($FMNH_2$ or $FADH_2$)

The function of the flavin is thus to transport reducing equivalents from the donor (substrate and/or coenzyme such as NADH) to the oxidant (acceptor). Flavins also react readily with molecular oxygen to form a flavin hydroperoxide,[12] which has been suggested as an intermediate in reactions mediated by flavoprotein oxygenases. In the holoenzyme, the flavin moiety is covalently bound to the apoenzyme. In most cases, the precise mode of attachment is not known. An exception is thiamine dehydrogenase, in which it has been shown[13] that the FAD prosthetic group is covalently bound to the protein via the 8α position of the flavin to the N(1) position of the imidazole ring of a histidine residue, as shown below.

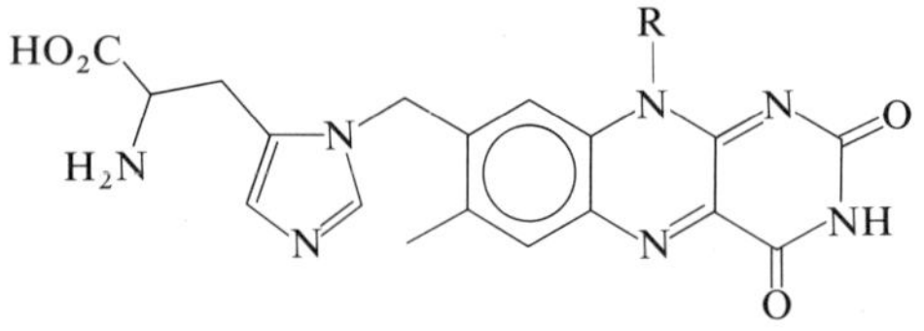

D. Hemes

Hemes are the prosthetic groups of *hemoproteins*, which are ubiquitous in nature and participate in oxygen transport (hemoglobin, myoglobin), electron transport (cytochromes *c*), and oxygen redox chemistry (cytochromes *P*-450 and peroxidases). A *heme* is an iron–porphyrin complex (in most instances iron protoporphyrin IX), which in its ferrous form is called *ferroprotoporphyrin IX* or simply heme. Ferroprotoporphyrin IX is the prosthetic group of numerous important enzymes and of the oxygen transport proteins hemoglobin and myoglobin.

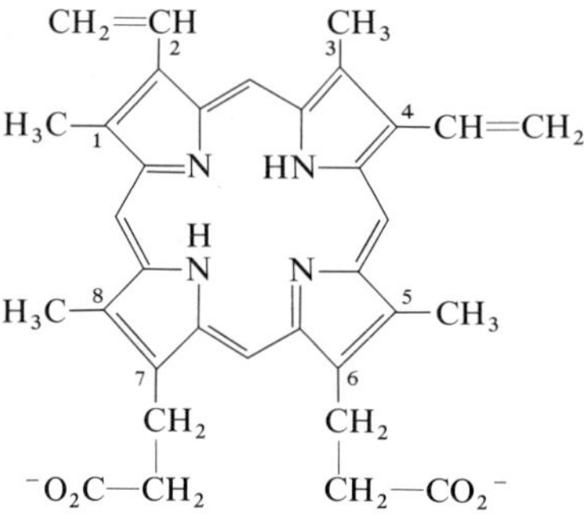

Protoporphyrin IX

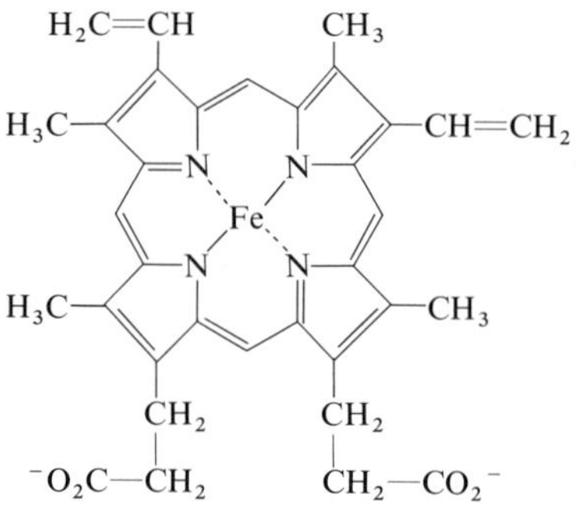

Ferroprotoporphyrin IX
(Proto)heme (IX)

Hemes are oxidized by O_2 or other oxidizing agents to the corresponding iron (III) derivatives, which are called *ferrihemes* or *hemins*. These iron(III) complexes have a residual positive charge, which in the isolated complex is neutralized by an anion. If the counterion is chloride, the resulting complex is designated as *hemin chloride* or *chloroheme*. In hemes, the iron is coordinated to four nitrogen atoms and capable of axially binding two additional ligands. In this process, the binding of the fifth ligand generally facilitates the coordination of the sixth ligand. The resulting octahedral iron(II) complexes are known as *hemochromes*. With additional ligands, hemins readily form hexacoordinate octahedral complexes called *hemi-*

chromes. When the heme prosthetic group is incorporated into the apoenzyme, the resulting hemoprotein either may have a hemochrome structure,

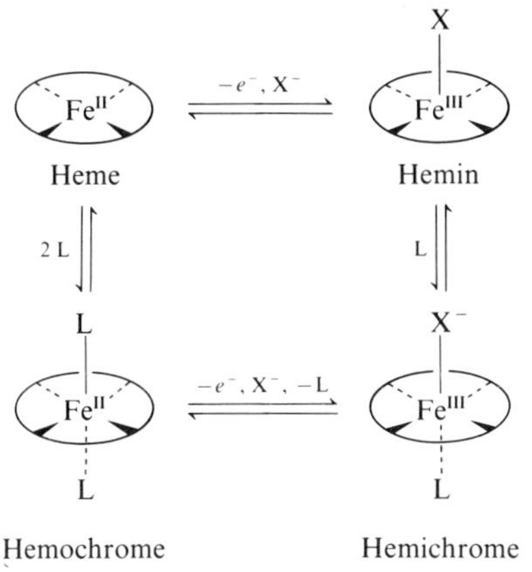

Heme $\underset{}{\overset{-e^-,\,X^-}{\rightleftharpoons}}$ Hemin (7)

2 L ⇅ L ⇅

Hemochrome $\overset{-e^-,\,X^-,\,-L}{\rightleftharpoons}$ Hemichrome (8)

in which both the fifth and sixth coordination sites are occupied by nitrogen atoms from the amino acid residues of the protein, or may be of the so-called open type, in which at least one site is not occupied by a nitrogen ligand. *Catalases* and *peroxidases* are generally believed to be of the open type. Among the side-chain groups that coordinate to the iron in hemoproteins, the imidazole moiety of histidine and the thiol group of cysteine are the most prevalent. In particular, the imidazole group of histidine is of great importance in the chemistry of enzymes. At neutral pH, imidazole (with a pK_a of about 6) can act as a proton donor or acceptor. *Hemoglobin* (Hb) is the heme-containing oxygen transport protein in the red blood cells of vertebrates. Although Hb is not an enzyme, it has many of the properties of enzymes. Thus, it is capable of catalyzing many oxidations characteristic of the peroxidases and oxygenases, e.g., oxidations with H_2O_2 and hydroxylation of aniline to *p*-aminophenol with dioxygen.[14]

The prosthetic group of hemoglobin contains ferroprotoprophyrin IX coordinated to the imidazole nitrogen of a histidine residue in the protein globin. The sixth coordination site is available for coordination with dioxygen. In the five-coordinate deoxyhemoglobin, the iron is displaced out of the porphyrin ring towards the histidine ligand. In the octahedral oxygenated form (oxyHb), however, the nitrogens and the iron are all coplanar.

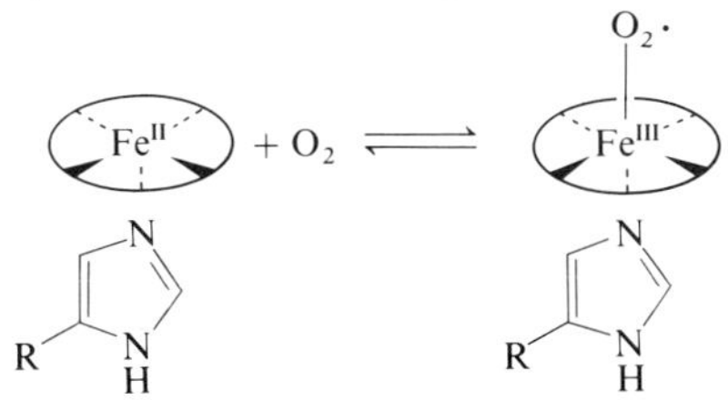

Fe^{II} + O_2 ⇌ Fe^{III}–$O_2\cdot$ (9)

A *cytochrome* is a hemoprotein whose characteristic mode of action is electron and/or hydrogen transport by virtue of a reversible valency change of the iron. The iron(II) form is designated as a ferrocytochrome, and the iron(III) form is a ferricytochrome. Cytochromes are not enzymes but are protein coenzymes that are essential, as transporters of redox equivalents, to the catalytic activity of many oxidoreductases. Cytochromes are classified on the basis of the structure of the heme prosthetic group. For example, cytochromes *b* contain protoheme as the prosthetic group. The heme prosthetic groups of cytochromes *a* and *c* are shown below.

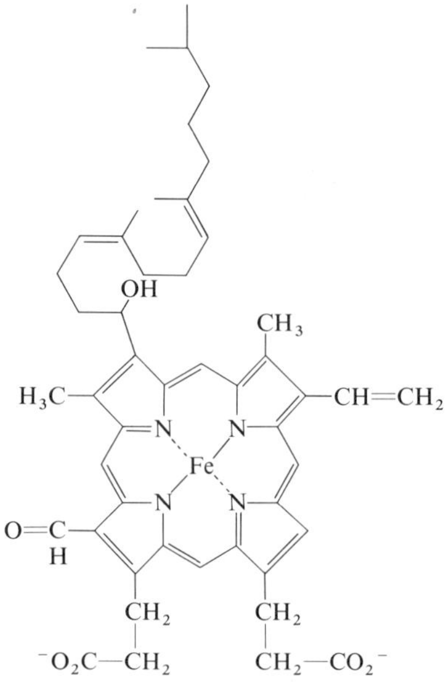

Heme *a*
(prosthetic group of
class *a* cytochromes)

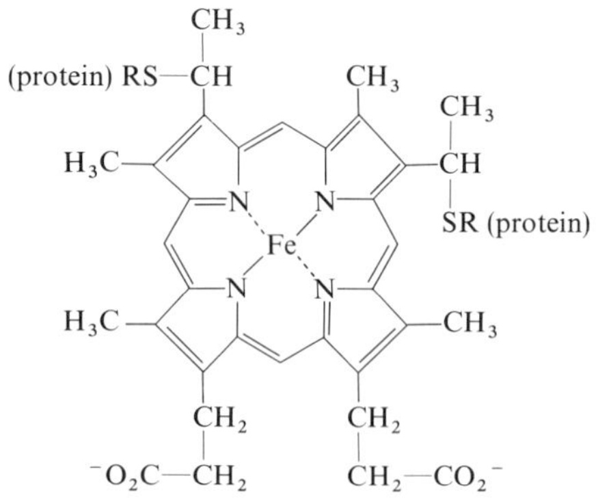

Heme *c*
(prosthetic group of
class *c* cytochromes)

Most cytochromes are low-spin hexacoordinate complexes, the fifth and sixth ligands usually provided by the protein. They are therefore generally incapable of interacting with external ligands such as O_2, CO, or CN^-. Exceptions are known, however, among them being cytochrome *P*-450, the terminal electron carrier of numerous monooxygenases of mammalian liver microsomes, adrenal mitochondria, and bacterial systems. The cytochrome isolated from camphor-oxidizing microorganisms has been the most thoroughly studied.[15–18] The prosthetic group is ferriprotoporphyrin IX, which in the enzyme is thought to contain a cysteinyl mercaptide and, possibly, histidinyl nitrogen as axial ligands. The reduced ferrocytochrome forms a complex with dioxygen, which is reasonably stable at low temperatures. These monooxygenases are composite, multicomponent protein systems. The ultimate hydrogen donor (cosubstrate) is NADH or NADPH. Hydrogen is initially transferred from the cosubstrate to the flavin group of a flavoprotein reductase. The next link in the chain of electron transfer is an iron–sulfur protein coenzyme (*vide infra*), leading to the terminal electron carrier cytochrome *P*-450. The complete reaction scheme is discussed in more detail later (Section III).

E. Protein Coenzymes

There is a group of proteins that take part in enzymatic transformations but possess no catalytic activity in themselves. Protein coenzymes have low molecular weights compared to most enzymes. Those employed by oxidoreductases can be divided into four groups: the cytochromes and iron–sulfur proteins, which have already been discussed, flavoproteins, and thiol coenzymes.

Ferredoxins are iron–sulfur protein coenzymes that mediate electron transport in many important biological processes. The prosthetic groups of these proteins contain iron coordinated with sulfide and cysteinyl mercaptide. For example, plant ferredoxin (MW ~11,000) contains two iron atoms each ligated tetrahedrally by two bridging sulfide groups and two cysteinyl mercaptides. *Rubredoxin* and *putidaredoxin* are related iron–sulfur protein coenzymes utilized by many bacterial monooxygenases.

Several flavoproteins can be classified as protein coenzymes. A low-molecular-weight flavoprotein, *flavodoxin*, has been isolated from several species of microorganisms. It contains FAD as a prosthetic group and serves as an electron carrier in redox reactions. Its function is often interchangeable with that of ferredoxin. *Thioredoxins* are low-molecular-weight proteins having thiol groups at their active sites which transfer redox equivalents via successive dithiol–disulfide cycles.

F. Nonheme Iron, Copper, and Molybdenum

In addition to the hemoprotein enzymes, there are also many oxidoreductases that contain nonheme iron. Examples are the dioxygenases pyrocatechase and metapyrocatechase. Copper ions form the prosthetic group of the oxygen transport protein hemocyanin and of several important oxidases[19,20] and oxygenases[21]. In some cases the holoenzyme contains Cu(I), and in others Cu(II), and sometimes both. The mode of coordination of the apoenzyme to the copper is generally not known. Transfer of electrons from the substrate to dioxygen at the active site involves the Cu(I)/Cu (II) redox couple, and in some cases complex formation with dioxygen.

There are several examples of oxidoreductases that contain molybdenum as the prosthetic group.[22–25] The best known example is *xanthine oxidase*[25] (EC 1.2.3.2) isolated from cow's milk. It is a complex enzyme that contains iron, FAD, and molybdenum. It catalyzes the oxidation of xanthine to urate, the incorporated oxygen being derived from water and not from dioxygen.

$$\text{xanthine} + O_2 + H_2O \xrightarrow{[\text{xanthine oxidase}]} \text{urate (8-OH)} + H_2O_2 \quad (10)$$

Oxidations with molybdoenzymes are thought to involve the Mo(VI)/Mo(V) and/or Mo(VI)/Mo(IV) redox couple in a direct interaction with the substrate. Transfer of oxygen from an oxomolybdenum(VI) group to the substrate may be involved.

III. TYPES OF OXIDATIVE TRANSFORMATION

Enzymes that catalyze the dehydrogenation of primary substrates, employing a hydrogen acceptor such as NAD^+ or $NADP^+$ as the cosubstrate, are designated *dehydrogenases.* When molecular oxygen serves as the hydrogen acceptor, to form water or hydrogen peroxide as the coproduct, the enzymes are called *oxidases. Hydroperoxidases* (or *peroxidases*) are oxidoreductases that catalyze the oxidation of substrates with hydrogen peroxide (which can sometimes be replaced by an alkyl hydroperoxide), forming water as a coproduct. In 1955, a fourth group of enzymes were discovered which participate in a diverse group of important reactions involving the incorporation of dioxygen into organic substrates. They were discovered independently by Mason[26] and Hayaishi[27] are referred to as *oxygenases.*

Mason and co-workers[26] showed by ^{18}O labeling that the oxidation of 3,4-dimethylphenol to 4,5-dimethylcatechol catalyzed by a *phenolase* involved the incorporation of oxygen exclusively from dioxygen and not from water.

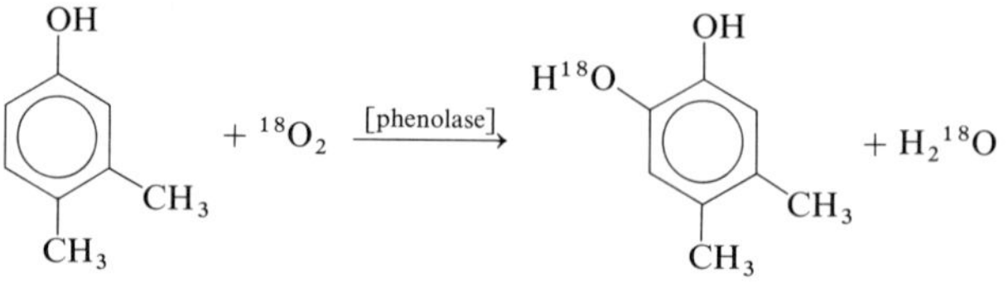

$$\text{3,4-dimethylphenol} + {}^{18}O_2 \xrightarrow{[\text{phenolase}]} \text{4,5-dimethylcatechol (}H^{18}O\text{)} + H_2{}^{18}O \quad (11)$$

This was the first demonstration of a *monooxygenase*, which catalyzes the incorporation of a single atom from dioxygen into a primary substrate, the second oxygen being reduced to water. Monooxygenases require the presence of a hydrogen donor (often NADH or NADPH) as cosubstrate, which is consumed according to the equation:

$$S + DH_2 + O_2 \longrightarrow SO + D + H_2O \qquad (12)$$

where S = substrate and DH_2 = donor. In some cases, the substrate itself, a reaction intermediate, or a product can function as the hydrogen donor, as in the example given in reaction 11. Since no cosubstrate is required, the enzyme is called an *internal monooxygenase* as opposed to an *external monooxygenase*, which requires the addition of a cosubstrate, such as NADH.

Hayaishi and co-workers[27] similarly demonstrated by ^{18}O labeling studies that the two oxygen atoms incorporated during oxidative cleavage of catechol to *cis,cis*-muconic acid, catalyzed by *pyrocatechase* (EC 1.13.11.1), were both derived from dioxygen.

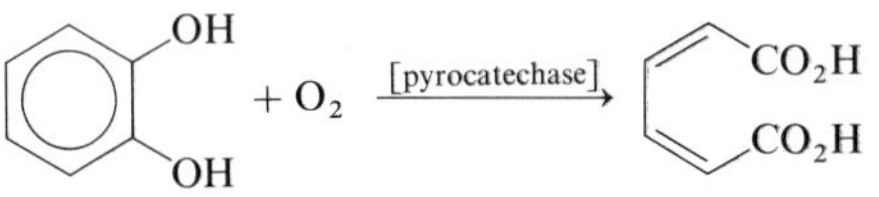

$$\text{catechol (OH, OH)} + O_2 \xrightarrow{[\text{pyrocatechase}]} \text{cis,cis-muconic acid } (CO_2H, CO_2H) \qquad (13)$$

This was the first example of a dioxygenase, which represents a group of enzymes catalyzing the general reaction:

$$S + O_2 \longrightarrow SO_2 \qquad (14)$$

where S = substrate and SO_2 = oxidized substrate. No cosubstrate is needed in these reactions.

Finally, a new group of enzymes, identified recently as *superoxide dismutases* (EC 1.15.1.1), catalyzes the disproportionation of superoxide anion to hydrogen peroxide and oxygen.

The various types of oxidative transformation mediated by oxidoreductases are summarized in Table I. We shall now examine these types in more detail, placing emphasis on the oxygenases owing to the enormous interest this diverse group of enzymes has attracted in recent years.

A. Dehydrogenases

More than a hundred different dehydrogenases have been identified and classified.[7,8] The majority of these employ NAD^+ or $NADP^+$ as the hydrogen acceptor according to the stoichiometry:

$$SH_2 + NAD^+ \longrightarrow S + NADH + H^+ \qquad (15)$$

TABLE I Oxidative Transformations Mediated by Oxidoreductases

Class	General reaction	Example
Dehydrogenase	$SH_2 + D \longrightarrow S + DH_2$	*Alcohol dehydrogenase*; alcohol : NAD oxidoreductase (EC 1.1.1.1) $>CHOH + NAD^+ \longrightarrow >C{=}O + NADH + H^+$
	$S + D + H_2O \longrightarrow SO + DH_2$	*Aldehyde dehydrogenase*; aldehyde : NAD oxidoreductase (EC 1.2.1.3) $RCHO + NAD^+ + H_2O \longrightarrow RCO_2H + NADH + H^+$
Oxidase	$SH_2 + O_2 \longrightarrow S + H_2O_2$	*Glycollate oxidase*; glycollate : oxygen oxidoreductase (EC 1.1.3.1) $HOCH_2CO_2^- + O_2 \longrightarrow OHCCO_2^- + H_2O_2$
	$S + H_2O + O_2 \longrightarrow SO + H_2O_2$	*Aldehyde oxidase*; aldehyde : oxygen oxidoreductase (EC 1.2.3.1) $RCHO + H_2O + O_2 \longrightarrow RCO_2H + H_2O_2$ L-*Amino-acid oxidase*; L-Amino acid : oxygen oxidareductase (EC 1.4.3.2) $\text{L-RCH(NH}_2\text{)CO}_2\text{H} + H_2O + O_2 \longrightarrow RCOCO_2H + NH_3 + H_2O_2$
	$2\,SH_2 + O_2 \longrightarrow 2S + 2H_2O$	*p-Diphenol oxidase*; *p*-diphenol : oxygen oxidoreductase (EC 1.10.3.1) $2\,HO{-}C_6H_4{-}OH + O_2 \longrightarrow 2\,O{=}C_6H_4{=}O + 2\,H_2O$
Peroxidase	$SH_2 + H_2O_2 \longrightarrow S + 2H_2O$	*Peroxidase*; donor : hydrogen-peroxide oxidoreductase (EC 1.11.1.7) $\text{Donor} + H_2O_2 \longrightarrow \text{oxidized donor} + 2\,H_2O$ *Catalase*; hydrogen peroxide : hydrogen-peroxide oxidoreductase (EC 1.11.1.6) $H_2O_2 + H_2O_2 \longrightarrow O_2 + 2\,H_2O$

		Chloroperoxidase; 3-oxoadipate : hydrogen-peroxide oxidoreductase (adding chlorine) (EC 1.11.1.10) 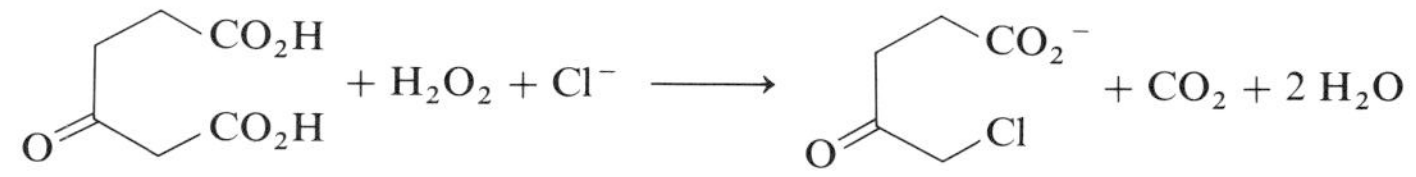
		CO_2H, CO_2H, O $+ H_2O_2 + Cl^- \longrightarrow$ CO_2^-, Cl, O $+ CO_2 + 2\ H_2O$
Dioxygenase	$S + O_2 \longrightarrow SO_2$	*Tryptophan oxygenase*; L-tryptophan: oxygen oxidoreductase (EC 1.13.11.11) 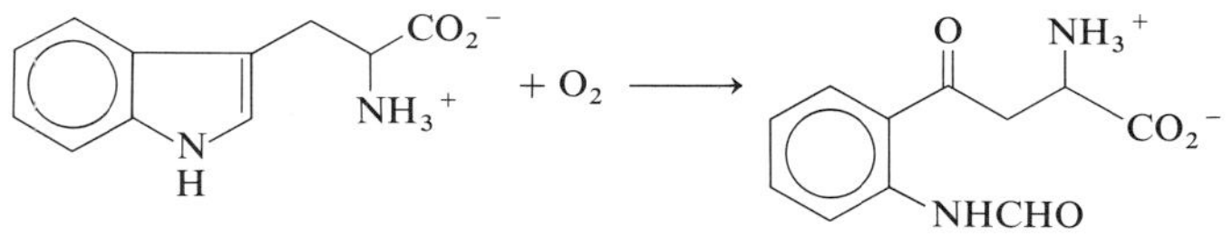
		CO_2^-, NH_3^+, N, H $+ O_2 \longrightarrow$ O, NH_3^+, CO_2^-, NHCHO
		Lipoxygenase; linoleate : oxygen oxidoreductase (EC 1.13.11.12)
		$RCH{=}CHCH_2CH{=}CHR' + O_2 \longrightarrow RCH{=}CHCH{=}CHCH(O_2H)R'$
		Prolyl hydroxylase; proline, 2-oxo-glutarate : oxygen oxidoreductase (EC 1.14.11.2) 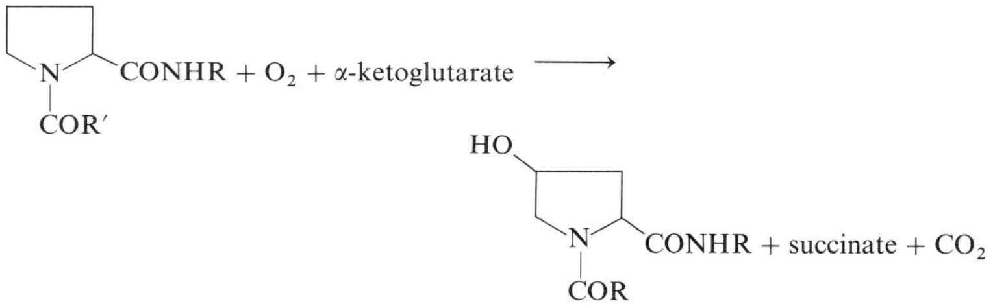
		N, COR′, CONHR $+ O_2 + \alpha$-ketoglutarate $\longrightarrow$ HO, N, COR, CONHR + succinate $+ CO_2$
Monooxygenase	$S + DH_2 + O_2 \longrightarrow SO + D + H_2O$	*Camphor 5-monooxygenase*; camphor, reduced NAD : oxygen oxidoreductase (EC 1.14.15.1) 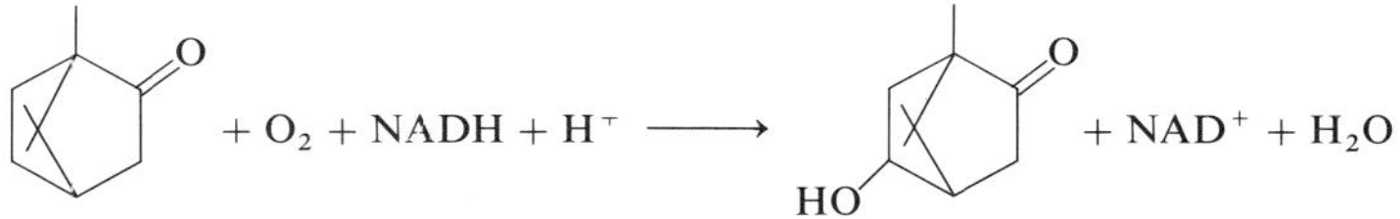
		O $+ O_2 + NADH + H^+ \longrightarrow$ O, HO $+ NAD^+ + H_2O$

(*continued*)

TABLE I *(continued)*

Class	General reaction	Example
		Squalene epoxidase; squalene, reduced NADP : oxygen oxidoreductase (EC 1.14.99.7) squalene + O_2 + NADPH + $H^+ \longrightarrow$ squalene oxide + $NADP^+$ + H_2O
		Dopamine hydroxylase; 3,4-dihydroxyphenylalanine, ascorbate : oxygen oxidoreductase (EC 1.14.17.1) [3,4-dihydroxyphenyl ring: OH, OH, $CH_2CH_2NH_2$] + O_2 + ascorbate $\longrightarrow$ [3,4-dihydroxyphenyl ring: OH, OH, $HOCHCH_2NH_2$] + H_2O + dehydroascorbate
Superoxide dismutase		*Superoxide dismutase* (EC 1.15.1.1) $2\,O_2^{\overline{\cdot}} + 2\,H^+ \longrightarrow O_2 + H_2O_2$

The best known example is the *alcohol dehydrogenase* (EC 1.1.1.1) isolated from bakers' yeast, which catalyzes the dehydrogenation of ethanol and other primary alcohols to the corresponding aldehydes. Hydrogen transfer to NAD proceeds stereospecifically to one side of the molecule, as shown by deuterium labeling studies.

$$\text{(R–N}^{+}\text{ pyridinium–CONH}_2) + CH_3CD_2OH \longrightarrow \text{(R–N 4-D,H-dihydropyridine–CONH}_2) + CH_3CDO + H^+ \qquad (16)$$

This important subject has been reviewed by Popjak[28]. *Aldehyde dehydrogenase* (EC 1.2.1.3) catalyzes the oxidation of aldehydes to carboxylic acids according to the general equation:

$$RCHO + NAD^+ + H_2O \longrightarrow RCO_2H + NADH + H^+ \qquad (17)$$

B. Oxidases

Oxidases catalyze the oxidation of substrates by employing dioxygen as the hydrogen acceptor. Typical examples are *xanthine oxidase* (reaction 10) and aldehyde oxidase (EC 1.2.3.1), which catalyzes the oxidation of aldehydes to carboxylic acids according to the stoichiometry:

$$RCHO + H_2O + O_2 \longrightarrow RCO_2H + H_2O_2 \qquad (18)$$

Both of these enzymes are molybdoenzymes. Cytochrome *c* oxidase (EC 1.9.3.1), a copper protein, catalyzes the oxidation of ferrocytochrome *c* to ferricytochrome *c* according to eq 19.[29]

$$4\,Fe^{II}(cyt) + 4\,H^+ + O_2 \longrightarrow 4\,Fe^{III}(cyt) + 2\,H_2O \qquad (19)$$

C. Peroxidases

The *(hydro)peroxidases*[30–33] are a group of hemiproteins that are widely distributed in mammalian liver, microorganisms, and especially plants. Peroxidases (EC 1.11.1.7), the most studied example of which is horseradish peroxidase (HRP), catalyze the general reaction:

$$ROOH + SH_2 \longrightarrow ROH + H_2O + S \qquad (20)$$

The enzyme is not substrate specific, and it can catalyze the oxidation of phenols, aromatic and aliphatic amines, hydroquinones, enediols, leuco dyes, cytochrome *c*, and ferrocyanide. In the case of *catalase*[34] (EC 1.11.1.6), the substrate (SH_2) in eq 20 is a second molecule of H_2O_2 and the reaction is:

$$H_2O_2 + H_2O_2 \longrightarrow O_2 + 2\,H_2O \qquad (21)$$

Catalase also catalyzes the oxidation of primary alcohols to aldehydes according to eq 22.

$$RCH_2OH + H_2O_2 \longrightarrow RCHO + 2\,H_2O \qquad (22)$$

Cytochrome *c* peroxidasc (EC 1.9.3.1) catalyzes the oxidation of ferrocytochrome *c* to ferricytochrome *c*.

$$2\,Fe^{II}(cyt) + H_2O_2 + 2\,H^+ \longrightarrow 2\,Fe^{III}(cyt) + 2\,H_2O \qquad (23)$$

The enzyme *chloroperoxidase* (EC 1.11.1.10) catalyzes the oxidative chlorination of certain substrates (e.g., 3-oxoacids, cyclic 1,3-diketones, and substituted phenols; see Table I for an example).

All catalases and most peroxidases contain ferriprotoporphyrin IX in a high-spin state, as the prosthetic group. In the course of reaction, all of them form one or more of a series of intermediate complexes identified by their magnetic and optical properties.[31,33] For example, HRP initially forms an intermediate, designated as compound I, which is the same irrespective of whether H_2O_2 or RO_2H is used.[35] It is now generally accepted [31] that compound I is formally an oxoiron(V) porphyrin, as first suggested by George.[36] Compound I reacts with the substrate to form compound II and the substrate radical (HS·). Compound II is formally an oxoiron(IV) porphyrin, and it reacts with a second substrate molecule to regenerate HRP. The full reaction scheme is shown below.

Scheme I:

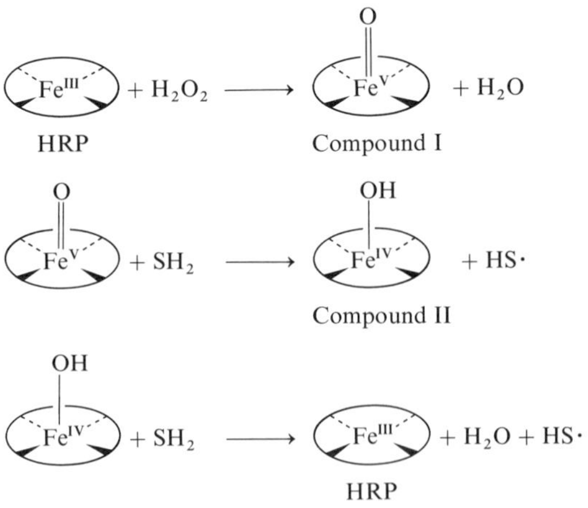

$$Fe^{III}\ (\text{HRP}) + H_2O_2 \longrightarrow O{=}Fe^{V}\ (\text{Compound I}) + H_2O \qquad (24)$$

$$O{=}Fe^{V} + SH_2 \longrightarrow HO{-}Fe^{IV}\ (\text{Compound II}) + HS\cdot \qquad (25)$$

$$HO{-}Fe^{IV} + SH_2 \longrightarrow Fe^{III}\ (\text{HRP}) + H_2O + HS\cdot \qquad (26)$$

$$2\,HS\cdot \longrightarrow S + SH_2 \quad (\text{or } HS{-}SH) \qquad (27)$$

In some cases, HRP catalyzes the hydroxylation of aromatic substrates (compare reactions with the *P*-450 enzymes described later). Catalase also forms an oxoiron(V) intermediate (compound I), which reacts directly with H_2O_2 or RCH_2OH to regenerate catalase and produce O_2 or RCHO. Since there is no evidence for free radical intermediates, Hamilton[37] suggested an ionic mechanism.

Scheme II:

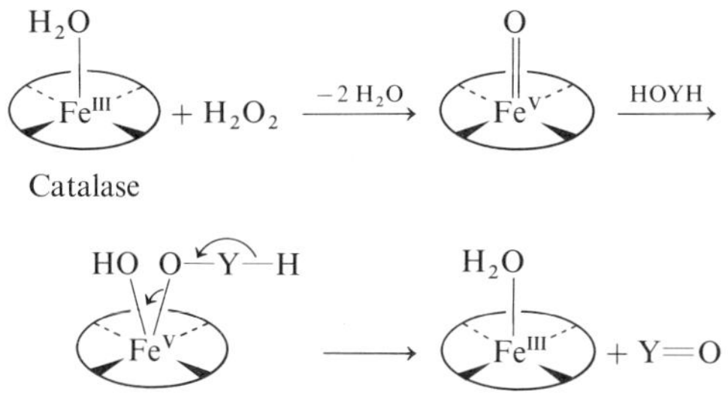

H_2O–Fe^{III} (Catalase) + H_2O_2 $\xrightarrow{-2\,H_2O}$ O=Fe^{V} $\xrightarrow{HOYH}$

HO O–Y–H on Fe^{V} $\longrightarrow$ H_2O–Fe^{III} + Y=O

where Y = O or RCH. Chloroperoxidase[38] also initially forms an oxoiron(V) porphyrin, which presumably affords the chlorinated product via ligand transfer as shown in Scheme III.

Scheme III:

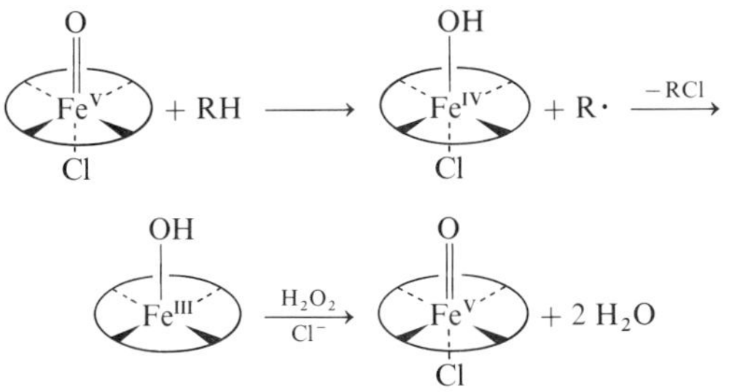

O=Fe^{V}–Cl + RH $\longrightarrow$ HO–Fe^{IV}–Cl + R· $\xrightarrow{-RCl}$

HO–Fe^{III} $\xrightarrow[Cl^-]{H_2O_2}$ O=Fe^{V}–Cl + 2 H_2O

D. Oxygenases

Oxygenases, an extremely important group of enzymes, are widely distributed in animals, plants, and microorganisms.[39–48] In animals, they play a vital role in the biosynthesis and metabolism of amino acids, fatty acids, porphyrins, prostaglandins, vitamins, and steroids. Oxygenases have been isolated from liver, kidney, adrenal gland, brain, and other tissues. Monooxygenases contained in mammalian liver microsomes, for example, mediate transformations of foreign compounds and several natural metabolites into nontoxic products. In adrenal mitochondria, they play a major role in steroidal hormone biosynthesis. Monooxygenases also regulate the biosynthesis and catabolism of biogenetic amines, such as dopamine, adrenaline, noradrenaline, and serotonine, involved in the specific functions of nerve cells. In plants, oxygenases participate in the biosynthesis, transformation, and degradation of many cell constitutents. They are involved, for example, in the metabolism of phenolic compounds and the biosynthesis of alkaloids.

Aerobic microorganisms, such as *Pseudomonas Mycobacteria* and *Nocardia* are rich sources of oxygenases. Almost all the oxygenases that have been crystallized are of microbial origin. In contrast to mammalian oxygenases, those of microbial origin are often easily obtained in large quantities. Numerous microorganisms are known to hydroxylate selectively a wide variety of hydrocarbons.[2–6] Many of these reactions have potential synthetic value (see later discussion for examples).

Dioxygenases are defined as enzymes that catalyze reactions in which both atoms of molecular oxygen are incorporated into the substrate(s). In most cases, one substrate accepts both oxygen atoms, and the enzyme is referred to as an intramolecular dioxygenase (EC subclass 1.13.11).

Intramolecular dioxygenase

$$\underset{\text{Substrate}}{S + O_2} \longrightarrow \underset{\text{Oxidized substrate}}{SO_2} \tag{28}$$

In others, a cosubstrate is employed which accepts one of the oxygen atoms, and the enzyme is designated as an intermolecular dioxygenase (EC 1.14.11).

Intermolecular dioxygenase

$$S + S' + O_2 \longrightarrow SO + S'O \tag{29}$$

The *catechol dioxygenases*[49] of microbial origin are a group of nonheme iron, intramolecular dioxygenases that catalyze the oxidative cleavage of catechol and its derivatives. *Pyrocatechase* (EC 1.13.11.1), which catalyzes the oxidative cleavage of catechol to *cis*, *cis*-muconic acid, has already been mentioned (see eq 13). *Metapyrocatechase* (EC 1.13.11.2), on the other hand, catalyzes oxidative cleavage of catechol via a different pathway.

$$C_6H_4(OH)_2 + O_2 \xrightarrow{[\text{metapyrocatechase}]} OHC{-}CH{=}CH{-}CH{=}C(OH){-}CO_2H \tag{30}$$

Similarly, *protocatechuate 3,4-dioxygenase* (EC 1.13.11.3) and *protocatechuate 4,5-dioxygenase* (EC 1.13.11.8) catalyze the analogous oxidative cleavage reactions of protocatechuic acid.

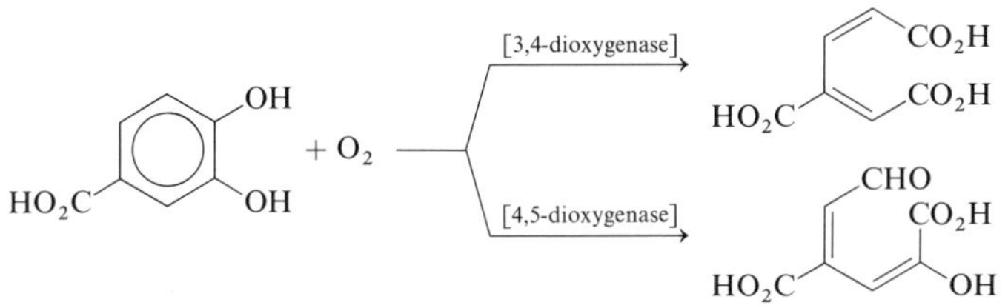

[3,4-dioxygenase] (31)

[4,5-dioxygenase] (32)

All of the above enzymes contain nonheme ferrous ion as the prosthetic group. *Tryptophan 2,3-dioxygenase* (EC 1.13.11.11), on the other hand, contains both hemin and copper ion as prosthetic groups.[50] This enzyme catalyzes the oxidative cleavage of L-tryptophan to formylkynurenine, which is the first step in the metabolic conversion of L-tryptophan to nicotinic acid and pyridine nucleotides.

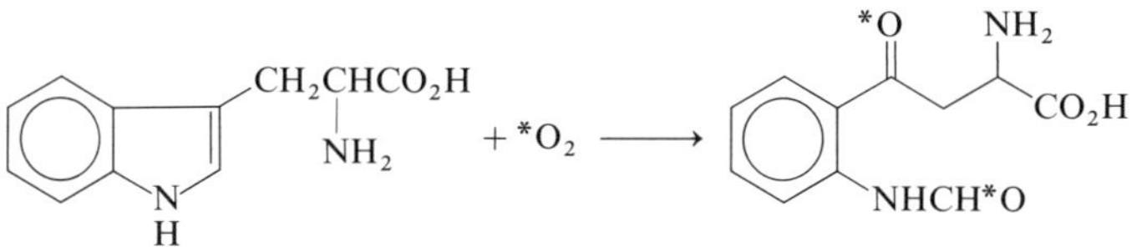

(33)

The active site contains ferriprotoporphyrin IX and cuprous ion.[50]

Quercetinase[21] is a copper-containing fungal dioxygenase that catalyzes the oxidative cleavage of the heterocyclic ring of quercetin.

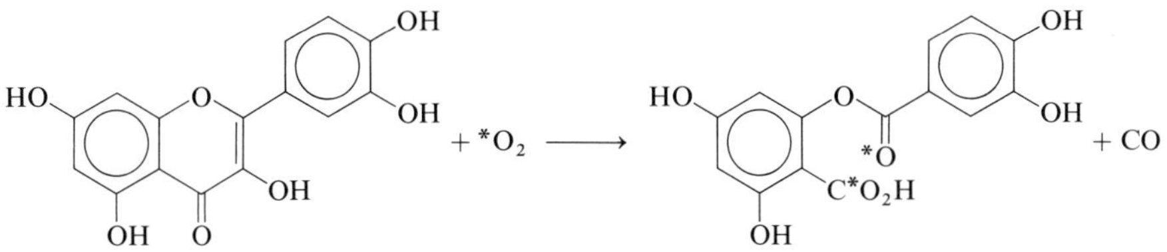

(34)

Oxidations catalyzed by another dioxygenase, *lipoxygenase*[51] (EC 1.13.11.12), are worthy of mention because of the close similarity to classical autoxidations. Lipoxygenases catalyze the oxidation of unsaturated fatty acids containing a *cis,cis*-1,4-pentadiene unit to an allylic hydroperoxide.

$$R_1CH{=}CHCH_2CH{=}CHR_2 + O_2 \longrightarrow R_1CH{=}CHCH{=}CH\underset{\displaystyle OOH}{\underset{|}{C}}HR_2 \qquad (35)$$

The biosynthesis of prostaglandins from unsaturated fatty acids involves the successive participation of both a dioxygenase and a monooxygenase. The mechanism shown below has been suggested[51] for the conversion of 8,11,14-eicosatrienoic acid to PGE_1 and $PGF_1\alpha$.

Scheme IV:

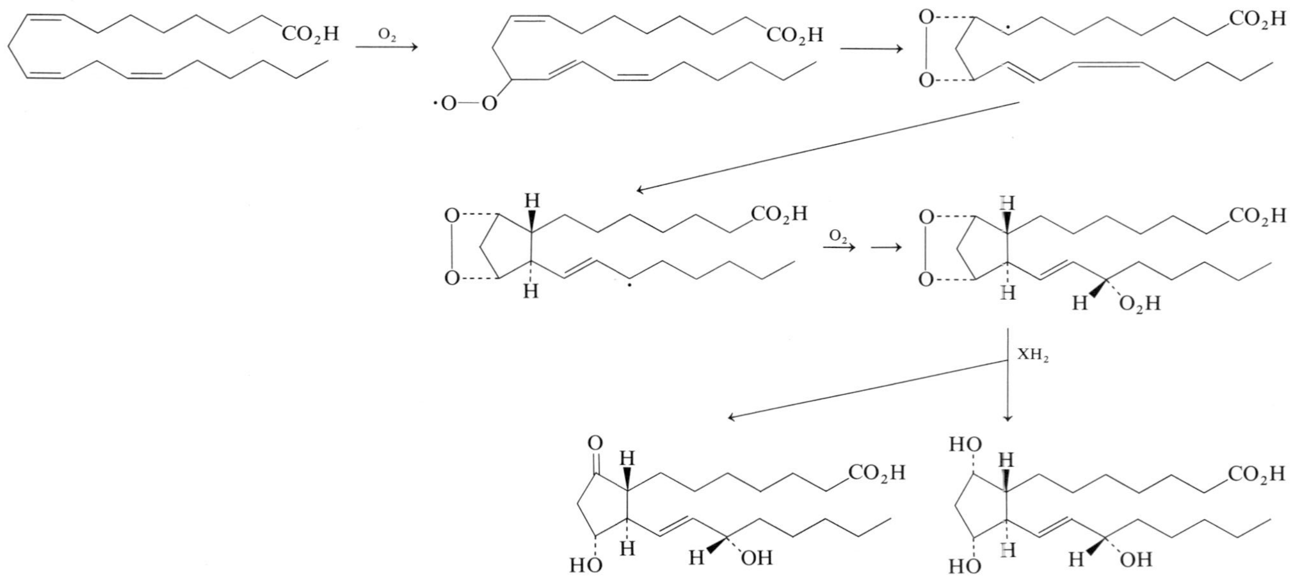

CO_2H $\xrightarrow{O_2}$ $\cdot O—O$ CO_2H ⟶ O O CO_2H

O O H H CO_2H $\xrightarrow{O_2}$ ⟶ O O H H CO_2H H O_2H

XH_2

O H HO H CO_2H H OH

PGE_1

HO H HO H CO_2H H OH

$PGF_{1\alpha}$

Interestingly, the formation of a hydroperoxy epidioxide has recently been demonstrated in the autoxidation of a hydroperoxide of methyl linolenate shown below [R = $MeO_2C(CH_2)_6$—].[52,52a]

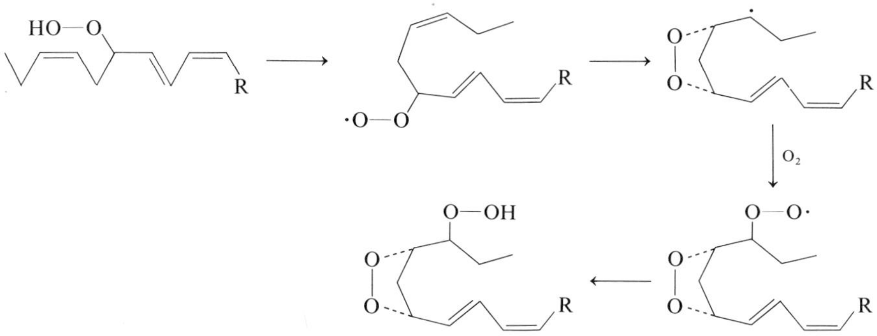

In this case, the intermediate carbon-centered radical reacts with dioxygen rather than undergoing an intramolecular addition to the double bond.

The *α-ketoglutarate-coupled dioxygenases* are a group of intermolecular dioxygenases that employ α-ketoglutarate as the cosubstrate,[53] which is converted to succinate. *Proline hydroxylase* (EC 1.14.11.2), which converts a proline moiety to a 4-hydroxyproline derivative in plants, is an example of this class of enzymes.

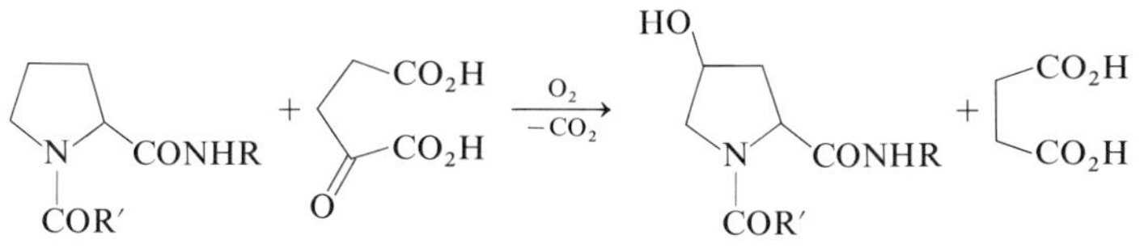

(36)

Internal monooxygenases (EC 1.13.12), the simplest type of monooxygenases, catalyze the incorporation of one oxygen atom into the substrate and the concomitant reduction of the second oxygen atom to water; the hydrogens are derived from the substrate itself. The overall reaction is expressed by the equation:

$$SH_2 + O_2 \longrightarrow SO + H_2O \tag{37}$$

The *internal flavin monooxygenases*[54] catalyze the oxidative decarboxylation of several amino acids. For example, *lysine 2-monooxygenase* (EC 1.13.12.2) from *Pseudomonas* catalyzes the conversion of L-lysine to δ-aminovaleramide.

$$H_2N(CH_2)_4\underset{\displaystyle NH_2}{\underset{|}{C}}HCO_2H + O_2 \longrightarrow H_2N(CH_2)_4CONH_2 + CO_2 + H_2O$$

Tyrosinase[21] (EC 1.10.3.1), the oldest of all known oxygenases, is a copper-containing internal monooxygenase that catalyzes the insertion of molecular oxygen into the ortho position of a variety of phenols. Interestingly, the hydrogens are provided by the catechol, which is formed as an intermediate in the reaction.

$$\text{R-C}_6\text{H}_4\text{OH} + O_2 \longrightarrow \text{R-}o\text{-benzoquinone} + H_2O \quad (38)$$

The *external monooxygenases* comprise a much larger group of enzymes than do the internal monooxygenases and require a cosubstrate, such as

TABLE II Reactions Effected by Monooxygenases

Type of reaction	Substrate	Product
Saturated C—H hydroxylation	$R_1R_2R_3C{-}H$	$R_1R_2R_3C{-}OH$
Aromatic hydroxylation	ArH	ArOH
Epoxidation of a double bond	$R_1CH{=}CHR_2$	$R_1CH{-}CHR_2$ (epoxide, O bridging)
Lactonization	$R_1C(=O)R_2$	$R_1C(=O)OR_2$
Hydroxylation of a nitrogen atom[164]	R_1R_2NH	$R_1R_2N{-}OH$
Formation of an amine oxide	$R_1R_2R_3N$	$R_1R_2R_3N{-}O$
Formation of a sulfoxide[163]	R_1R_2S	$R_1R_2S{=}O$
Deamination	$RCH(NH_2)CH_3$	$RCOCH_3 + NH_3$
N-Dealkylation	$R_1NHCH_2R_2$	$R_1NH_2 + R_2CHO$
O-Dealkylation	$ROCH_3$	$ROH + CH_2O$
S-Dealkylation	$RSCH_3$	$RSH + CH_2O$

NADH or NADPH, which functions as the hydrogen donor:

$$S + NADH + H^+ + O_2 \longrightarrow SO + NAD^+ + H_2O \quad (39)$$

This group of enzymes catalyzes a wide variety of oxidative transformations, which are summarized in Table II. It is this group of important enzymes that has largely stimulated the extensive studies of biomimetic oxygenations to be described later (Section V).

Liver microsomes contain numerous monooxygenases that mediate many of the transformations listed in Table II. The most common of these is the hydroxylation of aliphatic C—H bonds, which constitutes the most important reaction in the microbial transformations of simple hydrocarbons, terpenes, and steroids.[2–6] Although the active enzyme has not been isolated from the microorganism in many cases, these systems almost certainly involve monooxygenases. Oxygenation at nonactivated carbon, which occurs in many cases with high regio- and stereoselectivity, represents the most interesting practical application of microbial oxidations.[2–6] The regio- and stereospecific hydroxylation of steroid molecules by microorganisms is used for the commercial production of a variety of steroidal hormones. Many of the transformations mediated by microorganisms cannot be accomplished by simple chemical means. For example, alkane 1-monooxygenase (EC 1.14.15.3) from *Pseudomonas oleovorans* effects regiospecific hydroxylation of *n*-alkanes at the terminal methyl group.[55] The same microorganism is also capable of stereospecific epoxidation of isolated carbon–carbon double bonds.[56] The first step in the biosynthesis of cholesterol from squalene is the regiospecific epoxidation catalyzed by squalene 2,3-epoxidase[57] (EC 1.14.99.7).

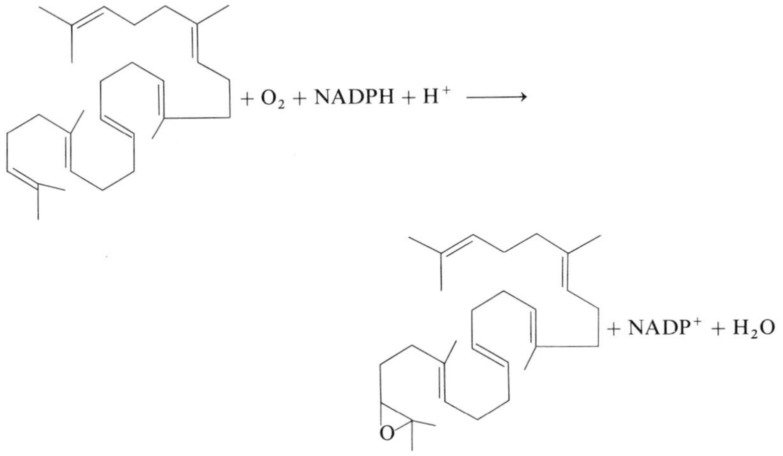

$$\text{[squalene]} + O_2 + NADPH + H^+ \longrightarrow$$

$$\text{[2,3-oxidosqualene]} + NADP^+ + H_2O \quad (40)$$

The second most common reaction catalyzed by monooxygenases is the hydroxylation of aromatic systems, which is a key step in the metabolism of aromatic amino acids. There are several examples of flavoprotein monooxygenases that employ NADH or NADPH as the cosubstrate and contain FAD as the prosthetic group.[54] For example, *salicylate hydroxylase* (EC 1.14.13.1) and *p-hydroxybenzoate hydroxylase* (EC 1.14.13.2) contain FAD as the prosthetic group and catalyze the oxidative decarboxylation of salicylic acid,

$$\text{C}_6\text{H}_4(\text{CO}_2\text{H})(\text{OH}) + \text{NADH} + \text{H}^+ + {}^*\text{O}_2 \longrightarrow \text{C}_6\text{H}_4({}^*\text{OH})(\text{OH}) + \text{NAD}^+ + \text{CO}_2 + \text{H}_2{}^*\text{O} \quad (41)$$

and the hydroxylation of *p*-hydroxybenzoate.[54]

$$p\text{-HO-C}_6\text{H}_4\text{-CO}_2\text{H} + \text{NADPH} + \text{H}^+ + {}^*\text{O}_2 \longrightarrow 3\text{-}{}^*\text{OH-}4\text{-HO-C}_6\text{H}_3\text{-CO}_2\text{H} + \text{NADP}^+ + \text{H}_2{}^*\text{O} \quad (42)$$

Dopamine β-monooxygenase[58] (EC 1.14.17.1.) is a copper-containing monooxygenase that employs ascorbate as the cosubstrate and catalyzes the hydroxylation of dopamine to noradrenaline.[58a]

$$(\text{HO})_2\text{C}_6\text{H}_3\text{CH}_2\text{CH}_2\text{NH}_2 + \text{ascorbate} + {}^*\text{O}_2 \longrightarrow (\text{HO})_2\text{C}_6\text{H}_3\text{CH}({}^*\text{OH})\text{CH}_2\text{NH}_2 + \text{dehydroascorbate} + \text{H}_2{}^*\text{O} \quad (43)$$

Several iron-containing hydroxylases require both a pyridine nucleotide and a pterin as cofactors.[59] The latter acts as an intermediate hydrogen donor between the reduced pyridine nucleotide cosubstrate on the one hand and the monooxygenase on the other. For example, *phenylalanine hydroxylase*[58,59], (EC 1.14.16.1) obtained from mammalian liver employs a pterin as a shuttle between a reductase and the monooxygenase proper. The ultimate hydrogen donor (cosubstrate) is NADPH, as shown in Scheme V.

Scheme V:

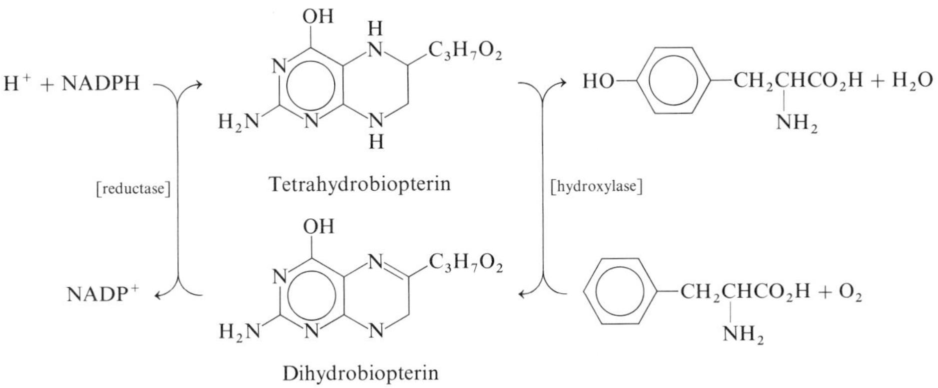

H^+ + NADPH → [reductase] → $NADP^+$

Tetrahydrobiopterin

Dihydrobiopterin

[hydroxylase]

HO–C_6H_4–$CH_2CH(NH_2)CO_2H + H_2O$

C_6H_5–$CH_2CH(NH_2)CO_2H + O_2$

An interesting feature of this reaction is the migration of the *p*-hydrogen of the substrate to the meta position of the tyrosine product. This effect, which is termed the NIH shift (because it was discovered at the National Institute of Health), is a common feature of many aromatic hydroxylations catalyzed by monooxygenases (*vide infra*).

Monooxygenases containing the *cytochrome P*-450 constitute a significant group of enzymes that have been extensively studied in recent years.[14–18,60–69] As mentioned earlier, cytochrome *P*-450 is a protein containing ferriprotoporphyrin IX as the prosthetic group. Cytochrome *P*-450 is a component of many monooxygenases in mammalian liver microsomes, adrenal mitochondria, and numerous bacteria. Examples of *P*-450-containing monooxygenases include many steroid hydroxylases, oxygenases mediating drug and carcinogen metabolism, and bacterial monooxygenases catalyzing the hydroxylation of aromatic and aliphatic hydrocarbons. The bacterial monooxygenase *camphor 5-oxygenase* of *Pseudomonas putida*, examined by Gunsalus and co-workers,[15–17,63–66] catalyzes the hydroxylation of camphor.

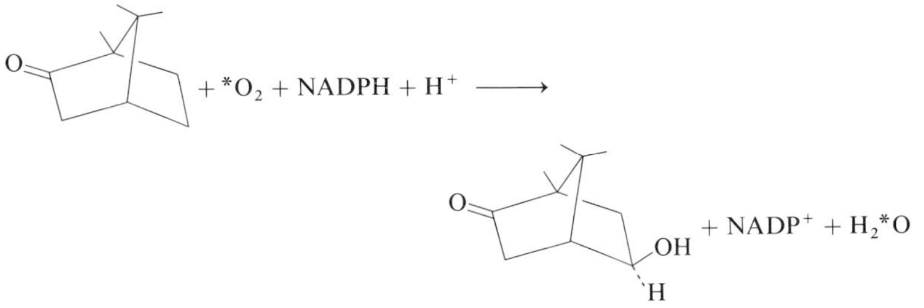

$$\text{camphor} + {}^*O_2 + \text{NADPH} + H^+ \longrightarrow \text{5-exo-hydroxycamphor} + \text{NADP}^+ + H_2{}^*O \quad (44)$$

Camphor 5-oxygenase is a multicomponent enzyme that contains cytochrome P-450$_{cam}$, a flavoprotein, and an iron–sulfur protein, *putidaredoxin*. The ultimate hydrogen donor is NADPH, and the hydrogens (electrons plus protons) are transferred via the flavoprotein and putidaredoxin to the cytochrome, which activates molecular oxygen and mediates oxygen transfer to the substrate. The sequence of events is depicted in Scheme VI.

Scheme VI:

NADPH + H^+	P-$FADH_2$	2 P-$(Fe^{II}S)_2$	P-450(Fe^{II})	O_2 →
				P-450($Fe^{V}O$) → H_2O
$NADP^+$	P-FAD	2 P-$(Fe^{III}S)_2$	P-450(Fe^{III})	RH → ROH

In the first step of the reaction cycle, the FAD-containing flavoprotein undergoes a two-electron reduction by NADPH to afford $FADH_2$ (reduced flavin). The latter reduces the oxidized form of the putidaredoxin (which contains two ferric ions coordinated to sulfide and a cysteinyl mercaptide) to the ferrous form. The reduced putidaredoxin then transfers electrons to the cytochrome P-450. The final steps in the reaction cycle are the activation of molecular oxygen by the P-450 and subsequent oxygen transfer to the substrate. Cytochrome P-450$_{cam}$ has been isolated in crystalline form,[64] and the mechanism of oxygen activation and transfer to substrate has been thoroughly studied.[17] The reaction sequence involves six well-defined steps.[17,67]

1. The formation of an enzyme–substrate complex with the ferricytochrome P-450 containing six-coordinate Fe(III). There is much evidence in favor of a cysteinyl mercaptide (RS) group as the fifth, axial ligand. The other axial ligand is unknown, although Ullrich[67] favors a hydroxyl group of a serine residue. Complex formation is accompanied by a transition from a hexacoordinate, low-spin complex to a pentacoordinate, high-spin heme complex with displacement of the sixth axial ligand. The lipophilic substrate presumably resides in a hydrophobic pocket of the active site, and complex formation is accompanied by a conformational change in the protein.
2. One-electron reduction of the ferricytochrome to the ferrocytochrome, the electron being provided by putidaredoxin. This step requires the neutralization of an anion (probably mercaptide) by a proton. The ferrocytochrome is also pentacoordinate, which leaves a vacant site.
3. Complexation with dioxygen to form a superoxoferricytochrome that is analogous to superoxohemoglobin (oxyHb).

4. One-electron reduction of the superoxoferricytochrome by putidaredoxin, yielding an electrophilic oxygen species thought to be formally a perferrylcytochrome [O=Fe(V)].

5. Oxygen transfer from the perferrylcytochrome to the C—H bond or π system of the substrate.

6. Dissociation of the product from the enzyme, which is facilitated by the decreased hydrophobicity of the product compared to the substrate. The endogenous sixth, axial ligand is restored, and the cycle can continue, as depicted in Scheme VII.

Scheme VII:

L–Fe^{III}–RS $\xrightarrow[-L]{\text{substrate (R'H)}}$ Fe^{III}–RS $\xrightarrow{e^-, H^+}$ Fe^{II}–RSH $\xrightarrow{O_2}$ ·O–O–Fe^{III}–RSH $\xrightarrow{e^-, H^+}$

HO–O–Fe^{III}–RSH $\xrightarrow{-H_2O}$ O=Fe^{V}–RS $\xrightarrow{R'H}$ Fe^{III}–RS + R′OH $\xrightarrow{L}$ L–Fe^{III}–RS

The mechanism for oxygen transfer from the oxoiron(V) porphyrin to the substrate (R′H) is discussed in more detail in the following section. It is likely that other cytochrome *P*-450-containing bacterial monooxygenases and those derived from mammalian liver microsomes[60] and adrenal cortex mitochondria[68,69] involve analogous multicomponent enzyme systems and reaction pathways.

E. Superoxide Dismutases

Superoxide dismutases mediate the removal of the cytotoxic superoxide ion from biological systems by catalyzing the reaction:

$$2\,O_2^{\cdot -} + 2\,H^+ \longrightarrow H_2O_2 + O_2 \tag{45}$$

It is perhaps not surprising that superoxide is prevalent in biological systems when one considers that the first step in the binding of dioxygen by many metalloenzymes involves the formation of a superoxo complex.

$$M^{(n-1)+} + O_2 \rightleftharpoons \cdot O{-}O{-}M^{n+} \tag{46}$$

Ligand displacement by nucleophiles leads to free superoxide. Knowles and co-workers[70] showed that superoxide is formed when reduced xanthine oxidase, which contains molybdenum(V), reacts with dioxygen. McCord and Fridovich[71] subsequently found that a copper enzyme, erythrocuprein, was responsible for scavenging the superoxide generated by *xanthine oxidase*.[71a] Since then, the superoxide dismutases have been studied extensively,[72–75] and both iron- and manganese-containing superoxide dismutases have been identified. Fee and McClune[75] have pointed out that simple inorganic complexes of Cu, Fe, and Mn catalyze the disproportionation in eq 45, sometimes even more efficiently than the enzymes. Moreover, reaction 45 is fast even in the absence of a catalyst. In terms of classical autoxidation systems, the superoxide dismutases can be considered to be very efficient scavengers of the superoxide anion-radical.

IV. MECHANISMS OF ENZYMATIC OXIDATION

The mechanisms of enzymatic oxidations have been the subject of much interest and speculation ever since the study of biological oxidation processes was initiated by Lavoisier more than 200 years ago. The early development of modern biochemistry was marked by attempts to classify all biological oxidations under one all-encompassing mechanism. For example, Wieland[76] believed that all biological oxidations consisted of a series of dehydrogenations. As noted earlier, it was thought until 1955 that the function of dioxygen was solely that of an electron acceptor. When oxygen was incorporated into the substrate it was thought to be derived from water. In the last two decades, however, it has become increasingly apparent that the majority of oxidizing enzymes are oxygenases, which incorporate molecular oxygen directly into the substrate. It is now appreciated that no single mechanism is applicable to all enzymatic oxidations and that many of the mechanistic pathways are interrelated.

It is perhaps not surprising that nature has developed oxygen-utilizing enzymes for the metabolism of hydrophobic compounds such as aromatic hydrocarbons. In contrast, hydrophilic substrates, such as purines, are usually hydroxylated by employing water in a hydration–dehydrogenation sequence. For example, xanthine oxidase probably involves hydration and subsequent dehydrogenation mediated by oxomolybdenum(VI).

$$\text{(xanthine hydrate, N-bound } Mo^{VI}{=}O\text{; C(H)(OH))} \longrightarrow \text{(uric acid)—OH} + Mo^{IV}OH \qquad (47)$$

The O=Mo(VI) moiety is then regenerated by oxidation of the reduced Mo(IV) with dioxygen; i.e., the latter is a terminal oxidant. In enzymatic reactions catalyzed by oxygenases, in addition to its role as oxygen acceptor, the substrate usually acts as an *allosteric regulator*; i.e., dioxygen binds to the active site only in the presence of the substrate. A conformational change in the protein resulting from complexation of the substrate can lead to the creation of a vacant site (compare the reaction cycle in Scheme VII for *P*-450 enzymes).

Since enzymatic systems and metal catalysis involve the same basic components (i.e., transition metal ions and dioxygen or hydroperoxides), similar mechanisms involving either homolytic or heterolytic pathways are to be expected in which metal–substrate, metal–dioxygen, and metal–hydroperoxide interactions occur, as described in Chapters 3–7. Thus, in enzymatic systems, the same types of reaction intermediates, such as free radicals, metal–peroxide complexes, and oxometal species, are likely to be encountered as in the *in vitro* systems. However, it is worth emphasizing that the substrate is isolated at the active site of the enzyme, where reactive intermediates, e.g., free radicals, cannot easily diffuse away. (This situation is equivalent to a high local concentration of the metal catalyst.) As a result, free radical chain pathways involving radical–substrate interactions are less favored than in the *in vitro* systems, as is observed. Moreover, enzymatic oxidations take place under conditions in which normal autoxidations proceed very slowly, if at all.

A. Free Radicals as Intermediates

Since the prosthetic group of oxidoreductases contains metals such as iron and copper or organic groups (flavins) that readily undergo one-electron transfer, it is reasonable that free radicals will be formed as transient intermediates in many cases. Homolytic mechanisms have often been rejected for enzymatic oxidations, however, on the grounds that they are not consistent with the observed retention of configuration at an asymmetric C—H bond. However, this stereochemical result is possible if the reactants are so oriented on an enzyme that C—H bond fission and C—O bond formation must occur on the same face of the molecule.[77] Indeed, retention of configuration in the cage reactions of free radical pairs has been demonstrated.[78] In cases where the substrate forms a stable radical, one-electron oxidation of the substrate is the initial step in the reaction. For example, the first step in the oxidation of ascorbic acid catalyzed by the copper enzyme *ascorbate oxidase* (EC 1.10.3.3.) is thought to be a one-electron oxidation by copper(II).[37]

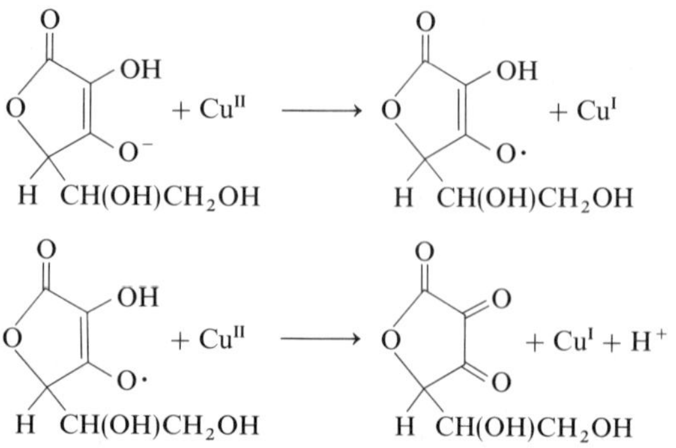

$$\text{ascorbate monoanion (O=C–O ring, C(OH)=C(O}^-\text{), H, CH(OH)CH}_2\text{OH)} + Cu^{II} \longrightarrow \text{ascorbyl radical (C(OH)=C(O}\cdot\text{))} + Cu^{I} \quad (48)$$

$$\text{ascorbyl radical (C(OH)=C(O}\cdot\text{), H, CH(OH)CH}_2\text{OH)} + Cu^{II} \longrightarrow \text{dehydroascorbic acid (C(=O)–C(=O), H, CH(OH)CH}_2\text{OH)} + Cu^{I} + H^+$$

This step is followed by the regeneration of the Cu(II) enzyme by oxidation with dioxygen:

$$4\,Cu^{I} + O_2 + 4\,H^+ \longrightarrow 4\,Cu^{II} + 2\,H_2O \quad (49)$$

In the oxidation of galactose catalyzed by *galactose oxidase* (EC 1.1.3.9), on the other hand, the substrate does not form a stable radical and a heterolytic mechanism is preferred.[37] A reasonable mechanism is one involving β-hydride elimination from the Cu(II) alcoholate followed by insertion of dioxygen into a Cu—H bond.

$$RCH_2OCu^{II}X \longrightarrow RCH{=}O + XCu^{II}H \xrightarrow{O_2} XCu^{II}O_2H \quad (50)$$

$$XCu^{II}O_2H \xrightarrow{RCH_2OH} RCH_2OCu^{II}X + H_2O_2 \quad (51)$$

where R = $\overline{CH(OH)CH(OH)CH(OH)CH(OH)OCH}$—.

Hamilton and co-workers[79] have presented an alternative mechanism involving a Cu(I)/Cu(III) redox couple in the oxidation of galactose mediated by galactose oxidase, e.g.,

$$Cu^{I} + O_2 \longrightarrow O_2Cu^{II} \xrightarrow{2\,H^+} Cu^{III} + H_2O_2$$

$$Cu^{III} + {>}C(H)(OH) \xrightarrow{-2\,H^+} {>}C{=}O + Cu^{I}, \qquad \text{etc.}$$

It was noted that the paucity of well-characterized copper(III) complexes is not so much due to the high redox potentials [depending on the ligands, the E^0 for the Cu(II)/Cu(III) couple can be less than that for a Fe(II)/Fe(III) couple], but to the fact that copper(III) complexes are kinetically unstable—indeed, the property required for catalysis. The participation of a Cu(I)/Cu(III) couple has also been proposed in the copper-catalyzed oxidation of ascorbic acid by dioxygen[80] (compare ascorbate oxidase discussed above.)

B. Hemoproteins and the Role of Oxoiron Intermediates

The hemoproteins fulfill a multiplicity of roles in nature. (The participation of heme proteins in oxidation and in the activation of dioxygen has been recently reviewed by Chang and Dolphin.[81]) *Hemoglobin*[82] reversibly binds dioxygen but does not activate it. *Cytochrome c oxidase*[83] catalyzes the reduction of oxygen. *Cytochromes P-450* (monooxygenase) and *tryptophan oxygenase* (dioxygenase) catalyze the activation and insertion of dioxygen into organic substrates. Finally, *peroxidases* catalyze the transfer of oxygen from a hydroperoxide to a substrate. Interaction with dioxygen is limited to ferrohemes containing a vacant sixth coordination site (hemoglobin and oxygenases). The ferrihemes, on the other hand, can form an active oxidant by reaction with H_2O_2, as in the peroxidases. The type of interaction is further determined by the protein, which controls the axial ligand modes of the heme group, both the number and the type being variable. Williams[84] has pointed out that the essential difference between the prosthetic group of an oxygen carrier (hemoglobin) and that of an oxygen activator (cytochrome *P*-450) is the presence in the latter of a second electron donor site. In cytochrome *P*-450, this is a cysteinyl mercaptide (RS) group; the axial ligand in hemoglobin is a histidinyl imidazole group. The active intermediate is now believed to be formally an oxoiron(V) heme formed as shown in Scheme VIII (compare Scheme VII).[67]

Scheme VIII:

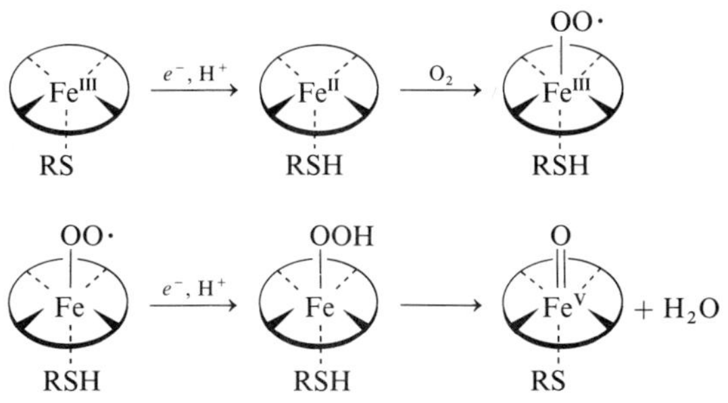

Fe^{III} (RS) $\xrightarrow{e^-,\ H^+}$ Fe^{II} (RSH) $\xrightarrow{O_2}$ Fe^{III}–OO· (RSH)

Fe–OO· (RSH) $\xrightarrow{e^-,\ H^+}$ Fe–OOH (RSH) $\longrightarrow$ Fe^{V}=O (RS) $+ H_2O$

The exact role of the mercaptide group is not clearly defined, although the facile one-electron change ($RS^- \rightarrow RS\cdot + e$) is undoubtedly important in mediating electron transfer. Although it is now generally accepted that the active oxidant in *P*-450 monooxygenases, catalase, peroxidase, and chloroperoxidase,[85] is an oxoiron(V) heme (protoporphyrin IX), they are formed via totally different pathways (Scheme IX).

Scheme IX:

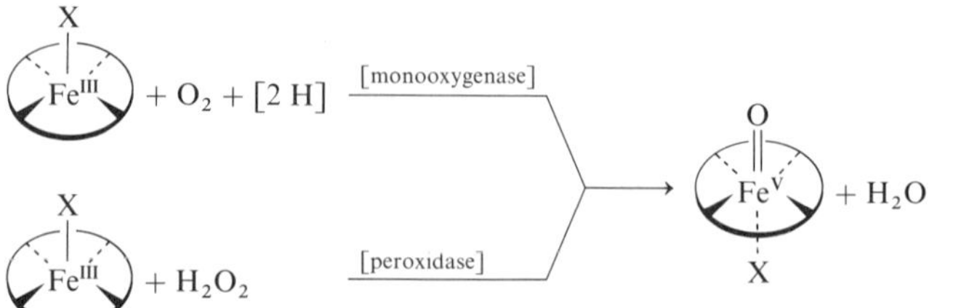

$$\left(\overset{X}{\underset{}{\mathrm{Fe^{III}}}}\right) + O_2 + [2\,H] \xrightarrow{[\text{monooxygenase}]} \left(\overset{O}{\underset{X}{\mathrm{Fe^{V}}}}\right) + H_2O \tag{52}$$

$$\left(\overset{X}{\mathrm{Fe^{III}}}\right) + H_2O_2 \xrightarrow{[\text{peroxidase}]}$$

Several recent investigations have confirmed the formal equivalence of the intermediates in Scheme IX. Following the initial findings of Hrycay and co-workers[86–88] that cytochrome *P*-450 can function as a peroxidase, several groups[89–92] have demonstrated that liver microsomal *P*-450 cytochrome can catalyze the hydroxylation of hydrocarbon substrates by H_2O_2 or RO_2H in the absence of NADPH, oxygen, and cytochrome *P*-450 reductase (i.e., via the peroxidase pathway in Scheme IX). In addition to hydroperoxides and peracids, other single-oxygen donors such as chlorite,[90] periodate,[90] and especially iodosobenzene[93] can also be used as the oxidizing agent, the oxoiron(V) cytochrome being generated according to:

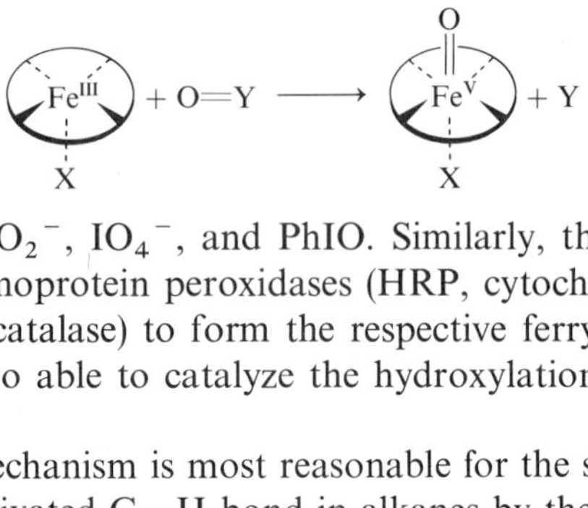

$$\left(\underset{X}{\mathrm{Fe^{III}}}\right) + O{=}Y \longrightarrow \left(\overset{O}{\underset{X}{\mathrm{Fe^{V}}}}\right) + Y \tag{53}$$

where O=Y is ClO_2^-, IO_4^-, and PhIO. Similarly, these oxygen donors also react with hemoprotein peroxidases (HRP, cytochrome *c* peroxidase, chloroperoxidase, catalase) to form the respective ferryl(V) intermediates. Peroxidases are also able to catalyze the hydroxylation of aromatic compounds.[31]

A free radical mechanism is most reasonable for the subsequent hydroxylation of an unactivated C—H bond in alkanes by the ferryl(IV) oxidant. Indeed, Groves and co-workers[94] have presented evidence in favor of transient radical intermediates in the hydroxylation of norbornane catalyzed by the liver microsomal cytochrome *P*-450 system. Thus, the large isotope effect ($k_H/k_D = 11.5 \pm 1$) and a significant amount of epimerization are consistent with rate-limiting hydrogen abstraction followed by fast ligand transfer, in which the intermediate alkyl radical is only partially constrained in the enzyme–substrate cage.

Scheme X:

$$O{=}Fe^{V} + RH \longrightarrow [HOFe^{IV}\,R\cdot]$$

$$[HOFe^{IV}\,R\cdot] \longrightarrow Fe^{III} + ROH$$

The authors noted that the large kinetic isotope effect is similar to those observed[95,96] for alkane oxidations by the well-characterized high-valent oxo complexes of manganese and chromium. The mechanism in Scheme X for hydroxylation by cytochrome *P*-450 monooxygenases was proposed by Hamilton.[37,97–99] He noted the remarkable similarity of many monooxygenase-catalyzed reactions to carbene and nitrene reactions and suggested that these enzymes react via an oxygen atom transfer or "oxenoid" mechanism. He further predicted that there should be a group of electrophilic oxenoid reagents O═X capable of transferring an oxygen atom to organic substrates such as alkanes, alkenes, and aromatic compounds, i.e.,

$$\mathrm{O{=}X + S \longrightarrow SO + X} \tag{54}$$

Before elaborating further the possible reactions of the electrophilic oxoiron (V) porphyrin intermediate, it is appropriate to consider the possible canonical structures of this oxenoid species. Since various structures can be written in which the odd electron and charges are delocalized onto the ligands, e.g.,

$$\underset{\substack{|\\ \mathrm{P}}}{^{-}\mathrm{O{-}Fe^{+}}} \longleftrightarrow \underset{\substack{|\\ \mathrm{P}}}{\mathrm{O{=}Fe^{V}}} \longleftrightarrow \underset{\substack{|\\ \mathrm{P}^{\dot{+}}}}{\mathrm{O{=}Fe^{IV}}} \longleftrightarrow \underset{\substack{|\\ \mathrm{P}}}{\cdot\mathrm{O{-}Fe^{IV}}}$$

(where P = porphyrin), both heterolytic and homolytic pathways are possible in the reactions of this intermediate with C—H bonds or π systems. With substrates containing an activated C—H bond, a heterolytic mechanism is probably favored, as in the oxidation of primary alcohols by the oxorion (V) intermediate of catalase in Scheme II.

$$\mathrm{O{=}Fe^{V} + RCH_2OH \longrightarrow \begin{matrix}\mathrm{HO}\\ \mathrm{RCH_2O}\end{matrix}\!\!>\!Fe^{V} \longrightarrow Fe^{III} + RCHO + H_2O} \tag{55}$$

In addition to catalyzing the hydroxylation of aliphatic C—H bonds, cytochrome *P*-450 monooxygenases also mediate, *inter alia*, the epoxidation of alkenes and the hydroxylation of aromatic substrates (see Section III). A[2 + 2]-cycloaddition can be envisaged for the stereospecific epoxidation of alkenes, analogous to that proposed by Sharpless and co-workers[100] for the epoxidation of alkenes by oxochromium(VI) reagents, as discussed in Chapter 6.

$$\begin{matrix}\mathrm{O{=}Fe^{V}}\\ \mathrm{C{=}C}\\ \mathrm{R}\quad\quad\mathrm{R}\end{matrix} \longrightarrow \begin{matrix}\mathrm{O{-}Fe^{V}}\\ |\quad\ |\\ \mathrm{C{-}C}\\ \mathrm{R}\quad\quad\mathrm{R}\end{matrix} \longrightarrow \mathrm{Fe^{III}} + \begin{matrix}\mathrm{O}\\ \mathrm{C{-}C}\\ \mathrm{R}\quad\quad\mathrm{R}\end{matrix} + \mathrm{H_2O} \tag{56}$$

Indeed, Sharpless and Flood[100a] noted the many similarities between the reactions of monooxygenases and traditional oxometal oxidants such as O═Cr(VI). Possible heterolytic and homolytic mechanisms for the reaction of oxometal reagents with C—H bonds and π systems have been discussed in Chapter 6.

The validity of the ferryl postulate is supported by recent studies of Groves and co-workers,[101] who showed that the oxidation of cyclohexene to cyclohexene oxide and cyclohexenol by cumene hydroperoxide is catalyzed by cytochrome *P*-450. It is noteworthy that a catalytic system containing chlorotetraphenylporphyrinatoiron(III) and iodosobenzene gave the same products. Deuterium labeling indicated that the cyclohexenol is formed with partial equilibration of an allylic intermediate, and a mechanism involving a caged radical pair as in Scheme X was suggested. At this juncture, it is not clear whether the same radical pair is involved in the formation of epoxide.

Much of the current understanding of the mechanism of hydroxylation of aromatic hydrocarbons catalyzed by cytochrome *P*-450 monooxygenases has resulted from the discovery of the NIH shift. In the course of studies on *phenylalanine hydroxylase* from liver, Udenfriend and co-workers[102,103] showed that the labeled hydrogen of [4-^{3}H]phenylalanine undergoes a 1,2-shift during enzymatic hydroxylation to produce [3-^{3}H]tyrosine in more than 90% yield. Similarly, the chlorine atom of 4-chlorophenylalanine undergoes a 1,2-shift to afford 3-chlorotyrosine,

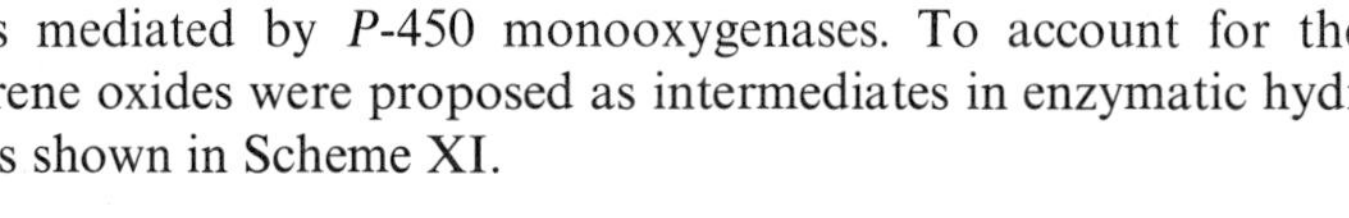

$$X\text{—}C_6H_4\text{—}CH_2CH(NH_2)CO_2H \xrightarrow[\text{[enzyme]}]{O_2} HO\text{—}C_6H_3(X)\text{—}CH_2CH(NH_2)CO_2H$$

where X = D, ^{3}H, and Cl. Subsequently, extensive studies by Jerina and co-workers[104–108] revealed that this intramolecular 1,2-shift of the group being displaced by hydroxyl is a characteristic feature of aromatic hydroxylations mediated by *P*-450 monooxygenases. To account for the NIH shift, arene oxides were proposed as intermediates in enzymatic hydroxylations, as shown in Scheme XI.

Scheme XI:

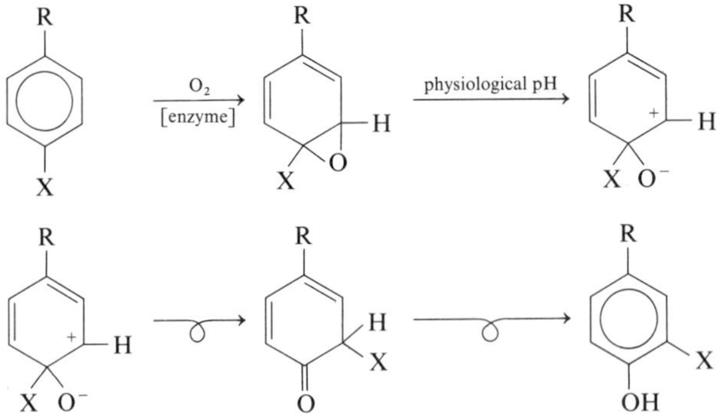

Boyland[109] was the first to suggest that the initial step in aromatic hydroxylation involved the formation of a transient arene oxide. The intermediacy of arene oxides has now been firmly established by detailed mechanistic studies.[110–113] Thus, the nonenzymatic rearrangement of 3,4-[4-^{2}H]toluene oxide gave 4-hydroxytoluene with the same amount of deuterium retention (in the 3 position) as that observed in the enzymatic hydroxylation of [4-^{2}H]toluene by liver microsomes.[105] 1,2-Naphthalene oxide could actually be isolated as the initial product of the enzymatic hydroxylation of naphthalene.[105] Arene oxides as obligatory intermediates provide the basis for explaining ortho/meta/para isomer ratios observed in enzymatic hydroxylations.[104] For example, the absence of *m*-cresol in the metabolites of toluene is consistent with the fact that none of the three possible toluene oxide intermediates rearranges to this product.[104] Other products formed in the metabolism of aromatic substrates have been shown to be derived from transient arene oxides. The latter have also been implicated as causative agents in the carcinogenicity of certain hydrocarbons.[104]

The formation of arene oxides in the biochemical oxidations involves either a homolytic or heterolytic attack of the electrophilic oxoiron(V) porphyrin on the aromatic π system. A heterolytic mechanism is depicted in Scheme XII.

Scheme XII:

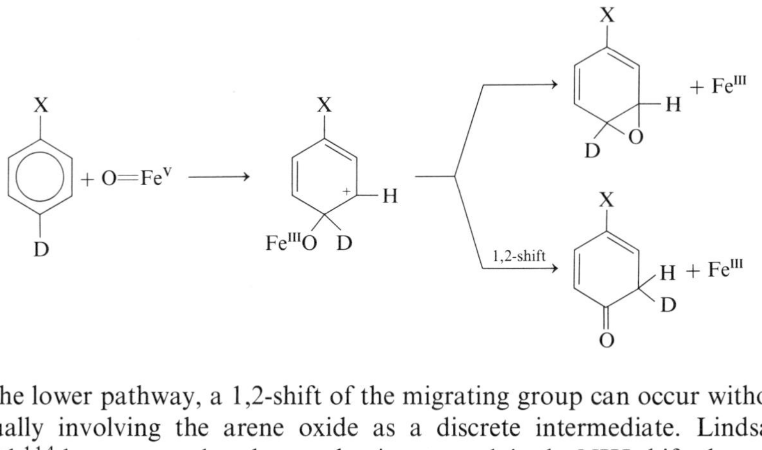

In the lower pathway, a 1,2-shift of the migrating group can occur without actually involving the arene oxide as a discrete intermediate. Lindsay-Smith[114] has proposed such a mechanism to explain the NIH shift observed in model systems, which are described in the next section.

Another interesting reaction is the recently reported formation of an iron porphyrin–ylide complex upon the reaction of the microsomal cytochrome *P*-450 with 1,3-benzodioxole in the presence of O_2 and NADPH.[115] The formation of the stable ylide complex with cytochrome *P*-450 and model iron porphyrins could explain the severe inhibition of the detoxifying monooxygenases of insects. (1,3-Benzodioxole derivatives are widely used as insecticide synergists.) The ylide formation involves the formal replacement of an oxo ligand by a carbene ligand. A possible mechanism is outlined in Scheme XIII.

Scheme XIII:

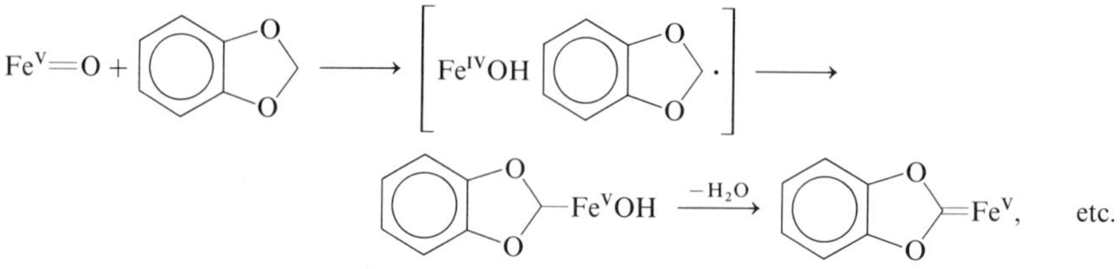

$$Fe^{V}{=}O + (\text{1,3-benzodioxole}) \longrightarrow [Fe^{IV}OH \; (\text{benzodioxol-2-yl})^{\cdot}] \longrightarrow$$

$$(\text{benzodioxol-2-yl}){-}Fe^{V}OH \xrightarrow{-H_2O} (\text{benzodioxol-2-ylidene}){=}Fe^{V}, \quad \text{etc.}$$

Analogous iron(IV) ylides have been recently prepared from the reaction of iron(II) porphyrins with Cl_3CO_2Et or Cl_3CCN,[116]

$$Fe^{II}P + Cl_3CX \longrightarrow (Cl)(X)C{=}Fe^{IV}P + 2Cl^-$$

where $X = CO_2Et$ and CN and $P =$ porphyrin.

Another oxygenase containing ferriprotoporphyrin IX as its prosthetic group is *tryptophan 2,3-dioxygenase* (EC 1.13.11.11), which catalyzes the oxidative cleavage of tryptophan to formylkynurenine, described in reaction 33. In contrast to the reactions with heme-containing monooxygenases discussed above, dioxygen is apparently not cleaved to an oxoiron intermediate. It is probably significant that tryptophan dioxygenase also contains copper ion in the prosthetic group but no cysteinyl mercaptide ligand. Evidence suggests that the catalytically active species contains Cu(I)/Fe(III) or Cu(II)/Fe(II) and that both the copper and iron are involved in the complexation of tryptophan and dioxygen.[50] Another significant observation[117] is the reaction of tryptophan with singlet oxygen to afford formylkynurenine via the hydroperoxide.

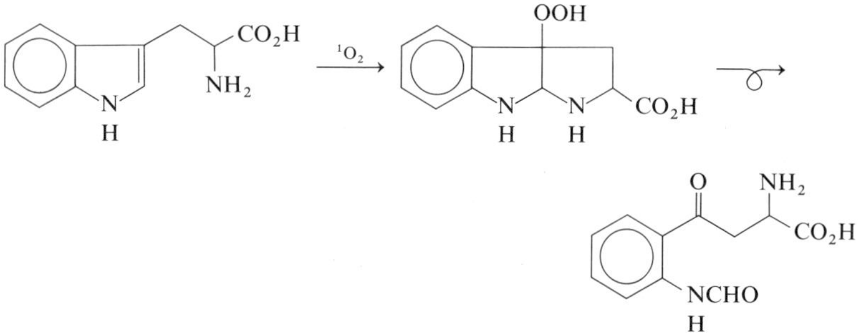

$$\text{tryptophan} \xrightarrow{^1O_2} \text{hydroperoxide (OOH, } CO_2H\text{)} \longrightarrow \text{formylkynurenine (NHCHO, } NH_2, CO_2H\text{)} \qquad (57)$$

It is possible that singlet oxygen formed by the decomposition of a bimetallic μ-peroxo complex is the active oxygenating species in the enzymatic system, e.g.,

Scheme XIV:

$$Fe^{III}OOCu^{II} \longrightarrow \longrightarrow Fe^{II} + {}^{1}O_2 + Cu^{I}$$

$$Fe^{II} + Cu^{I} + O_2 \longrightarrow Fe^{III}OOCu^{II}, \quad \text{etc.}$$

It is worth noting that, in the model system involving a cobalt(II) complex (see the following section), the reaction appears to involve oxygen insertion into an alkylcobalt(III) intermediate as the key step.

C. Nonheme Iron, Copper, and Flavin-Containing Oxygenases

A number of nonheme iron- and copper-containing oxygenases mediate the hydroxylation and oxidative cleavage of phenolic substrates. For example, the catechol dioxygenases[49,97] *pyrocatechase* and *metapyrocatechase* are nonheme iron enzymes that catalyze the oxidative cleavage of catechol described in reactions 13 and 30, respectively. Many of these reactions bear marked resemblances to the corresponding reactions with singlet oxygen and/or base-catalyzed autoxidations [118] and are generally considered to proceed via hydroperoxide intermediates. Hamilton[97] proposed a Baeyer–Villiger rearrangement of the cyclohexadienyl hydroperoxide to explain the formation of *cis,cis*-muconic acid and hydroxymuconic acid semialdehyde in the cleavage of catechol by pyrocatechase and metapyrocatechase, respectively.

Scheme XV:

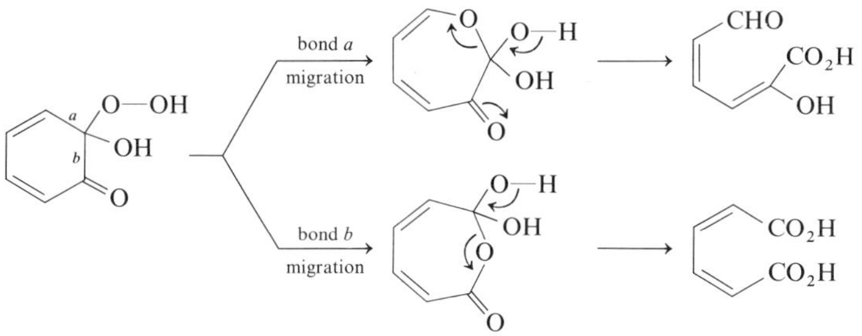

bond *a* migration

bond *b* migration

The hydroperoxide could be formed by an attack of a superoxoiron(III) complex on catechol.[97] In this context, two important points are worth noting: (1) Phenolic substrates readily undergo one-electron oxidation by Fe(III), and (2) nucleophilic displacement of superoxide by protic substrates such as phenols is expected to be facile, i.e.,

$$ArOH + O_2Fe^{III} \longrightarrow ArOFe^{III} + HO_2\cdot \qquad (58)$$

Reaction 58 could be followed by oxygen insertion into the aryloxyiron(III) intermediate (see Chapter 4 for a discussion of similar reactions with cobalt complexes), probably via a free radical chain pathway.

Scheme XVI:

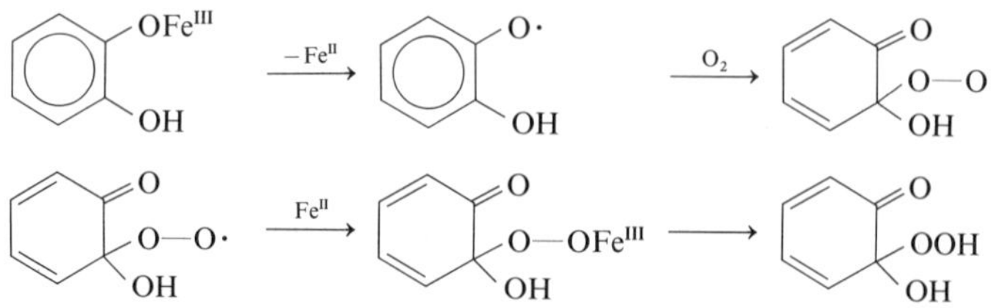

Cleavage could also occur directly via a Baeyer–Villiger rearrangement of the alkylperoxyiron(III) intermediate. These mechanisms are, however, highly speculative, and experimental verification of the various steps is needed.

The copper-containing monooxygenases *tyrosinase* and *dopamine hydroxylase* catalyze the hydroxylation of phenols in reaction 38 and dopamine in reaction 43, respectively. *Quercetinase* is a copper-containing dioxygenase (reaction 34). The mechanisms of these reactions are poorly understood.[21,118] These enzymes generally contain more than one copper atom per molecule, often both Cu(II) and Cu(I).[118a] Model studies of oxidations of phenolic substrates with copper complexes have generally revealed that the initial step is one-electron oxidation of the substrate by Cu(II), as discussed in Chapter 4.

We suggest in Scheme XVII that *hydroxylations* of phenolic substrates mediated by nonheme iron and copper (and, probably, flavin) monooxygenases proceed via a fundamentally different mechanism from the hydroxylation of nonactivated aromatic rings by the heme monooxgenases,

Scheme XVII:

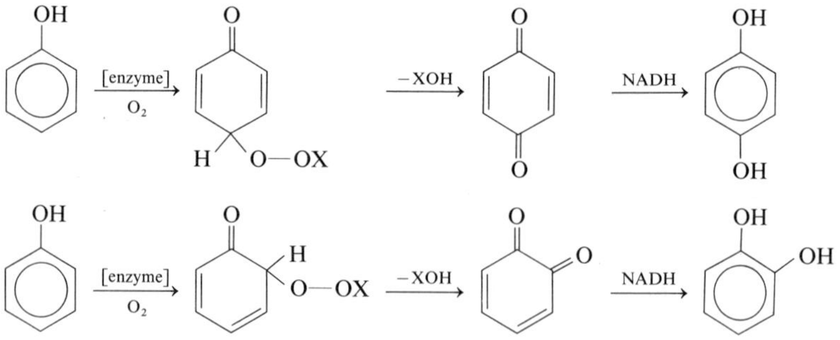

where X = H, Fe(III), Cu(II), or flavin. The intermediate cyclohexadienyl hydroperoxide is essentially the same as that proposed in the reactions with dioxygenases above. Oxygen incorporation precedes the cleavage of the O—O bond in this case, in contrast to the *P*-450 monooxygenases. The cosubstrate (NADH or another donor) is utilized in a subsequent reduction step. Thus, the distinction between a monooxygenase and a dioxygenase may sometimes be quite arbitrary and may be *determined by the fate of the organoperoxy intermediate.* Future investigations of these enzymes may therefore lead to a reclassification of the types on a mechanistic basis. We note that the above mechanism also explains the absence of meta isomers in these enzymatic hydroxylations.

Recent work implicates the flavin hydroperoxide[119] and/or the derived oxaziridine[120] shown below as the enzyme-bound active oxidants (electrophiles) in reactions catalyzed by flavin oxygenases.

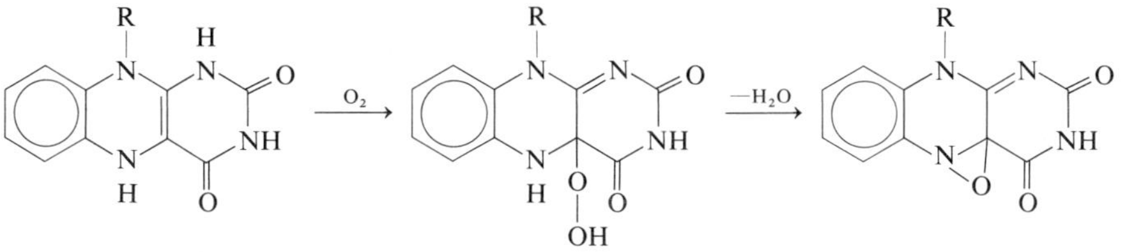

The hydroperoxide moiety above is flanked by two electron-withdrawing substituents that could confer electrophilic properties on it comparable to those of an organic peracid. Bruice and co-workers [119] have shown that model flavin hydroperoxides are able to oxidize tertiary amines to the corresponding *N*-oxides at ambient temperature, a reaction characteristic of flavin monooxygenases and simple organic peracids.

$$\text{flavin—O}_2\text{H} + \text{R}_3\text{N} \longrightarrow \text{flavin—OH} + \text{R}_3\text{NO}$$

In enzymatic oxidations, the resulting hydroxyflavin is presumably reduced to the flavin (and water) by the cosubstrate. Model flavin hydroperoxides were also shown[119,120] to effect other oxidations characteristic of flavin mono- and dioxygenases.

V. BIOMIMETIC OXYGENATIONS

Interest in the study[43,44,97,118] of chemical models that mimic oxygenases has developed for two reasons: first, to provide a basis for understanding enzymatic oxidations and, second, to develop simple, catalytic systems that, under mild conditions, exhibit the high selectivities characteristic of enzymatic oxidations. It is not surprising, therefore, that most studies have concentrated on chemical models for the cytochrome *P*-450 monooxygenases, which mediate the selective hydroxylation of alkanes and arenes and the epoxidation of alkenes.

A. Monooxygenases

Chemical models for monooxygenases can be divided into three basic types: (1) oxygenase, consisting of $Fe^{II} + O_2 +$ hydrogen donor; (2) peroxidase, consisting of $Fe^{III} + H_2O_2$ (or another single-oxygen donor); and (3) oxometal, consisting of oxenoid reagents (X=O).

1. Oxygenase Route

Udenfriend and co-workers[121] observed that a mixture of Fe(II), EDTA, ascorbic acid, and dioxygen at neutral pH is capable of hydroxylating arenes to phenols under mild conditions. [The EDTA is not necessary at lower pH,[37] at which Fe(II) is soluble. It may also lower the redox potential of the Fe(II)/Fe(III) couple.[122]] Udenfriend's reagent also hydroxylates alkanes to alcohols and epoxidizes olefins, albeit in low yields.[37,123,124] The ascorbic acid can be replaced by a variety of hydrogen donors such as enediols,[121] reduced nicotinamide coenzymes,[125] cysteine,[126] thiosalicylic acid,[127] and the pyrimidine[124] and tetrahydropteridine[128] shown below. A tetrahydropteridine derivative is the cofactor in phenylalanine hydroxylase (see Scheme V.):

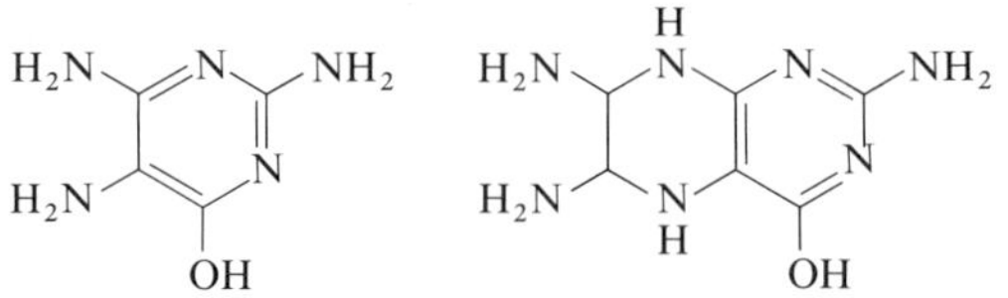

Mimoun and Seree de Roch[129] have recently described several model systems, consisting of a low-valent transition metal, dioxygen, and a hydrogen donor, which hydroxylate alkanes and arenes. For example, $FeCl_2$ was used in conjunction with hydrazobenzene, *o*-phenylenediamine, or 8-hydroxyquinoline.

Despite numerous investigations,[37,122,130,131] the mechanism of hydroxylation by the Udenfriend and related reagents remains unresolved. In one rationale, dioxygen is initially reduced to H_2O_2, and hydroxylation involves hydroxyl radicals formed by reaction of H_2O_2 with Fe(II) (see Fenton's reagent in Chapter 3, Section I,A). However, Fenton's reagent and Udenfriend's reagent afford different isomer distributions in the hydroxylation of arenes,[131] indicating that the hydroxyl radical is not the (only) hydroxylating species in the Udenfriend system. The latter probably involves competing pathways for hydroxylation, since variations in reaction conditions, especially the metal concentration, lead to changes in isomer distributions.

The fact that the Udenfriend reagent does not induce the NIH shift[132] has generally been taken as evidence of a mechanism different from that involved in enzymatic hydroxylation. However, recent results have shed new light on this controversy. Lindsay-Smith and co-workers[114] have reported that the hydroxylation of deuteriated aromatic compounds with Fenton's reagent in an aprotic solvent (CH_3CN, CH_3NO_2) produces NIH shifts that are comparable to those observed with cytochrome *P*-450 enzymes. The extent of deuterium migration is related to the amount of water in the system. These results, taken together with those of Groves and co-workers[133] (see Chapters 3 and 6), support the formation of oxoiron intermediates from Fenton's reagent in aprotic solvents. They both favor a mechanism involving oxoiron(V) or oxoiron(IV) complexes as the active oxidant in the model systems, i.e., the same as that presented above for the *in vivo* system.

Scheme XVIII:

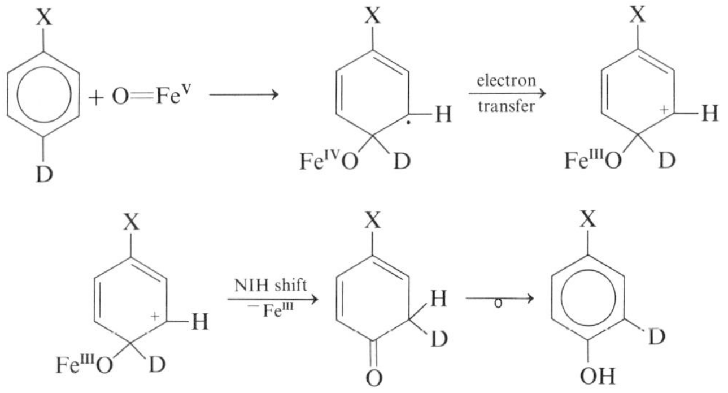

The diminution in the NIH shift with increased proportions of water in the solvent results from the competing process:[114]

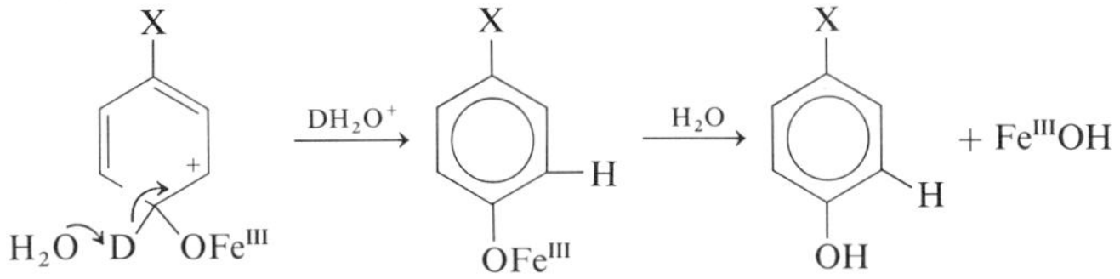

The mechanism for the formation of O═Fe(V) species from Fe(II), O_2, and DH_2 has been discussed earlier. In the enzymatic system, the oxoiron(V) heme resides in a hydrophobic pocket of the *P*-450 cytochrome, allowing the intramolecular 1,2-shift to compete favorably with the reaction of the attached cyclohexadienyl intermediate with water.

Hemin (ferriprotoporphyrin IX) catalyzes the *p*-hydroxylation of aniline by dioxygen at pH > 11 in the presence of NADH or NADPH as the cosubstrate.[134] This constitutes an example of a biomimetic oxygenation mediated by a prosthetic group in the absence of the apoenzyme. The requirement for both dioxygen and a reductant is typical of a monooxygenase. The involvement of HO_2^- was suggested on the basis of an observed maximum rate at pH 11.7, which corresponds to the pK_a of hydrogen peroxide. Furthermore, the addition of H_2O_2 resulted in a reduction of the induction period. The addition of superoxide dismutase resulted in an inhibition, suggesting that superoxide $O_2^{\cdot-}$ was involved, as shown in the following mechanism.

$$HOFe^{III} + NADH + H^+ \longrightarrow Fe^{II} + NAD^+ + H_2O$$

$$Fe^{II} + O_2 \longrightarrow [O_2Fe] \xrightarrow{NADH} HOOFe^{III} + NAD^+$$

$$HOOFe^{III} + C_6H_5\text{—}NH_2 \longrightarrow HO\text{—}C_6H_4\text{—}NH_2 + HOFe^{III}$$

The active oxidant is presented as hydroperoxoiron (III) species or, equivalently, its oxoiron(V) rearrangement product O=Fe(V)OH.

Patin and Mignani[135] have utilized the template effect of a juxtaposed xanthate group to effect selective oxidation of ring A in cholestanol by dioxygen in the presence of iron(II) perchlorate.

$$O_2 + \text{(3-O-xanthate steroid, } S{=}C(SCH_3)O\text{—)} \xrightarrow[HOAc,\ 120^\circ C]{[Fe(ClO_4)_2]} \text{(diacetoxy steroid, AcO, AcO)}$$

(45% yield)

This oxidation is reminiscent of the many stereoselective and regioselective hydroxylations of steroids catalyzed by monooxygenases [see eq 1, for example]. In a similar fashion, the oxidation of adamantane favors the secondary position when it is carried out with a biomimetic system modeling hydroxylase.[136]

$$O_2 + \text{adamantane} \xrightarrow[HOCH_2CH_2SH/py]{[(SalenFe^{III})_2O]} \text{1-adamantanol (OH)} + \text{2-adamantanol (OH)} + \text{2-adamantanone (O)}$$

1 : 2.4 : 0.8

2. *Peroxidase Route*

Hamilton and Friedman[37,97,137] have described the hydroxylation of aromatic compounds with the model system consisting of Fe(III), H_2O_2, and catechol, which also involves an oxoiron(V) species as the putative oxidant formed by a process such as:

$$\mathrm{HO{-}O{-}Fe^{III}} \longrightarrow \mathrm{O{=}Fe^{V}(OH)}$$

Groves and co-workers[138] and Chang and Kuo[139] have recently described chemical models consisting of a porphyrinatoiron(III) chloride and iodosoarenes as the single-oxygen donor. Thus, tetraphenylporphyrinatoiron(III) chloride in combination with iodosobenzene effected stereospecific epoxidation of olefins in good yield,[135] e.g.,

$$\text{cis-PhCH=CHPh} \xrightarrow{[(\mathrm{TPP})\mathrm{Fe^{III}}],\ \mathrm{PhIO}} \text{cis-stilbene oxide (H, H; Ph, Ph)} \quad (60)$$

82% yield

Cyclohexane afforded cyclohexanol in 8% yield with the same reagent. The low yields obtained with the less reactive alkane are due to a competing intramolecular hydroxylation of the porphyrin ligand. This side reaction was more serious with the system consisting of octaethylporphyrinatoiron(III) chloride and iodosobenzene studied by Chang and Kuo.[139] Both groups observed regioselective, intramolecular hydroxylation of an aliphatic side chain in an appropriately substituted porphyrin complex. These reactions probably involve an oxoiron(V) porphyrin intermediate, formed as shown below.

$$(\mathrm{Fe^{III}})\text{–Cl} + \mathrm{ArIO} \longrightarrow \mathrm{O{=}}(\mathrm{Fe^{V}})\text{–Cl} + \mathrm{ArI} \quad (61)$$

The mechanisms of oxygen transfer to alkanes and alkenes have been discussed in the preceding section.

3. Oxometal Route

Sharpless and Flood[100a] observed reactions with oxo transition metal oxidants that are characteristic of monooxygenases, such as the stereospecific epoxidation of olefins and arene hydroxylation. For example, the oxidation of tritium-labeled naphthalene to naphthoquinone with chromyl chloride and acetate was accompanied by tritium migration. Heteroaromatic amine oxides, such as pyridine *N*-oxide, have also been shown [140–142] to transfer oxygen to alkanes, alkenes, and arenes upon photolysis. Retention of configuration was observed in alkane hydroxylation, and the NIH shift in arene hydroxylation.[97] The active oxidant probably has significant diradical character and may be formally considered to be O—N(IV), analogous to O—Fe(IV) discussed earlier.

Finally, the system consisting of Sn(II) and O_2 oxidizes alkanes to the alcohols[129,143] at ambient temperatures, and it produces the NIH shift during the hydroxylation of arenes.[104] The superoxotin(III) complex OOSn(III) was suggested to be the active oxidant.[143] Other low-valent metal ions such as Ti(III), V(III), and Cu(I),[122,129] having a redox potential of approximately 0.15 V, are also effective in promoting *stoichiometric* hydroxylations, although usually in low yield, e.g.,[129]

$$VCl_3 + RH + O_2 \longrightarrow ROH + VOCl_3$$

B. Dioxygenases

Nishinaga and co-workers[144–149] have found that Schiff base complexes of cobalt(II), such as Co(Salen) and Co(Salpr), catalyze many of the reactions characteristic of dioxygenases. The mechanism of these reactions has been discussed in detail in Chapter 4, and they probably involve the free radical chain autoxidation of an organocobalt(III) intermediate. Initiation proceeds via nucleophilic displacement by the substrate on a superoxocobalt(III) or a μ-peroxocobalt(III) complex:

Scheme XIX:

$$SH + LCo^{III}O_2 \longrightarrow LCo^{III}S + \cdot O_2H \qquad (62)$$

$$LCo^{III}S \longrightarrow LCo^{II} + S\cdot \qquad (63)$$

$$S\cdot + O_2 \longrightarrow SO_2\cdot \qquad (64)$$

$$LCo^{II} + SO_2\cdot \longrightarrow LCo^{III}O_2S \qquad (65)$$

$$LCo^{III}O_2S + SH \longrightarrow LCo^{III}S + SO_2H \longrightarrow \text{products} \qquad (66)$$

where SH = protic substrate, and L = Salen, Salpr. [In eq 62, LCo(III)O_2 may be $LCoO_2CoL$; then $HO_2\cdot$ will be $LCoO_2H$.] Scheme XIX is analogous to the mechanism proposed for catechol dioxygenases in Scheme XVI. For example, Co(Salpr) catalyzes the oxygenation of 4-alkyl-2,6-di-*tert*-butylphenols at room temperature to produce the corresponding *p*-quinols in quantitative yield.[146] The reaction proceeds via the alkylperoxycobalt(III) complex, which can be isolated in some cases.[144–146]

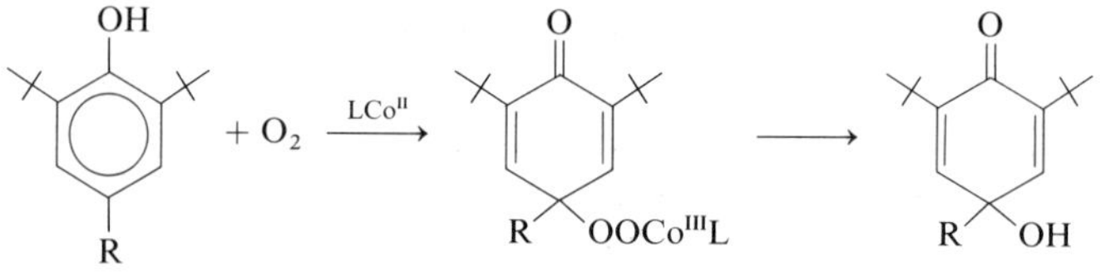

OH … R $+ O_2 \xrightarrow{LCo^{II}}$ O … R, $OOCo^{III}L$ $\longrightarrow$ O … R, OH (67)

Interestingly, when R in eq 67 is an aryl group, the corresponding *o*-peroxy complex is obtained,[147]

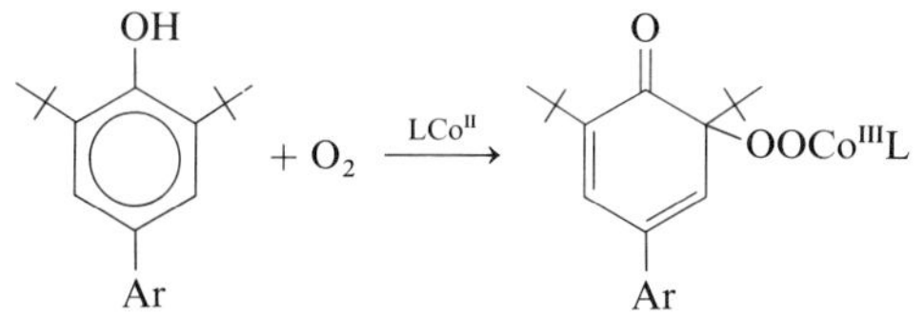

$$\text{2,6-di-}t\text{-Bu-4-Ar-phenol} + O_2 \xrightarrow{LCo^{II}} \text{2,6-di-}t\text{-Bu-4-Ar-6-(OOCo}^{III}\text{L)-cyclohexa-2,4-dienone} \tag{68}$$

which is analogous to the cyclohexadienylperoxyiron(III) intermediate proposed for the catechol dioxygenases in Scheme XVI. Similarly, Co(Salpr) and/or Co(Salen) were shown to catalyze the oxidative cleavage of indole derivatives[148] and 3-hydroxyflavones,[149] which are models for tryptophan dioxygenase in reaction 33 and quercetinase in reaction 34, respectively.

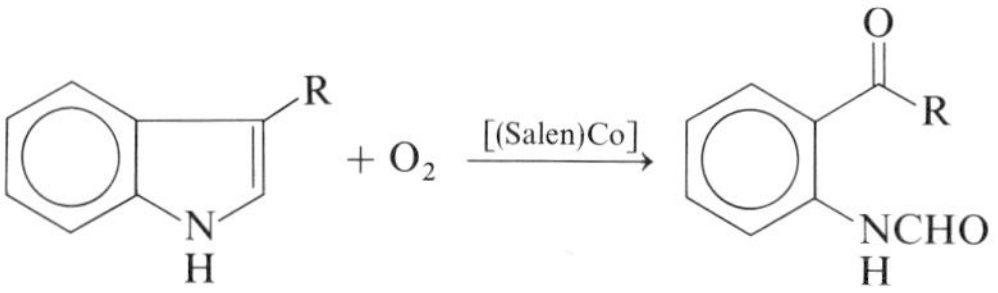

$$\text{3-R-indole} + O_2 \xrightarrow{[(Salen)Co]} \text{2-(NHCHO)C}_6\text{H}_4\text{COR} \tag{69}$$

$$\text{R-3-hydroxyflavone (2-Ar)} + O_2 \xrightarrow{[(Salen)Co]} \text{R-C}_6\text{H}_3\text{(OCOAr)(CO}_2\text{H)} + CO \tag{70}$$

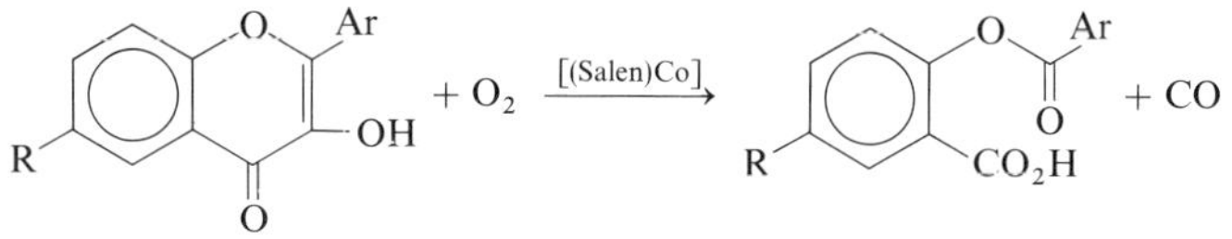

The reactions catalyzed by the cobalt complexes discussed above provide valuable insights into the mechanisms of oxidations mediated by dioxygenases. However, they are not true enzyme models since the oxygenases that they mimic contain copper and/or iron as prosthetic groups. It is interesting, therefore, that oxidative cleavage of aromatic molecules has also been effected with dioxygen in the presence of simple iron and copper complexes. For example, the oxidative cleavage of 3-methylindole by dioxygen (reaction 69, R = CH_3) in the presence of CuCl/pyridine has been described.[150] (Oxidative cleavage of catechols by CuCl/pyridine/O_2 has been discussed in Chapter 4.) Similarly, iron(II)–bipyridine complexes, in the presence of pyridine, catalyze the oxidative cleavage of 3,5-di-*tert*-butylcatechol by dioxygen.[151]

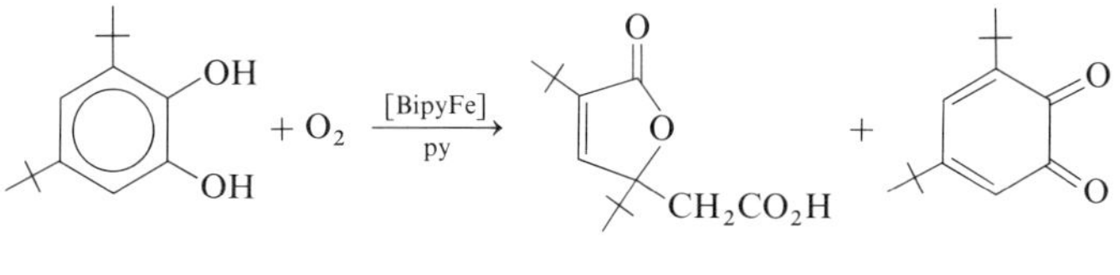

$$\text{3,5-di-}t\text{-Bu-catechol} + O_2 \xrightarrow[\text{py}]{[BipyFe]} \text{lactone (}t\text{-Bu, }t\text{-Bu, CH}_2\text{CO}_2\text{H)} + \text{3,5-di-}t\text{-Bu-}o\text{-benzoquinone}$$

C. Oxidases

There has been much interest recently in the study of models for copper- and/or iron-containing oxidases.[152–155] Thus, the catalytic oxidation of alcohols to aldehydes by dioxygen in the presence of cuprous chloride–bipyridine complexes, L_2CuCl, has been described and the following mechanism suggested.[155]

$$Cu^I + O_2 \longrightarrow \cdot O_2Cu^{II} \xrightarrow{\text{>C(H)OH}} \text{>C=O} + H_2O_2 + Cu^I$$

$$H_2O_2 + \text{>C(H)(OH)} \xrightarrow{Cu^I} \text{>C=O} + 2\,H_2O$$

The initial oxygenation of copper(I) is involved in the production of hydrogen peroxide with concomitant oxidation of the alcohol. This step is followed by the copper(I)-catalyzed oxidation of a second alcohol by the hydrogen peroxide formed as a coproduct in the first step. (Compare the mechanisms discussed earlier for galactose oxidase, for which this reaction is a model.) The second stage was demonstrated in independent experiments with the alcohol in combination with H_2O_2 and catalytic amounts of L_2CuCl.

In summary, the reactions observed with chemical models lend support to the mechanistic postulates outlined in the preceding section. Oxidations catalyzed by heme-containing monooxygenases involve oxoiron(V)–porphyrin complexes, which are formed by the reductive rearrangement of a superoxoiron(III) porphyrin. In the case of peroxidases, the active oxoiron(V) porphyrin is derived from Fe(III) and H_2O_2. Dioxygenases, on the other hand, appear to involve the direct reaction of the substrate with a superoxo or μ-peroxo complex of iron(III) and copper(II) (or a peroxyflavin species); nonheme iron- and copper-containing monooxygenases probably also participate in similar pathways. Singlet oxygen may also be formed, possibly via extrusion from a peroxo or μ-peroxo complex (or flavin hydroperoxide).

VI. IMMOBILIZED ENZYMES

Although work on biomimetic oxygenation provides insights into enzymatic oxidation mechanisms, it is unlikely that simple chemical models can be devised which mimic the combination of high rates, substrate selectivities, and regio- and stereospecificities observed with enzymes. The apoenzyme is of vital importance in determining the type of reaction taking place at the prosthetic group, since it influences the size and nature of the active site. This is amply illustrated by comparing the *P*-450 monooxygenases with the peroxidases, both of which contain iron protoporphyrin IX as the prosthetic group. In the former, the active site contains the heme in a hydrophobic pocket readily accessible to a nonpolar hydrocarbon substrate but not to anions. Peroxidases, in contrast, are generally glycoproteins (carbohydrate-containing), and the active site is hydrophilic, which explains why they employ H_2O_2 and oxidize polar substrates.

An alternative approach to devising chemical models for enzymes is to make enzymes themselves more attractive for large-scale synthesis by immobilization on suitable supports. Recent advances in enzyme immobilization have led to many improvements in enzymatic conversions.[156–158]

It is worth recalling that many oxidoreductases require the use of stoichiometric amounts of electron (hydrogen) donors, such as NADH (*vide supra*). There is currently much interest, therefore, in the immobilization of such coenzymes.[159] Immobilizing enzymes can have many advantages over using soluble enzymes[156–158]: (1) The enzymes can be reused; (2) continuous processing becomes practical; (3) There is improved stability to pH, temp, etc.; (4) product purity is greater; (5) enzyme activity is sometimes enhanced; and (6) effluent problems and handling of materials are minimized. Indeed, it is widely believed that biocatalysis will play an important role in the future production of chemicals. Microbial epoxidation[160] may, for example, be exploited in the future for the production of the bulk chemical propylene oxide and for the synthesis of a variety of chiral epoxides for use in organic synthesis.

An interesting enzymatic process for the production of propylene oxide has recently been reported.[161] In the two-stage process, propylene is treated with a chloride salt, hydrogen peroxide and the enzyme *chloroperoxidase* to yield the chlorohydrin. The latter is then converted to the epoxide and chloride, which is recycled, using another enzyme, *chlorohydrin epoxidase*.

$$CH_3CH{=}CH_2 + Cl^- + H_2O_2 \xrightarrow{[\text{chloroperoxidase}]} CH_3CH(OH){-}CH_2Cl + HO^-$$

$$CH_3CH(OH){-}CH_2Cl \xrightarrow{[\text{chlorohydrin epoxidase}]} CH_3\overset{\displaystyle O}{CH{-}CH_2} + HCl$$

The hydrogen peroxide used in the first step can be produced chemically or biologically (by enzymatic oxidation of glucose, for example).

Finally, the interesting use of a protein (bovine serum albumin) as a phase transfer catalyst for the oxidation of formaldehyde dithioacetals with aqueous sodium periodate,

$$RSCH_2SR + NaIO_4 \xrightarrow[H_2O]{[\text{protein}]} RSCH_2S(=O)R$$

has recently been described.[162]

REFERENCES

1. D. H. Peterson and H. C. Murray, *J. Am. Chem. Soc.* **74**, 1871 (1952).
2. G. S. Fonken and R. A. Johnson, "Chemical Oxidations with Microorganisms." Dekker, New York, 1972.
3. R. A. Johnson, *in* "Oxidation in Organic Chemistry" (W. S. Trahanovsky, ed.), Part C, p. 131. Academic Press, New York, 1978.
4. K. Kieslich, *Synthesis* **120**, 147 (1969).
5. K. Kieslich, "Microbial Transformations of Non-Steroid Cyclic Compounds." Thieme, Stuttgart, 1976.
6. W. Charney and H. L. Herzog, "Microbial Transformation of Steroids." Academic Press, New York, 1967.
7. "Enzyme Nomenclature 1978." Academic Press, New York, 1979.
8. T. E. Barman, "Enzyme Handbook," Vols. 1 and 2. Springer-Verlag, Berlin and New York, 1969; Suppl. I, 1974.
9. H. Beinert, *in* "The Enzymes" (P. D. Boyer, H. Lardy, and K. Myrbäck, eds.), 2nd rev. ed., Vol. 2, pp. 339–416. Academic Press, New York, 1960.
10. A. Ehrenberg and P. Hemmerich, *in* "Biological Oxidations" (T. P. Singer, ed.), pp. 239–262. Wiley (Interscience), New York, 1968. See also T. C. Bruice, *Acc. Chem. Rsch.* **13**, 256 (1980).
11. P. Hemmerich, *Prog. Chem. Org. Nat. Prod.* **33**, 451 (1976).
12. J. W. Hastings, C. Balny, C. LePeuch, and P. Douzou, *Proc. Natl. Acad. Sci. U.S.A.* **70**, 3468 (1973); also see H. W. Orf and D. Dolphin, *ibid.* **71**, 2646 (1974).
13. D. E. Edmondson and C. Gomez-Moreno, *in* "Mechanisms of Oxidizing Enzymes" T. P. Singer and R. N. Ondarza, eds.), Elsevier, Amsterdam, 1977.
14. J. J. Mieyal, *Bioorg. Chem.* **4**, 315 (1978).
15. (a) Y. Ishimura, V. Ullrich, and J. A. Peterson, *Biochem. Biophys. Res. Commun.* **42**, 140 (1971).
    (b) I. C. Gunsalus, J. D. Lipscomb, V. Marshall, H. Frauenfelder, E. Münck, and E. Greenbaum, *Biochem. J.* **125**, 5P (1972).
16. M. Sharrock, E. Münck, P. G. Debrunner, V. Marshall, J. D. Lipscomb, and I. C. Gunsalus, *Biochemistry* **12**, 258 (1973).
17. I. C. Gunsalus, J. R. Meeks, J. D. Lipscomb, P. Debrunner, and E. Münck, *in* "Molecular Mechanisms of Oxygen Activation" (O. Hayaishi, ed.), p. 561. Academic Press, New York, 1974.
18. R. W. Estabrook, J. B. Schenkman, W. Cammer, R. Remmer, D. Y. Cooper, S. Narasimhulu and O. Rosenthal, *in* "Biological and Chemical Aspects of Oxygenases" (K. Block and O. Hayaishi, eds.), p. 153. Maruzen, Tokyo, 1966.
19. B. G. Malmström, *in* "Oxidases and Related Redox Systems" (T. E. King, H. S. Mason, and M. Morrison, eds.), Vol. 1, p. 207. Wiley, New York, 1965.
20. E. Frieden, J. A. McDermott, and S. Osaki, *in* "Oxidases and Related Redox Systems" (T. E. King, H. S. Mason, and M. Morrison, eds.) vol. 1, p. 240. Wiley, New York, 1965.
21. W. H. Vanneste and A. Zuberbühler, in "Molecular Mechanisms of Oxygen Activation" (O. Hayaishi, ed.), p. 371. Academic Press, New York, 1974.
22. E. I. Stiefel, W. E. Newton, G. D. Watt, K. L. Hadfield, and W. A. Bulen, *Adv. Chem. Series* **162**, 353 (1977); W. H. Orme-Johnson, G. S. Jacob, M. T. Henzl, and B. A. Averill, *ibid.* p. 389.
23. J. T. Spence, *Coord. Chem. Rev.* **8**, 475 (1969).
24. R. C. Bray and J. C. Swann, *Struc. Bonding* (*Berlin*) **11**, 107 (1972).

25. R. C. Bray, G. Palmer, and H. Beinert, *in* "Oxidases and Related Redox Systems" (T. King, H. S. Mason, and M. Morrison, eds.), Vol. 1, p. 359. Wiley, New York, 1965.
26. H. S. Mason, W. L. Fowlks, and E. Peterson, *J. Am. Chem. Soc.* **77**, 2914 (1955); H. S. Mason, *Science* **125**, 1185 (1957).
27. O. Hayaishi, M. Katagari, and S. Rothberg, *J. Am. Chem. Soc.* **77**, 5450 (1955).
28. C. Popjak, *in* "The Enzymes" (P. D. Boyer, ed.), 3rd ed., Vol. 2, p. 134. Academic Press, New York, 1970.
29. P. Nicholls and B. Chance, *in* "Molecular Mechanisms of Oxygen Activation" (O. Hayaishi, ed.), p. 479. Academic Press, New York, 1974.
30. B. C. Saunders, A. G. Holmes-Siedle, and B. P. Stark, "Peroxidases". Butterworth, London, 1964.
31. H. B. Dunford and J. S. Stillman, *Coord. Chem. Rev.* **19**, 187 (1976).
32. P. Nicholls, *in* "Oxygenases" (O. Hayaishi, ed.), p. 273. Academic Press, New York, 1962.
33. (a) I. Yamazaki, *in* "Molecular Mechanisms of Oxygen Activation" (O. Hayaishi, ed.), p. 535. Academic Press, New York, 1974.
    (b) H. B. Dunford and B. B. Hasinoff, *Adv. Chem. Series* **100**, 413 (1971).
34. P. Nichols and G. R. Schonbaum, *in* "The Enzymes" (P. D. Boyer, H. Lardy, and K. Myrbäck, eds.), 2nd rev. ed., Vol. 8, p. 147, Academic Press, New York, 1963.
35. K. G. Paul, P. I. Ohlsson, and S. Wold, *Acta Chem. Scand.* **33**, 747 (1979).
36. P. George, *in* "Currents in Biochemical Research" (D. E. Green, ed.), p. 338. Wiley (Interscience), New York, 1956.
37. G. A. Hamilton, *Adv. Enzymol.* **32**, 55 (1969).
38. D. R. Morris and L. P. Hager, *J. Biol. Chem.* **241**, 1763 (1966); L. P. Hager, D. R. Morris, F. S. Brown, and H. Eberwein, *J. Biol. Chem.* **241**, 1769 (1966); see also ref. 85.
39. O. Hayaishi, ed., "Oxygenases." Academic Press, New York, 1962.
40. K. Bloch and O. Hayaishi, eds., "Biological and Chemical Aspects of Oxygenases." Maruzen, Tokyo, 1966.
41. (a) O. Hayaishi, *Annu. Rev. Biochem.* **38**, 21 (1969).
    (b) O. Hayaishi, ed., "Molecular Mechanisms of Oxygen Activation." Academic Press, New York, 1974.
42. O. Hayaishi and M. Nozaki, *Science* **164**, 389 (1969).
43. A. A. Akhrem, D. I. Metelitsa, and M. E. Skurko, *Russ. Chem. Rev.* **44**, 398 (1975).
44. V. Ullrich, *Angew. Chem. Int. Ed.* **84**, 701 (1972).
45. J. P. Vandecasteele, *Kiñet. Katal.* (Eng. Trans.) **14**, 95 (1973).
46. T. E. King, H. S. Mason, and M. Morrison, eds., "Oxidases and Related Redox Systems," Vols. I and II. Wiley, New York, 1965.
47. H. S. Mason, *Annu. Rev. Biochem.* **38**, 21 (1969).
48. *Adv. Chem. Series* **77**, 170-307 (1968).
49. M. Nozaki, *in* "Molecular Mechanisms of Oxygen Activation" (O. Hayaishi, ed.), p. 135. Academic Press, New York, 1974.
50. P. Feigelson and F. O. Brady, *in* "Molecular Mechanisms of Oxygen Activation" (O. Hayaishi, ed.), p. 87. Academic Press, New York, 1974.
51. N. A. Porter, J. R. Nixon, and D. W. Gilmore, *ACS Symp. Series* **69**, 90 (1978); see also the following: M. Hamberg, B. Samuelsson, I. Björkhem, and H. Danielsson, *in* "Molecular Mechanisms of Oxygen Activation" (O. Hayaishi, ed.), p. 30. Academic Press, New York, 1974; S. Bergstrum, *Science* **158**, 382 (1967); B. Samuelsson, *J. Am. Chem. Soc.* **87**, 3011 (1965).
52. H. W. S. Chan, J. A. Mathew, and D. T. Coxon, *J. Chem. Soc., Chem. Commun.* p. 235 (1980).

52a. See also N. A. Porter, B. A. Weber, H. Weenen, and J. A. Khan, *J. Am. Chem. Soc.* **102**, 5597 (1980). N. A. Porter, A. N. Roe, and A. T. McPhail, *Ibid.*, **102**, 7574 (1980).
53. M. T. Abbott and S. Udenfriend, *in* "Molecular Mechanisms of Oxygen Activation" (O. Hayaishi, ed.), p. 167. Academic Press, New York, 1974.
54. M. S. Flashner and V. Massey, *in* "Molecular Mechanisms of Oxygen Activation" (O. Hayaishi, ed.), p. 245. Academic Press, New York, 1974; S. Yamamoto, Y. Maki, T. Nakazawa, Y. Kajita, H. Takeda, M. Nozaki, and O. Hayaishi, *Adv. Chem. Series* **77**, 177 (1968).
55. E. J. McKenna and M. J. Coon, *J. Biol. Chem.* **245**, 3882 (1970).
56. P. L. Kumler and P. J. DeJong, *J. Chem. Educ.* **52**, 475 (1975); S. W. May and R. D. Schwartz, *J. Am. Chem. Soc.* **96**, 4031 (1974).
57. E. J. Corey, W. E. Russey, and P. R. Ortiz de Montellano, *J. Am. Chem. Soc.* **88**, 4750 (1966); E. E. van Tamelen, J. D. Willett, R. B. Clayton, and K. E. Lord, *ibid.* p. 4752.
58. S. Kaufman, *in* "Biological and Chemical Aspects of Oxygenases" (K. Bloch and O. Hayaishi, eds.), p. 261. Maruzen, Tokyo, 1966; T. Nagatsu, *ibid.* p. 273; S. Kaufman, *Adv. Chem. Series* **77**, 172 (1968).
58a. For sulfoxidation, see S. W. May and R. S. Phillips, *J. Am. Chem. Soc.* **102**, 5981 (1980).
59. S. Kaufman and D. B. Fisher, *in* "Molecular Mechanisms of Oxygen Activation" (O. Hayaishi, ed.), p. 285. Academic Press, New York, 1974.
60. S. Orrenius and L. Ernster, *in* "Molecular Mechanisms of Oxygen Activation" (O. Hayaishi, ed.), p. 215. Academic Press, New York, 1974.
61. R. W. Estabrook, J. Baron, J. Peterson, and Y. Ishimura, *in* "Biological Hydroxylation Mechanisms" (G. S. Boyd and R. M. S. Smellie, eds.), pp. 159–185. Academic Press, New York, 1972.
62. V. Ullrich and H. Staudinger, *in* "Biochemie des Sauerstoffs" (H. Staudinger and B. Hess, eds.), pp. 229–248. Springer-Verlag, Berlin and New York, 1968.
63. I. C. Gunsalus, J. D. Lipscomb, V. Marshall, H. Frauenfelder, E. Greenbaum, and E. Münck, *in* "Biological Hydroxylation Mechanisms" (G. S. Boyd and R. M. S. Smellie, eds.), Academic Press, New York, 1972.
64. C. A. Yu and I. C. Gunsalus, *Biochem. Biophys. Res. Commun.* **40**, 1431 (1970).
65. I. C. Gunsalus, C. A. Tyson, and J. D. Lipscomb, *in* "Oxidases and Related Redox Systems" (T. E. King, H. S. Mason, and M. Morrison, eds.), Vol. 2, p. 583. Wiley, New York, (1973).
66. I. C. Gunsalus, P. W. Trudgill, and D. W. Cushman, *in* "Biological and Chemical Aspects of Oxygenases" (K. Bloch and O. Hayaishi, eds.), p. 339. Academic Press, New York, 1974; D. W. Cushman, R. L. Tsai, and I. C. Gunsalus, *Biochem. Biophys. Res. Commun.* **26**, 577 (1972).
67. V. Ullrich, *Top. Curr. Chem.* **83**, 68 (1979); see also *J. Mol. Catal.* **7**, 159 (1980); J. A. Peterson, *in* "Fundamental Research in Homogeneous Catalysis" (M. Tsutsui, ed.), Vol. 3, Plenum, New York, 1979; C. Caron, A. Mitschler, G. Riviere, L. Ricard, M. Schappacher, and R. Weiss, *J. Am. Chem. Soc.* **101**, 7401 (1979), and references therein.
68. K. Suzuki and T. Kimura, *Biochem. Biophys. Res. Commun.* **19**, 340 (1965); T. Kimura and K. Suzuki, *J. Biol. Chem.* **242**, 485 (1967).
69. D. Y. Cooper, S. Narasimhulu, and O. Rosenthal, *Adv. Chem. Series* **77**, 220 (1968).
70. P. F. Knowles, J. F. Gibson, F. M. Pick, and R. C. Bray, *Biochem. J.* **111**, 53 (1969).
71. J. M. McCord and I. Fridovich, *J. Biol. Chem.* **244**, 6049 (1969).
71a. See also D. R. Graham, L. E. Marshall, K. A. Reich, and D. S. Sigman, *J. Am. Chem. Soc.* **102**, 5419 (1980).

72. (a) I. Fridovich, *Accts. Chem. Rsch.* **5**, 321 (1972).
(b) I. Fridovich, *in* "Molecular Mechanisms of Oxygen Activation" (O. Hayaishi, ed.), p. 453. Academic Press, New York, 1974.
(c) I. Fridovich, *Am. Sci.* **63**, 54 (1975).
(d) I. Fridovich, *Adv. Enzymol.* **41**, 35 (1974).
73. J. M. McCord, B. B. Keele, and I. Fridovich, *Proc. Natl. Acad. Sci. U.S.A.* **68**, 1024 (1971).
74. S. J. Lippard, A. R. Burger, K. Ugurbil, J. S. Valentine, and M. W. Pantoliano, *Adv. Chem. Series*, **162**, 251 (1977).
75. J. A. Fee and G. J. McClune, *in* "Mechanisms of Oxidizing Enzymes" (T. P. Singer and R. N. Ondarza, eds.), p. 273. Elsevier, Amsterdam, 1978.
76. H. Wieland, "On the Mechanism of Oxidation." Yale Univ. Press, New Haven, Connecticut, 1932.
77. D. F. Jones and R. Howe, *J. Chem. Soc. C* pp. 2801, 2809, 2816, 2821, 2827 (1968).
78. T. Koenig and J. M. Owens, *J. Am. Chem. Soc.* **96**, 4054 (1974); **95**, 8485 (1973); J. P. Engström and F. D. Greene, *J. Org. Chem.* **37**, 968 (1972).
79. G. A. Hamilton, P. K. Adolf, J. deJersey, G. C. DuBois, C. R. Dyrkacz, and R. D. Libby, *J. Am. Chem. Soc.* **100**, 1899 (1978); see also G. A. Hamilton, R. D. Libby, and R. C. Hartzell, *Biochem. Biophys. Res. Commun.* **55**, 333 (1973).
80. R. F. Jameson and N. J. Blackburn, *J. Chem. Soc., Dalton* p. 1596 (1976).
81. C. K. Chang and D. Dolphin, *Bioorg. Chem.* **4**, 37 (1978).
82. J. H. Wang, *in* "Oxygenases" (O. Hayaishi, ed.), p. 470. Academic Press, New York, 1962.
83. K. Okunuki, *in* "Oxygenases" (O. Hayaishi, ed.), p. 409. Academic Press, New York, 1962; B. G. Malmström, *Adv. Chem. Series* **162**, 173 (1977).
84. R. J. P. Williams, *Struc. Bonding (Berlin)* **8**, 123 (1970); *Chimia* **24**, 155 (1970). R. J. P. Williams, *Biochem. J.* **117**, 14P (1970).
85. L. P. Hager, D. L. Doubek, R. M. Silverstein, J. H. Hargis, and J. C. Martin, *J. Am. Chem. Soc.* **94**, 4364 (1972).
86. E. G. Hrycay and P. J. O'Brien, *Arch. Biochem. Biophys.* **153**, 480 (1972).
87. E. G. Hrycay and R. A. Prough, *Arch. Biochem. Biophys.* **165**, 331 (1974).
88. E. G. Hrycay, H. G. Jonen, A. Y. H. Lu, and W. Levin, *Arch. Biochem. Biophys.* **166**, 145 (1975).
89. G. D. Nordblum, R. E. White, and M. J. Coon, *Arch. Biochem. Biophys.* **175**, 524 (1976).
90. E. Hrycay, J. A. Gustafsson, M. Ingelman-Sundberg, and L. Ernster, *Biochem. Biophys. Res. Commun.* **66**, 209 (1975).
91. A. D. Rahimtula and P. J. O'Brien, *Biochem. Biophys. Res. Commun.* **60**, 440 (1974); **62**, 268 (1975).
92. F. F. Kadlubar, K. C. Morton, and D. M. Ziegler, *Biochem. Biophys. Res. Commun.* **54**, 1255 (1973).
93. F. Lichtenberger, W. Nastainczyk, and V. Ullrich, *Biochem. Biophys. Res. Commun.* **70**, 939 (1976). See also L. T. Burka, A. Thorsen, and F. P. Goengerich, *J. Am. Chem. Soc.* **102**, 7615 (1980).
94. J. T. Groves, G. A. McClusky, R. E. White, and M. J. Coon, *Biochem. Biophys. Res. Commun.* **81**, 154 (1978).
95. K. B. Wiberg, *in* "Oxidation in Organic Chemistry" (K. B. Wiberg, ed.), pp. 36, 109. Academic Press, New York, 1965.
96. R. Stewart, *in* "Oxidation Mechanisms," pp. 50–53. Benjamin, New York, 1964.
97. G. A. Hamilton, *in* "Molecular Mechanisms of Oxygen Activation" (O. Hayaishi, ed.), p. 405. Academic Press, New York, 1974.

98. G. A. Hamilton, *Prog. Bioorg. Chem.* **1**, 83 (1971).
99. G. A. Hamilton, J. R. Giacin, T. M. Hellman, M. E. Snook, and J. W. Weller, *Ann. N.Y. Acad. Sci.* **212**, 4 (1973).
100. K. B. Sharpless, A. Y. Teranishi, and J. E. Bäckvall, *J. Am. Chem. Soc.* **99**, 3120 (1977).
100a. K. B. Sharpless and T. C. Flood, *J. Am. Chem. Soc.* **93**, 2316 (1971).
101. J. T. Groves, O. F. Akinbote, and G. E. Avaria, *in* "Microsomes and Drug Oxidations" (M. J. Coon, ed.), Academic Press, New York, 1980.
102. G. Guroff, J. W. Daly, D. Jerina, J. Renson, B. Witkop, and S. Udenfriend, *Science* **157**, 1524 (1967).
103. J. Daly, G. Guroff, D. Jerina, S. Udenfriend, and B. Witkop, *Adv. Chem. Series* **77**, 270 (1968).
104. D. M. Jerina, *Chemtech* **3**, 120 (1973).
105. D. M. Jerina, J. W. Daly, and B. Witkop, *J. Am. Chem. Soc.* **90**, 6523, 6525 (1968).
106. D. M. Jerina, N. Kaubisch, and J. W. Daly, *Proc. Natl. Acad. Sci. U.S.A.* **68**, 2545 (1971); *Biochemistry* **11**, 3080 (1972).
107. D. M. Jerina, J. W. Daly, B. Witkop, P. Zaltzman-Nirenberg, and S. Udenfriend, *Biochemistry* **9**, 147 (1970).
108. D. R. Boyd, J. W. Daly, and D. M. Jerina, *Biochemistry* **11**, 1961 (1972).
109. E. Boyland, *Spec. Publ.—Chem. Soc.* **5**, 40 (1950).
110. (a) G. J. Kasparek and T. C. Bruice, *J. Am. Chem. Soc.* **94**, 198 (1972).
(b) G. J. Kasparek, T. C. Bruice, H. Yagi, N. Kaubisch, and D. M. Jerina, *J. Am. Chem. Soc.* **94**, 7876 (1972).
(c) G. J. Kasparek, P. Y. Bruice, T. C. Bruice, H. Yagi, and D. M. Jerina, *J. Am. Chem. Soc.* **95**, 6041 (1973).
111. P. Y. Bruice, T. C. Bruice, H. G. Selander, H. Yagi, and D. M. Jerina, *J. Am. Chem. Soc.* **96**, 6814 (1974).
112. E. A. Fehnel, *J. Am. Chem. Soc.* **94**, 3961 (1972).
113. D. L. Whelan and A. M. Ross, *J. Am. Chem. Soc.* **96**, 3678 (1974).
114. L. Castle, J. R. Lindsay-Smith, and G. V. Buxton, *J. Mol. Catal.* **7**, 235 (1980).
115. D. Mansuy, J. P. Battioni, J. C. Chottard, and V. Ullrich, *J. Am. Chem. Soc.* **101**, 3971 (1979).
116. D. Mansuy, P. Guérin, and J. C. Chottard, *J. Organometal. Chem.* **171**, 195 (1979); D. Mansuy, J. C. Chottard, M. Lange, and J. P. Battioni, *J. Mol. Catal.* **7**, 215 (1980); D. Mansuy, *Pure Appl. Chem.* **52**, 681 (1980).
117. M. Kakagawa, S. Kato, S. Kataoka, and T. Hino, *J. Am. Chem. Soc.* **101**, 3136 (1979).
118. T. Matsuura, *Tetrahedron* **33**, 2869 (1977), and references cited therein.
118a. E.g. see R. S. Himmelwright, N. C. Eickman, C. D. Lubien, K. Lerch and E. I. Solomon, *J. Am. Chem. Soc.* **102**, 7339 (1980).
119. C. Kemal and T. C. Bruice, *J. Am. Chem. Soc.* **101**, 1635 (1979); *ibid.* **102**, 498 (1980); see also C. Walsh, *Accts. Chem. Rsch.* **13**, 148 (1980).
120. W. H. Rasteller, T. R. Gadek, J. P. Tane, and J. W. Frost, *J. Am. Chem. Soc.* **101**, 2228 (1979). S. Muto and T. C. Bruice, *Ibid.*, **102**, 4472, 7559 (1980).
121. S. Udenfriend, C. T. Clark, J. Axelrod, and B. B. Brodie, *J. Biol. Chem.* **208**, 731 (1954).
122. M. B. Dearden, C. R. Jefcoate, and J. R. Lindsay-Smith, *Adv. Chem. Series* **77**, 260 (1968).
123. G. A. Hamilton, *J. Am. Chem. Soc.* **86**, 3391 (1964).
124. G. A. Hamilton, R. J. Workman, and L. Woo, *J. Am. Chem. Soc.* **86**, 3390 (1964).
125. H. Staudinger, B. Kerekjarto, V. Ullrich, and Z. Zubrzycki, *in* "Oxidases and Related Redox Systems" (T. King, H. S. Mason, and M. Morrison, eds.), p. 815. Wiley, New York, 1965.

126. T. Gertsen, *Biochem. Z.* **336**, 251 (1962).
127. V. Ullrich, *Z. Naturforsch., B: Anorg. Chem., Org. Chem., Biochem., Biophys., Biol.* **24B**, 699 (1969).
128. A. Bobst and M. Viscontini, *Helv. Chim. Acta* **49**, 884 (1966); M. Viscontini, *Angew. Chem.* **80**, 492 (1968); *Fortschr. Chem. Forsch.* **9**, 605 (1968); M. Viscontini, H. Leidner, G. Mattern, and T. Okada, Helv. Chim. Acta **49**, 1911 (1966).
129. H. Mimoun and I. Seree de Roch, *Tetrahedron* **31**, 777 (1975).
130. V. Ullrich and H. Staudinger, *in* "Biological and Chemical Aspects of Oxygenases" (K. Bioch and O. Hayaishi, eds.), p. 235. Maruzen, Tokyo, 1966.
131. R. O. C. Norman and G. K. Radda, *Proc. Chem. Soc., London* p. 138 (1962).
132. D. M. Jerina, J. W. Daly, W. Landis, B. Witkop, and S. Udenfriend, *J. Am. Chem. Soc.* **89**, 3347 (1967).
133. J. T. Groves and G. A. McClusky, *J. Am. Chem. Soc.* **98**, 859 (1976); J. T. Groves and M. van der Puy, *ibid.* p. 5274; **96**, 5290 (1974).
134. P. A. Adams, M. C. Berman, and D. A. Baldwin, *J. Chem. Soc., Chem. Commun.* p. 856 (1979).
135. H. Patin and C. Mignani, *J. Chem. Soc., Chem. Commun.* p. 685 (1979).
136. I. Tabushi, T. Nakajima, and K. Seto, *Tetrahedron Lett.* **21**, 2565 (1980).
137. G. A. Hamilton and J. P. Friedman, *J. Am. Chem. Soc.* **85**, 1008 (1963).
138. J. T. Groves, T. E. Nemo, and R. S. Myers, *J. Am. Chem. Soc.* **101**, 1032 (1979).
139. C. K. Chang and M. S. Kuo, *J. Am. Chem. Soc.* **101**, 3413 (1979).
140. J. W. Weller and G. A. Hamilton, *J. Chem. Soc., Chem. Commun.* p. 1390 (1970).
141. D. M. Jerina, D. R. Boyd, and J. W. Daly, *Tetrahedron Lett.* p. 457 (1970).
142. U. Frommer and V. Ullrich, *Z. Naturforsch., B: Anorg. Chem., Org. Chem., Biochem., Biophys., Biol.* **24B**, 583 (1969).
143. N. Z. Muradov, A. E. Shilov, and A. A. Shteinman, *Kinet. Katal.* **13**, 1219 (1972); A. E. Shilov and A. A. Shteinman, *ibid.* **14**, 117 (1973).
144. A. Nishinaga, H. Tomita, T. Shimizu, and T. Matsuura, *in* "Fundamental Research in Homogeneous Catalysis" (Y. Ishii and M. Tsutsui, eds.), p. 241. Plenum, New York, 1978.
145. A. Nishinaga and H. Tomita, *J. Mol. Catal.* **7**, 179 (1980).
146. A. Nishinaga, K. Watanabe, and T. Matsuura, *Tetrahedron Lett.* p. 1291 (1974).
147. A. Nishinaga, K. Nishizawa, H. Tomita, and T. Matsuura, *J. Am. Chem. Soc.* **99**, 1287 (1977).
148. A. Nishinaga, *Chem. Lett.* p. 273 (1975).
149. A. Nishinaga, T. Tojo, and T. Matsuura, *J. Chem. Soc., Chem. Commun.* p. 896 (1974).
150. J. Tsuji and H. Takayanagi, *Chem. Lett.* p. 65 (1980).
151. T. Funabiki, H. Sakamoto, S. Yoshida, and K. Tarama, *J. Chem. Soc., Chem. Commun.* p. 754 (1979).
152. R. H. Petty, B. R. Welch, L. J. Wilson, L. A. Bottomley, and K. M. Kadish, *J. Am. Chem. Soc.* **102**, 611 (1980).
153. T. Mashiko, J. C. Marchon, D. T. Musser, C. A. Reed, M. E. Kastner, and W. R. Scheidt, *J. Am. Chem. Soc.* **101**, 365 (1979).
154. M. J. Gunter, L. N. Mander, G. M. McLaughlin, K. S. Murray, K. J. Berry, P. E. Clark, and D. A. Buckingham, *J. Am. Chem. Soc.* **102**, 1470 (1980).
155. M. Munakata, S. Nishibayashi, and H. Sakamoto, *J. Chem. Soc., Chem. Commun.* p. 219 (1980).
156. M. L. Sinnott, *Chem. Br.* **15**, 293 (1979).
157. P. Brodelius, *Adv. Biochem. Eng.* **10**, 99 (1978).
158. W. R. Vieth and K. Venkatasubramanian, *Chemtech* **3**, 667 (1973).
159. S. S. Wang and C. K. King, *Adv. Biochem. Eng.* **12**, 119 (1979).

160. H. Ohta and H. Tetsukawa, *J. Chem. Soc., Chem. Commun.* p. 849 (1978); *Agric. Biol. Chem.* **43**, 2099 (1979). S. W. May, *Enzyme Microb. Technol.* **1**, 15 (1979).
161. Cited in *Chemical Week* (July 30, 1980, p. 15) and *European Chemical News* (August 4, 1980, p. 22).
162. K. Ogura, M. Fujita, and H. Iida, *Tetrahedron Lett.* **21**, 2233 (1980).
163. See Y. Watanabe, T. Iyanagi and S. Oae, *Tetrahedron Lett.* **21**, 3685 (1980).
164. S. Ball and T. C. Bruice, *J. Am. Chem. Soc.* **102**, 6498 (1980).

ADDITIONAL READING

H. R. Mahler and E. H. Cordes, "Biological Chemistry," 2nd ed., Harper Row, New York, 1971.

K. G. Scrimgeour, "Chemistry and Control of Enzyme Reactions." Academic Press, New York, 1977.

C. J. Gray, "Enzyme-Catalyzed Reactions. "Van Nostrand-Reinhold, Princeton, New Jersey, 1971.

L. E. Bennett, Metalloprotein redox reactions. *Prog. Inorg. Chem.* **18**, 1 (1973).

W. A. Pryor, ed., "Free Radicals in Biology," Vols. 1 and 2. Academic Press, New York, 1976; Vol. 3, 1977; Vol. 4, 1980.

A. M. Michelson, J. M. McCord, and I. Fridovich, "Superoxide and Superoxide Dismutases." Academic Press, New York, 1977.

R. P. Hanzlik, "Inorganic Aspects of Biological and Organic Chemistry." Academic Press, New York, 1976.

G. L. Eichhorn and L. G. Marzilli, eds., "Advances in Inorganic Biochemistry," Vol. 1. Am. Elsevier, New York, Vol. I 1979. Includes superoxide dismutase, copper-containing oxidases, cytochrome *P*-450, hemerythrin, etc.

E. E. van Tamelen, ed., "Biorganic Chemistry," Vol. 4. Academic Press, New York, 1978. Includes electron transfer and energy conversion, cofactors, oxygen activation by heme proteins, redox chemistry of NAD(P).

J. B. Jones, C. J. Sih, and D. Perlman, eds., "Applications of Biochemical Systems in Organic Chemistry," Parts 1 and 2. Wiley, New York, 1976.

I. Chibata, Ed., "Immobilized Enzymes", Halstead Press, New York, 1978.

Part Two

Synthetic Methodology for Metal-Catalyzed Oxidations

Chapter 9

Olefins

Olefins are particularly important building blocks in organic synthesis, and their conversions to oxygen-containing derivatives via a wide variety of oxidative transformations constitute prime examples of the broad synthetic utility of metal-catalyzed oxidations.

I. AUTOXIDATION

Before discussing the various catalytic oxidative transformations of olefins, it is appropriate to consider their reactions under classical autoxidation conditions. Olefins are generally labile substrates that can readily undergo autoxidation under relatively mild conditions in the absence of catalysts.[1–6] Indeed, the more reactive olefins rapidly deteriorate in air at ambient temperatures, unless an antioxidant is added. Individual reactivities of olefins are determined by the ratio of the rate constants for chain propagation and chain termination and are defined by the quantity $k_p/(2k_t)^{1/2}$, which is referred to as the oxidizability of the substrate, as discussed in Chapter 2. The propagation rate constant k_p is roughly related to the C—H bond strength, and the termination rate constant k_t increases in the order tertiary < secondary < primary alkylperoxy radicals (see Chapter 2). The oxidizabilities of several olefins are listed in Table I.

The most reactive olefins are generally those containing conjugated unsaturation (e.g., styrene and cyclohexa-1,3-diene) as well as cyclohexene and 2,3-dimethyl-2-butene. Terminal olefins (e.g., propylene and 1-octene) are generally the least reactive; an internal olefin is about ten times as reactive as its terminal isomer (e.g., 1-octene compared to 2-octene). The low rate of oxidation of 1,4-pentadiene and 1,3-butadiene compared to that of other conjugated olefins is attributable to the high termination rate constants for the unhindered, largely primary, allylic peroxy radical.

A typical reaction scheme for the early stages of an olefin autoxidation is depicted in Scheme I for propylene (primary products in boxes).

Scheme I:

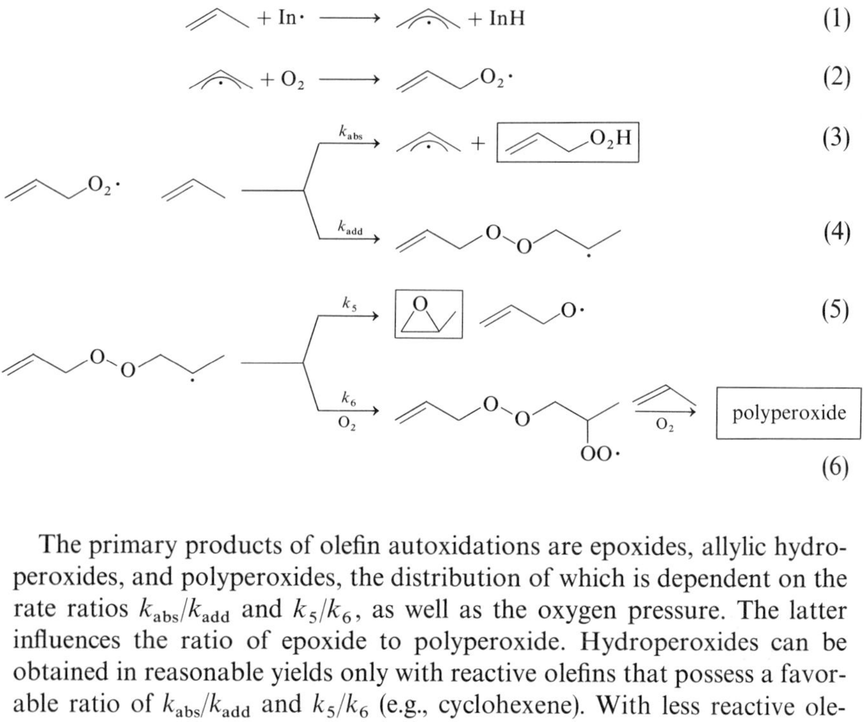

+ In· ⟶ + InH (1)

$+ O_2$ ⟶ O_2· (2)

k_{abs} ⟶ + O_2H (3)

O_2·

k_{add} ⟶ O O · (4)

k_5 ⟶ O O· (5)

O O ·

k_6 O_2 ⟶ O O OO· O_2 ⟶ polyperoxide (6)

The primary products of olefin autoxidations are epoxides, allylic hydroperoxides, and polyperoxides, the distribution of which is dependent on the rate ratios k_{abs}/k_{add} and k_5/k_6, as well as the oxygen pressure. The latter influences the ratio of epoxide to polyperoxide. Hydroperoxides can be obtained in reasonable yields only with reactive olefins that possess a favorable ratio of k_{abs}/k_{add} and k_5/k_6 (e.g., cyclohexene). With less reactive olefins such as propylene and 1-octene, the intermediate hydroperoxides are not stable at the temperatures required for a reasonable rate of oxidation. They undergo homolytic decomposition to afford alcohols and carbonyl

TABLE I Absolute Rate Constants for Olefin Autoxidations at 30°C[a]

Olefin	$k_p/(2k_t)^{1/2} \times 10^3$ ($M^{-1/2}$ sec$^{-1/2}$)	k_p (M^{-1} sec^{-1})	$2k_t \times 10^{-6}$ (M^{-1} sec^{-1})
Cyclohexa-1,3-diene	100	810	66
Methyl linoleate	21	62	8.8
α-Methylstyrene	13	10	0.6
Styrene	6.3	41	42
2,3-Dimethyl-2-butene	3.2	2.6	0.64
Cyclohexene	2.3	5.4	5.6
Cyclopentene	2.8	7.0	6.2
3-Heptene	0.54	1.4	6.4
1,4-Pentadiene	0.42	14	1080
1-Octene	0.062	1.0	260

[a] From Howard.[2]

compounds, e.g.,

$$\text{CH}_2{=}\text{CHCH}_2\text{O}_2\text{H} \xrightarrow{\Delta} \text{CH}_2{=}\text{CHCH}_2\text{O}\cdot + \text{HO}\cdot \longrightarrow \text{CH}_2{=}\text{CHCH}_2\text{OH}, \quad \text{CH}_2{=}\text{CHCHO}, \quad \text{etc.} \tag{7}$$

Polyperoxides also undergo thermal cleavagc to carbonyl compounds, e.g.,

$$[\text{polyperoxide}]_n \xrightarrow{\Delta} n\,\text{CH}_3\text{CHO} + n\,\text{HCHO} \tag{8}$$

The aldehydes undergo rapid, further autoxidation, affording acylperoxy radicals and peracids, which can react with the olefin to produce more epoxide.

$$\text{R}\dot{\text{C}}\text{O} + \text{O}_2 \longrightarrow \text{RCO}_3\cdot \tag{9}$$

$$\text{RCO}_3\cdot + \text{CH}_2{=}\text{CHCH}_3 \longrightarrow \text{RCO}_3\text{CH}_2\dot{\text{C}}\text{HCH}_3 \longrightarrow \text{CH}_3\text{-epoxide} + \text{RCO}_2\cdot \tag{10}$$

$$\text{RCO}_2\cdot \longrightarrow \text{R}\cdot + \text{CO}_2, \quad \text{etc.} \tag{11}$$

It is readily apparent from Scheme I that olefin autoxidations generally lead to highly complex reaction mixtures, especially at conversions higher than a few percent. The high selectivities to hydroperoxides claimed for many olefins in the early literature[6] can be attributed largely to the lack of reliable analytical techniques for the analysis of such complex mixtures. Even at very low olefin conversions, autoxidations are generally quite unselective and are thus of limited synthetic value. A few examples are given in Table II. It is noteworthy that large amounts of polymeric material are formed in all the examples.

TABLE II AIBN-Initiated Autoxidation of Olefins[a]

Olefin	Temp (°C)	Conversion (%)	Selectivity (%) Epoxide[b]	RO_2H	Carbonyls[c]	Residue[d]	Total
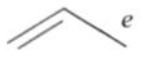 [e]	110	3.0	32 (48)	9	20[f]	14	91
	70	3.5	5 (9)	56	6	24	95
	70	4.7	15 (23)	23	11	33	90
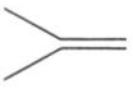	80	3.9	22 (24)	3	29	41	97
	50	4.8	15 (16)	44	11	22	93
	90	3.6	6 (6)	34	13	42	95

[a] Data from van Sickle *et al.*[5]
[b] Figures in parentheses include glycol derivatives.
[c] Aldehydes and ketones.
[d] Mostly polymeric peroxides and hydroperoxides.
[e] In benzene solution using 1,1-azobis(cyclohexane-1-carbonitrile) as the initiator.
[f] Includes 10% CO.

Early studies of olefin autoxidations were carried out in connection with mechanistic interest in the aerial deterioration of rubber and unsaturated fats. Further interest was stimulated by the possibility of employing the autoxidation of propylene for the production of propylene oxide. Although selectivities to propylene oxide in the region 40–50% are attainable at low conversions, its separation from the plethora of by-products represents a formidable obstacle. Furthermore, little advantage accrues from the use of transition metals (e.g., Co, Mn) as autoxidation catalysts in these reactions.[7] Although they increase the rate and eliminate induction periods, they have little effect on the selectivity. Indeed, epoxide yields are often lower in the presence of metal catalysts. Autoxidation of olefins can be productive in a limited context. For example, the autoxidation of methylpentene at temperatures in the range 45°–75°C and 10–11 bar oxygen pressure in the presence of a cobaltic naphthenate catalyst (0.2 g liter^{-1}) affords the α,β-unsaturated ketone.[7a] Subsequent hydrogenation of mesityl oxide affords methyl isobutyl ketone in an overall yield of approximately 90%.

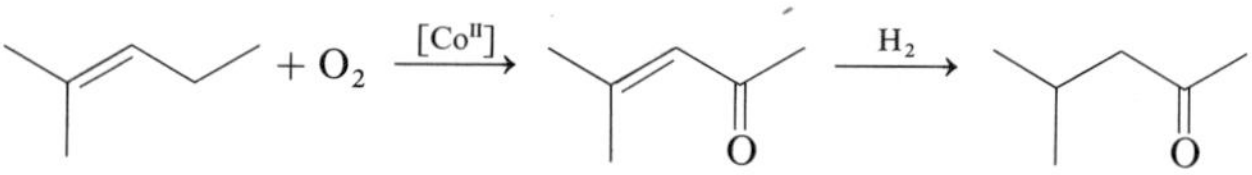

$$+ O_2 \xrightarrow{[Co^{II}]} \quad \xrightarrow{H_2}$$

The initial step represents a selective allylic oxidation under autoxidation conditions (see Section III).

II. EPOXIDATION

$$\rangle C{=}C\langle \Longrightarrow \rangle C\overset{O}{\triangle}C\langle$$

Ethylene oxide is prepared commercially by the gas-phase oxidation of ethylene with air or oxygen over a supported silver catalyst at elevated temperatures (ca. 250°C).[8] Unfortunately, the oxidation of propylene and higher olefins under these conditions leads to very low yields of epoxides. The silver-catalyzed autoxidation is successful only with ethylene for the following reasons. (1) The absence of labile C—H bonds renders the radical chain propagation steps via hydrogen abstraction extremely unfavorable. (2) Polyperoxide formation by homolytic addition is slow compared to that of styrene, for example. Moreover, the low stability of the intermediate adduct (i.e., $ROOCH_2CH_2\cdot$) makes unimolecular cleavage to epoxide more favorable than polyperoxide formation via the bimolecular reaction with dioxygen (especially at higher temperatures).

Propylene oxide is now manufactured on a large scale by the metal-catalyzed epoxidation of propylene with an alkyl hydroperoxide produced by hydrocarbon autoxidation. Processes employing molybdenum catalysts with both *tert*-butyl hydroperoxide (TBHP),

$$CH_3CH{=}CH_2 + (CH_3)_3COOH \xrightarrow{[Mo]} CH_3\overset{O}{CH{-}CH_2} + (CH_3)_3COH \qquad (12)$$

and ethylbenzene hydroperoxide (EBHP) are used.[9]

$$CH_3CH{=}CH_2 + PhCH(CH_3)OOH \xrightarrow{[Mo]} CH_3\overset{O}{CH{-}CH_2} + PhCH(CH_3)OH \qquad (13)$$

A heterogeneous process involving titanium(IV) on a silica support as the catalyst has also been developed for the latter.[10] The catalyst is prepared by impregnating silica with $TiCl_4$ or an organotitanium compound, followed by calcination. Particularly effective catalysts are obtained when the deposition of titanium(IV) on silica is followed by further treatment with an organic silylating agent to remove residual Brønsted acidity arising from SiOH groups. This catalyst is at least as selective as the best homogeneous molybdenum catalysts. It has the added advantage of being completely insoluble in the reaction medium, which makes it eminently suitable for a continuous, fixed-bed operation.[10] In this mode, propylene is converted to propylene oxide in 93–94% selectivity, based on EBHP at 96% conversion of the latter. By comparison, the homogeneous molybdenum catalyst affords 90% selectivity at 92% EBHP conversion under comparable conditions.

The *tert*-butyl alcohol coproduct in eq 12 can be recycled by dehydration followed by hydrogenation or converted to methyl *tert*-butyl ether, which is an excellent high-octane component for gasoline.

$$(CH_3)_3COH + CH_3OH \longrightarrow (CH_3)_3COCH_3 + H_2O \qquad (14)$$

The methylphenylcarbinol coproduct in reaction 13 is dehydrated to styrene, which is sold or recycled.

$$PhCH(CH_3)OH \longrightarrow PhCH{=}CH_2 + H_2O \qquad (15)$$

A process has been described for the coproduction of ethylene and propylene oxides, in a variable ratio, by the epoxidation of mixtures of ethylene and propylene with TBHP.[11]

Alkyl hydroperoxides in combination with homogeneous (Mo, V) and heterogeneous (TiO_2/SiO_2) catalysts form a versatile group of reagents for the epoxidation of olefins in general.[12,13] The reactions are performed in hydrocarbon solvents at moderate temperatures (generally 80°–120°C) and afford epoxides in very high selectivities. The mechanism and factors governing the catalytic activity have been discussed in Chapter 3. Many industrial applications have been envisaged[14,15] for the hydroperoxide process, such as the production of long-chain primary alcohols from α-olefins by epoxidation and subsequent hydrogenation over a modified nickel catalyst.[16] Indeed, epoxides are versatile chemical intermediates that can be converted to a variety of products, as illustrated in Scheme II.

Scheme II:

$$RCH_2\overset{O}{\overset{\|}{C}}CH_3$$

$$\uparrow [H^+]$$

$$\begin{array}{c} RCH{=}CHCH_2OH \\ \text{and/or} \\ RCH(OH)CH{=}CH_2 \end{array} \xleftarrow{[Li_2HPO_4]} RCH_2\overset{O}{CH{-}CH_2} \xrightarrow[{[Ni]}]{H_2} RCH_2CH_2CH_2OH$$

$$\downarrow [AlPO_4]$$

$$RCH_2CH_2CHO$$

Propylene oxide, for example, is rearranged to allyl alcohol over basic lithium phosphate[17,18] and to propionaldehyde over aluminum phosphate.[18] Epoxides are selectively rearranged to allylic alcohols in the presence of aluminum alkoxide catalysts.[19,20]

One industrial application envisaged by Farberov[14] is the conversion of isopentane to isoprene according to Scheme III:

Scheme III:

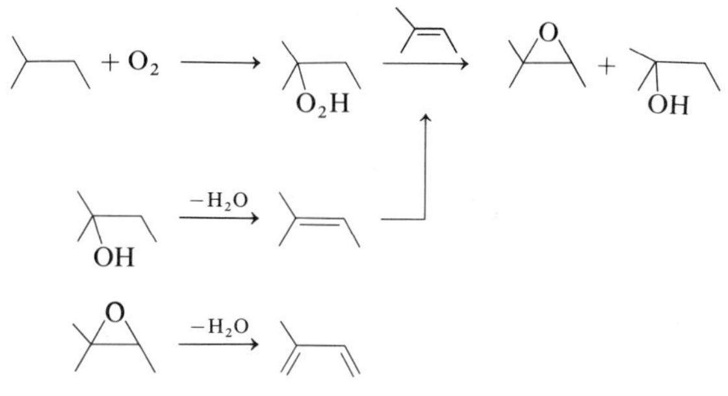

$$\text{isopentane} + O_2 \longrightarrow \text{tert-amyl hydroperoxide } (O_2H) \xrightarrow{\text{2-methyl-2-butene}} \text{2-methyl-2,3-epoxybutane} + \text{tert-amyl alcohol } (OH)$$

$$\text{tert-amyl alcohol } (OH) \xrightarrow{-H_2O} \text{2-methyl-2-butene}$$

$$\text{2-methyl-2,3-epoxybutane} \xrightarrow{-H_2O} \text{isoprene}$$

Net reaction:

$$\text{isopentane} + O_2 \longrightarrow \text{isoprene} + 2\,H_2O$$

Epoxidation of cyclohexenes can be utilized for the production of phenols.[15]

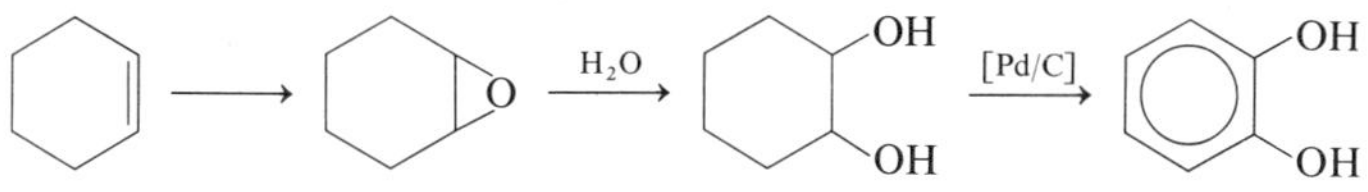

$$\text{cyclohexene} \longrightarrow \text{cyclohexene oxide} \xrightarrow{H_2O} \text{cyclohexane-1,2-diol} \xrightarrow{[Pd/C]} \text{catechol} \quad (16)$$

$$\text{4-vinylcyclohexene} \longrightarrow \text{4-vinylcyclohexene oxide} \xrightarrow{H_2O} \text{4-vinylcyclohexane-1,2-diol (HO, OH)} \xrightarrow{[Pd/C]} \text{4-ethylcatechol (HO, OH)} \quad (17)$$

$$\text{4-vinylcyclohexene oxide} \xrightarrow[H_2]{[Pd/C]} \text{4-ethylphenol (OH)} + \text{3-ethylphenol (HO)} \quad (18)$$

$$\text{cyclohexene} \longrightarrow \text{cyclohex-2-enyl hydroperoxide } (O_2H) \longrightarrow \text{2,3-epoxycyclohexanol (OH)} \xrightarrow{H_2O} \text{cyclohexane-1,2,3-triol (OH, OH, OH)} \xrightarrow{[Pd/C]} \text{pyrogallol (OH, OH, OH)} \quad (19)$$

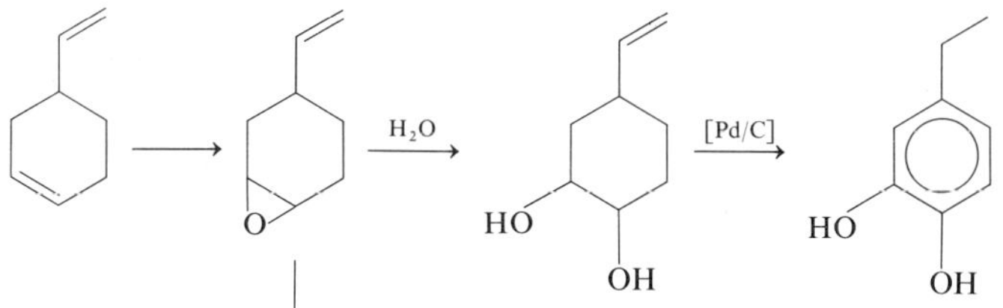

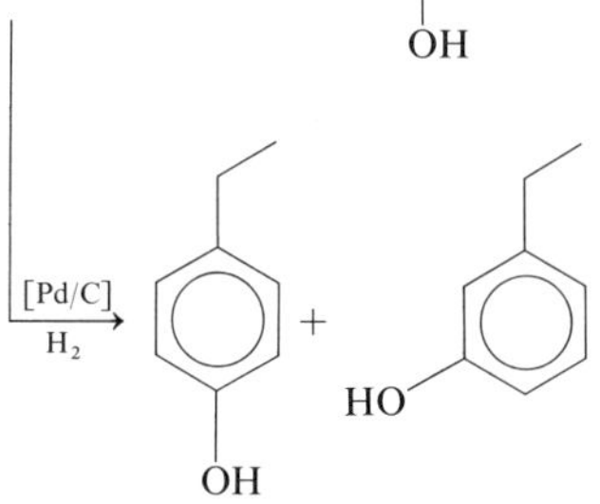

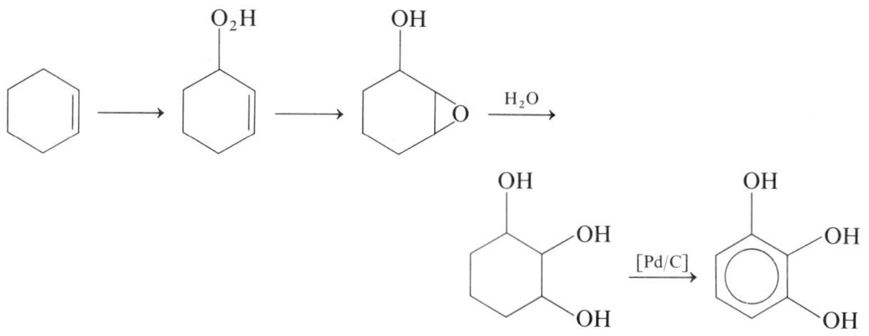

tert-Butyl hydroperoxide offers many advantages over the traditional organic peracids used in organic synthesis.[13] It is thermally more stable and is less sensitive to contamination by metals than either organic peracids or hydrogen peroxide and is therefore safer to handle. Furthermore, in contrast to peracids and H_2O_2, TBHP is unreactive toward most organic compounds *in the absence of catalysts.* Thus, TBHP/metal catalyst reagents are particularly useful for the epoxidation of acid-sensitive olefins and those containing a functional group that reacts with a peracid. An example of the latter is the selective epoxidation of citral with TBHP.[10]

CHO $\xrightarrow[\text{[Ti}^{IV}\text{/SiO}_2]]{\text{TBHP}}$ CHO (20)

Sheng and co-workers[21,22] have described the epoxidation of a wide variety of olefins with metal catalyst/alkyl hydroperoxide reagents. Sharpless and Verhoeven[13] have reported a detailed procedure for epoxidations with TBHP/$Mo(CO)_6$ which affords 85–95% isolated yields with a variety of nonfunctionalized olefins.

Relative reactivities of double bonds are enhanced by increasing alkyl substitution. This is reflected in the selective monoepoxidation of nonconjugated dienes, for example.[15,21,22] Similar regioselectivities are observed in epoxidations with organic peracids.

$\xrightarrow[\text{[Mo}^{VI}]]{\text{TBHP}}$ O (21)

$\xrightarrow[\text{[Mo}^{VI}]]{\text{TBHP}}$ O (22)

Conjugated dienes are generally less reactive than the corresponding compounds containing isolated double bonds (e.g., isoprene is less reactive than 2-methyl-2-butene).[21,22] They selectively form the monoepoxide, e.g.,[15,21,22]

$\xrightarrow[\text{[Mo}^{VI}]]{\text{TBHP}}$ O + O (23)

$$\xrightarrow[\text{[Mo}^{\text{VI}}\text{]}]{\text{TBHP}} \quad (24)$$

Electron-withdrawing groups retard the rate of epoxidation. For example, allyl chloride is about one-tenth as reactive as propylene toward TBHP/Mo(VI).[23] Acrylic esters and acrylonitrile cannot be epoxidized with these reagents.[22] However, epoxidation is not seriously impeded when the electron-withdrawing group is sufficiently removed from the double bond. 4-Cyanohexene, for example, afforded the corresponding epoxide in 88% yield.[15]

CN $\xrightarrow[\text{[Mo}^{\text{VI}}\text{]}]{\text{TBHP}}$ O CN (25)

Tolstikov and co-workers[24] used *tert*-amyl hydroperoxide (TAHP) with molybdenum catalysts for the selective epoxidation of a variety of terpenes, e.g.

$\xrightarrow[\text{[Mo}^{\text{VI}}\text{]}]{\text{TAHP}}$ O (26)

$\xrightarrow[\text{[Mo}^{\text{VI}}\text{]}]{\text{TAHP}}$ O (27)

$\xrightarrow[\text{[Mo}^{\text{VI}}\text{]}]{\text{TAHP}}$ O (28)

$\xrightarrow[\text{[Mo}^{\text{VI}}\text{]}]{\text{TAHP}}$ O (29)

The stereochemistry of these epoxidations of nonfunctionalized olefins is determined by steric constraints, preferential attack of the double bond occurring from the less shielded face of the substrate molecule. The same stereoselectivity is observed with organic peracids. Thus, in the epoxidation of nonfunctionalized olefins, only minor differences in regio- and stereoselectivity are observed, and expected, between the metal–hydroperoxide reagents and organic peracids. By contrast, olefins containing functional groups can produce completely different regio- and stereoselectivities with the metal–hydroperoxide reagents. Orientation of the olefin by coordination to the metal catalyst through a functional group can result in preferential transfer of oxygen to a particular double bond in a diene (regioselectivity) or to a particular face of the substrate (stereoselectivity). For example, quite remarkable results have been obtained in the epoxidations of unsaturated alcohols with the metal–hydroperoxide reagents.[12,13]

List and Kuhnen[25] first reported that allylic alcohols give excellent yields of epoxides with a combination of cumene hydroperoxide and V_2O_5. Subsequently, Sheng and Zajacek[22] noted, in their study of epoxidations of functionalized olefins, that allylic alcohols gave unexpected results. With simple olefins, molybdenum catalyzed epoxidations are $\sim 10^2$ times faster than vanadium-catalyzed epoxidations, whereas with allylic alcohols vanadium gives higher rates and better yields than molybdenum. The difference is reflected in the different regioselectivities observed in the epoxidation of the dienol below.[22]

OH —TBHP [catalyst]→ O OH + OH O (30)

$VO(acac)_2$	4	:	1
$Mo(CO)_6$	1	:	1

The exceptional reactivity of allylic alcohols toward the vanadium(V)–hydroperoxide reagent is due to an efficient intramolecular oxygen transfer from the coordinated alkyl hydroperoxide to the double bond of an allylic alcohol coordinated through its hydroxyl group (see Chapter 3). That vanadium catalysts in particular are able to cause such rate accelerations can be attributed to the strong coordination of alcohol ligands to vanadium(V).

The exceptionally facile epoxidation of allylic alcohols by vanadium(V)–TBHP has been exploited for the regioselective epoxidation of geraniol and linaloöl to the previously unknown monoepoxides in eqs 31 and 32, respectively.[26]

OH —TBHP $[VO(acac)_2]$→ OH O (31)

93% (98%)

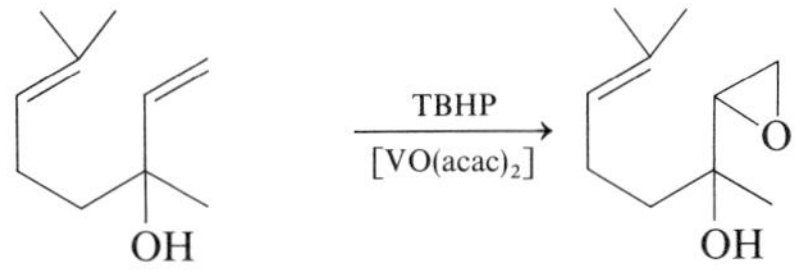

$$\xrightarrow[\text{[VO(acac)}_2\text{]}]{\text{TBHP}} \qquad (32)$$

OH OH

84% (95%)

(The numbers in parentheses refer to the isomeric purity.) Such regioselectivities are not possible with other epoxidizing agents. The reactivity of the $VO(acac)_2$–TBHP reagent is underscored by the fact that reactions 31 and 32 proceed readily at room temperature. High regioselectivities are also observed with homoallylic alcohols, as in the synthesis of the prostaglandin intermediate in eq 33.[27]

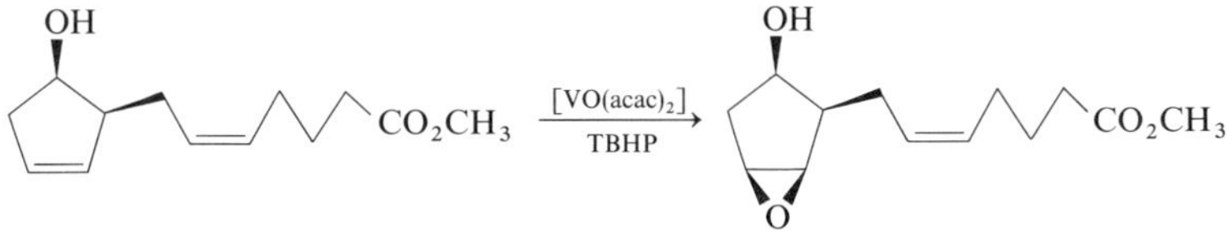

$$\text{CO}_2\text{CH}_3 \xrightarrow[\text{TBHP}]{\text{[VO(acac)}_2\text{]}} \text{CO}_2\text{CH}_3 \qquad (33)$$

91%

The transition metal–hydroperoxide reagents also exhibit remarkable stereoselectivities resulting from the preferential syn transfer of oxygen within the ternary metal–hydroperoxide–substrate complex (see the example above).[13] A comparison[26] of the stereoselectivities obtained with $Mo(CO)_6$–TBHP, $VO(acac)_2$–TBHP, and perbenzoic acid (PBA) in the epoxidation of cyclohexen-3-ol revealed that the molybdenum- and vanadium-catalyzed reactions were virtually stereospecific. They also showed rate enhancements (compared to cyclohexene as a reference) of 4.5 and 200, respectively. On the other hand, PBA showed no rate enhancement and was less stereoselective:

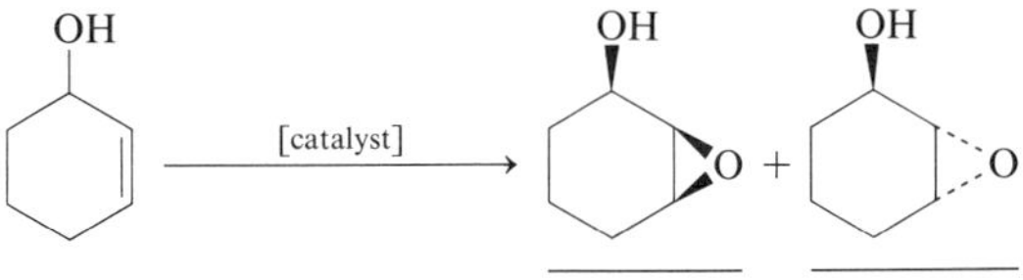

$$\xrightarrow{\text{[catalyst]}} \qquad (34)$$

	syn	anti
TBHP–$Mo(CO)_6$	98%	2%
TBHP–$VO(acac)_2$	98%	2%
PBA	92%	8%

Differences in stereoselectivity were more pronounced in the case of the homoallylic alcohol:

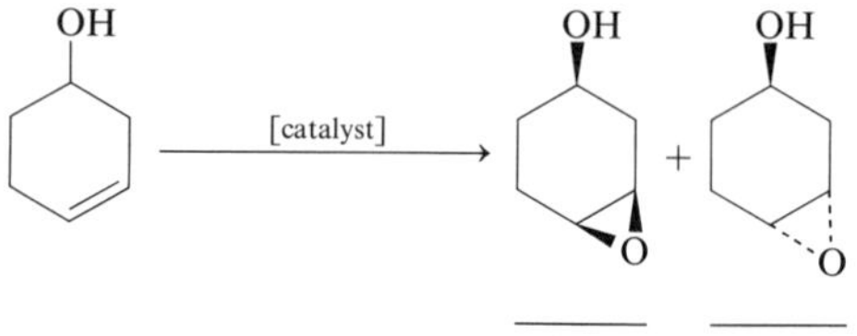

[catalyst] (35)

TBHP–Mo(CO)$_6$	98%	2%
TBHP–VO(acac)$_2$	98%	2%
PBA	60%	40%

The rate enhancements in eq 35 were ~10 for both molybdenum and vanadium. Thus, molybdenum is the catalyst of choice because of its much higher rate with the parent olefin (cyclohexene).

The selective epoxidation of allylic alcohols with the combination of TBHP and VO(acac)$_2$ has been exploited in a useful method for the stereoselective transposition of allylic alcohols.[28]

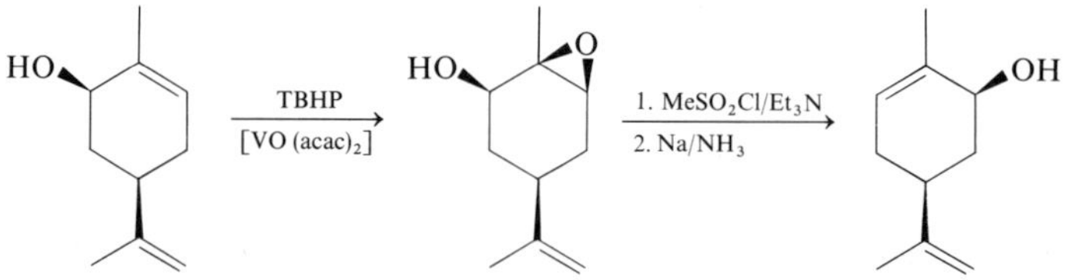

TBHP / [VO (acac)$_2$]; 1. MeSO$_2$Cl/Et$_3$N, 2. Na/NH$_3$ (36)

(−)-*cis*-Carveol (+)-*cis*-Carveol

Sharpless and co-workers[29] investigated the stereoselectivity of epoxidation of acyclic allylic alcohols. The *erythro*-epoxide was formed almost exclusively with TBHP–VO(acac)$_2$, in contrast to the corresponding reactions with *m*-chloroperbenzoic acid (MCPBA), which were virtually nonselective:

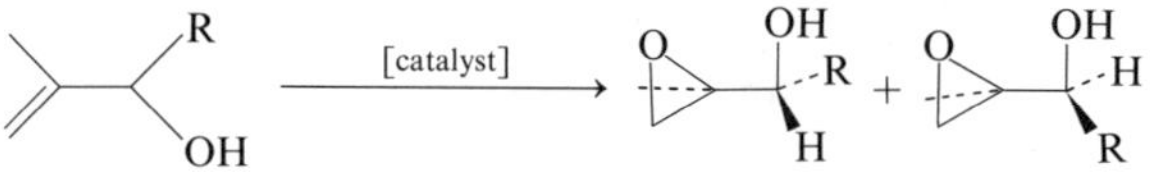

[catalyst] (37)

		threo	*erythro*
R = CH$_3$	TBHP–[V^V]	5%	95%
	TBHP–[Mo^{VI}]	16%	84%
	MCPBA	45%	55%
R = *n*-C$_4$H$_9$	TBHP–[V^V]	2%	98%
	TBHP–[Mo^{VI}]	16%	84%
	MCPBA	41%	59%

Breslow and Maresca[30] have reported examples of the template-directed remote epoxidation of double bonds which exploit the coordinating ability of the hydroxyl group to the metal catalyst in these systems, e.g.,

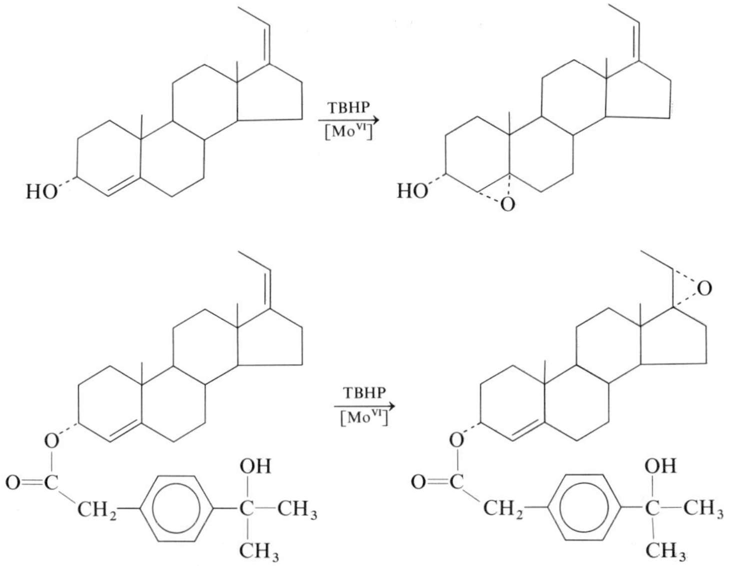

TBHP $[Mo^{VI}]$

TBHP $[Mo^{VI}]$

Sharpless and Verhoeven[13] have reviewed many more examples of the exploitation of metal-catalyzed epoxidations of unsaturated alcohols in the synthesis of complex organic molecules.

Allylic alcohols have also proved to be suitable substrates for effecting asymmetric epoxidations with alkyl hydroperoxide–metal catalyst reagents. Yamada and co-workers[31] employed a chiral molybdenum catalyst (see below) for the epoxidation of allylic alcohols with cumene hydroperoxide and obtained enantiomeric excesses of the order of 13–33%. Sharpless and co-workers[32] used TBHP in combination with a chiral complex of *N*-phenylcamphorylhydroxamic acid (see below) and achieved enantiomeric excesses in the order of 30–50%.

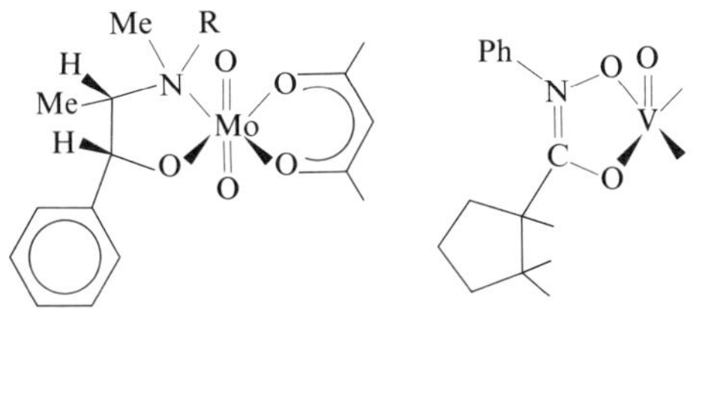

More recently, the same group[33] reported that epoxidations of allylic alcohols with TBHP–titanium isopropoxide, in the presence of (+)- or (−)-diethyl tartrate as the chiral ligand, affords enantiomeric excesses consistently greater than 90% and, in many cases (Table III), >95%. Although the method employs stoichiometric amounts of the titanium isopropoxide and the chiral ligand, it was noted that these reagents are commercially available at low to moderate cost. Another characteristic of this method is that, when a given tartrate enantiomer is employed, the epoxidic oxygen is delivered to the same enantioface of the olefin substrate regardless of the substitution pattern (see Scheme VIII, Chapter 3). The method has been applied to the synthesis of the chiral epoxy alcohols shown below,

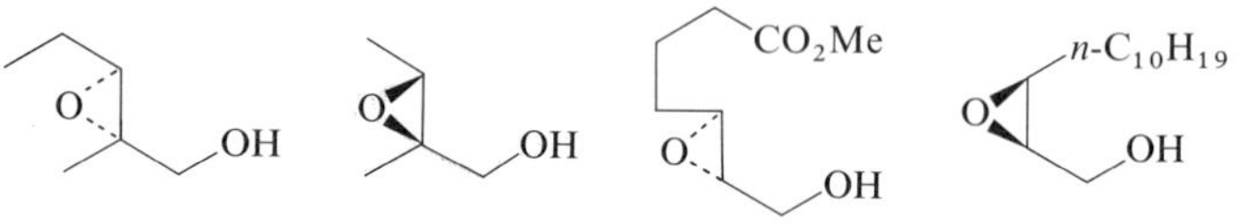

O OH   O OH CO_2Me O OH n-$C_{10}H_{19}$ O OH

which are key intermediates in the synthesis of the natural products, methymycin, erythromycin, leukotriene C-1, and (+)-dispalure (sex attractant of the gypsy moth).[33a]

In another study, Otsuka and co-workers[34] employed TBHP–$MoO_2(acac)_2$ in the presence of optically active diols, such as (+)-diethyl tartrate, for the asymmetric epoxidation of nonfunctionalized olefins. Perhaps not surprisingly, small enantiomeric excesses (<10%) were generally observed. The olefin substrate is not coordinated strongly to the metal catalyst in these systems, and asymmetric induction is expected to be more difficult.

The asymmetric epoxidation of olefins is an area that is likely to attract increasing attention in the future. Chiral epoxides constitute ideal building blocks for asymmetric syntheses, since subsequent reactions do not generally involve the chiral center.[35]

In all of the above examples of functional-group-mediated stereoselective epoxidation, the functional group involved was hydroxyl. Tolstikov and co-workers[36] have also described stereospecific epoxidations of unsaturated acetates, e.g.,

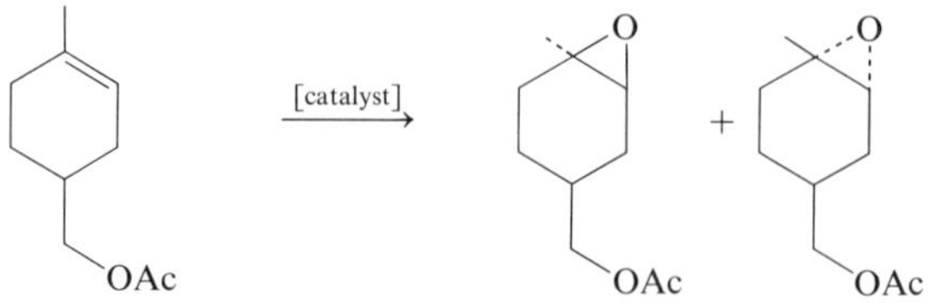

OAc $\xrightarrow{\text{[catalyst]}}$ O OAc + O OAc (38)

TAHP–$[Mo(CO)_6]$	100%	0%
PBA	50%	50%

TABLE III Asymmetric Epoxidation of Allylic Alcohols[a]

Allylic alcohol	Epoxy alcohol	% yield	% e.e.	Configuration
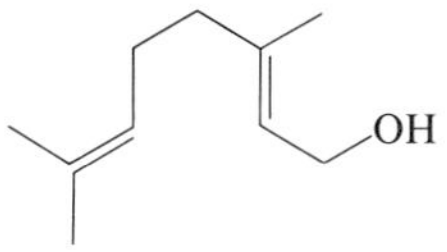 OH	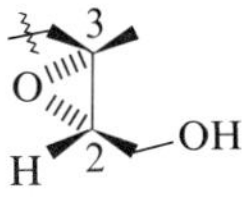 3, O, H, 2, OH	77	95	2(*S*), 3(*S*)
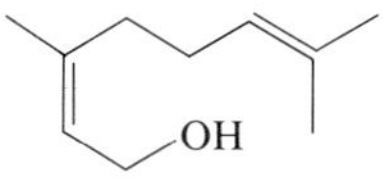 OH	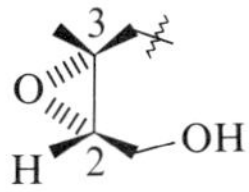 3, O, H, 2, OH	79	94	2(*S*), 3(*R*)
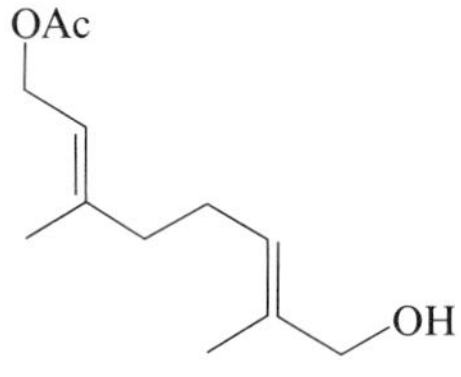 OAc, OH	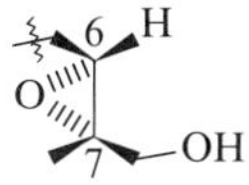 6, H, O, 7, OH	70	>95	6(*S*), 7(*S*)
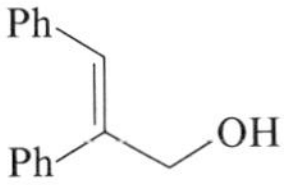 Ph, Ph, OH	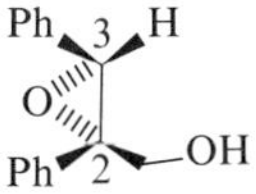 Ph, 3, H, O, Ph, 2, OH	87	>95	2(*S*), 3(*S*)
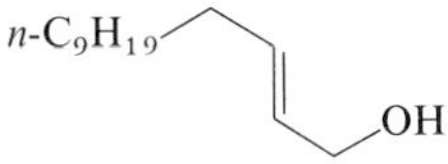 *n*-C_9H_{19}, OH	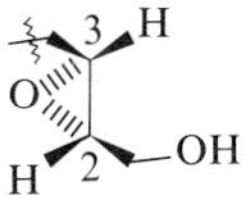 3, H, O, H, 2, OH	79	>95	2(*S*), 3(*S*)
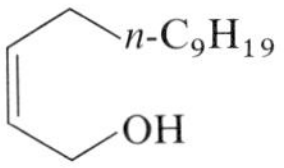 *n*-C_9H_{19}, OH	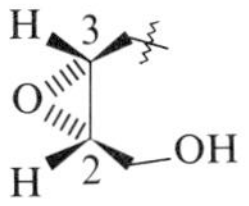 H, 3, O, H, 2, OH	82	90	2(*S*), 3(*R*)
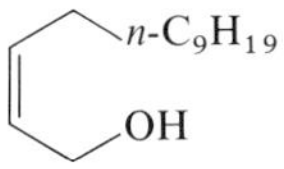 *n*-C_9H_{19}, OH	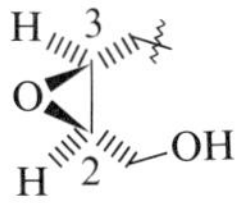 H, 3, O, H, 2, OH	80	90	2(*R*), 3(*S*)
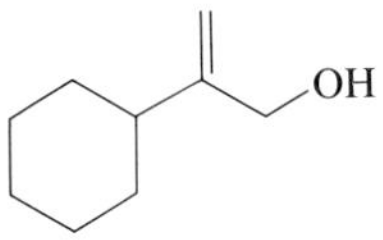 OH	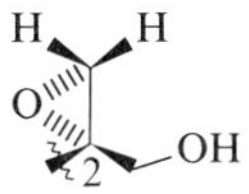 H, H, O, 2, OH	81	>95	2(*S*)

[a] From Katsuki and Sharpless.[33]

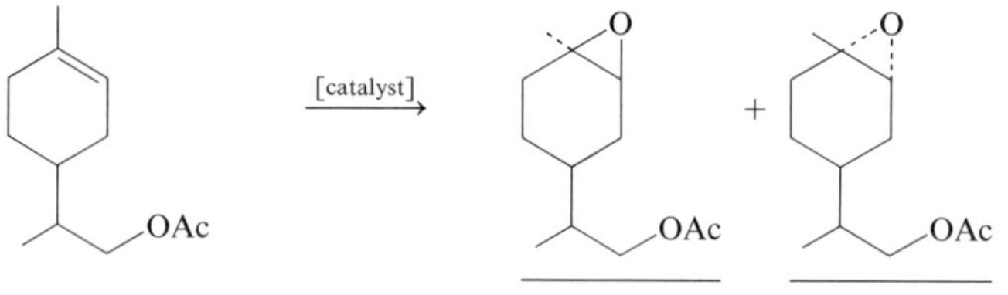

$$\xrightarrow{[\text{catalyst}]} \qquad (39)$$

	OAc (α-epoxide)	OAc (β-epoxide)
TAHP–[$Mo(CO)_6$]	100%	0%
PBA	50%	50%

Similar, surprisingly high stereoselectivities were reported for steroidal acetates.[37] These results are quite remarkable, especially when one considers that acetate groups are much less strongly coordinating than alcohol groups, and contrast with the lack of stereoselectivity observed in the molybdenum-catalyzed epoxidation of 1-cyclohexenyl acetate.[26]

Hydrogen peroxide–metal catalyst reagents (usually Mo, W, or Se) have also been used for the epoxidation of olefins, although they do not have the same broad synthetic utility as the alkyl hydroperoxide–metal reagents.[12] Reactions with H_2O_2 are generally carried out in the presence of water and/or a polar solvent, which seriously retard these reactions (see Chapter 3). Moreover, once formed, the epoxide is often readily hydrolyzed to the corresponding glycol. The H_2O_2–metal catalyst reagents are useful, however, for the epoxidation of certain water-soluble olefins such as allyl alcohol and α,β-unsaturated acids (as their salts). The epoxidation of allyl alcohol to glycidol using H_2O_2–Na_2WO_4 is applied industrially as an alternative to the traditional route via epichlorohydrin.

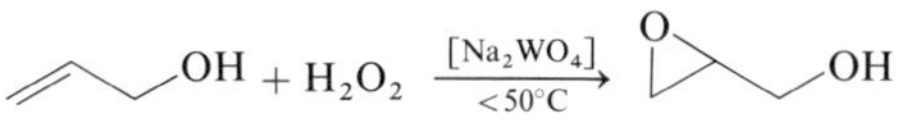

$$\text{CH}_2\text{=CHCH}_2\text{OH} + H_2O_2 \xrightarrow[<50^\circ C]{[Na_2WO_4]} \text{glycidol (OH)}$$

Hydrogen peroxide–metal combinations are the reagents of choice for the epoxidation of α,β-unsaturated acids (see Table IV for examples) which are inert to alkyl hydroperoxide–metal reagents. Certain of the more reactive nonfunctionalized olefins also give good epoxide yields with these reagents. Good yields of epoxides can be obtained with simple olefins when water is removed from the reaction medium. For example, propylene is epoxidized in 85% yield with H_2O_2 in the presence of molybdenum,[38] boron,[39] and arsenic[40] catalysts if the water is continuously removed by azeotropic distillation. Recently, a novel epoxidation of olefins with superoxide salts has been carried out in the presence of organic sulfur compounds.[40a]

TABLE IV Metal-Catalyzed Epoxidation of Olefins with Hydrogen Peroxide

Olefin	Catalyst	Solvent	Temp (°C)	Epoxide selectivity (%)[a]	Reference
 OH	$NaHWO_4$	H_2O	50	89	b
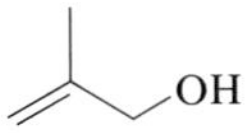 OH	$NaHWO_4$	H_2O	50	High	b
 HO—, —OH	$NaHWO_4$	H_2O	50	90	b
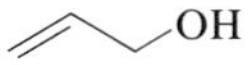 OH	H_2WO_4/Et_3N	H_2O	20	70	c
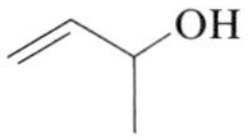 OH	H_2WO_4/Et_3N	H_2O	20	90	c
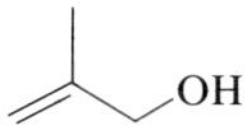 OH	H_2WO_4/Et_3N	H_2O	20	96	c
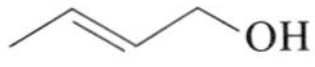 OH	H_2WO_4/Et_3N	H_2O	20	97	c
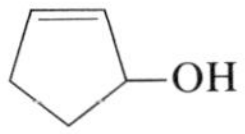 —OH	H_2WO_4/Et_3N	H_2O	20	90	c
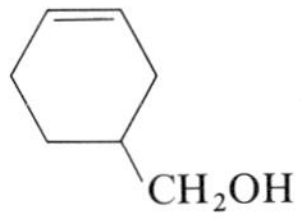 CH_2OH	H_2WO_4/Et_3N	H_2O	20	48	c
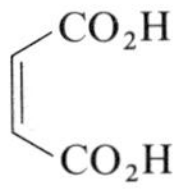 CO_2H, CO_2H	Resin–hydrogen tungstate	H_2O	60	95	d
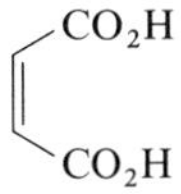 CO_2H, CO_2H	Na_2WO_4	H_2O	65	77	e
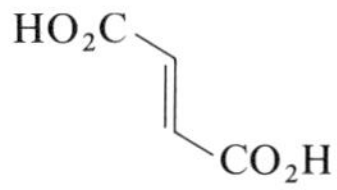 HO_2C, CO_2H	Na_2WO_4	H_2O	65	50	e
 CO_2H	Na_2WO_4	H_2O	65	50	e
	H_2WO_4	PrOH	25	89	f
	H_2MoO_4	$PrOH/H_2O$	25	91	f
	V_2O_5	$PrOH/H_2O$	25	90	f
	SeO_2	$PrOH/H_2O$	25	92	f

(continued)

TABLE IV (*continued*)

Olefin	Catalyst	Solvent	Temp (°C)	Epoxide selectivity (%)[a]	Reference
	H_2WO_4	PrOH	25	88	f
	SeO_2	PrOH	25	96	f
	$ArSeO_2H$	CH_2Cl_2	25	87	g
	H_2WO_4	PrOH	25	23	f
	SeO_2	PrOH	25	79	f
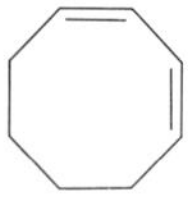	SeO_2	PrOH	25	83	f
	H_2WO_4	PrOH	25	78	f
	SeO_2	PrOH	25	97	f
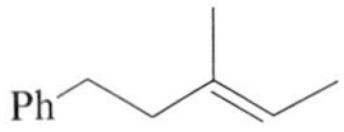 Ph	$ArSeO_2H$	CH_2Cl_2	25	94	g
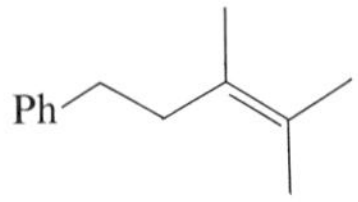 Ph	$ArSeO_2H$	CH_2Cl_2	25	91	g
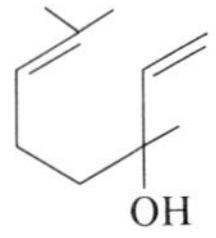 OH	$PhSeO_2H$[h]	CH_3OH	25	63[i]	j
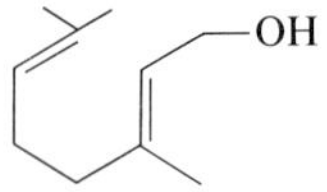 OH	$PhSeO_2H$[h]	CH_3OH	25	65[i]	j

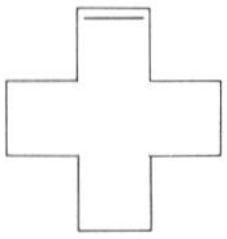

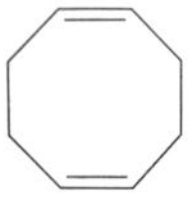

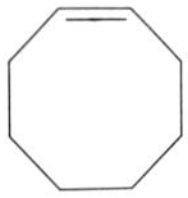

[a] The selectivities quoted are generally based on the olefin consumed. Selectivities based on the hydrogen peroxide consumed are generally lower than these values.

[b] G. J. Carlson, J. R. Skinner, C. W. Smith, and C. H. Wilcoxen, U.S. Patent 2, 833, 787 (1958) to Shell Development.

[c] H. C. Stevens and A. J. Kaman, *J. Am. Chem. Soc.* **87**, 734 (1965).

[d] G. G. Allen, U.S. Patent, 3, 156, 709 (1964) to DuPont.

[e] G. B. Payne and P. H. Williams, *J. Org. Chem.* **24**, 54 (1959).

[f] J. Itakura, H. Tanaka, and H. Ito, *Bul. Chem. Soc. Japan* **42**, 1604 (1969).

[g] H. Reich, F. Chow, and S. L. Peake, *Synthesis*, p. 299 (1978).

[h] Stoichiometric amounts of $PhSeO_2H$ were used.

[i] Selective epoxidation of the double bond farthest removed from the OH group was observed.

[j] P. A. Grieco, Y. Yokoyama, S. Gilman, and M. Nishizawa, *J. Org. Chem.* **42**, 2034 (1977).

III. ALLYLIC OXIDATION

$$\underset{\displaystyle |}{>C=C}-\overset{\displaystyle OR}{\overset{|}{C}H}- \Longleftarrow \underset{\displaystyle |}{>C=C}-CH_2- \Longrightarrow \underset{\displaystyle |}{>C=C}-\overset{\displaystyle O}{\overset{\|}{C}}-$$

Allylic oxidation lies in the realm of the oxometal reagents. The vapor-phase oxidation and ammoxiation of propylene to acrolein and acrylonitrile, respectively, over bismuth molybdate and related catalysts are well-known processes.[41,42] Selenium dioxide, on the other hand, is the reagent of choice for the laboratory-scale, stoichiometric oxidation of olefins to the corresponding allylic alcohols.[43] As discussed in Chapter 6, these reactions may proceed via a common mechanism involving an ene addition and sigmatropic rearrangement.

R, H, O, M^{n+}, O $\longrightarrow$ R, OH, M^{n+}, O $\longrightarrow$ R, O, $M^{(n-2)+}$, OH

Allylic alcohols are formed following the hydrolysis of the intermediate. The α,β-unsaturated carbonyl compounds result from the further oxidation of the allylic alcohol and/or directly from the intermediate by β-elimination.

H, O, $M^{(n-2)+}$, R, OH $\longrightarrow$ R, O $+ M^{(n-4)+} + H_2O$

$\downarrow H_2O$

R, OH $+ (HO)_2M^{(n-2)+} \xrightarrow{O_2M^{n+}}$ R, O $+ (HO)_2M^{(n-2)+}$

In catalytic oxidations, the oxidant (O_2M^{n+}) is regenerated by the reoxidation of the reduced catalyst with air or oxygen.

Adams[44] studied the oxidation of a variety of olefins with oxygen over bismuth molybdate at 460°C. Propylene afforded acrolein in 90–86% selectivity at 10–40% conversion. As shown in Table V, the selectivity decreases with increasing conversion owing to the further oxidation of the product. This reaction, because of its commercial importance, has been extensively studied,[45,46] and several mixed-oxide catalyst systems have been developed, including bismuth phosphomolybdates, uranium antimonates, and various multicomponent systems.[45]

TABLE V Oxidation of Olefins over Bismuth Molybdate at 460°C[a]

Olefin	Relative reactivity[b]	Olefin conversion (%)	Product	Selectivity (%)
	0.11	10 40 80	CHO	90 86 73
	1.00	40 80		95 90
	0.26 (cis) 0.19 (trans)	40 80		90 85
	0.50	40 70	CHO	72 72
	1.38	20 60		87 38
	0.43	5 15		88 79
	2.7	40		80
	4.2	35		60
	2.0	40		62

[a] From Adams.[44]
[b] Relative to 1-butene.

Isobutene is oxidized to methacrolein, albeit in lower selectivity. In all other cases, a side reaction involving oxidative dehydrogenation to conjugated dienes is favorable and becomes the predominant reaction:

$$\text{H} \quad \text{O—M}^{(n-2)+} \quad \text{OH} \longrightarrow \quad + (\text{HO})_2\text{M}^{(n-2)+} \qquad (40)$$

Thus, butenes are converted to 1,3-butadiene, and methylbutenes to isoprene. With olefins higher than C_5, oxidation is generally rather unselective,[44] as a result of the further oxidation of the initially formed products.

In the presence of ammonia, the same catalysts promote *ammoxidation* of propylene and isobutene to afford acrylonitrile and methacrylonitrile, respectively,[41,42,47] according to the stoichiometry in eq 41 (for a speculative mechanism, see Chapter 6).

$$CH_2{=}C(R)CH_3 + NH_3 + \tfrac{3}{2}\,O_2 \longrightarrow CH_2{=}C(R)CN + 3\,H_2O \qquad (41)$$

This transformation is the basis for several commercial processes for acrylonitrile production.[41,42]

Selenium dioxide complements the bismuth molybdate oxidations in a certain sense. Thus, olefins higher than propylene are stoichiometrically oxidized by SeO_2 to the corresponding allylic alcohols in moderate to good yields. When the reaction is carried out in acetic acid, the corresponding allylic acetates are obtained. These reactions have recently been reviewed by Rabjohn.[43] Terminal olefins yield primary allylic alcohols, with concomitant migration of the double bond.

$$CH_3(CH_2)_3CH{=}CH_2 \xrightarrow{[SeO_2]} CH_3(CH_2)_2CH{=}CHCH_2OH \qquad (42)$$

Disubstituted olefins are preferentially oxidized at a methylene group, e.g.,

$$CH_3CH{=}CHCH_2CH_3 \xrightarrow{[SeO_2]} CH_3CH{=}CHCH(OH)CH_3 \qquad (43)$$

The mechanism, which appears to involve an ene addition followed by a sigmatropic rearrangement, has been discussed in Chapter 6. Unsaturated carbonyl compounds and, in some cases, conjugated dienes are also formed (compare the bismuth molybdate oxidations) via the decomposition of the allylseleninic acid:

$$CH_2{=}CHCH{=}CHCH_3 \xleftarrow{-Se(OH)_2} CH_3CH(OSeOH)CH{=}CHCH_3 \begin{cases} \xrightarrow[-Se(OH)_2]{H_2O} CH_3CH(OH)CH{=}CHCH_3 \\ \xrightarrow[-H_2O]{-Se} CH_3C(=O)CH{=}CHCH_3 \end{cases}$$

A serious drawback of SeO_2 oxidations in organic syntheses is the formation of colloidal selenium and organoselenium compounds, which are difficult to separate. In order to circumvent these problems, Umbreit and Sharpless[48] developed a method for the SeO_2-catalyzed allylic oxidation of olefins with TBHP. This method produces allylic alcohols in moderate to good yields (generally better than that obtained with SeO_2 alone) with a wide variety of olefins. For example, β-pinene was selectively oxidized in 86% yield.

$$\text{(methylenecyclohexane derivative)} \xrightarrow[\text{TBHP}]{[\text{SeO}_2]} \text{(allylic alcohol, OH)} \tag{44}$$

Imidoselenium(IV) reagents effect the *stoichiometric* allylic amination of olefins,[49] e.g.,

$$\text{RN}{=}\text{Se}{=}\text{NR} + \text{(olefin)} \longrightarrow \text{(allylic amine, NHR)}, \quad \text{etc.} \tag{45}$$

This amination is the stoichiometric equivalent of the ammoxidation of olefins over bismuth molybdate catalysts.

Another reaction that is commonly used for allylic substitution of olefins is the peroxy ester reaction,[50–52] in which an olefin is treated with a *tert*-butylperoxy ester (usually the acetate) in the presence of catalytic amounts of a copper salt. Good yields of allylic acetates are generally obtained (see Chapter 3 for a discussion of the mechanism). Terminal olefins generally afford the 3-acetoxy derivative.

$$RCH_2CH{=}CH_2 + t\text{-}BuO_2Ac \xrightarrow{[Cu^I]} \underset{\displaystyle OAc}{RCH}CH{=}CH_2 + t\text{-}BuOH \tag{46}$$

Internal olefins can give complex mixtures resulting from attack at different allylic positions, with or without rearrangement. A few examples are given in Table VI.

The perester can be replaced by the less expensive TBHP if the reaction is carried out in acetic acid as solvent.[53] Cyclohexene with TBHP in acetic acid in the presence of 1 mol % CuCl affords 3-acetoxycyclohexene in 88% yield.[53]

$$\text{(cyclohexene)} + t\text{-}BuO_2H + HOAc \xrightarrow{[Cu^I]} \text{(cyclohexenyl)}{-}OAc + t\text{-}BuOH + H_2O \tag{47}$$

Metal acetates, such as $Pb(OAc)_4$, $Tl(OAc)_3$, $Hg(OAc)_2$, and $Pd(OAc)_2$, also effect allylic acetoxylation of olefins.[54] The first three require stoichiometric amounts of oxidants and are of limited utility for large-scale synthesis. The palladium(II)-catalyzed oxidation of ethylene and terminal olefins to ketones will be presented later in Section VI. However, when applied to olefins branched at the double bond such as isobutene, in which ketone formation is impossible, it often results in allylic oxidation.[55] In aqueous solution, the primary products are allylic alcohols, which generally undergo further oxidation and afford complex mixtures.[55] In acetic acid, propylene, 1-butene, and 1-pentene give mixtures of enol acetates and allylic acetates,

TABLE VI Allylic Acetoxylation of Olefins with *t*-Butyl Peracetate/CuCl[a]

Olefin	Products		Yield (%)
1-Butene or *cis,trans*-2-butene	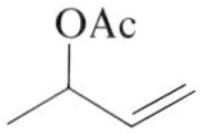 OAc	(89–99%)	70–85
	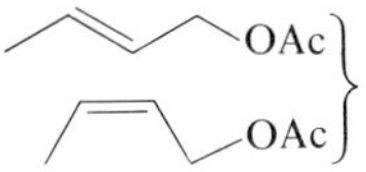 OAc, OAc	(11–6%)	
1-Pentene	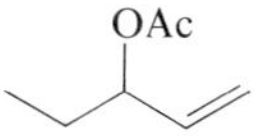 OAc	(89%)	88
	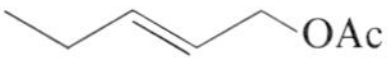 OAc	(11%)	
1-Octene	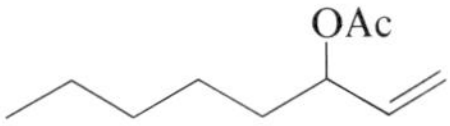 OAc	(87%)	89
	 OAc	(13%)	
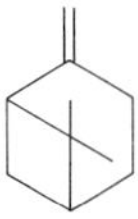	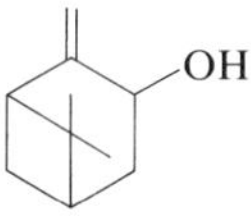 OH	(90%)[b]	85–90
	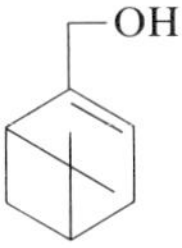 OH	(8%)[b]	
2-Pentene	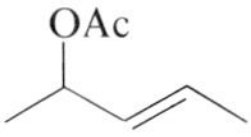 OAc	(68%)	89
	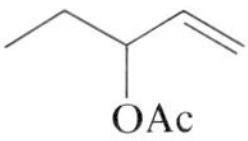 OAc	(29%)	
	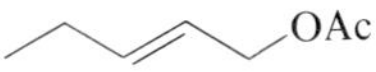 OAc	(3%)	
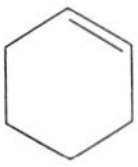	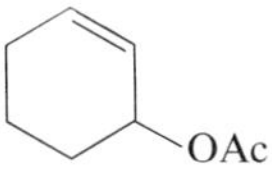 OAc		95
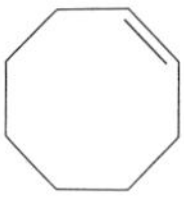	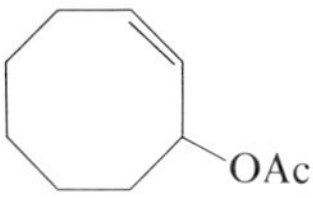 OAc		38
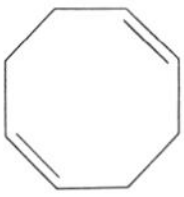	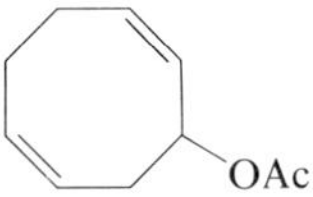 OAc	(71%)	96
	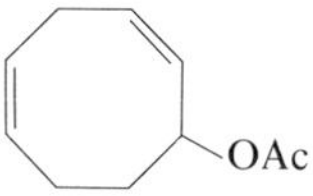 OAc	(29%)	

[a] From Rawlinson and Sosnovsky.[50]

[b] Obtained after hydrolysis.

whereas internal olefins and higher terminal olefins afford predominantly allylic acetates.

$$\mathrm{-\overset{H}{\underset{|}{C}}-\overset{H}{\underset{|}{C}}=\overset{H}{C}-} \xrightarrow{[Pd(OAc)_2]} \begin{cases} \mathrm{>C=\overset{|}{C}-\overset{H}{\underset{OAc}{C}}-} \\ \mathrm{-\overset{H}{\underset{|}{C}}-\underset{|}{C}=C<_{OAc}} \end{cases}$$

However, virtually all of the examples in the literature[55] involve the use of stoichiometric quantities of the palladium(II) as the oxidant. The reaction does not appear to have been adequately examined as a catalytic method (e.g., using O_2 and copper salts to regenerate the oxidant) for the preparation of allylic acetates. It should be mentioned, however, that these reactions are often complicated by the palladium-catalyzed olefin isomerization and the formation of palladium(II)–π-allyl complexes.

The allylic oxidation of olefins with a stoichiometric mixture of $Mn(OAc)_3$ and KBr in acetic acid has been reported.[56] Cyclohexene afforded 3-acetoxycyclohexene in 83% yield. As noted in Chapter 5, surprisingly few studies have been made of olefin autoxidations in the presence of *high concentrations* of Co(III) and Mn(III) acetates, with or without bromide ion. By analogy with the widespread use of these reagents for benzylic oxidation, one might expect them to have synthetic utility for allylic oxidation.

IV. GLYCOL (ESTER) FORMATION

$$\mathrm{>C=C<} \Longrightarrow \mathrm{>\overset{HO}{\overset{|}{C}}-\overset{OH}{\overset{|}{C}}<}$$

The classical *catalytic* method for converting olefins to the corresponding glycols utilizes H_2O_2 in the presence of catalytic amounts of acidic metal oxides (Milas reagents).[57] Osmium tetroxide generally gives the best results and reacts directly with the olefin to form a cyclic osmate ester,

$$\mathrm{>C=C<} + \mathrm{OsO_4} \longrightarrow \mathrm{\begin{matrix} C-O \\ | \\ C-O \end{matrix}\!\!>Os(=O)_2}$$

followed by H_2O_2 cleavage to regenerate OsO_4 (see Chapter 3).[58]

$$\mathrm{\begin{matrix} C-O \\ | \\ C-O \end{matrix}\!\!>Os(=O)_2} \xrightarrow{H_2O_2} \mathrm{\begin{matrix} C-OH \\ | \\ C-OH \end{matrix}} + \mathrm{OsO_4} + \mathrm{H_2O} \qquad (48)$$

The reaction is usually conducted in *tert*-butyl alcohol as solvent and affords exclusively the *cis*-glycol. Hydrogen peroxide can be replaced by TBHP,[59] $NaClO_3$,[60] NaOCl,[61] and amine oxides such as *N*-methylmorpholine oxide.[62] Although catalytic osmylation is the most reliable method for converting a variety of olefins to *cis*-glycols, overoxidation to α-ketols often leads to lower yields. The latter can be overcome by the use of *N*-methylmorpholine oxide as the stoichiometric oxidant.[62] For example, *cis*-1,2-cyclohexanediol was prepared in 89–90% isolated yield using this procedure.[62]

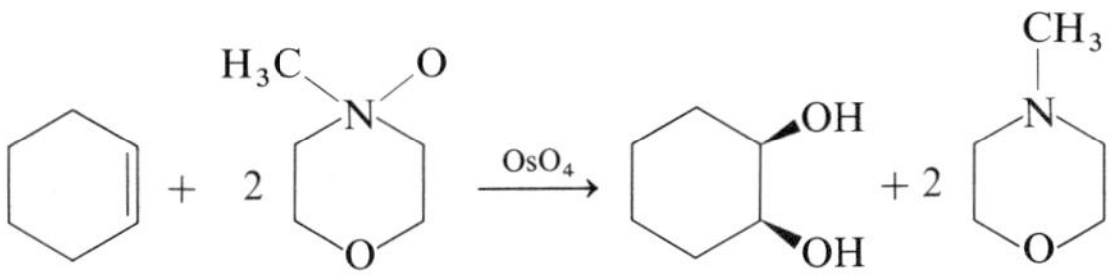

$$\text{cyclohexene} + 2\ \text{N-methylmorpholine N-oxide} \xrightarrow{OsO_4} \textit{cis}\text{-1,2-cyclohexanediol} + 2\ \text{N-methylmorpholine} \tag{49}$$

Potassium oleate was converted, in 92% isolated yield, to *erythro*-9,10-dihydroxystearic acid using a combination of OsO_4 and NaOCl.[61]

$$CH_3(CH_2)_7CH{=}CH(CH_2)_7CO_2K \longrightarrow CH_3(CH_2)_7CH(OH){-}CH(OH)(CH_2)_7CO_2K \tag{50}$$

Stoichiometric hydroxylations with OsO_4 are generally carried out in the presence of pyridine, which accelerates the reaction and gives improved yields.[58] Presumably this involves coordination to the osmium at some point along the reaction pathway. An interesting recent development is the use of chiral bases to produce asymmetric induction in the hydroxylation of olefins with OsO_4.[63] Enantiomeric excesses observed were generally in the range 60–80% when optically active quinuclidine bases, such as dihydroquinine acetate, were used.

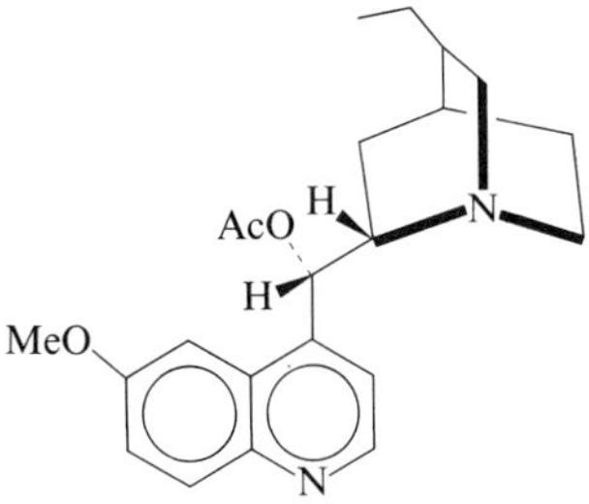

Dihydroquinine acetate

The mechanism shown below was proposed to account for the results (see Chapter 6).

$$O_2(O=)Os(L) \underset{}{\overset{-L}{\rightleftharpoons}} OsO_4 \xrightarrow{>C=C<} O_2(O=)Os(C=C)$$

$$O_2(O=)Os(C=C) \xrightarrow{L} L(O=)(O)Os\text{–}O\text{–}C\text{–}C \xrightarrow{L} L_2(O=)_2Os(O\text{–}C)(O\text{–}C)$$

The synthetic utility of these results would be considerably enhanced if the method could be made catalytic.

Alkaline potassium permanganate is also commonly used for the stoichiometric conversion of olefins to *cis*-glycols, but the method has less synthetic utility than catalytic osmylation.

Ethylene is converted to ethylene glycol diacetate using molecular oxygen with a combination metal–halide catalyst in acetic acid.[64] Very high (95%) selectivities have been claimed for the system consisting of TeO_2 and bromide salts (see Chapter 5 for a discussion of possible mechanisms).

$$CH_2{=}CH_2 + 2\,HOAc + \tfrac{1}{2}\,O_2 \xrightarrow{[TeO_2/Br^-]} AcOCH_2CH_2OAc + H_2O \qquad (51)$$

The SeO_2 oxidation of ethylene in acetic acid at 100–125°C in the presence of mineral acids such as HCl also affords ethylene glycol diacetate.[43,65] Styrene and butadiene similarly undergo 1,2-addition.[66] Although of undoubted commercial importance for the production of ethylene glycol, these methods are presumably limited to ethylene and do not represent general methods for converting olefins to glycol esters. With higher olefins, the severe conditions employed are expected to lead to complex mixtures of oxidation products.

Under certain conditions, the palladium(II)-catalyzed oxidation of ethylene and terminal olefins leads predominantly to the formation of glycol mono- and/or diacetates, as described in Chapter 7. A process for the manufacture of ethylene and propylene glycol monoacetates in 90–95% yield, using molecular oxygen and a catalyst combination using palladium(II) and nitrate in acetic acid, has been developed.[67] The Teijin process for the direct conversion of ethylene to ethylene glycol via oxychlorination has been discussed in Chapter 5.

The OsO_4-catalyzed oxyamination of olefins with chloramine salts[68] was discussed in Chapter 6.

$$>C{=}C< + ArSO_2NClNa \xrightarrow{[OsO_4]} -\underset{ArSO_2NH}{\overset{|}{C}}-\underset{OH}{\overset{|}{C}}-, \quad \text{etc.} \qquad (52)$$

V. OXIDATIVE CLEAVAGE

$$>C{=}C< \Longrightarrow >C{=}O + O{=}C<$$

Ozone is one of the most commonly used reagents for oxidative cleavage of double bonds.[69] Certain nucleophilic olefins also undergo oxidative cleavage with singlet oxygen via a dioxetane intermediate.[70] A discussion of these reactions is beyond the scope of this book. However, oxometal reagents, in addition to being useful allylic oxidants, can also effect oxidative cleavage of double bonds. A characteristic feature of oxometal reagents that effect the oxidative cleavage of double bonds is the presence of a *cis*-dioxometal functionality. Reaction with a double bond can produce cleavage via a [4 + 2]- or a [2 + 2]-cycloaddition, as discussed in Chapter 6. When the reduced form of the oxidant is reoxidized by molecular oxygen (i.e., in a catalytic oxidation), the overall reaction has been described as *olefin–oxygen metathesis*[71] and is *formally equivalent* to cleavage with singlet oxygen. Unfortunately, oxidative cleavage of olefins over metal oxide catalysts, at elevated temperatures in the gas phase, almost always leads to complete combustion to CO_2, CO, and H_2O, since the cleavage products (aldehydes, ketones, etc.) are generally more reactive than the reactant olefins. However, one example of a selective cleavage has been reported, namely, the oxidation of ethylene to formaldehyde with 76% selectivity at 0.4% conversion over an iron molybdate catalyst at 350°–400°C.[71]

$$CH_2{=}CH_2 + O_2 \xrightarrow{[Fe_xMoO_y]} 2\,HCHO \tag{53}$$

Ruthenium tetroxide in carbon tetrachloride was first introduced as an oxidizing agent by Djerassi and Engle.[72] Ruthenium tetroxide is an extremely powerful oxidant that can cleave double bonds at ambient temperatures.[73] The synthetic utility of RuO_4 increased considerably with the finding that it can be used in catalytic amounts in conjunction with NaOCl as the primary oxidant.[74] Thus, either $RuCl_3$ or RuO_2 catalyzes the oxidation of a variety of substrates with NaOCl at ambient temperatures.[74,75] Olefins undergo smooth oxidative cleavage to ketones or carboxylic acids in an aqueous alkaline solution of NaOCl containing catalytic quantities of Ru salts,[74] e.g.,

$$CH_3(CH_2)_9CH{=}CH_2 \xrightarrow[NaOH]{[RuCl_3],\ NaOCl} \underset{90\%}{CH_3(CH_2)_9CO_2H}, \text{ etc.} \tag{54}$$

$$CH_3(CH_2)_5\underset{\displaystyle (CH_2)_3CH_3}{\underset{|}{C}}{=}CH_2 \xrightarrow[NaOH]{[RuCl_3],\ NaOCl} \underset{83\%}{CH_3(CH_2)_5\overset{\displaystyle O}{\overset{\|}{C}}(CH_2)_3CH_3}, \text{ etc} \tag{55}$$

Cyclohexene similarly affords adipic acid in 86–95% yield.[75] Apparently the aldehydes formed as the initial products of cleavage of 1,2-disubstituted olefins undergo further rapid oxidation to the corresponding carboxylic acids. Foglia and co-workers[61] found that the addition of a base such as NaOH is required to circumvent the competing chlorination reactions. Potassium oleate was converted to a mixture of azelaic and pelargonic acids in more than 90% yield [61] after neutralization.

$$CH_3(CH_2)_7CH{=}CH(CH_2)_7CO_2K \xrightarrow{[RuO_2],\ NaOCl} CH_3(CH_2)_7CO_2K + KO_2C(CH_2)_7CO_2K \tag{56}$$

The same authors reported that the addition of a phase transfer catalyst (Bu_4NBr) led to further improvements in the oxidative cleavage of water-immiscible acyclic olefins.[76] Thus, 1-pentadecene and *cis*-9-octadecene afforded the corresponding carboxylic acids in 100 and 84% yields, respectively, in mixtures of CH_2Cl_2 and H_2O at ambient temperatures,

$$\mathrm{R(H)C{=}C(H)R'} \xrightarrow[NaOH,\ Bu_4NBr]{[RuO_2],\ NaOCl} RCO_2H + R'CO_2H \tag{57}$$

where $R = C_{13}H_{27}$, $R' = H$, and $R = R' = C_8H_{17}$.

Cyclic olefins are oxidized to the corresponding dicarboxylic acids using aqueous nitric acid in the presence of a vanadium(V) catalyst (usually V_2O_5 or NH_4VO_3).[77] The reaction is carried out in the liquid phase in the temperature range 50°–100°C. Cyclohexene, for example, afforded adipic acid in 68% yield, together with minor amounts of the lower dibasic acids.[78,79] Wilhoit[80] improved the yields by using 25–50% aqueous nitric acid in conjunction with an OsO_4–V_2O_5 catalyst at 70°–90°C. Cyclohexene afforded adipic acid in 89% yield. Similarly, cyclododecene and cyclooctene were oxidized to 1,12-dodecanedioic and 1,8-octanedioic acids in 86 and 82% yield, respectively.

$$\text{cyclododecene} \xrightarrow[[OsO_4/V_2O_5]]{HNO_3} HO_2C(CH_2)_{10}CO_2H \tag{58}$$

$$\text{cyclooctene} \xrightarrow[[OsO_4/V_2O_5]]{HNO_3} HO_2C(CH_2)_6CO_2H \tag{59}$$

It should be noted, however, that virtually stoichiometric amounts of OsO_4 are used, and the reaction probably involves initial hydroxylation by OsO_4

followed by oxidative cleavage of the resulting glycol by the combination of HNO_3 and V_2O_5.[79] This situation is analogous to the oxidative cleavage of olefins with sodium periodate ($NaIO_4$) using catalytic amounts of OsO_4.[81] The latter effects hydroxylation of the olefin, and the periodate is responsible for the cleavage of the glycol formed and reoxidation of the reduced osmium. Indeed, an alternative approach to oxidative cleavage is a two-step procedure involving the initial conversion of the olefin to the glycol followed by cleavage of the glycol with one of several reagents (see Chapter 12 for glycol cleavage reactions).

Parshall[82] has reported the oxidative cleavage of cyclododecene to 1,12-dodecanedioic acid, in 30% yield, using a mixture of Re_2O_7 and H_2O_2 in acetic acid at ambient temperature. The use of Re_2O_7 as an oxidant, in combination with an inexpensive terminal oxidant such as H_2O_2 or NaOCl, warrants further investigation.

The oxidative cleavage of *n*-butenes to acetic acid is achieved in a rather novel way. The mixture of 1-butene and 2-butene is first converted to *sec*-butyl acetate, which is then subjected to liquid-phase autoxidation at 200°C and 60 bar in the absence of a catalyst.[83]

$$\text{2-butene} + \text{1-butene} + CH_3CO_2H \xrightarrow{H^+} CH_3CH_2CH(O_2CCH_3)CH_3$$

$$CH_3CH_2CH(O_2CCH_3)CH_3 \xrightarrow{O_2} 3\ CH_3CO_2H, \quad \text{etc.}$$

The yield of acetic acid based on *n*-butenes is 60%. This procedure lacks broad utility as a method for the selective oxidative cleavage of double bonds.

VI. OXIDATIVE KETONIZATION

$$\text{H}\!-\!C\!=\!C\!-\!\text{H} \Longrightarrow \text{H}\!-\!C(\text{H})\!-\!C(\!=\!O)\!-$$

Ethylene is oxidized smoothly to acetaldehyde by aqueous solutions of $PdCl_2$ at ambient temperature.

$$CH_2{=}CH_2 + H_2O + PdCl_2 \longrightarrow CH_3CHO + Pd + HCl \quad (60)$$

In the presence of $CuCl_2$ the palladium is reoxidized,

$$Pd + 2\ CuCl_2 \longrightarrow PdCl_2 + 2\ CuCl \quad (61)$$

and the $CuCl_2$ can in turn be regenerated by dissolved oxygen.

$$2\,CuCl + \tfrac{1}{2}\,O_2 + 2\,HCl \longrightarrow 2\,CuCl_2 + H_2O \tag{62}$$

The overall reaction is catalytic in $PdCl_2$ and $CuCl_2$ and constitutes the Wacker process for the oxidation of acetaldehyde.[55]

$$CH_2{=}CH_2 + \tfrac{1}{2}\,O_2 \xrightarrow{[PdCl_2/CuCl_2]} CH_3CHO \tag{63}$$

The oxidation is carried out at 100°–110°C and 10 bar to produce acetaldehyde in about 95% yield (see Chapter 7 for the mechanism). Fixed-bed processes have also been described.[55,84] A major problem in the Wacker process is the extremely corrosive nature of the aqueous solutions of $PdCl_2$ and $CuCl_2$, which necessitates the use of costly titanium alloys as materials of construction.[84]

Propylene and 1-butene or 2-butene are similarly oxidized by dioxygen with the $PdCl_2/CuCl_2$ catalyst to produce acetone and methyl ethyl ketone, respectively, both in about 90% yield. The rates of oxidation decrease in the order ethylene > propylene > butenes, although the differences are small, being roughly 9 : 2 : 1. Olefins branched at the double bond cannot form ketones and undergo allylic oxidation. Terminal olefins of higher molecular weight generally afford low yields of ketones under the standard Wacker conditions, since mixtures of isomeric ketones are often formed as a result of the palladium-catalyzed isomerization of the olefin; i.e., the oxidation is not sufficiently fast to compete effectively with side reactions. However, the effective oxidation of the higher terminal olefins has been realized by using alcohols,[85,86] DMF,[87] or sulfolane[88] as cosolvents. Thus, Lloyd and Luberoff[86] found that most monosubstituted and 1,2-disubstituted alkenes were readily oxidized to the corresponding ketones in aqueous methanol or ethanol solutions. The extent of isomerization was markedly dependent on the reaction temperature. For example, 1-octene, which generally produces a mixture of 2-, 3-, and 4-octanones, gave a 97% yield of 2-octanone at 30°C, 85% at 60°C, but only 62% at 90°C. The oxidation of cyclohexene with $PdCl_2$ and $CuCl_2$ in aqueous ethanol afforded cyclohexanone in 95% yield.[86] The amount of isomeric aldehydes formed in the oxidation of terminal olefins generally did not exceed a few percent.

Clement and Selwitz[87] found aqueous DMF to be an excellent medium for these reactions. For example, 1-dodecene was oxidized to 2-dodecanone in greater than 80% yield, and 10-undecenoic acid afforded 10-ketoundecanoic acid in 83% yield.

$$CH_2{=}CH(CH_2)_8CO_2H + O_2 \xrightarrow[DMF\text{–}H_2O]{[PdCl_2/CuCl_2]} CH_3\underset{\underset{O}{\|}}{C}(CH_2)_8CO_2H \tag{64}$$

Tsuji and co-workers[89] have incorporated the oxidation of terminal olefins to methyl ketones using dioxygen and the $PdCl_2/CuCl$ catalyst in DMF or methanol as a key step in the synthesis of a variety of natural products. A few examples are shown below.

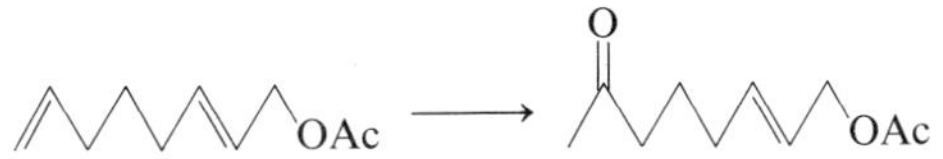

OAc ⟶ OAc (65)

CO_2R ⟶ CO_2R (66)

RO ⟶ RO (67)

Cuprous chloride was found to be superior to cupric chloride as the co-catalyst, since it minimized the chlorination of the ketone product. In a typical procedure,[90] 10 mmol of olefin was treated with 10 mmol of CuCl, 2 mmol of $PdCl_2$, and oxygen at room temperature for 20 hr. (Note that the system is not truly catalytic.) Similarly, McQuillin and Parker[91] reported the oxidation of the diene in eq 68 to the enone in "good" yield.

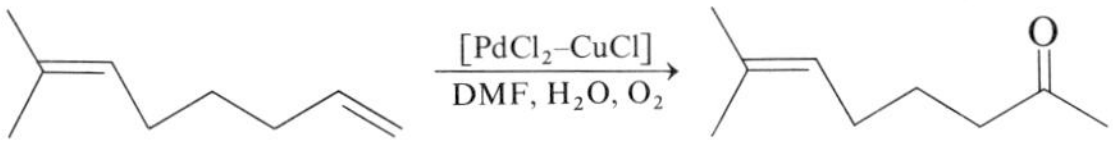

$$\xrightarrow[\text{DMF, H}_2\text{O, O}_2]{[\text{PdCl}_2\text{–CuCl}]} \qquad (68)$$

It is worth noting, however, that the amounts of palladium and copper chlorides used in these syntheses are generally prohibitive for commercial production. Moreover, if smaller amounts of catalysts are used, the more forcing conditions required will lead to increased isomerization and other side reactions.

Fujimoto and Kunugi[92] obtained excellent yields of ketones in the oxidation of olefins with mixtures of oxygen and steam over a $PdCl_2$ catalyst supported on charcoal at 105°C. Thus, propylene, 1-butene, and cyclohexene afforded acetone (89%), methyl ethyl ketone (93%), and cyclohexanone (84%), respectively.

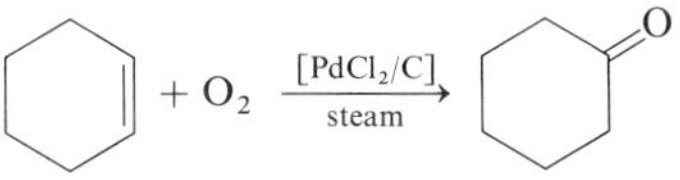

$$+ \text{O}_2 \xrightarrow[\text{steam}]{[\text{PdCl}_2/\text{C}]} \qquad (69)$$

Apparently no cocatalyst is needed, since the active charcoal support is capable of catalyzing the reoxidation of the palladium catalyst. This method may have general synthetic utility for the conversion of olefins to ketones.

The oxidation of α,β-unsaturated carbonyl compounds under the usual conditions (see above) in DMF using $PdCl_2$–CuCl with O_2 was found to be very slow.[93] Tsuji and co-workers[93] found that α,β-unsaturated esters and α,β-unsaturated ketones are efficiently oxidized to the corresponding β-keto esters and β-diketones, respectively, using Na_2PdCl_4 as the catalyst and TBHP or H_2O_2 as the stoichiometric oxidant (see Table VII). Suitable solvents are 50% aqueous acetic acid, isopropanol, and *N*-methylpyrrolidone. A 1 : 1 mixture of 30% H_2O_2 and acetic acid is particularly effective.

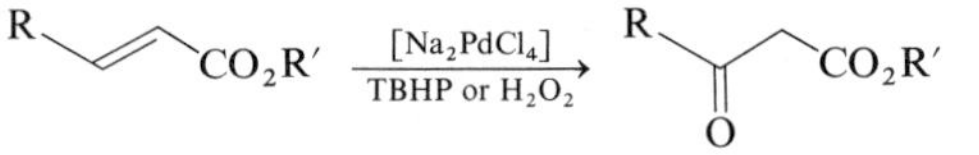

$$\text{R}\diagup\!\!=\!\!\diagdown\text{CO}_2\text{R}' \xrightarrow[\text{TBHP or H}_2\text{O}_2]{[\text{Na}_2\text{PdCl}_4]} \text{R}\diagdown\underset{\text{O}}{\overset{}{\text{C}}}\diagup\diagdown\text{CO}_2\text{R}' \qquad (70)$$

TABLE VII Oxidation of α,β-Unsaturated Carbonyl Compounds with Na_2PdCl_4/TBHP[a]

Substrate[b]	Product	Yield %[c]	(%)[d]
 CO_2Me	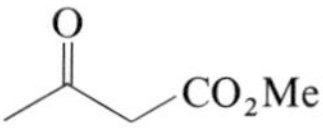 O, CO_2Me		(83)
 CO_2Me	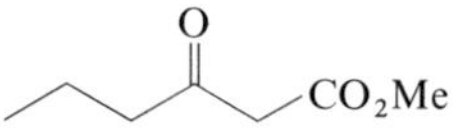 O, CO_2Me	78	(96)
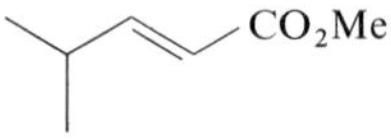 CO_2Me	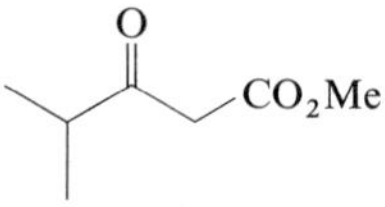 O, CO_2Me	64	(96)
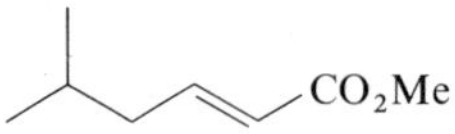 CO_2Me	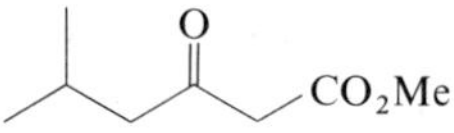 O, CO_2Me	68	(80)
 CO_2Me	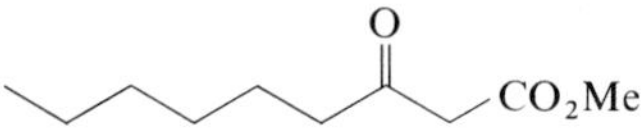 O, CO_2Me	75	
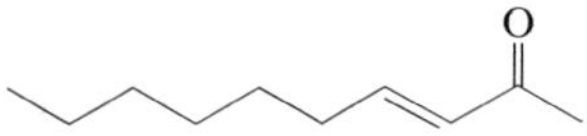 O	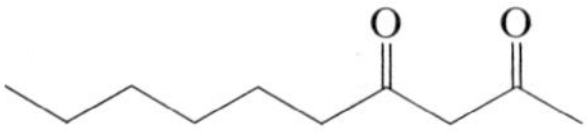 O O	59	
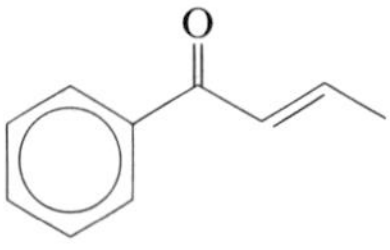 O	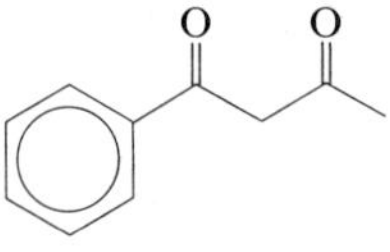 O O	59	

[a] From Tsuji *et al.*[93]

[b] Carried out in 50% aqueous acetic acid for 3 hr at 50°C using Na_2PdCl_4 (0.2 equiv) and *t*-butyl hydroperoxide (1.2–2.0 equiv).

[c] Isolated yields.

[d] Determined by gas–liquid chromatography using internal standards.

$$\mathrm{R\text{-}CH{=}CH\text{-}C(O)\text{-}R'} \xrightarrow[\text{TBHP or } H_2O_2]{[Na_2PdCl_4]} \mathrm{R\text{-}C(O)\text{-}CH_2\text{-}C(O)\text{-}R'} \qquad (71)$$

This system was not suitable for the selective oxidation of simple terminal olefins, such as 1-octene, because it caused extensive double bond migration. The reaction pathway would involve reoxidation of palladium by H_2O_2 or TBHP in a traditional Wacker sequence or peroxymetallation of the olefin as proposed by Mimoun for the oxidation of simple olefins to methyl ketones with $Pd(O_2CCF_3)_2$/TBHP[94] or $Pd(O_3SCH_3)_2/H_2O_2$[95] (see Chapter 7 for a detailed discussion of the mechanism).

An alternative procedure for converting olefins to ketones involves a hydration–dehydrogenation sequence, i.e.,

$$\mathrm{(H)C{=}C} + H_2O \longrightarrow \mathrm{(H)(HO)C{-}C(H)} \xrightarrow{-[2H]} \mathrm{O{=}C{-}C{-}H}$$

When oxygen is employed as the hydrogen acceptor in the second step, the reaction is often referred to as *oxyhydration* and the overall reaction is equivalent to the Wacker process, i.e.,

$$\mathrm{(H)C{=}C} + \tfrac{1}{2}\,O_2 \longrightarrow \mathrm{O{=}C{-}C{-}H} \qquad (72)$$

We have mentioned in Chapter 6 the vapor-phase oxidation of propylene with H_2O–O_2 mixtures over an SnO_2/MoO_3 catalyst at 100°–160°C to afford acetone in 90% selectivity[96] by an oxyhydration mechanism. (This oxidation contrasts with the formation of acrolein, via allylic oxidation, over a bismuth molybdate catalyst.) The reaction is reminiscent of certain reactions mediated by molybdoenzymes (discussed in Chapter 8) which are also thought to involve oxyhydration.

Other transition metal oxides, such as TiO_2, Fe_2O_3, and Cr_2O_3, also form active oxyhydration catalysts in combination with MoO_3, but they are less effective than the preferred SnO_2/MoO_3 combination.[96,97] The latter has also been used for the conversion of 1-butene and 2-butene to methyl ethyl ketone, and 2-pentene to a mixture of methyl propyl ketone and diethyl ketone.[96]

Finally, the recently discovered rhodium-catalyzed oxidation of terminal olefins to methyl ketones with molecular oxygen[98] is noteworthy since it proceeds by a different mechanism than the palladium-catalyzed oxidation (see Chapter 4 for a discussion of the mechanism). Rhodium catalysis appears to offer an alternative method for the mild, selective oxidation of terminal olefins to ketones in the liquid phase.

VII. OXIDATIVE NUCLEOPHILIC SUBSTITUTION

$$\gt C{=}C\lt_{H} \Longrightarrow \gt C{=}C\lt_{X}$$

If the Wacker reaction is conducted in an acetic acid medium, vinyl acetate is produced in what is formally an oxidative nucleophilic substitution of hydrogen by acetate, i.e.,[55,84]

$$CH_2{=}CH_2 + HOAc + \tfrac{1}{2}\,O_2 \xrightarrow{[Pd^{II}/Cu^{II}]} CH_2{=}CHOAc + H_2O \qquad (73)$$

Acetaldehyde is also produced by the accompanying hydrolysis of the vinyl acetate by the water produced. The ratio of acetaldehyde to vinyl acetate can be varied by changing the water content of the reaction medium. The commercial processes[41,55,84,99,100] generally operate in the range 100°–130°C and 30–35 bar. A fixed-bed process involving palladium supported on charcoal appears to be preferred in practice.[99] Corrosion in the vinyl acetate process is the same as that encountered in the Wacker process. The reaction does not have general synthetic utility, since the oxidation of higher olefins generally leads to complex mixtures. Lower terminal olefins give mainly enol acetates, analogous to the reaction in eq 73, with the 1,2-addition products being the major by-products in the presence of cupric salts.[101] Internal olefins afford mainly allylic acetates, as discussed above. Reactions with higher olefins are further complicated by olefin isomerization and the formation of π-allyl complexes.

Other nucleophiles can also be used in conjunction with a Pd(II) catalyst.[55] For example, the oxidation of olefins with dioxygen over a $PdCl_2/CuCl_2$ catalyst in the presence of alcohols produces the corresponding ketals.[86]

$$RCH{=}CHR + 2\,R'OH + \tfrac{1}{2}\,O_2 \longrightarrow RCH_2CR(OR')_2 + H_2O \qquad (74)$$

When the nucleophile is a carbanion, such as malonate, oxidative nucleophilic substitution provides a method for producing a new C—C bond.[55]

The oxidative coupling of olefins,[55] which occurs under certain conditions in the presence of Pd(II) salts, is another reaction that is formally an oxidative substitution of hydrogen by a carbanion. For example, vinyl acetate reacts with $Pd(OAc)_2$ to give, *inter alia*, 1,4-diacetoxy-1,3-butadiene.[102]

$$2\ CH_2{=}CHOAc + Pd(OAc)_2 \longrightarrow AcOCH{=}CH{-}CH{=}CHOAc + Pd + 2\,HOAc \qquad (75)$$

The reaction can be made catalytic in Pd(II) by using $Cu(OAc)_2$ as the primary oxidant.[102] Similarly, α-methylstyrene affords the corresponding diene.[103]

$$2\,\mathrm{PhC(CH_3){=}CH_2} \xrightarrow{[\mathrm{Pd^{II}/Cu^{II}}]} \mathrm{(Ph)(H_3C)C{=}CH{-}CH{=}C(Ph)(CH_3)}, \quad \text{etc.} \qquad (76)$$

A related reaction is the Pd(II)-mediated oxidative coupling of olefins to arenes, exemplified by the formation of styrene from ethylene and benzene. The reaction can be made catalytic in Pd(II) by employing a silver or copper cocatalyst and dioxygen.[104]

$$\mathrm{PhH + C_2H_4 + \tfrac{1}{2}\,O_2 \xrightarrow[\mathrm{Cu(OAc)_2}]{[\mathrm{Pd(OAc)_2},} PhCH{=}CH_2 + H_2O} \qquad (77)$$

This oxidative coupling appears to be a general reaction, and it has been used for the *oxidative arylation* of a variety of olefins.[105]

$$\mathrm{ArH} + \;\rangle\mathrm{C{=}C}\langle^{}_{\mathrm{H}} \longrightarrow \;\rangle\mathrm{C{=}C}\langle^{}_{\mathrm{Ar}} \qquad (78)$$

VIII. OXIDATIVE CARBONYLATION

$$\rangle\mathrm{C{=}C}\langle^{}_{\mathrm{H}} \Longrightarrow \;\rangle\mathrm{C{=}C}\langle^{}_{\mathrm{CO_2H}}$$

A process for the oxidative carbonylation of ethylene to acrylic acid has been described[106] which involves a Pd(II)/Cu(II) catalyst together with carbon monoxide and dioxygen (see Chapter 7).

$$\mathrm{CH_2{=}CH_2 + CO + \tfrac{1}{2}\,O_2 \xrightarrow{[\mathrm{Pd^{II}/Cu^{II}}]} CH_2{=}CHCO_2H} \qquad (79)$$

The scope of this reaction as a general method for converting olefins to α,β-unsaturated carboxylic acids does not appear to have been investigated fully. Propylene is reported[106] to afford crotonic acid.

IX. OXIDATIVE DIMERIZATION

$$2\;\rangle\mathrm{C{=}C}\langle^{\mathrm{H}} + \tfrac{1}{2}\,\mathrm{O_2} \Longrightarrow \;\rangle\mathrm{C{=}\overset{|}{C}{-}\underset{|}{C}{=}C}\langle + \mathrm{H_2O}$$

or

$$\rangle\mathrm{C{=}CCH_3} + \tfrac{1}{2}\,\mathrm{O_2} \Longrightarrow \;\rangle\mathrm{C{=}\underset{|}{C}CH_2CH_2\underset{|}{C}{=}C}\langle + \mathrm{H_2O}$$

We have already mentioned the Pd(II) catalysis of the oxidative dimerization of olefins to 1,3-dienes. This reaction formally represents a Pd(II)-mediated nucleophilic substitution of a vinylic hydrogen by a vinyl carbanion.

Oxidative dimerization of propylene to 1,5-hexadiene occurs over certain oxometal catalysts in the vapor phase and in the absence of dioxygen. The oxometal compound is reduced stoichiometrically but reoxidized by dioxygen in a separate step. The overall reaction is given by:

$$2\ \text{propylene} + \tfrac{1}{2}O_2 \longrightarrow \text{1,5-hexadiene} + H_2O \qquad (80)$$

The oxidative coupling is reported to occur over Tl(I) and Tl(III) oxides, In(III) oxide,[107] Bi_2O_3/SnO_2,[108] Bi_2O_3/P_2O_5,[109] and Bi_2O_3 alone.[110] Swift and co-workers[110] reported high selectivities (76%) to 1,5-hexadiene using Bi_2O_3 at temperatures between 475° and 520°C. The reaction probably involves initial formation of an allyl radical followed by dimerization.

X. CONJUGATED DIENES

In addition to the usual oxidative transformations (allylic attack, 1,2-addition) of olefins, conjugated dienes readily undergo 1,4-oxidative processes. This special characteristic of conjugated dienes will be considered in this section. The simplest member, 1,3-butadiene, a very important petrochemical building block, is obtained in large quantities from naphtha cracking. Autoxidation of 1,3-butadiene affords a polyperoxide via the following propagation sequence.[111]

$$ROO\cdot + \text{butadiene} \longrightarrow ROO\text{–}CH_2CH{=}CH\text{–}\dot{C}H_2 \text{ (allyl)}$$

$$ROO\text{(allyl)} \xrightarrow{O_2} RO_2\text{–}CH_2CH{=}CHCH_2\text{–}O\text{–}O\cdot \xrightarrow{\text{butadiene}}$$

$$RO_2\text{–}CH_2CH{=}CHCH_2\text{–}O\text{–}O\text{–}CH_2CH{=}CH\text{–}\dot{C}H_2, \quad \text{etc.}$$

The reaction, catalyzed by Rh(I) and Ir(I) complexes, was suggested to involve the addition of coordinated dioxygen to coordinated butadiene to form a metal-bonded alkylperoxy radical:[112]

$$(\text{butadiene})M^{I} + O_2 \longrightarrow (\text{butadiene})M^{II}OO\cdot \longrightarrow (\text{allyl–}CH_2OO\cdot)M^{II} \qquad (81)$$

Hydrogenation of the polyperoxide affords 1,4-butanediol, an important industrial monomer that is prepared via the addition of 2 mol of formaldehyde to acetylene.

$$HC{\equiv}CH + 2\,H_2CO \xrightarrow{[Cu]} HO\text{–}CH_2C{\equiv}CCH_2\text{–}OH \xrightarrow{H_2} HO(CH_2)_4OH \qquad (82)$$

Much effort has been devoted in recent years to the production of 1,4-butanediol via the selective oxidation of butadiene.[113] In a study of various palladium-containing catalysts the Pd/Te/charcoal system was found to be the most effective for liquid-phase 1,4-diacetoxylation of butadiene (compare 1,2-diacetoxylation of ethylene with palladium catalysts).

$$CH_2{=}CH{-}CH{=}CH_2 + 2\,HOAc + \tfrac{1}{2}\,O_2 \xrightarrow{[Pd/Te/C]} AcOCH_2CH{=}CHCH_2OAc + H_2O \quad (83)$$

An alternative approach[114] involves reaction with cupric acetate/lithium bromide/dioxygen in the catalytic sequence below (compare halide catalysis in the diacetoxylation of ethylene). The net reaction is equivalent to eq 83.

$$CH_2{=}CH{-}CH{=}CH_2 + 2\,CuBr_2 \longrightarrow BrCH_2CH{=}CHCH_2Br + 2\,CuBr \quad (84)$$

$$BrCH_2CH{=}CHCH_2Br + 2\,Cu(OAc)_2 \longrightarrow AcOCH_2CH{=}CHCH_2OAc + 2\,CuBr_2 \quad (85)$$

$$2\,CuBr + 2\,HOAc + \tfrac{1}{2}\,O_2 \longrightarrow Cu(OAc)_2 + CuBr_2 + H_2O \quad (86)$$

An interesting application of oxidative 1,4-diacetoxylation of a conjugated diene in organic synthesis is the following conversion of tiglic aldehyde to 4-acetoxytiglic aldehyde.[115]

$$CH_3CH{=}C(CH_3)CHO \xrightarrow{Ac_2O} CH_2{=}CH{-}C(CH_3){=}CHOAc \xrightarrow[{[Pd/Cu]}]{O_2,\ HOAc} AcOCH_2CH{=}C(CH_3)CH(OAc)_2$$

$$AcOCH_2CH{=}C(CH_3)CH(OAc)_2 \xrightarrow[-Ac_2O]{} AcOCH_2CH{=}C(CH_3)CHO$$

XI. ACETYLENES

In contrast to olefins, the autoxidation of acetylenes has not been studied in any detail. In principle, propagation, by analogy with olefins, should involve competition between addition and hydrogen abstraction:

$$ROO\cdot + {-}\overset{H}{\underset{|}{\overset{|}{C}}}{-}C{\equiv}C{-} \longrightarrow \begin{cases} {>}CH{-}C(RO_2){=}\dot{C}{<} \\ {>}\dot{C}{-}C{\equiv}C{-} + RO_2H \end{cases}$$

However, addition to the triple bond, forming an unstable vinylic radical, is less favorable than addition to the double bond. Hydrogen abstraction, on the other hand, proceeds at about the same rate as with the corresponding olefins.[116] For example, the propagation rate constant (k_p) per reactive hydrogen for 5-decyne is 0.7 M^{-1} sec^{-1} compared to 0.35 M^{-1} sec^{-1} for 3-heptene [D(R—H) = 83 kcal mol^{-1} in both cases].[116] Autoxidation of acetylenes proceeds, therefore, mainly via an abstraction mechanism forming α-acetylenic hydroperoxides as the primary products.

The autoxidation of straight-chain octynes occurs mainly at the carbon–hydrogen bond adjacent to the triple bond to yield acetylenic hydroperoxides and ketones.[116a] Approximately 2–7% yields of α,β-unsaturated ketones were formed by a competing oxidation at the triple bond via putative oxirene intermediates.

$$-\overset{|}{\underset{\substack{| \\ H}}{C}}-C{\equiv}C- + ROO\cdot \begin{cases} \xrightarrow[(95\%)]{\text{abstraction}} -\overset{|}{\underset{\bullet}{C}}-C{\equiv}C- + RO_2H \\ \xrightarrow[(5\%)]{\text{addition}} -\overset{|}{\underset{\substack{| \\ H}}{C}}-\underset{\bullet}{C}{=}\underset{\substack{| \\ OOR}}{C}- \end{cases}$$

$$-\underset{\substack{| \\ H}}{C}-\underset{\bullet}{C}{=}\underset{\substack{| \\ OOR}}{C}- \xrightarrow{-RO\cdot} \left[-\underset{\substack{| \\ H}}{C}-\underset{\diagdown O \diagup}{C{=}C}-\right] \longrightarrow -\overset{|}{C}{=}\underset{\substack{| \\ H}}{C}-\underset{\substack{\| \\ O}}{C}-$$

The α-methylene group of 2-octyne is oxidized under these conditions eight times faster than the methyl group.

Acetylenes undergo oxidation by selenium dioxide in a reaction analogous to the allylic oxidation of olefins by this reagent.[117]

$$-\overset{|}{\underset{\substack{| \\ H}}{C}}-C{\equiv}C- \xrightarrow{SeO_2} -\overset{|}{\underset{\substack{| \\ OH}}{C}}-C{\equiv}C- \qquad (87)$$

Oxidation of acetylenes to acetylenic alcohols with TBHP in the presence of catalytic amounts of SeO_2, in dichloroethane at 25°C, has recently been reported.[118] Internal acetylenes showed a marked tendency to undergo α,α′-dihydroxylation:

$$RCH_2C{\equiv}CCH_2R \xrightarrow[{[SeO_2]}]{\text{excess TBHP}} R\underset{\substack{| \\ OH}}{C}HC{\equiv}C\underset{\substack{| \\ OH}}{C}HR \qquad (88)$$

The reactivity sequence for acetylenes is $CH_2 \simeq CH > CH_3$, allowing for selective monohydroxylation in the case of CH_2 or CH versus CH_3. Relative rate data indicated that acetylenes are slightly more reactive than the corresponding olefins. Acetylenes bearing one methylene and one methine substituent afforded the enynone as the major product at 80°C, e.g.,

$$\text{C}_6\text{H}_{11}\text{—C}\equiv\text{C—CH}_2\text{CH}_3 \xrightarrow[\text{[SeO}_2\text{], 80°C}]{\text{3 mol TBHP}} \text{(1-hydroxycyclohexyl)—C}\equiv\text{C—CH(OH)CH}_3 \; (18\%) + \text{(cyclohexenyl)—C}\equiv\text{C—COCH}_3 \; (52\%) \quad (89)$$

Acetylenes possessing no active methylene groups undergo oxidation at the triple bond with SeO_2.[117] For example, diphenylacetylene is oxidized to benzil in 35% yield.

$$\text{PhC}\equiv\text{CPh} \xrightarrow{\text{oxidant}} \text{PhCOCOPh} \quad (90)$$

oxidant	yield
SeO_2	35%
$[RuO_2]$, NaOCl	83%

In general carbon–carbon triple bonds are more resistant to electrophilic attack than the corresponding double bonds. Consequently, more drastic conditions are generally required to attack acetylenes, and formation of cleavage products is more likely. Various oxometal reagents such as RuO_4,[119] $KMnO_4$,[120] and OsO_4[58] attack acetylenes at the triple bond to form α-diketones. For example, NaOCl in the presence of RuO_2 as catalyst is an effective reagent for reaction 90, benzil being obtained in 83% yield together with a small amount of benzoic acid formed via oxidative cleavage.

The catalytic oxidation of alkynes to α-diketones using OsO_4 with $KClO_3$,[121] TBHP,[58], *N*-methylmorpholine oxide,[58] or hydrogen peroxide[122] as the stoichiometric oxidant has been described, e.g.,

$$\text{PhC}\equiv\text{CCH}_3 \xrightarrow[\text{[OsO}_4\text{]}]{\text{TBHP}} \text{PhC(=O)—C(=O)CH}_3 \quad (62\%) \quad (91)$$

Terminal acetylenes afford carboxylic acids via oxidative cleavage.

Cyclic osmate esters have been isolated as intermediates in the reaction of OsO_4L (L = quinuclidine) with acetylenes.[123]

$$\text{R—C}\equiv\text{C—R}' + 2\,\text{OsO}_4\text{(L)} \longrightarrow \text{L}\rightarrow\text{OsO}_2\text{(O}_2\text{C(R)—C(R}'\text{)O}_2\text{)OsO}_2\leftarrow\text{L} \quad (92)$$

Hydrolysis of these complexes with sodium sulfite yields the α-diketone from internal acetylenes and carboxylic acids from terminal acetylenes.

Terminal acetylenes undergo oxidative dimerization when treated with copper(II) salts in the presence of an amine (usually pyridine) in an organic solvent (usually methanol).[124]

$$2\ \mathrm{RC{\equiv}CH} \xrightarrow[\text{pyridine}]{\mathrm{CuX_2}} \mathrm{RC{\equiv}C{-}C{\equiv}CR} \tag{93}$$

In the presence of dioxygen, the reaction is catalytic in copper. For example, bubbling oxygen through a solution of phenylacetylene and cuprous chloride catalyst in pyridine at 30°–40°[125] or stoichiometric oxidation with cupric acetate in pyridine–methanol (1 : 1) at reflux temperature affords diphenyldiacetylene in 70–86% yield.[126]

$$\mathrm{PhC{\equiv}CH} \xrightarrow[\text{or } [\mathrm{Cu^I/pyridine}],\ \mathrm{O_2}]{\mathrm{Cu^{II}/pyridine}} \mathrm{PhC{\equiv}C{-}C{\equiv}CPh} \tag{94}$$

The method has been applied to the synthesis of a wide variety of acetylenic compounds.[124,127] The reaction presumably involves the following steps.

$$\mathrm{RC{\equiv}CH + R'_3N} \rightleftharpoons \mathrm{RC{\equiv}C^- + R'_3NH^+} \tag{95}$$

$$\mathrm{RC{\equiv}C^- + Cu^{II}} \longrightarrow \mathrm{RC{\equiv}C\cdot + Cu^{I}} \tag{96}$$

$$2\ \mathrm{RC{\equiv}C\cdot} \longrightarrow \mathrm{RC{\equiv}C{-}C{\equiv}CR} \tag{97}$$

Tsuji and co-workers[128] have recently reported the facile, oxidative carbonylation of terminal acetylenes catalyzed by $PdCl_2$ in the presence of stoichiometric amounts of $CuCl_2$.

$$\mathrm{RC{\equiv}CH + CO + R'OH + 2\ CuCl_2} \xrightarrow{[\mathrm{Pd^{II}}]} \mathrm{RC{\equiv}CCO_2R' + 2\ CuCl + 2\ HCl} \tag{98}$$

The reaction proceeds at 1 bar and 25°C in alcohol solvent in the presence of a stoichiometric amount of sodium acetate (to neutralize the HCl formed). Yields in the range 58–74% were obtained with a variety of terminal acetylenes. The reaction is analogous to the oxidative carbonylation of olefins (*vide supra*).

REFERENCES

1. F. R. Mayo, *Accts. Chem. Rsch.* **1**, 193 (1968).
2. J. A. Howard, *Adv. Free-Radical Chem.* **4**, 49 (1972).
3. W. F. Brill and B. J. Barone, *J. Org. Chem.* **29**, 140 (1964).
4. M. Hession, K. Jones, and H. M. E. Steiner, *J. Chem. Soc. A* p. 350 (1971).
5. D. E. van Sickle, F. R. Mayo, R. M. Arluck, and M. G. Syz, *J. Am. Chem. Soc.* **89**, 967 (1967).

6. K. R. Hargrave and A. L. Morris, *Trans. Faraday Soc.* **52**, 89 (1956).
7. However, see K. Kaneda, K. Jitukawa, T. Itoh and S. Teranishi, *J. Org. Chem.* **45**, 3004 (1980).
7a. R. N. Lacey and K. Allison, U.S. Patent 3, 258, 491 (1966) to B.P.
8. P. A. Kilty and W. M. H. Sachtler, *Catal. Rev.* **10**, 1 (1974).
9. R. Landau, G. A. Sullivan, and D. Brown, *Chemtech* **9**, 602 (1979).
10. R. A. Sheldon, *J. Mol. Catal.* **7**, 107 (1980), and references cited therein.
11. M. I. Farberov, B. N. Bobylev, and D. I. Epstein, *Proc. Acad. Sci. USSR* **226**, 28 (1976).
12. R. A. Sheldon, *Aspects Homogeneous Catal.* **4**, 3.
13. K. B. Sharpless and T. R. Verhoeven, *Aldrichim. Acta* **12**, 63 (1979).
14. M. I. Farberov, *Sov. Chem. Ind.* **11** (1), 7 (1979).
15. G. A. Tolstikov, V. P. Yurev, and U. M. Dzhemilev, *Russ. Chem. Rev.* **44**, 319 (1975).
16. B. N. Bobylev, M. I. Farberov, and S. A. Kesarev, *Sov. Chem. Ind.* **11** (3), 153 (1979).
17. W. I. Denton, U.S. Patent 2,986,585 (1961) to Olin Mathieson. See also S. S. Srednev, S. I. Kryukov and M. I. Farberov, *Zh. Org. Khim.* **12**, 1882 (1976).
18. T. Imanaka, Y. Okamoto, and S. Teranishi, *Bul. Chem. Soc. Japan* **45**, 1353 (1972); see also J. B. Moffat, *Catal. Rev.* **18**, 199 (1979).
19. E. H. Eschinasi, *J. Org. Chem.* **35**, 1598 (1970).
20. S. Terao, M. Shiraishi, and K. Kato, *Synthesis* p. 467 (1979).
21. M. N. Sheng, J. G. Zajacek, and T. N. Baker, *Prepr., Div. Pet. Chem., Am. Chem. Soc.* E19 (1970).
22. M. N. Sheng and J. G. Zajacek, *J. Org. Chem.* **35**, 1839 (1970).
23. R. A. Sheldon, *Recl. Trav. Chim. Pays Bas* **92**, 253 (1973).
24. V. P. Yurev, J. A. Gailyunas, Z. G. Isaeva, and G. A. Tolstikov, *Bul. Acad. Sci. USSR, Div. Chem. Sci.* **24**, 885 (1974).
25. F. List and L. Kuhnen, *Erdöel & Kohle, Erdgas, Petrochem.* **20**, 192 (1967).
26. K. B. Sharpless and R. C. Michaelson, *J. Am. Chem. Soc.* **95**, 6136 (1973).
27. M. Kobayashi, S. Kurozumi, T. Toru, and S. Ishimoto, *Chem. Lett.* p. 1341 (1976).
28. A. Yasuda, H. Yamamoto, and H. Nozaki, *Bul. Chem. Soc. Japan* **52**, 1757 (1979).
29. B. E. Rossiter, T. R. Verhoeven, and K. B. Sharpless, *Tetrahedron Lett.* p. 4733 (1979); S. Tanaka, H. Yamamoto, H. Nozaki, K. B. Sharpless, R. C. Michaelson, and J. D. Cutting, *J. Am. Chem. Soc.* **96**, 5254 (1974).
30. R. Breslow and L. M. Maresca, *Tetrahedron Lett.* p. 623 (1977); p. 887 (1978).
31. S. Yamada, T. Mashiko, and S. Terashima, *J. Am. Chem. Soc.* **99**, 1988 (1977).
32. R. C. Michaelson, R. E. Palermo, and K. B. Sharpless, *J. Am. Chem. Soc.* **99**, 1990 (1977).
33. T. Katsuki and K. B. Sharpless, *J. Am. Chem. Soc.* **102**, 5974 (1980).
33a. B. E. Rossiter, T. Katsuki and K. B. Sharpless, *J. Am. Chem. Soc.* **103**, 464 (1981).
34. K. Tani, M. Hanafusa, and S. Otsuka, *Tetrahedron Lett.* p. 3017 (1979).
35. See J. J. Baldwin, A. W. Raab, K. Mensler, B. H. Arison, and D. E. McClure, *J. Org. Chem.* **43**, 4876 (1978), and references cited therein.
36. V. P. Yurev, I. A. Gailyunas, L. V. Spirikin, and G. A. Tolstikov, *J. Gen. Chem. USSR* **45**, 2269 (1975).
37. G. A. Tolstikov, U. M. Dzhemilev, V. P. Yurev, and S. F. Rafikov, *Proc. Acad. Sci. USSR, Chem. Sect.* **208**, 45 (1973); G. A. Tolstikov, V. P. Yurev, I. A. Gailyunas, and S. F. Rafikov, *ibid.* **214**, 23 (1974).
38. J. P. Schirmann and S. Y. Delavarenne, Ger. Patent 2,752,626 (1978) to Ugine Kuhlmann.
39. J. P. Schirmann, M. Pralus, and S. Y. Delavarenne, Ger. Patent 2,803,757 (1977) to Ugine Kuhlmann.
40. J. P. Schirmann and S. Y. Delavarenne, Ger. Patent 2,803,791 (1977) to Ugine Kuhlmann.
40a. S. Oae and T. Takata, *Tetrahedron Lett.* **21**, 3689 (1980).

41. T. Dumas and W. Bulani, "Oxidation of Petrochemicals: Chemistry and Technology." Applied Science, London, 1974.
42. D. J. Hucknall, "Selective Oxidation of Hydrocarbons." Academic Press, London, 1974.
43. N. Rabjohn, *Org. React.* **24**, 261 (1976).
44. C. Adams, *Ind. Eng. Chem.* **61**, 30 (1969).
45. G. W. Keulks, L. D. Krenzke, and T. M. Notermann, *Adv. Catal.* **27**, 183 (1978).
46. G. W. Keulks, *in* "Catalysis in Organic Synthesis" (G. V. Smith, ed.), p. 109. Academic Press, New York, 1977.
47. B. C. Gates, J. R. Katzer, and G. C. A. Schuit, "Chemistry of Catalytic Processes," p. 325. McGraw-Hill, New York, 1979.
48. M. A. Umbreit and K. B. Sharpless, *J. Am. Chem. Soc.* **99**, 5526 (1977).
49. K. B. Sharpless, T. Hori, L. K. Truesdale, and C. O. Dietrich, *J. Am. Chem. Soc.* **98**, 269 (1976); K. B. Sharpless and S. P. Singer, *J. Org. Chem.* **41**, 2504 (1976).
50. D. J. Rawlinson and G. Sosnovsky, *Synthesis* p. 1 (1972).
51. G. Sosnovsky and S. O. Lawesson, *Angew. Chem. Int. Ed.* **3**, 269 (1964).
52. G. Sosnovsky, *Tetrahedron* **21**, 871 (1965).
53. C. Walling and A. Zavitsas, *J. Am. Chem. Soc.* **85**, 2084 (1963).
54. D. J. Rawlinson and G. Sosnovsky, *Synthesis* p. 567 (1973); W. Kitching, *Organometal. React.* **3**, 319 (1972).
55. R. Jira and W. Freiesleben, *Organometal. React.* **3**, 1 (1972).
56. J. R. Gilmore and J. M. Mellor, *J. Chem. Soc. C* p. 2355 (1971).
57. N. A. Milas and S. Sussman, *J. Am. Chem. Soc.* **58**, 1302 (1936); **59**, 2345 (1937); N. A. Milas, *J. Am. Chem. Soc.* **59**, 2342 (1937).
58. For a recent review, see M. Schroder, *Chem. Rev.* **80**, 187 (1980).
59. A. Byers and W. J. Hickinbottom, *J. Chem. Soc.* p. 286 (1948); T. L. Gresham and T. R. Steadman, *J. Am. Chem. Soc.* **71**, 737 (1949); K. B. Sharpless and K. Akashi, *ibid.* **98**, 1986 (1976); U.S. Patent 4,049,724 (1975) to Atlantic Richfield.
60. K. A. Hofmann, *Ber. Dtsch. Chem. Ges.* **45**, 3329 (1912).
61. T. A. Foglia, P. A. Barr, A. J. Malloy, and M. J. Costanzo, *J. Am. Oil Chem. Soc.* **54**, 870A (1977).
62. V. van Rheenen, R. C. Kelly, and D. F. Cha, *Tetrahedron Lett.* p. 1973 (1976); V. van Rheenen, D. Y. Cha, and W. M. Hartley, *Org. Synth.* **58**, 43 (1978); see also W. P. Schneider and A. V. McIntosh, U.S. Patent 2,769,824 (1957); R. Ray and D. S. Matteson, *Tetrahedron Lett.* **21**, 449 (1980).
63. S. G. Hentges and K. B. Sharpless, *J. Am. Chem. Soc.* **102**, 4263 (1980).
64. J. Kollar, U.S. Patents 3,689,535 (1972); 3,668,239 (1972); 3,778,468 (1973); J. R. Valbert, U.S. Patent 3,715,388 (1973) all to Halcon International.
65. D. H. Olson, *Tetrahedron Lett.* p. 2053 (1966).
66. K. A. Javaid, N. Sonoda, and S. Tsutsumi, *Bul. Chem. Soc. Japan* **42**, 2056 (1969); *Ind. End. Chem. Prod. Res. Dev.* **9**, 87 (1970).
67. A. Mitsutani, *Chem. Econ. Eng. Rev.* **5**(3), 32 (1973).
68. K. B. Sharpless, A. O. Chong, and K. Oshima; *J. Org. Chem.* **41**, 177 (1976); E. Herranz and K. B. Sharpless, *ibid.* **43**, 2544 (1978); **45**, 2710 (1980); E. Herranz, S. A. Biller, and K. B. Sharpless, *J. Am. Chem. Soc.* **100**, 3596 (1978). See also J. E. Backvall and E. E. Bjorkman, *J. Org. Chem.* **45**, 2893 (1980).
69. P. S. Bailey, "Ozonation in Organic Chemistry," Vol. 1. Academic Press, New York, 1978.
70. C. S. Foote, *Accts. Chem. Rsch.* **1**, 104 (1968); C. Mumford, *Chem. Br.* **15**, 402 (1978); H. H. Wasserman and R. W. Murray, eds., "Singlet Oxygen." Academic Press, New York, 1979.

71. L. Kim, J. H. Raley, and C. S. Bell, *Recl. Trav. Chim. Pays-Bas* **96**, M136 (1977).
72. C. Djerassi and R. R. Engle, *J. Am. Chem. Soc.* **75**, 3838 (1953).
73. P. N. Rylander, *Engelhard Ind., Tech. Bull.* **9**, 135 (1969); D. G. Lee and M. van der Engh, *in* "Oxidation in Organic Chemistry" (W. Trahanovsky, ed.), Part B, p. 186. Academic Press, New York, 1973.
74. K. A. Keblys and M. Dubeck, U.S. Patent 3,409,649 (1968) to Ethyl Corp.
75. S. Wolfe, S. K. Hasan, and J. R. Campbell, *J. Chem. Soc. Chem. Commun.* p. 1420 (1970); see also H. Gopal and A. J. Gordon, *Tetrahedron Lett.* p. 2941 (1971).
76. T. A. Foglia, P. A. Barr, and A. J. Malloy, *J. Am. Oil Chem. Soc.* **54**, 858A (1977).
77. Y. Ogata, *in* "Oxidation in Organic Chemistry" (W. S. Trahanovsky, ed.), Part C, p. 295. Academic Press, New York, 1978.
78. J. F. Franz, J. F. Herber, and W. S. Knowles, *J. Org. Chem.* **30**, 1488 (1965); *Chem. Ind. (London)* p. 250 (1961).
79. G. Gut and W. Lindenmann, *Chimia* **22**, 307 (1968); G. Gut, R. V. Falkenstein, and A. Guyer, *ibid.* **19**, 581 (1965).
80. E. D. Wilhoit, U.S. Patent 3,461,160 (1969) to Du Pont.
81. R. Pappo, D. S. Allen, R. U. Lemieux, and W. S. Johnson, *J. Org. Chem.* **21**, 478 (1956).
82. G. W. Parshall, U.S. Patent 3,646,130 (1972) to Du Pont.
83. W. Schwerdtel, *Hydrocarbon Process.* **49**(11), 117 (1970).
84. B. C. Gates, J. R. Katzer, and G. A. Schuit, "Chemistry of Catalytic Processes." McGraw-Hill, New York, 1969.
85. A. A. Grigorev, M. Y. Klimenko, and I. I. Moiseev, USSR Patent 189,415 (1966); *Chem. Abstr.* **67**, 63,795 (1967).
86. W. G. Lloyd and B. J. Luberoff, *J. Org. Chem.* **34**, 3949 (1969).
87. W. H. Clement and C. M. Selwitz, *J. Org. Chem.* **29**, 241 (1964).
88. D. R. Fahey and A. E. Zuech, *J. Org. Chem.* **39**, 3276 (1974).
89. J. Tsuji, *Pure Appl. Chem.* **51**, 1235 (1979), and references cited therein; see also J. Tsuji, T. Yamaka, and T. Mandai, *Tetrahedron Lett.* p. 3741 (1979).
90. J. Tsuji, I. Shimizu, and K. Yamamoto, *Tetrahedron Lett.* p. 2975 (1976).
91. F. J. McQuillin and D. G. Parker, *J. Chem. Soc. Perkin I* p. 809 (1974).
92. K. Fujimoto and T. Kunugi, *Catal., Proc. Int. Congr., 5th, 1972* Vol. 1, p. 445 (1973); K. Fujimoto, H. Takeda, and T. Kunugi, *Ind. Eng. Chem. Prod. Res. Dev.* **13**, 237 (1974).
93. J. Tsuji, H. Nagashima, and K. Hori, *Chem. Lett.* p. 257 (1980).
94. H. Mimoun, R. Charpentier, A. Mitschler, J. Fischer, and R. Weiss, *J. Am. Chem. Soc.* **102**, 1047 (1980).
95. M. Roussel and H. Mimoun, *J. Org. Chem.* **45**, 5381 (1980).
96. S. Tan, Y. Moro-oka, and A. Ozaki, *J. Catal.* **17**, 125, 132 (1970).
97. Y. Moro-oka, Y. Takita, S. Tan, and A. Ozaki, *Bul. Chem. Soc. Japan* **41**, 2820 (1968); Y. Moro-oka, Y. Takita, and A. Ozaki, *J. Catal.* **23**, 183 (1971).
98. H. Mimoun, M. M. Perez Machirant, and I. Seree de Roch, *J. Am. Chem. Soc.* **100**, 5437 (1978).
99. H. Krekeler and H. Schmitz, *Chem.-Ing.-Tech.* **40**, 785 (1968); H. Krekeler and W. Kronig, *World Pet. Congr., Proc., 7th, 1967* Vol. 5, pp. 41–48 (1967).
100. S. A. Miller, *in* "Ethylene and its Industrial Derivatives" (S. A. Miller, ed.), p. 946. Benn, London, 1969.
101. P. M. Henry, *J. Org. Chem.* **32**, 2575 (1967).
102. C. F. Kohll and R. van Helden, *Recl. Trav. Chim. Pays-Bas* **84**, 1263 (1965).
103. R. Hüttel and M. Bechter, *Angew. Chem.* **71**, 456 (1959); R. Hüttel, J. Kratzer, and M. Bechter, *Chem. Ber.* **94**, 766 (1961).

104. R. S. Shue, *J. Catal.* **26**, 112 (1972).
105. I. Moritani and J. Fujiwara, *Synthesis* p. 524 (1973); M. Watanabe, M. Yamamura, I. Moritani, Y. Fujiwara, and A. Sonoda, *Bul. Chem. Soc. Japan* **47**, 1035 (1974), and references cited therein.
106. D. M. Fenton and K. L. Olivier, *Chemtech* **2**, 220 (1972); D. M. Fenton, K. L. Olivier, and G. Biale, *Prepr., Div. Pet. Chem., Am. Chem. Soc.* **14**(4), C77 (1969).
107. D. L. Trimm and L. A. Doerr, *J. Catal.* **23**, 49 (1971).
108. T. Sakamoto, M. Egashira, and T. Seiyama, *J. Catal.* **16**, 407 (1970).
109. T. Seiyama, M. Egashira, T. Sakamoto, and I. Aso, *J. Catal.* **24**, 76 (1972).
110. H. E. Swift, J. E. Buzik, and J. A. Ondrey, *J. Catal.* **21**, 212 (1971).
111. C. T. Handy and H. S. Rothrock, *J. Am. Chem. Soc.* **80**, 5036 (1958).
112. F. Mares and R. Tang, *J. Org. Chem.* **43**, 4361 (1978).
113. K. Takehira, H. Mimoun, and I. Seree de Roch, *J. Catal.* **58**, 155 (1979), and references cited therein.
114. P. R. Stapp, *J. Org. Chem.* **44**, 3216 (1979).
115. H. M. Weitz and R. Fischer, Ger. Patent 2,819,592 (1979) to BASF.
116. S. Korcek, J. H. B. Chenier, J. A. Howard, and K. U. Ingold, *Can. J. Chem.* **50**, 2285 (1972).
116a. W. Pritzkow, R. Radeglia, and W. Schmidt-Renner, *Neftekhimiya* **19**, 885 (1979).
117. N. Rabjohn, *Org. React.* **5**, 331 (1949).
118. B. Chabaud and K. B. Sharpless, *J. Org. Chem.* **44**, 4202 (1979).
119. H. Gopal and A. J. Gordon, *Tetrahedron Lett.* p. 2941 (1971).
120. N. A. Khan and M. S. Newman, *J. Org. Chem.* **17**, 1063 (1952).
121. L. Bassignani, A. Brandt, V. Caciagli, and L. Re, *J. Org. Chem.* **43**, 4245 (1978).
122. N. A. Milas, U.S. Patent 2,347,358 (1944).
123. M. Schroder, A. J. Nielson, and W. P. Griffith, *J. Chem. Soc., Dalton* p. 1607 (1979).
124. G. Eglinton and W. McCrae, *Adv. Org. Chem.* **4**, 225 (1963).
125. A. S. Hay, *J. Org. Chem.* **25**, 1275 (1960); G. Eglinton and A. R. Galbraith, *J. Chem. Soc.* p. 889 (1959).
126. I. D. Campbell and G. Eglinton, *Org. Synth.* **45**, 39 (1965).
127. F. Sondheimer, R. Wolovsky, and Y. Gaoni, *J. Am. Chem. Soc.* **82**, 754, 755 (1960).
128. J. Tsuji, M. Takahashi, and T. Takahashi, *Tetrahedron Lett.* **21**, 849 (1980).

Chapter 10

Aromatic Hydrocarbons

Aromatic hydrocarbons constitute one of the major groups of basic building blocks of the petrochemical industry.[1] The most important source is the catalytic reforming of naphtha, which affords a mixture of benzene, toluene, ethylbenzene, and xylenes, together with C_9 and C_{10} aromatic hydrocarbons. Oxidation is probably the most widely used unit process in the conversion of aromatic hydrocarbons to such industrial chemicals as phenol, terephthalic acid, and acetone.[2]

I. AUTOXIDATION

Autoxidations of alkyl-substituted benzenes involve predominantly attack at the reactive benzylic C—H bonds. The primary products are benzylic hydroperoxides, formed via a classical free radical mechanism involving the chain propagation sequence in eqs 1 and 2,

$$\mathrm{Ar\overset{\diagup}{\underset{\diagdown}{C}}\cdot + O_2 \longrightarrow Ar\overset{|}{\underset{|}{C}}{-}O{-}O\cdot} \quad (1)$$

$$\mathrm{Ar\overset{|}{\underset{|}{C}}{-}O{-}O\cdot + Ar\overset{|}{\underset{|}{C}}{-}H \xrightarrow{k_p} Ar\overset{|}{\underset{|}{C}}{-}O_2H + Ar\overset{\diagup}{\underset{\diagdown}{C}}\cdot} \quad (2)$$

as well as the chain termination step in eq 3.

$$2\ \mathrm{Ar\overset{|}{\underset{|}{C}}{-}O{-}O\cdot} \xrightarrow{k_t} \text{nonradical products} \tag{3}$$

These reactions are generally less complicated than olefin autoxidations since the aromatic moiety, in contrast to the olefin double bond, is inert to alkylperoxy radicals. Indeed, aromatic compounds that contain no alkyl substituents do not react with dioxygen under the usual autoxidation conditions. The oxidizabilities $k_p/(2k_t)^{1/2}$ of alkylaromatic hydrocarbons decrease significantly in the order tertiary > secondary > primary benzylic C—H bonds.[3] This trend is due largely to the significant increase in rate of termination, in the order primary > secondary > tertiary alkylperoxy radicals (see Table I).

Alkylaromatic compounds without benzylic C—H bonds (e.g., *tert*-butylbenzene) are, of course, unreactive toward autoxidation. Those compounds containing tertiary benzylic C—H bonds (e.g., cumene) or secondary benzylic C—H bonds (e.g., tetralin and ethylbenzene) generally undergo smooth autoxidation at moderate temperatures to afford the corresponding hydroperoxides in high selectivities. The autoxidation of cumene to cumene hydroperoxide is the first step in the well-known commercial process for the coproduction of phenol and acetone.[4]

$$\mathrm{PhCH(CH_3)_2 + O_2 \longrightarrow PhC(CH_3)_2O_2H} \tag{4}$$

$$\mathrm{PhC(CH_3)_2O_2H \xrightarrow{[H^+]} PhOH + (CH_3)_2CO} \tag{5}$$

The autoxidation of cumene is carried out at 90°–110°C in the absence of a metal catalyst. The solution of cumene hydroperoxide in cumene is concentrated and then subjected to acid-catalyzed decomposition to afford phenol in greater than 90% yields, based on cumene.

TABLE I Absolute Rate Constants for Autoxidations of Alkylaromatic Hydrocarbons at 30°C[a]

Substrate	$k_p/(2k_t)^{1/2} \times 10^3$ ($M^{-1/2}$ sec$^{-1/2}$)	k_p (M^{-1} sec^{-1})	$2k_t \times 10^{-6}$ (M^{-1} sec^{-1})
Cumene	1.50	0.18	0.015
Tetralin	2.30	6.4	7.6
Indane	1.70	4.8	8.2
Ethylbenzene	0.21	1.3	40
n-Butylbenzene	0.081	0.56	50
p-Xylene	0.049	0.84	300
o-Xylene	0.033	0.42	154
m-Xylene	0.028	0.48	300
Toluene	0.014	0.24	300

[a] From Howard.[3]

Ethylbenzene hydroperoxide, prepared by autoxidation of ethylbenzene, has achieved commercial importance as the oxidizing agent in the process for the coproduction of propylene oxide and styrene from propylene and ethylbenzene (see Chapter 9). These autoxidations are generally carried out in the absence of a metal catalyst, since cobalt, manganese, and copper complexes increase the rate of autoxidation at the expense of selectivity.* [Transition metal complexes increase the rate of initiation, as described in Chapter 3, but the resulting increased steady-state concentration of alkylperoxy radicals leads to faster termination and shorter chain lengths, i.e., a lower selectivity. Chain termination also results from the reaction of the alkylperoxy radicals with the reduced form of the catalyst (see Chapter 3).] With primary alkylbenzenes such as toluene and xylenes, the rate of termination is so high that it is not possible to achieve a high selectivity to hydroperoxide at reasonable rates of reaction. Thus, substantial amounts of aldehydes and alcohols are formed as primary products by the chain termination of the benzylperoxy radicals.

$$2\,ArCH_2O_2\cdot \longrightarrow ArCHO + O_2 + ArCH_2OH \qquad (6)$$

The aldehydes in turn undergo rapid autoxidation to produce the corresponding carboxylic acids. The addition of transition metal catalysts is, however, required to provide a useful rate of autoxidation of these substrates to carboxylic acids (*vide infra*).

Alkylaromatic hydrocarbons containing acidic C—H bonds (e.g., *p*-nitrotoluene) undergo base-catalyzed autoxidations in which the initiation step proceeds by a reaction of the carbanion directly with dioxygen.[5]

$$R^- + O_2 \longrightarrow R\cdot + O_2^{\cdot-} \qquad (7)$$

These reactions usually require strong bases, such as potassium *tert*-butoxide in dipolar aprotic solvents, (e.g., DMSO), and selectivities are generally not high. *p*-Nitrotoluene, for example, affords large amounts of products resulting from dimerization of the intermediate *p*-nitrobenzyl radicals, in addition to *p*-nitrobenzoic acid. However, Yamashita and co-workers[6] have recently reported that significant improvements accrue from the use of phase transfer catalysis (PTC) in these systems. Thus, *p*-nitrotoluene and fluorene are oxidized to *p*-nitrobenzoic acid and fluorenone, respectively, in high yields in a two-phase, benzene–water system at 40°C in the presence of KOH and a catalytic amount of trioctylmethylammonium chloride or 18-crown-6.

$$O_2N\text{–}C_6H_4\text{–}CH_3 + \tfrac{3}{2}O_2 \xrightarrow[\text{aq. KOH}/C_6H_6]{[\text{PTC}]} O_2N\text{–}C_6H_4\text{–}CO_2H + H_2O \qquad (8)$$

* In general, for a particular substrate, the selectivity to hydroperoxide decreases with increasing rate of autoxidation. Practical considerations generally necessitate a compromise between rate and selectivity.

$$\text{fluorene} + O_2 \xrightarrow[\text{aq. KOH}/C_6H_6]{[\text{PTC}]} \text{fluorenone} + H_2O \tag{9}$$

This technique provides a synthetically useful alternative to conventional metal-catalyzed autoxidation (*vide infra*) for the preparation of certain compounds, especially since deactivated alkylaromatic hydrocarbons such as *p*-nitrotoluene are extremely difficult to autoxidize to the corresponding carboxylic acids even in the presence of high concentrations of cobalt catalysts.

II. METAL-CATALYZED AUTOXIDATIONS OF ALKYLBENZENES TO CARBOXYLIC ACIDS

$$ArCH_2R \Longrightarrow ArCOR \Longrightarrow ArCO_2H$$

The traditional reagents for the conversion of methylbenzenes to the corresponding benzoic acids on the laboratory scale are potassium permanganate or chromyl compounds used in stoichiometric amounts. The industrial-scale oxidations of this type were generally carried out with stoichiometric amounts of nitric acid until the last decade. Even in the presence of small amounts of transition metal catalysts, such as cobalt salts, the autoxidation of methylbenzenes is an inefficient process. For example, the cobalt-catalyzed autoxidation of *p*-*tert*-butyltoluene to *p*-*tert*-butylbenzoic acid gave a 69% conversion and 67% selectivity in 4 hr at 168°C.[7] The oxidation of xylenes to the corresponding dicarboxylic acids, e.g., *p*-xylene to terephthalic acid, is even more difficult. The methyl group in the intermediate, *p*-toluic acid, is deactivated by the electron-withdrawing carboxyl group and is particularly resistant to further autoxidation.

$$p\text{-}CH_3C_6H_4CH_3 \xrightarrow{O_2} p\text{-}CH_3C_6H_4CO_2H \xrightarrow{O_2} p\text{-}HO_2CC_6H_4CO_2H \tag{10}$$

The presence of bromide has the significant effect of promoting the cobalt-catalyzed autoxidation of alkylbenzenes.[8] (Bromine atoms, rather than alkylperoxy radicals, are the principal chain carriers in the autoxidation mechanism, as described in Chapter 3.) These autoxidations are carried out in the liquid phase in acetic acid as solvent, typically at 195°–205°C and 30 bar using low concentrations (~0.001 *M*) of $Co(OAc)_2$ and NH_4Br. When this procedure is used, a variety of mono-, di-, and trimethylbenzenes can be oxidized to the corresponding mono-, di-, and tricarboxylic acids in good yields. The synergistic effect of bromide on the cobalt-catalyzed autoxidation of *p*-xylene is the basis of a commercial process for the manufacture of terephthalic acid.[9,10] Similarly, *o*-xylene and *m*-xylene afford phthalic and isophthalic acids. Trimethylbenzenes such as pseudocumene, mesitylene, and hemimellitine produce the corresponding tricarboxylic acids.

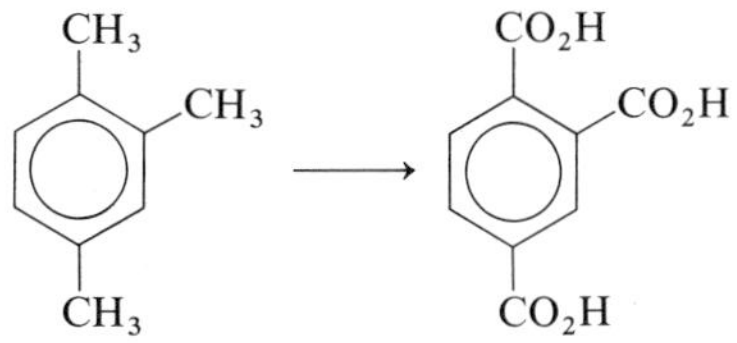

$$1,2,4\text{-}C_6H_3(CH_3)_3 \longrightarrow 1,2,4\text{-}C_6H_3(CO_2H)_3 \tag{11}$$

$$1,3,5\text{-}C_6H_3(CH_3)_3 \longrightarrow 1,3,5\text{-}C_6H_3(CO_2H)_3 \tag{12}$$

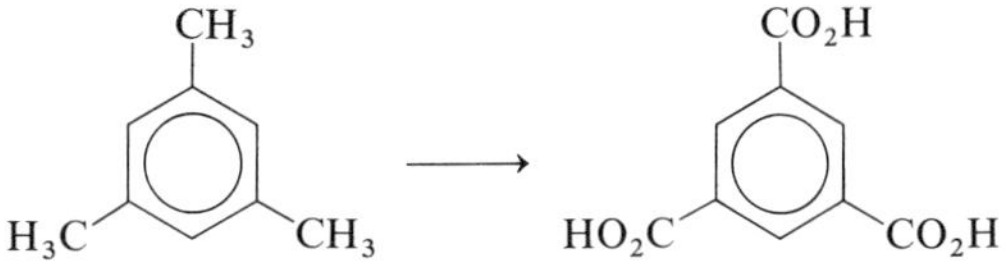

$$1,2,3\text{-}C_6H_3(CH_3)_3 \longrightarrow 1,2,3\text{-}C_6H_3(CO_2H)_3 \tag{13}$$

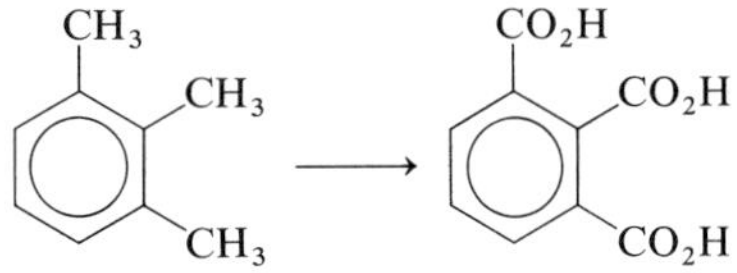

It was subsequently found that *p*-xylene could be smoothly oxidized to terephthalic acid in acetic acid when high concentrations of cobalt(II) acetate were used in combination with promoters such as bromide,[11] methyl ethyl ketone,[12,13] or ozone[14] at 110°C or less and 1 bar pressure. With sufficiently high cobalt concentrations (e.g., 0.4–0.5 mol per mole of xylene), smooth oxidation occurs even in the absence of a promoter.[15] Furthermore, the *cooxidation* of *p*-xylene and acetaldehyde (which is employed in stoichiometric amounts as a cosubstrate rather than as a promoter) can be efficiently carried out in the presence of high cobalt concentrations.[16] These reactions involve the direct reaction of the alkylaromatic substrate with cobalt(III) to produce the corresponding aldehyde by the following steps (see Chapter 3).

Scheme I:

$$ArCH_3 + Co^{III} \longrightarrow [ArCH_3]^{+\cdot} + Co^{II} \tag{14}$$

$$[ArCH_3]^{+\cdot} \longrightarrow ArCH_2\cdot + H^+ \tag{15}$$

$$ArCH_2\cdot + O_2 \longrightarrow ArCH_2O_2\cdot \tag{16}$$

$$ArCH_2O_2\cdot + Co^{II} \longrightarrow ArCHO + Co^{III}OH \tag{17}$$

$$ArCHO + \tfrac{1}{2}O_2 \longrightarrow ArCO_2H \tag{18}$$

TABLE II $Co(OAc)_2$-Catalyzed Oxidation of *p*-Xylene to Terephthalic Acid in Acetic Acid Solvent

$Co(OAc)_2$/xylene (molar ratio)	Promoter	Temp (°C)	TPA yield[a] (mol %)	Ref.
<0.001	Bromide	195–205	79	9
0.1	Methyl ethyl ketone	90–130	95	12, 13
0.4–0.5	None	<150	95	15
0.1	$ZrO(OAc)_2$	100	90	19
0.2	Bromide	90–110	72	14

[a] TPA, terephthalic acid. Yield differences in part reflect varying degrees of optimalization.

The function of the promoter is to mediate the smooth oxidation of the Co(II) catalyst to Co(III). In the absence of promoters, long induction periods can be observed during which Co(III) is formed slowly. Typical conditions of the most important terephthalic acid processes are summarized in Table II.

Synergistic effects have also been observed with mixed metal catalysts. For example, replacing 20% of the cobalt(II) by manganese(II) in the $Co(OAc)_2$/NaBr system resulted in a fivefold increase in rate.[17,18] More recently, it was found that the addition of zirconyl acetate led to a significant reduction in the induction period and an increase in the rate of oxidation of *p*-xylene and other alkylaromatic hydrocarbons catalyzed by $Co(OAc)_2$ in acetic acid at 100°C.[19]

The optimum molar ratio of $Co(OAc)_2$ to NaBr is 1:1, since the active catalyst is a bromocobalt acetate.[11]

$$Co(OAc)_2 + NaBr \longrightarrow Co(OAc)Br + NaOAc \tag{19}$$

When $CoBr_2$ is used as a catalyst, long induction periods are observed which are eliminated almost completely by the addition of an equimolar amount of a metal (Na, K, Ba, Zn, Co, Mn) acetate.[20] In the oxidation of mesitylene to mesitoic acid, the activity of $CoBr_2$[21] and $Co(OAc)_2$/NaBr[22] catalysts is increased significantly by the presence of amines such as *N,N*-diethylaniline and triethanolamine. The origin of the promoting effect of additives such as $ZrO(OAc)_2$ and amines is not clearly defined, but it is probably related to changes in monomer–dimer equilibria of the cobalt(II) complexes (see Chapter 3). The $Co(OAc)_2$/NaBr and $Co(OAc)_2$/MEK catalysts have broad synthetic utility for both the laboratory- and the industrial-scale oxidations of methylbenzenes to the corresponding carboxylic acids. The oxidations proceed under relatively mild conditions and in high yields with a wide variety of substrates (Table III). The carboxylic acid can in most cases be readily

TABLE III Cobalt-Catalyzed Oxidations of Methylbenzenes in Acetic Acid at 90°C

Catalyst	Substrate	Relative rate	Product (acid)	Yield (%)
$Co(OAc)_2$/MEK[a]	*p*-Xylene	1.00	Terephthalic	95
	m-Xylene	0.69	Isophthalic	90
	Toluene	1.00	Benzoic	89
	Chloro-*p*-xylene	0.65	Chloroterephthalic	75
	o-Xylene	0.22	*o*-Toluic	76
	p-Toluic acid	0.25	Terephthalic	92
Co(OAc)Br[b]	Toluene		Benzoic	86
	p-Chlorotoluene		*p*-Chlorobenzoic	88
	o-Chlorotoluene		*o*-Chlorobenzoic	86
	p-Xylene		Terephthalic	72
	m-Xylene		Isophthalic	67
	o-Bromotoluene		*o*-Bromobenzoic	91
	p-Ethyltoluene		4-Acetylbenzoic	78
	$H_3C-C_6H_4-O-C_6H_4-CH_3$		$HO_2C-C_6H_4-O-C_6H_4-CO_2H$	87
	$H_3C-C_6H_4-C(=O)-C_6H_4-CH_3$		$HO_2C-C_6H_4-C(=O)-C_6H_4-CO_2H$	76

[a] From Brill.[12]

[b] From Hay and Blanchard.[11]

crystallized out of the reaction mixture by cooling and the mother liquors recycled after makeup with fresh substrate.

The relative reactivities of primary alkyl (e.g., ethyl) and secondary alkyl (e.g., isopropyl) side chains is dependent on which catalyst system is employed. With a $Co(OAc)_2/NaBr$ catalyst, the oxidation involves chain transfer by bromine atoms, and cumene is significantly more reactive than toluene.[23] On the other hand, with either $Co(OAc)_2/MEK$ or $Co(OAc)_2/n$-butane (*vide infra*), the rate-limiting step is electron transfer oxidation of the substrate by Co(III), and toluene is significantly more reactive than cumene,[24,25] as shown in Table IV. Schulz and co-workers[25] utilized this rate difference for the selective autoxidation of *p*-cymene to *p*-isopropylbenzoic acid in the presence of a $Co(OAc)_2/n$-butane catalyst (since *n*-butane is a precursor for methyl ethyl ketone, it also functions as a promoter).

$$p\text{-}CH_3C_6H_4CH(CH_3)_2 + O_2 \xrightarrow[\text{HOAc, 100°C}]{[Co(OAc)_2/C_4H_{10}]} \underset{90\%}{p\text{-}HO_2CC_6H_4CH(CH_3)_2} + \underset{10\%}{p\text{-}HO_2CC_6H_4COCH_3} \quad (20)$$

Similarly, *p*-ethyltoluene afforded mainly *p*-ethylbenzoic acid (68%). In contrast, the autoxidation of *p*-cymene catalyzed by manganese (III) acetate afforded a mixture of *p*-toluic acid and *p*-methylacetophenone (the formation of which suggests a classical free radical pathway).[26] In the oxidation of 1,1-di-*p*-tolylethane with $Co(OAc)_2$/MEK(*n*-butane) in acetic acid at 105°C and 20 bar, the bridging ethylidene group survived intact to produce the dicarboxylic acid in eq 21 in 70% yield.[27]

TABLE IV Relative Reactivities of Alkylbenzenes toward Co(III) Oxidation

	Relative reactivity (per active hydrogen)		
Substrate	$Co(OAc)_2/NaBr$,[a] 60°C	$Co^{III}(OAc)_3$,[b] 65°C	$Co(OAc)_2/MEK(C_4H_{10})$[c] 105°C
Toluene	1.00	1.00	1.00
Ethylbenzene	8.3	1.3	1.8
Cumene	16.8	0.3	0.3

[a] Kamiya.[23]
[b] Heiba *et al.*[24]
[c] Onopchenko *et al.*[25]

$$H_3C-C_6H_4-CH(CH_3)-C_6H_4-CH_3 \xrightarrow[\text{[Co]}]{O_2} HO_2C-C_6H_4-CH(CH_3)-C_6H_4-CO_2H \quad (21)$$

The combination of $Co(OAc)_2$ and $Mn(OAc)_2$ is an excellent catalyst for the autoxidation of isopropylbenzenes to the corresponding carboxylic acids in acetic acid at moderate temperatures.[28] The reaction involves the initial formation of the corresponding acetophenone:

$$ArCH(CH_3)_2 \longrightarrow ArCOCH_3 \longrightarrow ArCO_2H$$

The Mn cocatalyst is probably important in the second step, since it is known that $Mn(OAc)_2$ is a better catalyst than $Co(OAc)_2$ for the autoxidation of acetophenones to the corresponding carboxylic acids.[29] The $Co(OAc)_2$/$Mn(OAc)_2$ catalyst can be used in an alternative process for terephthalic acid based on benzene and propylene by involving the autoxidation of diisopropylbenzene as a key feature.[28]

$$p\text{-}C_6H_4(CH(CH_3)_2)_2 + O_2 \xrightarrow[\substack{HOAc \\ 120°\text{–}150°C}]{[Co^{II}/Mn^{II}]} p\text{-}C_6H_4(CO_2H)_2 \quad (22)$$

These oxidations are not limited to simple alkylbenzenes. A variety of alkylpyridines can be oxidized selectively to the corresponding carboxylic acids using high concentrations of $Co(OAc)_3$ and dioxygen in acetic acid at 60°C,[30] e.g.,

$$3\text{-}CH_3C_5H_4N \longrightarrow 3\text{-}HO_2CC_5H_4N \quad (23)$$

$$4\text{-}CH_3C_5H_4N \longrightarrow 4\text{-}HO_2CC_5H_4N \quad (24)$$

An interesting recent development in the area of metal-catalyzed autoxidations of alkylbenzenes is the use of cobalt complexes that have been anchored on insoluble polymeric supports.[31] Thus, ethylbenzene was oxidized with dioxygen in the liquid phase at 70°–135°C in the absence of solvent with a catalyst consisting of cobalt(II) attached to a copolymer of diethylvinyl phosphonate and acrylic acid cross-linked by methylenediacrylamide, as in the proposed structure below.

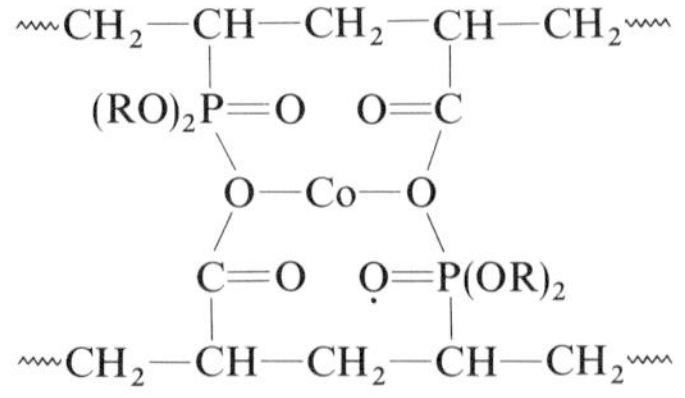

$$\begin{array}{c} \sim CH_2-CH-CH_2-CH-CH_2\sim \\ (RO)_2P{=}O \quad O{=}C \\ O-Co-O \\ C{=}O \quad O{=}P(OR)_2 \\ \sim CH_2-CH-CH_2-CH-CH_2\sim \end{array}$$

No induction period was observed, and the catalyst could be recycled several times with no apparent loss of activity. Unfortunately, the products of the autoxidation were not reported. However, this system is worthy of further investigation, since it provides for facile removal and recycling of the metal catalyst. It would be synthetically useful if it could be used for the efficient oxidation of alkylaromatic hydrocarbons to carboxylic acids, a reaction that usually requires the recycling of large amounts of cobalt catalyst.

The best known process for the heterogeneous oxidation of alkylaromatic compounds in the vapor phase is the oxidation of *o*-xylene to phthalic anhydride.[32]

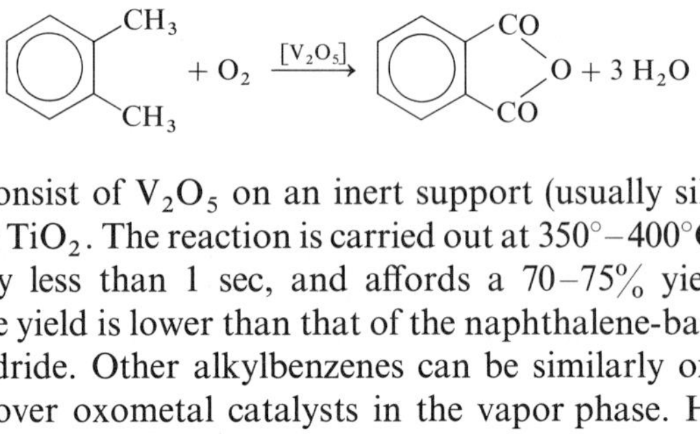

$$o\text{-}C_6H_4(CH_3)_2 + O_2 \xrightarrow{[V_2O_5]} C_6H_4(CO)_2O + 3\,H_2O \tag{25}$$

The catalyst consist of V_2O_5 on an inert support (usually silica) promoted with K_2SO_4 or TiO_2. The reaction is carried out at 350°–400°C, with contact times generally less than 1 sec, and affords a 70–75% yield of phthalic anhydride. The yield is lower than that of the naphthalene-based process for phthalic anhydride. Other alkylbenzenes can be similarly oxidized to carboxylic acids over oxometal catalysts in the vapor phase. However, these oxidations are generally much less selective than the liquid-phase processes discussed in the preceding section.

Methylbenzenes undergo ammoxidation to aromatic nitriles by reaction with a mixture of dioxygen and ammonia over oxometal catalysts at elevated temperatures. (This is analogous to the ammoxidation of propylene to acrylonitrile presented in Chapter 9.) For example, Hadley[33] reported the selective ammoxidation of *p*-xylene to terephthalonitrile with a V_2O_5 catalyst on an alumina support at 350°–400°C with yields in excess of 80%.

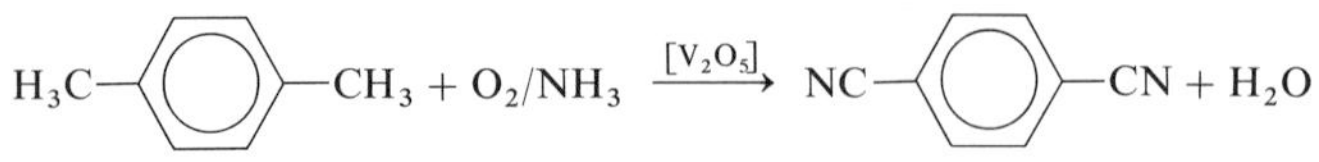

$$H_3C-C_6H_4-CH_3 + O_2/NH_3 \xrightarrow{[V_2O_5]} NC-C_6H_4-CN + H_2O \tag{26}$$

Methylpyridines were similarly converted to cyanopyridines,[34] e.g.,

$$\text{3-methylpyridine} + O_2/NH_3 \xrightarrow{[V_2O_5/Al_2O_3]} \text{3-cyanopyridine} \quad (82\%) \qquad (27)$$

The quantities of ammonia required in many of these reactions are considerably in excess of stoichiometric amounts owing to competition from the decomposition to dinitrogen. O'Donnell and co-workers[35] reported that high selectivities to aromatic nitriles (based on both the hydrocarbon and ammonia) are obtained over various oxometal (V, Cr, Mo, W) catalysts supported on Al_2O_3 when substantial amounts of water are added to the NH_3/O_2 feed.

A recently developed process for terephthalic acid manufacture is based on the two-stage conversion of *p*-xylene to terephthalonitrile, followed by hydrolysis.[36]

$$p\text{-}CH_3C_6H_4CH_3 \xrightarrow[2.\ O_2]{1.\ M_{ox},\ NH_3} p\text{-}NCC_6H_4CN \xrightarrow{H_2O} p\text{-}HO_2CC_6H_4CO_2H + NH_3 \qquad (28)$$

In the first stage, the *p*-xylene is treated with NH_3 over an oxometal catalyst to give terephthalonitrile and the reduced form of the catalyst.

$$ArCH_3 + NH_3 + 3\ O{=}M^{n+} \longrightarrow ArCN + 3\ H_2O + 3\ M^{(n-2)+} \qquad (29)$$

Thus, the oxometal reagent is consumed in stoichiometric amounts. In the second stage, the oxometal oxidant is regenerated by passing air at elevated temperatures over the reduced form. This procedure is referred[36] to as *oxidative ammonolysis* in order to distinguish it from direct ammoxidation with mixtures of dioxygen and ammonia. It should be emphasized, however, that the conversion of the substrate is the same in both cases (see Chapter 6 for a discussion of the mechanism). Toluene and *p*-xylene afford benzonitrile and terephthalonitrile, respectively, in a 90% yield. Similarly, other xylenes and alkylpyridines afford the corresponding dinitriles and cyanopyridines,[36] respectively. Interestingly, the concept of oxidative ammonolysis (i.e., the two-stage reaction) was the underlying theme in the initial studies of the conversion of propylene to acrylonitrile over oxometal catalysts.[37]

III. METAL-CATALYZED AUTOXIDATIONS OF ALKYLBENZENES TO ALDEHYDES AND KETONES

During the autoxidation of methylbenzenes, the initially formed benzaldehydes, being more reactive than the substrates, undergo further rapid oxidation to the corresponding carboxylic acid, i.e., $k_p \gg k'_p$ in the competition below.

$$RO_2\cdot \begin{cases} \xrightarrow[k_p]{ArCHO} Ar\dot{C}O + RO_2H \\ \xrightarrow[k'_p]{ArCH_3} ArCH_2\cdot + RO_2H \end{cases}$$

By contrast, in the presence of *high concentrations* of cobalt catalysts, the rate-limiting step is electron transfer oxidation of the substrate by Co(III). In this case, the intermediate benzaldehyde is less reactive than the substrate because of the electron-withdrawing effect of the carbonyl group, which increases the ionization potential of the molecule, i.e., $k_e > k'_e$:

$$XCo^{III} \begin{cases} \xrightarrow[k_e]{ArCH_3} [ArCH_3]^{+\cdot} + Co^{II} + X^- \\ \xrightarrow[k'_e]{ArCHO} [ArCHO]^{+\cdot} + Co^{II} + X^- \end{cases}$$

A consequence of the above dichotomy is that the selective oxidation of methylbenzenes to the corresponding benzaldehydes is feasible in the presence of high concentrations of cobalt catalysts under mild conditions. [We note, however, that this does not strictly apply to the $Co(OAc)_2/NaBr$ catalyst, in which the selectivity may be determined by the relative rates of hydrogen abstraction by bromine atoms.] Imamura and co-workers[38] have reported the selective oxidation of a variety of alkoxy- and aryloxytoluenes to the corresponding benzaldehydes using high concentrations (0.1–0.5 mol per mole substrate) of $Co(OAc)_2$ as catalyst in acetic acid at moderate temperatures (90°–120°C). Selectivities ranged typically between 50 and 75% at 40–80% conversion. When this procedure is used, *p*-methoxytoluene and *m*-phenoxytoluene can be efficiently oxidized to *p*-anisaldehyde and *m*-phenoxybenzaldehyde, respectively.[75]

$$p\text{-}MeO{-}C_6H_4{-}CH_3 + O_2 \xrightarrow[HOAc]{[Co(OAc)_2]} p\text{-}MeO{-}C_6H_4{-}CHO$$

$$m\text{-}PhO{-}C_6H_4{-}CH_3 + O_2 \xrightarrow[HOAc]{[Co(OAc)_2]} m\text{-}PhO{-}C_6H_4{-}CHO \qquad (30)$$

In the presence of strong acids (such as H_2SO_4 and CF_3CO_2H), the Co(III) and Mn(III) acetates oxidize alkylbenzenes at room temperature in acetic acid.[39] [The reaction proceeds via electron transfer with both Co(III) and Mn(III), and the strong acid considerably enhances the electrophilicity of the oxidant, as described in Chapter 3.] Toluene, for example, is oxidized by a stoichiometric amount of $Mn(OAc)_3/H_2SO_4$ in acetic acid in an oxygen atmosphere to give benzaldehyde in 71% yield, together with 24% benzyl acetate and 5% benzoic acid:

Scheme II:

$$PhCH_3 + AcOMn^{III} \longrightarrow [PhCH_3]^{+\cdot} + AcOMn^{II} \quad (31)$$

$$[PhCH_3]^{+\cdot} \longrightarrow PhCH_2\cdot + H^+ \quad (32)$$

$$PhCH_2\cdot + O_2 \longrightarrow PhCH_2O_2\cdot \quad (33)$$

$$PhCH_2O_2\cdot + Mn^{II} \longrightarrow PhCHO + HOMn^{III} \quad (34)$$

$$PhCH_2\cdot + AcOMn^{III} \longrightarrow PhCH_2OAc + Mn^{II} \quad (35)$$

Benzaldehyde is formed in a chain process by the sequence of reactions in eqs 31–34. The formation of benzyl acetate in reaction 35 results in chain termination, since the Mn(III) oxidant is not regenerated. Primary alkylbenzenes under the same conditions afford a mixture of ketones and benzylic acetates. Similarly, the autoxidation of ethylbenzene catalyzed by high concentrations of $Co(OAc)_3$ in acetic acid at 60°C afforded acetophenone as the major product, together with smaller amounts of α-phenylethyl acetate and the corresponding alcohol.[40] Autoxidation of primary alkylbenzenes in the presence of Co(OAc)Br under mild conditions also afforded ketones as the major product. The oxidation of tetralin, for example, at 40°C in acetic acid afforded α-tetralone.[11] Finally, there is a two-stage process for the production of benzaldehydes from methylbenzenes by oxidation with Mn(III) sulfate in sulfuric acid at 30°C and subsequent electrolytic regeneration of the Mn(III) oxidant.[41] *p*-Methoxytolune can be oxidized to *p*-anisaldehyde in very high selectivity using this process.

Under basic conditions, the *p*-cresols listed in Table V are readily oxidized with dioxygen to the corresponding *p*-hydroxybenzaldehydes in the presence of catalytic amounts of cobaltous chloride.[73]

$$HO\text{-}C_6HR^1R^2R^3\text{-}CH_3 + O_2 \xrightarrow[NaOH/MeOH]{[CoCl_2]} HO\text{-}C_6HR^1R^2R^3\text{-}CHO + H_2O$$

TABLE V Oxidation of *p*-Cresols to *p*-Hydroxybenzaldehydes[a]

R^1	R^2	R^3	Time (L) (hr)	Conversion (%)	Selectivity (%)
H	H	H	6	92	78
Cl	H	H	6	70	63
Br	H	H	6	65	66
Me	H	H	6	100	66
H	H	Me	6	100	62
Me	Me	H	4	100	59
t-Bu	H	H	6	100	58
t-Bu	t-Bu	H	0.5	100	52
EtO	H	H	6	100	61
MeO	MeO	H	2	100	55
Cl	Cl	H	6	60	70
Br	Br	H	6	58	68

[a] At 60°C and 1 bar O_2 pressure, typically with 50 mmol cresol, 0.5 mmol $CoCl_2$, 150 mmol NaOH in 16 ml MeOH[from Nishizawa *et al*].[73]

IV. METAL-CATALYZED OXIDATIONS OF ALKYLBENZENES TO BENZYLIC ACETATES

The oxidation of methylbenzenes to benzylic acetates can be effected with a mixture of $Mn(OAc)_3$ and KBr in acetic acid at 70°C.[42] The stoichiometric reaction involves a rate-limiting hydrogen abstraction by bromine atoms:

Scheme III:

$$Mn^{III} + Br^- \longrightarrow Mn^{II} + Br\cdot \qquad (36)$$

$$Br\cdot + ArCH_3 \longrightarrow ArCH_2\cdot + HBr \qquad (37)$$

$$ArCH_2\cdot + AcOMn^{III} \longrightarrow ArCH_2OAc + Mn^{II} \qquad (38)$$

When the autoxidation of alkylbenzenes catalyzed by Co(OAc)Br is carried out in the presence of NaOAc under anhydrous conditions, benzylic acetates are formed as the major products in more than stoichiometric amounts.[11] The *catalytic* oxidation of methylbenzenes to benzyl acetates has also been achieved using a $Pd(OAc)_2/Sn(OAc)_2$ catalyst in acetic acid at 100°C.[43]

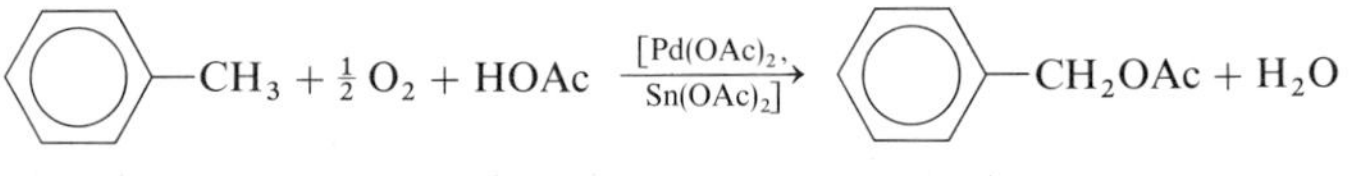

$$C_6H_5\text{—}CH_3 + \tfrac{1}{2}\,O_2 + HOAc \xrightarrow[Sn(OAc)_2]{[Pd(OAc)_2,} C_6H_5\text{—}CH_2OAc + H_2O \qquad (39)$$

The addition of a material with high surface area, such as charcoal, increased the rate substantially by dispersing the Pd(0) and facilitating its reoxidation.

Small amounts of benzaldehydes were also formed, presumably via hydrolysis and subsequent oxidation. The mechanism of this oxidation is fundamentally different from the corresponding oxidations with Co(III) and Mn(III) (described in Chapter 7). Thus polymethylbenzenes such as mesitylene and hexamethylbenzene were significantly less reactive than toluene.

Methylbenzenes are also oxidized to the corresponding benzyl acetates using peroxydisulfate as the stoichiometric oxidant in the presence of cupric acetate as the catalyst with added sodium acetate.[44] For example, *p*-methoxytoluene was converted to *p*-methoxybenzyl acetate in 77% yield in acetic acid. The following reaction pathway was suggested.

$$RC_6H_4CH_3 \xrightarrow[-SO_4^{2-}]{SO_4^{\cdot -}} [RC_6H_4CH_3]^{+\cdot} \xrightarrow{-H^+} RC_6H_4\dot{C}H_2$$

$$RC_6H_4\dot{C}H_2 \xrightarrow{Cu(OAc)_2} RC_6H_4CH_2OAc + CuOAc$$

Oxidation of toluene in acetic anhydride at 200°C affords benzylidene diacetate in yields up to 60% within 30 minutes.[74]

V. OXIDATIVE NUCLEAR SUBSTITUTION

$$ArH \Longrightarrow ArOH,\ ArO_2CR, \quad \text{etc.}$$

A simple chemical procedure for the selective hydroxylation of aromatic nuclei, to give the corresponding phenols, is highly desirable. Such a transformation may mimic the action of numerous monooxygenases that mediate the hydroxylation of aromatic substrates by a process that is tantamount to a cooxidation:

$$ArH + O_2 + SH_2 \xrightarrow{[\text{enzyme}]} ArOH + S + H_2O$$

There are basically two problems associated with the *in vitro* hydroxylation of aromatic nuclei. The first is related to the low reactivity of the aromatic nucleus toward alkylperoxy radicals. In contrast to olefins, which can react by both addition and abstraction mechanisms, alkylaromatic compounds react with alkylperoxy radicals only by hydrogen abstraction at the side chain. The second problem is that the phenolic product is significantly more reactive toward oxidation than the aromatic substrate from which it was derived.

Aromatic substrates undergo nuclear hydroxylation with Fenton's reagent, consisting of Fe(II) and H_2O_2. The formation of phenol involves the addition of the hydroxyl radical to an aromatic double bond, followed by oxidation of the intermediate hydroxycyclohexadienyl radical:

Scheme IV:

$$R\text{-}C_6H_5 + HO\cdot \longrightarrow R\text{-}C_6H_5(\cdot)(OH)(H) \tag{40}$$

$$R\text{-}C_6H_5(\cdot)(OH)(H) + Fe^{III} \longrightarrow R\text{-}C_6H_4\text{-}OH + Fe^{II} + H^+ \tag{41}$$

$$Fe^{II} + H_2O_2 \longrightarrow HOFe^{III} + HO\cdot \tag{42}$$

This reaction has been extensively studied.[45,46] In contrast to alkylperoxy radicals, the hydroxyl radical reacts predominantly by addition, even with alkyl-substituted aromatic compounds.[46] The yield of phenols increases with increasing Fe(III) concentration, as expected from Scheme IV. Higher yields of phenols result when Cu(II) is added to the system,[46] since it is more efficient than Fe(III) in the oxidation of the hydroxycyclohexadienyl radical in reaction 41. Phenol yields are also markedly influenced by the pH of the medium owing to the rapid, reversible, and acid-catalyzed protonation of the cyclohexadienyl radicals to give the corresponding cation-radical.

$$R\text{-}C_6H_5(\cdot)(OH)(H) + H^+ \longrightarrow R\text{-}C_6H_5^{+\cdot} + H_2O \tag{43}$$

According to Walling,[46] the products of side-chain oxidation are derived largely from subsequent formation of benzylic radicals by proton loss from the cation-radical. Furthermore, the reduction of the cation-radical by Fe(II) regenerates the starting material.

$$R\text{-}C_6H_5^{+\cdot} + Fe^{II} \longrightarrow R\text{-}C_6H_5 + Fe^{III} \tag{44}$$

The side reactions in eqs 43 and 44 lead to the inefficient use of hydrogen peroxide. Under optimal conditions (i.e., low $[H^+]$ and high $[Cu^{II}]$), good yields of phenols, based on both substrate and H_2O_2, have been reported.[46]

Aromatic compounds also undergo nuclear hydroxylation with the related $Fe(II)/S_2O_8^{2-}$ (peroxydisulfate) system, in which the reactive species is the sulfate anion-radical $SO_4^{\overline{\cdot}}$ formed in eq 45.

$$Fe^{II} + S_2O_8^{2-} \longrightarrow Fe^{III} + SO_4^{2-} + SO_4^{\overline{\cdot}} \qquad (45)$$

The aromatic substrate is oxidized by $SO_4^{\overline{\cdot}}$ to the aromatic cation-radical.[47] Hydration of the latter affords the hydroxycyclohexadienyl radical, which is the microscopic reverse of reaction 43. As with Fenton's reagent, significantly increased yields of phenols result in the presence of high Cu(II) concentrations and low H^+ concentrations.[46,47] Hydroxylation with H_2O_2 and $S_2O_8^{2-}$ are thus interrelated, as illustrated in Scheme V with toluene as the substrate.

Scheme V:

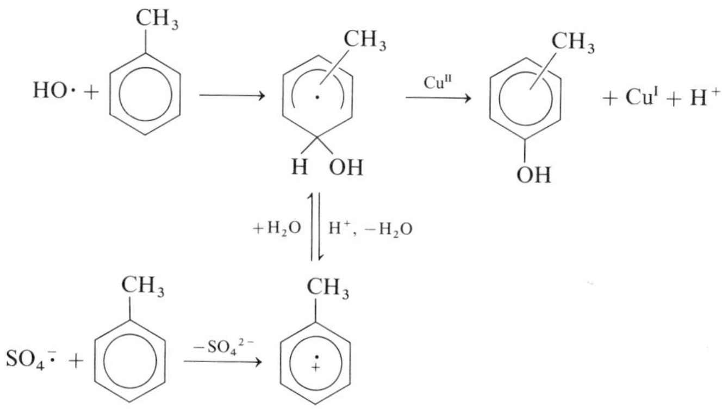

HO· + toluene ⟶ hydroxycyclohexadienyl radical (CH₃, H OH) $\xrightarrow{Cu^{II}}$ cresol (CH₃, OH) + Cu^I + H^+

$+H_2O$ ⇅ $H^+, -H_2O$

$SO_4^{\overline{\cdot}}$ + toluene $\xrightarrow{-SO_4^{2-}}$ toluene cation-radical

The distribution of the phenol isomers depends on the nature of the oxidant [Cu(II) or Fe(III)], on the pH, and on whether initial attack is by HO· or $SO_4^{\overline{\cdot}}$.[47] Iron(III) is a more selective electron transfer oxidant than copper(II) and gives rise to less of the meta isomer (i.e., thermodynamic control). At low acidities, the distribution of the phenol isomers is kinetically controlled in Fe(II)/H_2O_2 and determined by the point of hydroxyl radical attachment. At low acidities, kinetic control also prevails with Fe(II)/$S_2O_8^{2-}$, but hydration of the cation-radical is the determining factor. At high acidities, the hydroxycyclohexadienyl radicals can equilibrate with the cation-radical, and both systems afford the same thermodynamically controlled isomer distribution. Finally, it should be noted that hydroxylation of aromatic compounds with Fe(II)/H_2O_2 in aprotic media at low water concentrations may involve high-valent oxoiron complexes as the hydroxylating agents (see Chapters 6 and 8).

Udenfriend's reagent, consisting of a mixture of Fe(II), O_2, and ascorbic acid, as well as the related systems described in Chapter 8, also hydroxylates aromatic substrates. However, yields are generally poor to moderate, and the method does not have much synthetic utility.

Oxidative nuclear acetoxylation of aromatic substrates has been observed with peroxydisulfate in combination with either silver[48] or copper[49] salts in acetic acid as solvent. With silver, the reaction is believed to involve initially the formation of the cation-radical by electron transfer from the substrate to Ag(II). However, Cu(II) is not sufficiently powerful to oxidize the aromatic substrate, and the initial attack may involve $SO_4^{\cdot -}$. Naphthalene afforded acetoxynaphthalenes ($\alpha/\beta = 93/7$) in 90% selectivity, based on the converted naphthalene (39% on $S_2O_8^{2-}$).

$$\text{C}_{10}\text{H}_8 + S_2O_8^{2-} \xrightarrow[\text{HOAc}]{[\text{Cu}^{\text{II}}]} \text{C}_{10}\text{H}_7\text{OAc} + 2\,HSO_4^{-}$$

Interestingly, *p*-cymene afforded *p*-isopropylbenzyl acetate with the Cu(II)/$S_2O_8^{2-}$ reagent.

Nuclear acetoxylation occurs, in competition with oxidative coupling (see following section), in the reaction of aromatic substrates with palladium(II) acetate.[50–52] Nuclear acetoxylation is the preferred reaction,

$$\text{ArH} + \text{Pd(OAc)}_2 \longrightarrow \text{ArOAc} + \text{Pd} + \text{HOAc} \tag{46}$$

if excess of acetate ion is avoided and the reaction is performed under an oxygen atmosphere. (See Chapter 7 for a more detailed discussion of the conditions and the mechanism.) In the presence of oxygen, the reaction is catalytic in Pd(II), but the rates are too slow to be of practical value. The rate can be increased by the addition of cooxidants such as nitrate, dichromate, and persulfate.[52]

A process for the direct oxidation of benzene to phenol, using a combination of $PdCl_2$, $CuCl_2$, and dioxygen at 110°C and 3–6 bar, has been described.[53] Although a high selectivity (80%) was observed, the conversion was very low (2%).

Nuclear trifluoroacetoxylation of aromatic substrates can be effected with oxidants such as Co(III) and Pb(IV) in trifluoroacetic acid. The reactions involve cation-radicals as intermediates (see Chapter 4 for a detailed discussion). The aryl trifluoroacetate product is stable toward further oxidation owing to the strong deactivating effect of the electron-withdrawing trifluoroacetoxy group. The stoichiometric requirements of the reagents limit the synthetic utility of these reactions to small-scale procedures. However, they could possibly be employed in conjunction with the electrochemical regeneration of Co(III) or Pb(IV).

An alternative to the direct hydroxylation of benzene to phenol is the two-stage conversion of toluene to phenol via benzoic acid.[54–57] The facile oxidation of toluene to benzoic acid (*vide supra*) is followed by the second stage, in which the oxidation of molten benzoic acid is carried out with a mixture of air and steam at 230°–240°C in the presence of a cupric benzoate catalyst promoted with magnesium benzoate. The formation of phenol in about 90% yield proceeds according to the mechanism in Scheme VI.[54–60]

Scheme VI:

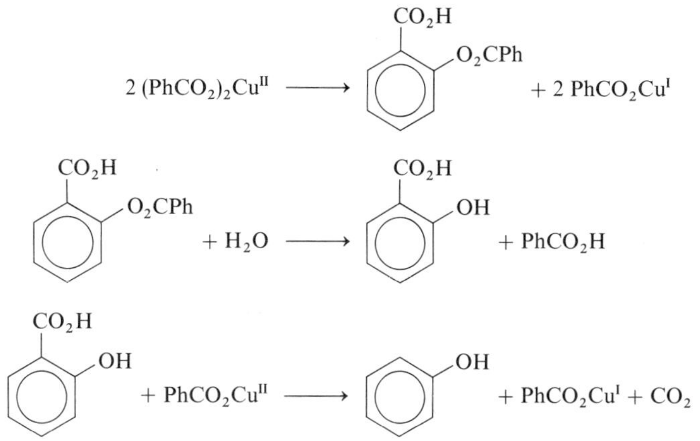

$$2\,(PhCO_2)_2Cu^{II} \longrightarrow \text{2-}(PhCO_2)C_6H_4CO_2H + 2\,PhCO_2Cu^{I} \qquad (47)$$

$$\text{2-}(PhCO_2)C_6H_4CO_2H + H_2O \longrightarrow \text{2-}(HO)C_6H_4CO_2H + PhCO_2H \qquad (48)$$

$$\text{2-}(HO)C_6H_4CO_2H + PhCO_2Cu^{II} \longrightarrow C_6H_5OH + PhCO_2Cu^{I} + CO_2 \qquad (49)$$

$$2\,PhCO_2Cu^{I} + 2\,PhCO_2H + \tfrac{1}{2}O_2 \longrightarrow 2\,(PhCO_2)_2Cu^{II} + H_2O \qquad (50)$$

In the absence of water, phenyl benzoate is formed by decarboxylation of the benzoylsalicylic acid intermediate.[55,58] Similarly, a variety of substituted benzoic acids afford the corresponding esters in which the ester group occupies the position ortho to the carboxyl group initially present. For example, *p*-toluic and *p*-*tert*-butylbenzoic acids afford the esters of *m*-cresol and *m*-*tert*-butylphenol, respectively,[58]

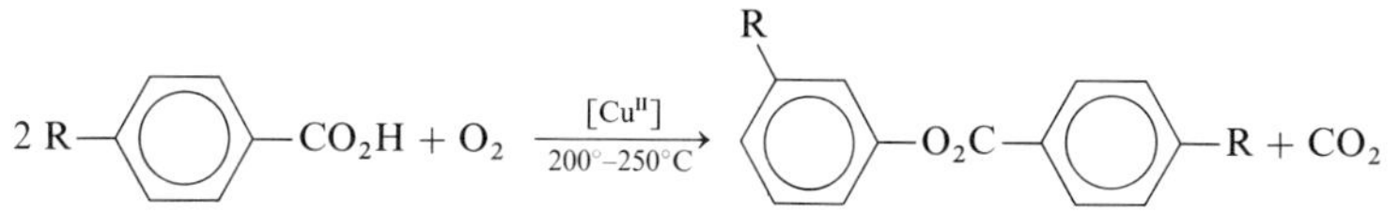

$$2\,R\text{-}C_6H_4\text{-}CO_2H + O_2 \xrightarrow[200^\circ\text{-}250^\circ C]{[Cu^{II}]} (m\text{-}R)C_6H_4\text{-}O_2C\text{-}C_6H_4\text{-}R + CO_2 \qquad (51)$$

where R = CH_3 and $(CH_3)_3C$. Hydrolysis affords the corresponding phenol in yields as high as 90%.[58] The reaction provides a means of preparing meta-substituted phenols, which may be otherwise difficult to obtain. The mechanism of the key oxidative rearrangement step in eq 47 has been discussed in some detail, and both heterolytic[58] and homolytic[55,60] mechanisms have been proposed. More recently an analogous vapor-phase process for the conversion of benzoic acid to phenol has been developed which gives selectivities approaching 90% at 50% conversion. A novel copper-containing catalyst is employed.[61]

VI. OXIDATIVE DIMERIZATION

$$2\,ArH \Longrightarrow Ar{-}Ar$$

The oxidative dimerization of aromatic hydrocarbons is catalyzed by Pd(II) salts.[62]

$$2\,ArH + \tfrac{1}{2}\,O_2 \xrightarrow{[Pd^{II}]} Ar{-}Ar + H_2O$$

Although the mechanism of the reaction has not been completely resolved, the first step probably involves electrophilic palladation[63–67] of the substrate since the reaction is promoted by strong Brønsted and Lewis acids, which increase the electrophilicity of the Pd(II).

$$ArH + X_2Pd^{II} \longrightarrow ArPd^{II}X + HX \qquad (52)$$

Both homolytic[62] and heterolytic mechanisms[63,67] have been suggested for subsequent conversion of the σ-arylpalladium(II) intermediate. For example, electrophilic addition to a second molecule of substrate may be followed by elimination of HPdX.[67]

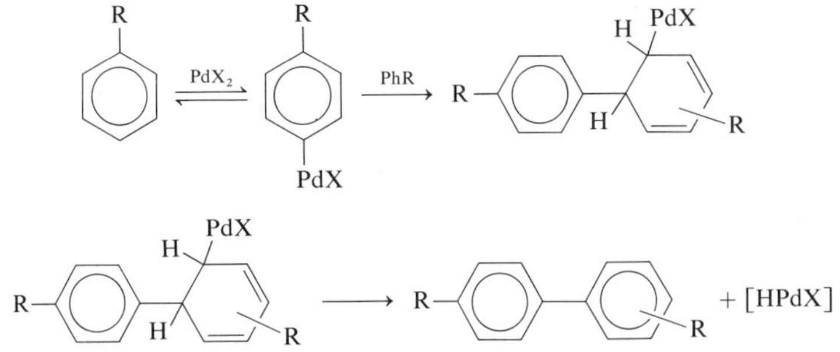

R PdX$_2$ R PhR R H PdX H R PdX

R H PdX H R R R + [HPdX]

Oxidative dimerization of electron-rich arenes to biaryls is also effected with thallium(III) trifluroacetate via the radical-cation intermediate formed by one-equivalent oxidation, as described in Chapter 5, Section I,B.[72] The procedure constitutes a convenient and useful method of synthesis for such isoquinoline alkaloids as ocoteine, in 46% yield,

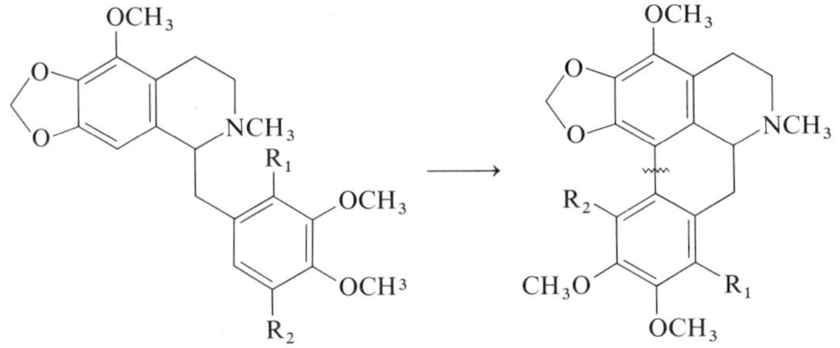

OCH$_3$ O O NCH$_3$ R$_1$ OCH$_3$ OCH$_3$ R$_2$ OCH$_3$ O O NCH$_3$ R$_2$ CH$_3$O R$_1$ OCH$_3$

and neolitsine, in 68% yield,

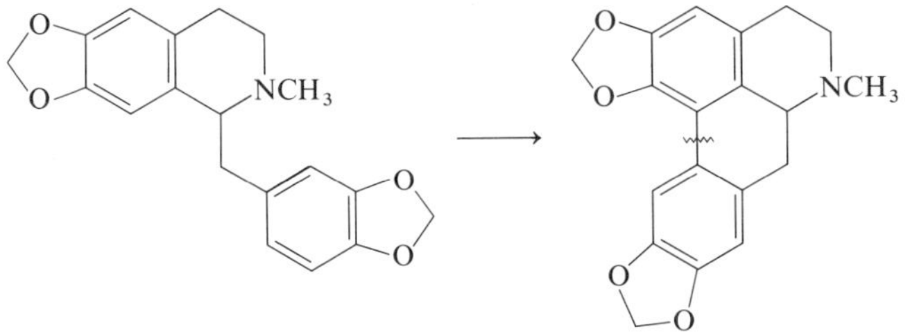

(The new biaryl bond is indicated by the wavy line.)

Toluene undergoes oxidative dimerization to stilbene over certain oxometal catalysts in the vapor phase (compare the conversion of propylene to 1,5-hexadiene described in Chapter 9). This forms the basis for converting toluene to styrene via oxidative dimerization and subsequent syn-proportionation of stilbene with ethylene.[68]

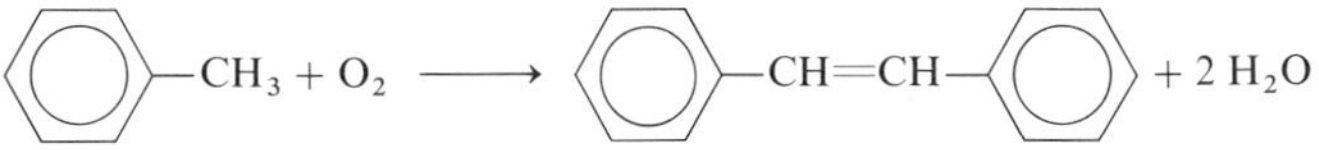

$$C_6H_5-CH_3 + O_2 \longrightarrow C_6H_5-CH{=}CH-C_6H_5 + 2\,H_2O \qquad (53)$$

$$C_6H_5-CH{=}CH-C_6H_5 + CH_2{=}CH_2 \longrightarrow 2\,C_6H_5-CH{=}CH_2 \qquad (54)$$

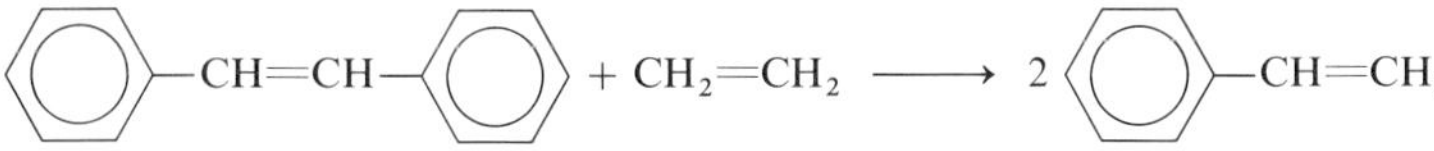

Reaction 53 presumably involves the formation of bibenzyl, by dimerization of benzyl radicals, and subsequent dehydrogenation.

VII. OXIDATIVE CLEAVAGE

The aromatic ring is a rather stable moiety which is generally cleaved only with powerful oxidants or under forcing conditions. (Certain dioxygenase enzymes mediate the cleavage of aromatic rings with molecular oxygen at ambient temperatures, as described in Chapter 8.) The vapor-phase oxidative cleavage of benzene and that of naphthalene over supported oxometal catalysts to give maleic anhydride and phthalic anhydride, respectively, are well-known commercial processes.[1,32] The production of maleic anhydride from benzene involves air oxidation over a MoO_3/P_2O_5 catalyst supported on alumina or silica at 425°–450°C. The yields obtained are in the range 67–72%.

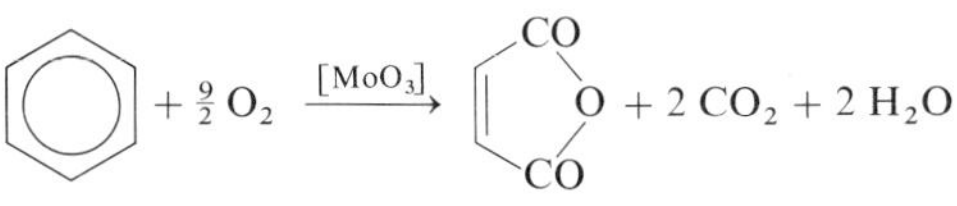

$$C_6H_6 + \tfrac{9}{2}\,O_2 \xrightarrow{[MoO_3]} \text{(CH–CO)}_2\text{O} + 2\,CO_2 + 2\,H_2O \qquad (55)$$

Catalysts employed for the naphthalene oxidation are generally based on V_2O_5 supported on silica. Yields of phthalic anhydride vary between 86 and 90% in the temperature range 360°–385°C.

$$C_{10}H_8 + \tfrac{9}{2}\,O_2 \xrightarrow{[V_2O_5]} C_6H_4(CO)_2O + 2\,CO_2 + 2\,H_2O \qquad (56)$$

Aromatic rings are oxidatively cleaved by RuO_4 at ambient temperatures (compare oxidative cleavage of olefins by RuO_4). The combination of NaOCl with a catalytic amount of RuO_2 has been used for the oxidative cleavage of naphthalene to phthalic acid[69] in 62–65% yields.

$$C_{10}H_8 \xrightarrow[CCl_4,\ H_2O]{[RuO_2],\ NaOCl} C_6H_4(CO_2H)_2 \qquad (57)$$

The oxidation of nitro- and hydroxy-substituted naphthalenes follows the expected pattern shown in eqs 58 and 59.

$$1\text{-}NO_2C_{10}H_7 \longrightarrow 3\text{-}NO_2C_6H_3(CO_2H)_2 \qquad (58)$$

$$1\text{-}HOC_{10}H_7 \longrightarrow C_6H_4(CO_2H)_2 \qquad (59)$$

Since alkanes are inert under these conditions, a synthetic method may be developed for the conversion of an aromatic ring to a carboxyl group by oxidation with sodium periodate and a catalytic amount of a ruthenium compound, e.g.,[70,71]

$$C_6H_{11}\text{–}C_6H_5 \xrightarrow[NaIO_4]{[RuO_4]} C_6H_{11}\text{–}CO_2H \qquad (60)$$

$$C_6H_5\text{–}CH_2NH_2 \xrightarrow[NaIO_4]{[RuCl_3]} HO_2CCH_2NH_2 \qquad (61)$$

Presumably, the periodate could be replaced by NaOCl in these reactions.

REFERENCES

1. M. Sittig, "Aromatic Hydrocarbons, Manufacture and Technology," Chem. Tech. Rev., No. 56. Noyes Data Corp., Park Ridge, New Jersey, 1976.
2. R. Landau and R. W. Simon, *Chem. Ind.* (*London*) p. 70 (1962).
3. J. A. Howard, *Adv. Free-Radical Chem.* **4**, 49 (1972).
4. W. T. Reichle, F. M. Konrad, and J. R. Brooks, *in* "Benzene and its Industrial Derivatives" (E. G. Hancock, ed.), Benn, London, 1975.
5. G. Sosnovsky, *in* "Organic Peroxides" (D. Swern, ed.), Vol. 3, p. 354. Wiley, New York, 1971.
6. J. Yamashita, S. Ishikawa, and H. Hashimoto, *ACS/CSJ Chem. Congr., 1979* Org. Chem. Div., Pap. No. 76 (1979).
7. R. M. Cole, A. W. Fairbairn, and K. D. Detling, *Chem. Eng. Sci.* **3**, Spec. Suppl., 67 (1954).
8. A Saffer and R. S. Barker, U.S. Patent 2,833,816 (1958) to Mid-Century Corp; for an interesting account of the historical development of the process, see R. Landau and A. Saffer, *Chem. Eng. Prog.* **64**(10), 20 (1968).
9. D. E. Burney, G. H. Weisemann, and N. Fragren, *Pet. Refiner* **38**(6), 186 (1959).
10. P. H. Towle and R. H. Baldwin, *Hydrocarbon Process* **43**(11), 149 (1964).
11. A. S. Hay and H. S. Blanchard, *Can. J. Chem.* **43**, 1306 (1965).
12. W. F. Brill, *Ind. Eng. Chem.* **52**, 837 (1960).
13. H. S. Bryant, C. A. Duval, L. E. McMakin, and J. I. Savoca, *Chem. Eng. Prog.*, **67**(9), 69 (1971).
14. A. S. Hay, J. W. Eustance, and H. S. Blanchard, *J. Org. Chem.* **25**, 616 (1960); see also G. A. Galstyan, V. A. Yakobi, M. M. Dvortsevoi, and T. M. Galstyan, *J. Appl. Chem. USSR* **51**, 123 (1978).
15. Y. Ichikawa, G. Yamashita, M. Tokashiki, and T. Yamaji, *Ind. Eng. Chem.* **62**(4), 38 (1970); Y. Ichikawa and Y. Takeuchi, *Hydrocarbon Process.* **51**(11), 103 (1972).
16. D. C. Hull, U.S. Patent 2,673,217 (1954) to Eastman Kodak.
17. D. A. S. Ravens, *Trans. Faraday Soc.* **55**, 1768 (1959).
18. Y. Kamiya, *Adv. Chem. Series* **76**, 192 (1968).
19. A. W. Chester, P. S. Landis, and E. J. Y. Scott, *Chemtech* **8**, 366 (1978).
20. F. F. Shcherbina and T. V. Lysukho, *Kinet. Katal. U.S.S.R.* **19**, 872 (1978).
21. M. Hronec and V. Vesely, *Col. Czech. Chem. Commun.* **42**, 1851 (1977).
22. M. Hronec and V. Vesely, *Col. Czech. Chem. Commun.* **40**, 2165 (1975).
23. Y. Kamiya, *J. Catal.* **33**, 480 (1974).
24. E. I. Heiba, R. M. Dessau, and W. J. Koehl, *J. Am. Chem. Soc.* **91**, 6830 (1969).
25. A. Onopchenko, J. G. D. Schulz, and R. Seekircher, *J. Org. Chem.* **37**, 1414 (1972); *J. Chem. Soc. Chem. Commun.* p. 939 (1971).
26. A. Onopchenko and J. G. D. Schulz, *J. Org. Chem.* **37**, 2564 (1972).
27. A. Onopchenko, J. G. D. Schulz, and R. E. Seekircher, *J. Org. Chem.* **37**, 2950 (1972).
28. J. P. Fortuin, M. J. Waale, and R. P. van Oosten, *Pet. Refiner* **38**(6), 189 (1959).
29. R. van Helden and E. C. Kooyman, *Recl. Trav. Chim. Pay-Bas* **80**, 57 (1961); H. J. den Hertog and E. C. Kooyman, *J. Catal.* **6**, 357 (1966).
30. J. D. V. Hanotier and M. G. S. Hanotier-Bridoux, Ger. Patent 2,242,386 (1974) to Labofina.
31. A. F. Efendiev, T. N. Shakhtaktinsky, L. F. Mustaeva, and H. L. Shick, *Ind. Eng. Chem. Prod. Res. Dev.* **19**(1), 75 (1980).
32. T. Dumas and W. Bulani, "Oxidation of Petrochemicals: Chemistry and Technology," pp. 53–64. Applied Science, London, 1974.
33. D. J. Hadley, U.S. Patent 2,846,462 (1958) to Distillers Co.; see also A. Farkas and R. Rosenthal, U.S. Patent 2,833,807 (1958) to Allied Chemical.

34. D. J. Hadley, U.S. Patent 2,839,535 (1958) to Distillers Co.
35. J. P. O'Donnell, R. M. Butler, and L. B. Simpson, U.S. Patent 3,462,476 (1969) to Esso.
36. M. C. Sze and A. P. Gelbein, *Hydrocarbon Process.* **55**(2), 103 (1976).
37. J. L. Callahan, R. K. Grasselli, E. C. Milberger, and H. A. Strecker, *Ind. Eng. Chem. Prod. Res. Dev.*, **9**(2), 134 (1970).
38. J. Imamura, M. Takehara, and K. Kizawa, Ger. Patent 2,605,678 (1976); J. Imamura, Y. Tobe and T. Yoshimoto, *ACS/CSJ Chem. Congr., 1979* Org. Chem. Div., Pap. No. 548 (1979); see also A. J. Chalk, S. A. Magennis, and W. E. Newman, *in* "Fundamental Research in Homogeneous Catalysis" (M. Tsutsui, ed.), Vol. 3, p. 445. Plenum, New York, 1979.
39. J. Hanotier, M. Hanotier-Bridoux, and P. de Radzitzky, *J. Chem. Soc., Perkin II*, p. 381 (1973); Br. Patent 1,206,268 (1970) to Labofina.
40. Y. Kamiya and M. Kashima, *Bul. Chem. Soc. Japan* **46**, 905 (1973).
41. D. H. Becking, U.S. Patent 3,985,809 (1976) to Oxy Metal Industries.
42. J. R. Gilmore and J. M. Mellor, *J. Chem. Soc. Chem. Commun.* p. 507 (1970).
43. D. R. Bryant, J. E. McKeon, and B. C. Ream, *J. Org. Chem.* **33**, 4123 (1968); *Tetrahedron Lett.* p. 3371 (1968); see also C. H. Bushweller, *ibid.* p. 6123.
44. A. Belli, C. Giordano, and A. Citterio, *Synthesis* p. 477 (1980).
45. G. Sosnovsky, *in* "Organic Peroxides" (D. Swern, ed.), Vol. 2, p. 269. Wiley (Interscience), New York, 1971; D. I. Metelitsa, *Russ. Chem. Rev.* **40**, 563 (1971).
46. C. Walling, *Accts. Chem. Rsch.* **8**, 125 (1975).
47. C. Walling, D. M. Camaioni, and S. S. Kim, *J. Am. Chem. Soc.* **100**, 4814 (1978), and references cited therein.
48. K. Nyberg and L. G. Wistrand, *J. Org. Chem.* **43**, 2613 (1978).
49. C. Giordano, A. Belli, A. Citterio, and F. Minisci, *J. Org. Chem.* **44**, 2314 (1979).
50. L. Eberson and L. Gomez-Gonzales, *J. Chem. Soc. Chem. Commun.* p. 263 (1971); *Acta Chem. Scand.* **27**, 1162 (1973).
51. L. Eberson and L. Gomez-Gonzales, *Acta Chem. Scand.* **27**, 1249, 1255 (1973).
52. L. Eberson and E. Jonsson, *Acta Chem. Scand., Series B* **B28**, 771 (1974); *J. Chem. Soc. Chem. Commun.* p. 885 (1974).
53. Br. Patent 1,141,238 (1969) to Lummus Corp.
54. W. W. Kaeding, *Hydrocarbon Process.* **43**(11), 173 (1964).
55. W. W. Kaeding, H. O. Kerlinger, and G. R. Collins, *J. Org. Chem.* **30**, 3754 (1965), and previous papers in this series.
56. W. W. Kaeding, R. O. Lindblom, and R. G. Temple, *Ind. Eng. Chem.* **53**, 805 (1961).
57. W. W. Kaeding, R. O. Lindblom, R. G. Temple, and H. I. Mahon, *Ind. Eng. Chem. Prod. Res. Dev.* **4**, 97 (1965).
58. W. G. Toland, *J. Am. Chem. Soc.* **83**, 2507 (1961).
59. D. M. Albright, C. Perlaky, and P. X. Masciantonio, *Ind. Eng. Chem. Prod. Res. Dev.*, **5**(1) 71 (1966).
60. W. Schoo, J. V. Veenland, J. C. van Velzen, T. J. de Boer, and F. L. J. Sixma, *Recl. Trav. Chim. Pays-Bas* **82**, 959 (1963).
61. A. P. Gelbein and A. S. Nislick, *Hydrocarbon Process.* **57**(11), 125 (1978).
62. H. Itatani and H. Yoshimoto, *Chem. Ind.* (*London*) p. 674 (1971).
63. R. van Helden and G. Verberg, *Recl. Trav. Chim. Pays-Bas* **84**, 1263 (1965).
64. J. M. Davidson and C. Triggs, *J. Chem. Soc. A* pp. 1324, 1331 (1968); *Chem. Ind.* (*London*) p. 457 (1966).
65. R. G. Brown, J. M. Davidson, and C. Triggs, *Prepr., Div. Pet. Chem., Am. Chem. Soc.* **14**(2), B23 (1969).
66. J. M. Davidson and C. Triggs, *Chem. Ind.* (*London*) p. 1361 (1967).

67. F. R. S. Clark, R. O. C. Norman, C. B. Thomas, and J. S. Willson, *J. Chem. Soc., Perkin I* p. 1289 (1974).
68. H. Wittcoff, *Chemtech* **8**, 239 (1978),
69. U. A. Spitzer and D. G. Lee, *J. Org. Chem.* **39**, 2468 (1974).
70. J. A. Caputo and R. Fuchs, *Tetrahedron Lett.* p. 4729 (1967).
71. D. C. Ayres, *J. Chem. Soc., Chem. Commun.* p. 440 (1975).
72. E. C. Taylor, J. G. Andrade, G. J. H. Rall, and A. McKillop, *J. Am. Chem. Soc.* **102**, 6513 (1980).
73. K. Nishizawa, K. Hamada and T. Aratani, Eur. Patent 12,939 (1979) to Sumitomo.
74. J. E. Lyons, R. W. Shinn and G. Suld, U.S. Patent 4,191,389 (1980) to Suntech.
75. See also P. Maggioni and F. Minisci, *La chimica e L'industria* (*Milan*) **61**, 101 (1979) for the oxidation of methylbenzene to the corresponding aldehyde with peroxydisulfate catalyzed by iron and copper.

Chapter 11

Alkanes

Alkanes are susceptible to homolytic attack, as exemplified in the metal-catalyzed autoxidation of *n*-butane to acetic acid and cyclohexane to adipic acid. Indeed, the oxidation of cyclohexane to cyclohexanone or adipic acid is of such importance as to merit an entire book devoted to this subject alone.[1]

I. FORMATION OF ALKYL HYDROPEROXIDE

$$-\overset{|}{\underset{|}{C}}-H + O_2 \Longrightarrow -\overset{|}{\underset{|}{C}}-O-OH$$

The initial products of the autoxidation of alkanes are generally the corresponding hydroperoxides formed by the free radical chain mechanism discussed in Chapter 2. The reactivities of various C—H bonds decrease in the order tertiary > secondary > primary. Indeed, alkanes containing a tertiary C—H bond can generally be selectively oxidized to the corresponding hydroperoxides with molecular oxygen in the temperature range between 100° and 140°C. The autoxidation is usually carried out without a metal catalyst, but a radical initiator such as di-*tert*-butyl peroxide or the hydroperoxide itself is added. The yield of hydroperoxide is related to the oxidizability of the substrate—higher rates of termination leading to lower yields of the hydroperoxide. Alkyl hydroperoxides are difficult to obtain selectively from the autoxidation of primary and secondary C—H bonds owing to the high rates of termination of primary and secondary alkylperoxy radicals. For a particular hydrocarbon, the selectivity to hydroperoxide is related to the *kinetic chain length* (KCL), which is inversely proportional to the rate of chain initiation (R_i).

$$\mathrm{KCL} = \frac{k_p[\mathrm{RH}]}{(2k_t)^{1/2}R_i^{1/2}} \tag{1}$$

When the kinetic chain length is high, the rate of oxidation is roughly equal to the rate of hydroperoxide formation. The selectivity to hydroperoxide is correspondingly high. For example, when KCL is 50, the predicted selectivity to hydroperoxide is 98% (i.e., 100 − 100/50). In practice, a compromise is struck between the rate and the selectivity; i.e., in order to achieve a practical and useful rate of oxidation, a certain degree of selectivity is sacrificed. Hydroperoxide selectivities also decrease with hydrocarbon conversion as a result of the increased competition from secondary processes as the alkane is depleted. Thus, hydrocarbon autoxidations are generally carried out to less than 20% conversion.

The autoxidation of isobutane to *tert*-butyl hydroperoxide (TBHP),

$$(CH_3)_3CH + O_2 \longrightarrow (CH_3)_3COOH \tag{2}$$

has been thoroughly studied,[2–7] since the product is an important initiator for radical polymerizations and an oxidant in metal-catalyzed epoxidations (see Chapters 3 and 9). Both liquid-[2,3,5–7] and vapor-phase[2,4,5] processes have been described for the selective oxidation of isobutane to TBHP. The reaction is carried out in the absence of a metal catalyst, usually between 100° and 140°C. At low rates of initiation and low conversions, selectivities higher than 90% can be achieved.[2] Under practical conditions, however, the liquid-phase oxidation of isobutane at 125°C (using di-*tert*-butyl peroxide as the initiator) produces TBHP in 75% yield, provided that the conversion of isobutane is limited to 8%. *tert*-Butyl alcohol (21%), acetone (2%), and isobutyl derivatives (1%) are formed as side products. The selectivity to TBHP drops to 64% at 20% conversion of isobutane.[3]

Hydrogen bromide (2–4% relative to isobutane) has been used to catalyze the vapor-phase autoxidation of isobutane.[4] Under these conditions, bromine atoms act as the chain transfer agents:

Scheme I:

$$\mathrm{-\overset{|}{\underset{|}{C}}-H + Br\cdot \longrightarrow -\dot{C}\!\!< + HBr} \tag{3}$$

$$\mathrm{-\dot{C}\!\!< + O_2 \longrightarrow -\overset{|}{\underset{|}{C}}-O-O\cdot} \tag{4}$$

$$\mathrm{-\overset{|}{\underset{|}{C}}-O-O\cdot + HBr \longrightarrow -\overset{|}{\underset{|}{C}}-O-OH + Br\cdot,\quad etc.} \tag{5}$$

Under these conditions, selectivities to TBHP of 70% were observed at conversions of isobutane as high as 90%.[4] Small amounts of bromo compounds are formed as impurities. Unfortunately, mixtures containing HBr are strongly corrosive and limit the materials of construction.

Other tertiary alkanes can be selectively autoxidized in the liquid phase to produce the corresponding hydroperoxide.[8,9] For example, a variety of saturated terpenes, listed in Table I, were oxidized to the corresponding hydroperoxides.[9] Alkanes containing two tertiary C—H bonds afford bis-hydroperoxides.[10,11] When the two relevant bonds are suitably juxtaposed, as in 2,4-dimethylpentane and 2,5-dimethylhexane, the formation of the bis-hydroperoxide occurs in high yields as a result of an intramolecular hydrogen abstraction.[10,11]

TABLE I ***tert*-Alkyl Hydroperoxides from the Liquid-Phase Autoxidation of Alkanes**

Substrate	Temp (°C)	Major product	Selectivity (%)	Ref.
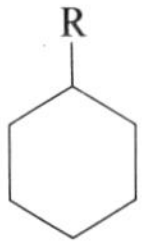 R	100	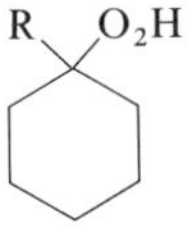 R, O_2H	—[a]	8
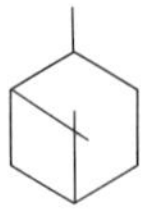	120	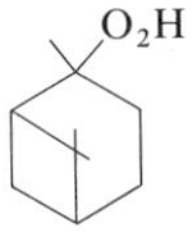 O_2H	96[b]	9
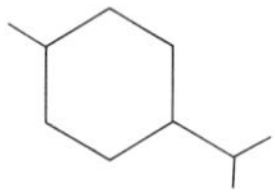	120	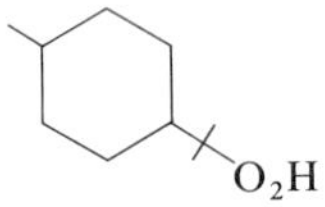 O_2H	94[b]	9
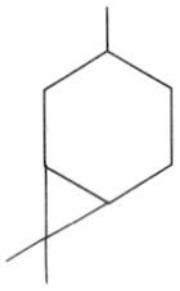	120	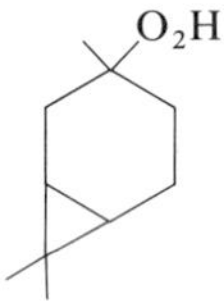 O_2H	95[b]	9
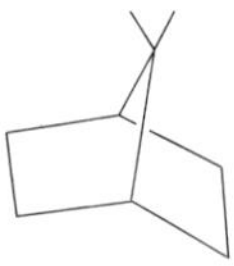	120	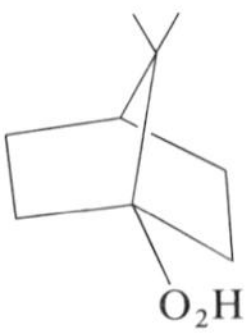 O_2H	96[b]	9

[a] R = Me, Et, *i*-Pr, *t*-Bu, cyclo-C_6H_{11}.

[b] The selectivity refers to total hydroperoxide; isomer distributions were not reported. The tertiary hydroperoxide is presumed to be the major product.

Scheme II:

$$-\underset{\underset{OO\cdot}{|}}{\overset{|}{C}}-CH_2-CH_2-\underset{\underset{H}{|}}{\overset{|}{C}}- \xrightarrow{\quad} -\underset{\underset{OOH}{|}}{\overset{|}{C}}-CH_2-CH_2-\dot{C}\langle \xrightarrow{O_2}$$

$$-\underset{\underset{OOH}{|}}{\overset{|}{C}}-CH_2-CH_2-\underset{\underset{OO\cdot}{|}}{\overset{|}{C}}- \xrightarrow{RH} -\underset{\underset{OOH}{|}}{\overset{|}{C}}-CH_2-CH_2-\underset{\underset{OOH}{|}}{\overset{|}{C}}- + R\cdot, \quad \text{etc.}$$

Selectivities to hydroperoxides can often be increased by carrying out the autoxidation in a polar solvent such as a nitrile (see Chapter 2).[12]

II. ALCOHOL AND KETONE FROM ALKANE

$$\rangle C\langle^{H}_{H} \Longrightarrow \rangle C\langle^{H}_{OH} \Longrightarrow \rangle C{=}O$$

tert-Alkyl hydroperoxides formed in the autoxidation of tertiary alkanes can be selectively reduced to the corresponding alcohol by aqueous sodium sulfite[13] or by catalytic hydrogenation.[14] *tert*-Alkyl hydroperoxides also decompose thermally to give the corresponding alcohol and dioxygen via a free radical chain process:

Scheme III:

$$ROOH \longrightarrow RO\cdot + HO\cdot \tag{6}$$

$$RO\cdot + RO_2H \longrightarrow RO_2\cdot + ROH \tag{7}$$

$$2\,RO_2\cdot \longrightarrow 2\,RO\cdot + O_2 \tag{8}$$

$$2\,RO_2\cdot \longrightarrow ROOR + O_2 \tag{9}$$

For long kinetic chain lengths, the overall reaction is:

$$2\,RO_2H \longrightarrow 2\,ROH + O_2 \tag{10}$$

For example, the thermal decomposition of TBHP (either neat at 100°C[15] or in chlorobenzene at 140°C[16]) yields *tert*-BuOH and oxygen in almost quantitative amounts. The chain decomposition of alkyl hydroperoxides can also be initiated by metal catalysts at ambient temperature, as described in Chapter 3. Thus, a solution of TBHP in chlorobenzene is completely decomposed within a few minutes at 25°C in the presence of a catalytic amount of a cobalt(II) salt.[17] The products consist of approximately 86% *tert*-butyl alcohol, 12% di-*tert*-butyl peroxide, and 93% oxygen. The decomposition of alkyl hydroperoxide to the corresponding alcohols is also catalyzed by selenium[18] and boron[19] compounds (*vide infra*).

The conversion of alkanes to alcohols should be accomplished in one step, in principle, by carrying out the autoxidation under conditions in which the hydroperoxide is simultaneously decomposed (i.e., high temperatures and metal catalysts). However, this method often leads to substantial amounts of ketone via β-scission of the intermediate alkoxy radicals, e.g.,

$$(CH_3)_3CO\cdot \longrightarrow (CH_3)_2C{=}O + CH_3\cdot \tag{11}$$

The formation of acetone from isobutane is facilitated at higher temperatures and in the presence of metal catalysts.

In the autoxidation of alkanes containing secondary C—H groups, ketone formation occurs at low conversions, predominantly via self-termination of secondary alkylperoxy radicals, e.g.,

$$2\ \text{C}_6\text{H}_{11}\text{—OO}\cdot \longrightarrow \text{C}_6\text{H}_{10}{=}\text{O} + \text{C}_6\text{H}_{11}\text{—OH} + O_2 \tag{12}$$

At higher conversions, the alcohol is oxidized further to the ketone, which is itself susceptible to further oxidation to, *inter alia*, carboxylic acids. It is possible, therefore, to obtain high selectivities to alcohols and ketones only at low conversions of alkane (usually $<10\%$).

Selectivities to alcohols can be improved by carrying out alkane autoxidations in the presence of stoichiometric amounts of boric acid (H_3BO_3), metaboric acid (HBO_2), or boric anhydride (B_2O_3). This effect was discovered by Bashkirov.[20] The boron compound reduces the intermediate hydroperoxides to the corresponding alkyl borates, dioxygen, and water,[21] e.g.,

$$6\ RO_2H + B_2O_3 \longrightarrow 2\,(RO)_3B + 3\ H_2O + 3\ O_2 \tag{13}$$

It is also conceivable that intermediate alkylperoxy radicals are intercepted by boron(III) compounds, e.g.,

$$(RO)_3B + RO_2\cdot \longrightarrow (RO_2)(RO)_2B + RO\cdot \tag{14}$$

The alkylperoxyboron(III) compounds are subsequently converted to alkyl borates. The alkyl borates are hydrolyzed in a later step to the corresponding alcohol and boric acid, which is recycled to the oxidation reactor.

The oxidation of cyclohexane to mixtures of cyclohexanol and cyclohexanone is extremely important industrially,[22,23] since these compounds are intermediates in the manufacture of nylon 6 and nylon 6,6. In the Dutch State Mine (DSM) process,[23] cyclohexane is oxidized by air in a batch operation at 155°C and 8–10 bar in the presence of a cobalt catalyst. A mixture of cyclohexanol and cyclohexanone (in a ratio of 1–2 : 1) is obtained with roughly 70% selectivity at approximately 10% cyclohexane conversion. The remainder (30%) consists of by-products such as *n*-butyric, *n*-valeric, succinic, glutaric, and adipic acids formed by ring cleavage of cyclohexanone and/or intermediate cyclohexyloxy radicals, e.g.,

$$\text{C}_6\text{H}_{11}\text{O}\cdot \xrightarrow{\sigma} \cdot\text{CH}_2\text{CH}_2\text{CH}_2\text{CH}_2\text{CH}_2\text{CHO} \xrightarrow{O_2} \text{carboxylic acids} \qquad (15)$$

The mixture of cyclohexanol and cyclohexanone can be dehydrogenated over a copper chromite catalyst to produce pure cyclohexanone.

The addition of boric acid to the cyclohexane oxidation results in an increase in the molar ratio of cyclohexanol and cyclohexanone to approximately 10 : 1 and in the overall selectivity to almost 90%.[22,24,25] Similarly, the air oxidation of cyclododecane carried to 30–35% conversion in the presence of boric acid or anhydride affords 80% cyclododecanol (as the borate ester), 10% cyclododecanone, and 10% by-products.[26,27] Dehydrogenation of the cyclododecanol–cyclododecanone mixture over copper chromite at 220°C affords pure cyclododecanone.[28]

III. CARBOXYLIC ACID FORMATION

$$R—CH_2—CH_2—R' \Longrightarrow RCO_2H + R'CO_2H$$

The alcohols and ketones formed as the primary products of alkane oxidations can be oxidized further to carboxylic acids by cleavage of C—C bonds, as shown in the simplified reaction sequence below.

$$RCH_2—CH_2R' \longrightarrow RCH(OH)—CH_2R' \longrightarrow RC(=O)CH_2R' \longrightarrow RCO_2H + R'CO_2H$$

The reaction can be carried out in multiple stages or in a single stage under conditions in which the alcohol and ketone intermediates are oxidized further. Under these forcing conditions, the formation of carboxylic acid can also occur by the fragmentation of intermediate alkoxy radicals.

$$R—CH(O\cdot)—CH_2R' \xrightarrow{\Delta} RCHO + R'CH_2\cdot \longrightarrow RCO_2H + R'CO_2H$$

Acetic acid is produced commercially by the liquid-phase autoxidation of *n*-butane[29] at 180°C and 60 bar in the presence of a cobalt (acetate) catalyst in acetic acid as solvent. Acetic acid is produced with a selectivity of 57% at greater than 90% conversion of *n*-butane.[29] The by-products include formic acid, acetaldehyde, methanol, acetone, methyl ethyl ketone, and various esters.

More recently, high concentrations ($\sim 0.2\ M$) of a cobalt(II) acetate catalyst have been employed with methyl ethyl ketone as a promoter for the autoxidation of *n*-butane in acetic acid at 100°–125°C and 20 bar pressure.[30] At butane conversions of approximately 80%, acetic acid is produced with 84% selectivity. The major by-products are propionic acid, *n*-butyric acid, and methyl ethyl ketone. Under these conditions (in contrast to the reaction at low cobalt concentrations), the reaction proceeds by the direct oxidation of the *n*-butane by Co(III). The sequence of steps may be that shown in Scheme IV (see Chapter 5):

Scheme IV:

$$n\text{-}C_4H_{10} + Co^{III} \longrightarrow CH_3\dot{C}HCH_2CH_3 + Co^{II} + H^+ \quad (16)$$

$$CH_3\dot{C}HCH_2CH_3 + O_2 \longrightarrow CH_3CH(OO\cdot)CH_2CH_3 \quad (17)$$

$$CH_3CH(OO\cdot)CH_2CH_3 + Co^{II} \longrightarrow CH_3COCH_2CH_3 + HOCo^{III} \quad (18)$$

The autoxidative process for the manufacture of acetic acid has been largely superseded by the rhodium-catalyzed carbonylation of methanol, i.e.,[39]

$$CH_3OH + CO \xrightarrow{[Rh]} CH_3CO_2H$$

Hanotier and co-workers[31] similarly found that alkanes are readily oxidized by Co(III) in acetic acid at ambient temperatures with strong acid promoters, such as sulfuric or trichloroacetic acid. (See Chapter 5 for a discussion of the effect of strong acids.) In the presence of dioxygen, ketones are formed in high yield. For example, *n*-heptane afforded a mixture of heptanones in 80% yield.

$$n\text{-}C_7H_{16} + O_2 \xrightarrow[Cl_3CCO_2H,\ 25^\circ C]{[Co^{III}]} \underset{78\%}{\text{2-heptanone}} + \underset{15\%}{\text{3-heptanone}} + \underset{7\%}{\text{4-heptanone}} \quad (19)$$

Interestingly, tertiary alkyl C—H bonds were oxidized in both systems[30,31] at significantly lower rates compared to secondary alkyl C—H bonds.[30,31] This effect is similar to that observed with alkylaromatic compounds, as described in Chapter 10.

The oxidation of cyclohexane to adipic acid can be achieved in either a two-stage or a single-stage process. In the two-stage process, the cyclohexane is initially oxidized to a mixture of cyclohexanol and cyclohexanone, which is then further oxidized to adipic acid in a separate step. For example, in the Du Pont process, nitric acid is used as the oxidant in the presence of Cu(II) and V(V) salts as catalysts at 75–80°C.[32,33] The reaction sequence is undoubtedly complex.[32] One possible pathway involves nitrosation of the cyclohexanone followed by oxidation of the resulting cyclohexane-1,2-dione or its monoxime with VO_2^+. The resulting VO^+ is reoxidized by nitric acid. The role of the Cu(II) is less clear.[32] (It may be involved in the nitrosation.)

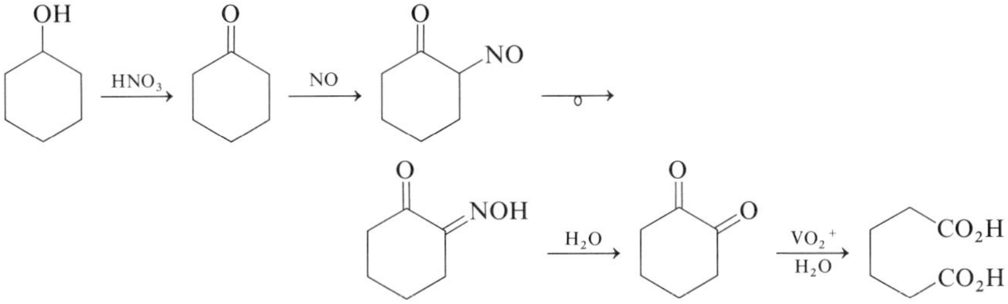

$$\text{cyclohexanol} \xrightarrow{HNO_3} \text{cyclohexanone} \xrightarrow{NO} \text{2-nitrosocyclohexanone} \longrightarrow \text{cyclohexane-1,2-dione monoxime} \xrightarrow{H_2O} \text{cyclohexane-1,2-dione} \xrightarrow[H_2O]{VO_2^+} HO_2C(CH_2)_4CO_2H$$

The yield of adipic acid is 92% under these conditions.[33]

An alternative process[34] involves the air oxidation of cyclohexanone,[34] or a mixture of cyclohexanol and cyclohexanone,[35] in acetic acid at 75–85°C in the presence of a manganese(II) acetate as the catalyst. The selectivity to adipic acid is 90% at conversions of 40–45%.[34]

When high concentrations of cobalt(II) acetate are employed as catalyst, the direct oxidation of cyclohexane to adipic acid can be carried out at 90°C in acetic acid.[36–38] (Compare the oxidation of *n*-butane to acetic acid.) Adipic acid is obtained in 70–75% selectivity at 80–85% conversion of cyclohexane. Similar to the oxidation of *n*-butane (*vide supra*), the reaction involves Co(III) as the chain transfer agent in a direct reaction with the cyclohexane substrate.[36]

Scheme V:

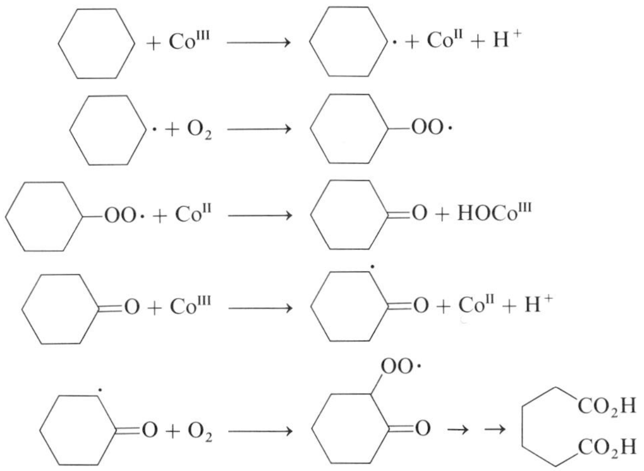

$$C_6H_{12} + Co^{III} \longrightarrow C_6H_{11}\cdot + Co^{II} + H^+ \tag{20}$$

$$C_6H_{11}\cdot + O_2 \longrightarrow C_6H_{11}OO\cdot \tag{21}$$

$$C_6H_{11}OO\cdot + Co^{II} \longrightarrow C_6H_{10}{=}O + HOCo^{III} \tag{22}$$

$$C_6H_{10}{=}O + Co^{III} \longrightarrow \cdot C_6H_9{=}O + Co^{II} + H^+ \tag{23}$$

$$\cdot C_6H_9{=}O + O_2 \longrightarrow \cdot OO\text{-}C_6H_9{=}O \rightarrow \rightarrow HO_2C(CH_2)_4CO_2H \tag{24}$$

1,12-Dodecanedioic acid is produced in a two-stage process from cyclododecane, which is analogous to the conversion of cyclohexane to adipic acid (*vide supra*).[32]

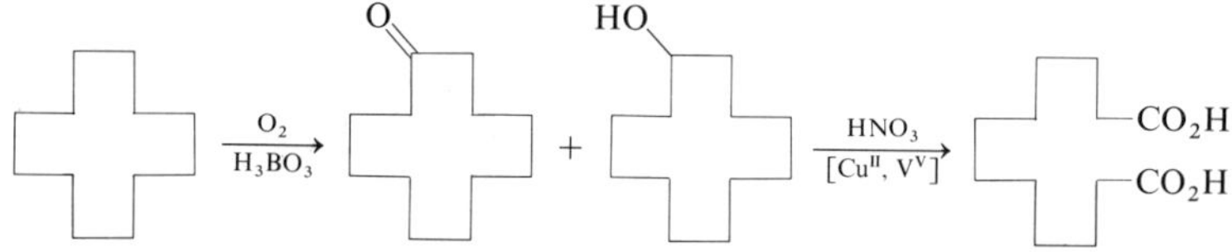

In this case, the single-stage oxidation is not successful, since cyclododecane is much less reactive than cyclohexane, and the oxidation with high concentrations of cobalt(II) acetate catalyst in acetic acid is unselective.[37,38]

Although much research has been devoted to the subject of alkane activation under mild conditions (see Chapter 7), no synthetically useful method has emerged, as yet, from this work.

REFERENCES

1. I. V. Berezin, E. T. Denisov, and N. M. Emanuel, "The Oxidation of Cyclohexane" (transl. by K. A. Allen). Pergamon, Oxford, 1966.
2. F. R. Mayo *Accts. Chem. Rsch.* **1**, 193 (1968).
3. D. E. Winkler and G. W. Hearne, *Ind. Eng. Chem.* **53**, 655 (1961); D. E. Winkler and G. W. Hearne, U.S. Patent 2,845,461 (1958) to Shell Development Co.
4. E. R. Bell, F. H. Dickey, J. H. Raley, F. F. Rust, and W. E. Vaughan, *Ind. Eng. Chem.* **41**, 2597 (1949).
5. D. L. Allara, T. Mill, D. G. Hendry, and F. R. Mayo, *Adv. Chem. Series* **76**, 40 (1968).
6. H. R. Grane, U.S. Patent 3,478,108 (1969) to Atlantic Richfield.
7. R. J. Harvey, U.S. Patent 3,449,217 (1969) to Halcon International.
8. W. Pritzkow and K. H. Gröbe, *Chem. Ber.* **93**, 2156 (1960).
9. W. Ester and A. Sommer, U.S. Patent 3,259,661 (1966) to Hibernia.
10. F. F. Rust, *J. Am. Chem. Soc.* **79**, 4000 (1957).
11. R. Criegee and P. Ludwig, *Erdoel & Kohle, Erdgas, Petrochem.* **15**, 523 (1962).
12. K. Tanaka and J. Imamura, *Chem. Lett.* p. 1347 (1974).
13. H. Hock and S. Lang, *Ber. Dtsch. Chem. Ges.* **75**, 313 (1942).
14. A. G. Davies, *J. Chem. Soc.* p. 3474 (1958).
15. N. A. Milas and D. M. Surgenor, *J. Am. Chem. Soc.* **68**, 205,643 (1946); see also R. Hiatt, T. Mill, and F. R. Mayo, *J. Org. Chem.* **33**, 1416 (1968), and subsequent papers in this series.
16. E. R. Bell, J. H. Raley, F. F. Rust, F. H. Seubold, and W. E. Vaughan, *Discuss. Faraday Soc.* **10**, 242 (1951).
17. R. Hiatt, K. C. Irwin, and C. W. Gould, *J. Org. Chem.* **33**, 1430 (1968).
18. D. T. Woodbridge, *J. Chem. Soc., Phys. Org.* p. 50 (1966).
19. P. F. Wolf, J. E. McKeon, and D. W. Cannell, *J. Org. Chem.* **40**, 1875 (1975).
20. A. N. Bashkirov, V. V. Kamzolkin, K. M. Sokova, and T. P. Andreyeva, *in* "The Oxidation of Hydrocarbons in the Liquid Phase" (N. M. Emanuel, ed.), p. 183. Pergamon, Oxford 1965.
21. H. Sakaguchi, Y. Kamiya, and N. Ohta, *Bul. Japan Pet. Inst.* **14**, 71 (1972).

22. S. A. Miller, *Chem. Process Eng.* (*London*) **50**(6), 63 (1969).
23. J. W. M. Steeman, S. Kaarsemaker, and P. J. Hoftyzer, *Chem. Eng. Sci.* **14**, 139 (1961).
24. M. Spielman, *AIChE J.* **10**, 496 (1964).
25. M. Sittig, "Combine Hydrocarbons and Oxygen for Profit," Chem. Process. Rev., No. 11, p. 185. Noyes Data Corp., Park Ridge, New Jersey, 1968.
26. A. N. Bashkirov, V. V. Kamzolkin, K. M. Sokova, T. P. Andreeva, L. I. Zakharkin, and V. V. Korneva, *Neftekhimiya* **1**, 527 (1961).
27. H. Graseman, *Erdoel & Kohle, Erdgas, Petrochem.* **22**, 751 (1969); F. Broich and H. Graseman, *ibid.* **18**, 360 (1965).
28. A. N. Bashkirov, V. V. Kamzolkin, and M. M. Potarin, *Neftekhimiya* **4**, 298 (1964).
29. R. P. Lowry and A. Aguilo, *Hydrocarbon Process.* **53**(11), 103 (1974).
30. A. Onopchenko and J. G. D. Schulz, *J. Org. Chem.* **38**, 909 (1973).
31. J. Hanotier, P. Camerman, M. Hanotier-Bridoux, and P. de Radzitzky, *J. Chem. Soc., Perkin II* p. 2247 (1972).
32. G. W. Parshall, *J. Mol. Catal.* **4**, 243 (1978).
33. A. F. Lindsay, *Chem. Eng. Sci.*, **3**, Suppl. 1, 78 (1954).
34. W. Fleming and W. Speer, U.S. Patent 2,005,183 (1935) to I. G. Farben Industries; see also H. W. Prengle and L. F. Hatch, *Hydrocarbon Process.* **49**(3), 106 (1970).
35. C. Parlant, I. Seree de Roch, and J. C. Balanceanu, *Bul. Soc. Chim. France* **11**, 2452 (1963).
36. A. Onopchenko and J. G. D. Schulz, *J. Org. Chem.* **38**, 3729 (1973).
37. K. Tanaka, *Hydrocarbon Process.* **53**(11), 114 (1974).
38. K. Tanaka, *Chemtech* **4**, 555 (1974).
39. D. Forster, *Adv. Organometal Chem.* **17**, 255(1979).

Chapter 12

Oxygen-Containing Compounds

The introduction of an oxygen-containing functionality into a hydrocarbon backbone results in a significant increase in the oxidizability of the adjacent C—H bond.[1,2] This point is illustrated in Table I for a series of α-substituted toluene derivatives.

I. ALCOHOLS

The oxidation of primary and secondary alcohols to aldehydes and ketones, respectively, is a reaction that is frequently used in organic synthesis. Stoichiometric oxidations with a variety of reagents such as chromium(VI) compounds (e.g., CrO_3/pyridine) and manganese dioxide are usually the

TABLE I Absolute Rate Constants for the Autoxidation of Some α-Substituted Toluenes at 30°C[a]

Substrate	$k_p/(2k_t)^{1/2} \times 10^3$ ($M^{-1/2}$ sec$^{-1/2}$)	k_p (M^{-1} sec^{-1})	$2k_t \times 10^{-6}$ (M^{-1} sec^{-1})
$PhCH_3$	0.014	0.24	300
$PhCH_2CH_3$	0.21	1.3	40
$PhCH_2OH$	0.85	4.8	32
$PhCH_2OCH_2Ph$	7.1	84	160
$PhCH_2COCH_2Ph$	0.55	3.3	36
$PhCHO$[b]	130	1900	210
$PhCH_2OAc$	0.31	4.6	220
$PhCH_2Cl$	0.42	3.0	50
$PhCH_2Br$	0.2	1.2	36

[a] From Howard.[2]
[b] Measured at 5°C.

methods of choice for laboratory-scale syntheses.[3] For example, oxometal salts such as pyridinium chlorochromate and pyridinium dichromate effect mild oxidations of carbohydrates as well as a variety of alcohols to the corresponding carbonyl compounds in methylene chloride at room temperature.[3a] The oxidation is promoted by insoluble inorganic compounds and by molecular sieves.[3b]

Alcohols can also be dehydrogenated to the corresponding carbonyl compounds over copper chromite or noble metal catalysts. This reaction is the microscopic reverse of the catalytic hydrogenation of carbonyl compounds and is often the method of choice for industrial-scale syntheses.

$$>C(H)(OH) \rightleftharpoons >C{=}O + H_2 \tag{1}$$

Since a thorough treatment of catalytic dehydrogenation is beyond the scope of this book, in the following discussion we shall be concerned mainly with metal-catalyzed oxidations of alcohols, usually employing molecular oxygen as the primary oxidant.

A. Autoxidation

The autoxidation of primary and secondary alcohols affords hydrogen peroxide and aldehydes or ketones, respectively, as the primary products. An unstable α-hydroxyalkyl hydroperoxide is an intermediate:[1]

Scheme I:

$$>C(H)(OH) \xrightarrow{[\text{initiation}]} >\dot{C}(OH) \qquad (2)$$

$$>\dot{C}(OH) + O_2 \longrightarrow >C(O{-}O\cdot)(OH) \qquad (3)$$

$$>C(O{-}O\cdot)(OH) + >C(H)(OH) \longrightarrow >C(O_2H)(OH) + >\dot{C}(OH) \qquad (4)$$

$$>C(O_2H)(OH) \longrightarrow >C{=}O + H_2O_2 \qquad (5)$$

$$>C(O{-}O\cdot)(OH) \longrightarrow >C{=}O + HO_2\cdot \qquad (6)$$

Hydrogen abstraction from the α-C—H bond of alcohols is more facile than that from the parent hydrocarbons, in accord with the lower C—H bond strengths [D(C—H) = 102 (CH_4), 92 (CH_3OH), 98 (C_2H_6), and 88 kcal mol^{-1} (CH_3CH_2OH)].[1] Abstraction from an α-C—H is also favored over that of the hydroxylic proton [e.g., D(O—H) = 102 kcal mol^{-1}].

Under the usual autoxidation conditions, the aldehydes derived from primary alcohols undergo further rapid oxidation to the corresponding carboxylic acids. The major products of ethanol autoxidation, for example, are acetic acid and ethyl acetate, the latter resulting from esterification of the ethanol by the acetic acid product.[1] On the other hand, the autoxidation of secondary alcohols affords high yields of ketones and hydrogen peroxide, forming the basis of a commercial process for the manufacture of hydrogen peroxide by the liquid-phase, peroxide-initiated autoxidation of isopropyl alcohol.[4] Hydrogen peroxide is reportedly produced in 98% yield. The acetone coproduct can be hydrogenated and recycled.

Tertiary alcohols are, of course, unreactive in autoxidation since they contain no labile C—H bond. It should be noted, however, that tertiary alcohols can react by prior dehydration to the corresponding olefin, which then undergoes oxidation, as in the recently reported[5] vapor-phase oxidation of *tert*-butyl alcohol to methacrylic acid over heterogeneous catalysts.

$$(CH_3)_3C{-}OH \xrightarrow{-H_2O} CH_2{=}C(CH_3)_2 \xrightarrow[\text{[catalyst]}]{O_2} CH_2{=}C(CH_3){-}CO_2H \qquad (7)$$

B. Catalytic Oxidation with Metal Salts

Transition metal ions can act as either catalysts or inhibitors of alcohol autoxidations. Inhibition results from scavenging of the intermediate α-hydroxyalkylperoxy radicals by the redox cycle:[1]

$$>C(OH)(O_2\cdot) + M^{n+} \longrightarrow >C{=}O + O_2 + M^{(n-1)+} + H^+ \tag{8}$$

$$>C(OH)(O_2\cdot) + M^{(n-1)+} \longrightarrow M^{n+} + >C(OH)(O_2^-) \tag{9}$$

Catalysis can result from chain initiation, which stems from the reaction of the metal species with H_2O_2 or with the alcohol substrate:

$$M^{n+} + H_2O_2 \longrightarrow M^{(n-1)+} + HO_2\cdot + H^+ \tag{10}$$

$$M^{(n-1)+} + H_2O_2 \longrightarrow M^{n+} + HO\cdot + HO^- \tag{11}$$

$$M^{n+} + >CHOH \longrightarrow M^{(n-1)+} + >\dot{C}OH + H^+ \tag{12}$$

For example, low concentrations of copper salts have been shown to inhibit the autoxidation of cyclohexanol.[6] On the other hand, phenanthroline–copper complexes catalyze the autoxidation of methanol to formaldehyde.[7]

The oxidation of secondary alcohols to ketones by Co(III) acetate in the presence of dioxygen[8] undoubtedly involves the direct, one-electron oxidation of the alcohol by Co(III). Although the overall result can be described adequately by equation 12, the reaction more likely involves the initial formation of an alkoxy radical via homolytic cleavage of an alkoxycobalt(III) intermediate:

Scheme II:

$$>CHOH + XM^{n+} \longrightarrow >CHOM^{n+} + HX \tag{13}$$

$$>CHOM^{n+} \longrightarrow >CHO\cdot + M^{(n-1)+} \tag{14}$$

$$>CHOH + >CHO\cdot \longrightarrow >CHOH + >\dot{C}OH \tag{15}$$

where M^{n+} = Co(III), Mn(III), etc. Whereas the oxidation in eq 14 is most likely to occur with strong, one-electron oxidants such as Co(III) and Mn(III), heterolytic pathways for the decomposition of an alkoxymetal intermediate are also possible:

Scheme III:

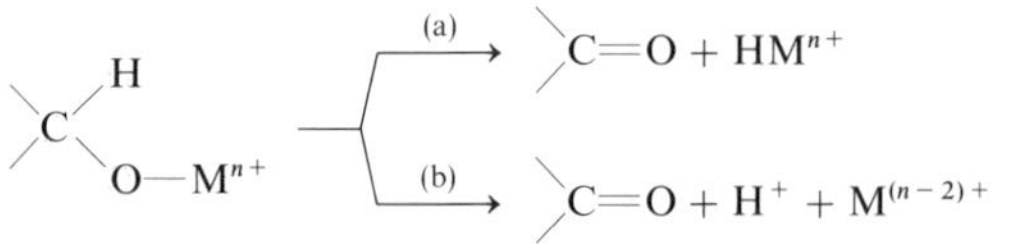

$$\begin{array}{l} \overset{\diagdown}{\underset{\diagup}{}}C\overset{H}{\diagup} \\ \quad \diagdown O{-}M^{n+} \end{array} \begin{array}{l} \xrightarrow{(a)} \;\rangle C{=}O + HM^{n+} \qquad (16) \\ \xrightarrow{(b)} \;\rangle C{=}O + H^{+} + M^{(n-2)+} \qquad (17) \end{array}$$

For example, the reduction of the noble metal salts of Pd(II) and Pt(II) by alcohols is a common method for preparing these metal hydrides (pathway a). Such a β-hydride elimination is presumably also involved in the Pd(II)-catalyzed oxidation of secondary alcohols to ketone by dioxygen at ambient temperature.[9] It is noteworthy that this reaction does not require Cu(II) as a cocatalyst for the regeneration of the Pd(II), the oxidant being regenerated by the reaction of O_2 with the palladium(II) hydride:

Scheme IV:

$$\rangle CHOPd^{II}X \longrightarrow \rangle C{=}O + HPd^{II}X \qquad (18)$$

$$HPd^{II}X + O_2 \longrightarrow HO_2Pd^{II}X \qquad (19)$$

$$HO_2Pd^{II}X + \rangle CHOH \longrightarrow \rangle CHOPd^{II}X + H_2O_2 \qquad (20)$$

A similar mechanism is probably involved in the oxidation of primary and secondary alcohols to aldehydes and ketones, respectively, with molecular oxygen over platinum supported on charcoal or reduced PtO_2 catalysts.[10,11] This oxidation is a useful method for the selective oxidation of primary alcohols to aldehydes, which is accomplished in *n*-heptane solution at temperatures in the range 40–60°C. When the oxidation is carried out in water at alkaline pH, the corresponding carboxylic acid is produced.[10,11] Secondary alcohols afford the corresponding ketones (see Table II for examples).

In pathway b of Scheme III, the metal undergoes a two-electron reduction without the formation of a metal hydride species. The oxidant is then regenerated as shown in Scheme V.

Scheme V:

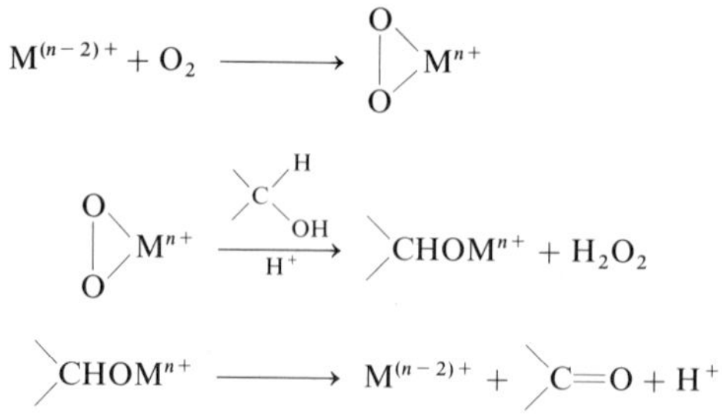

$$M^{(n-2)+} + O_2 \longrightarrow \begin{array}{c} O \\ | \\ O \end{array}\!\!\rangle M^{n+}$$

$$\begin{array}{c} O \\ | \\ O \end{array}\!\!\rangle M^{n+} \xrightarrow[H^{+}]{\rangle C\langle^{H}_{OH}} \rangle CHOM^{n+} + H_2O_2$$

$$\rangle CHOM^{n+} \longrightarrow M^{(n-2)+} + \rangle C{=}O + H^{+}$$

TABLE II Oxidation of Alcohols Catalyzed by Heterogeneous Platinum Catalysts[a]

Substrate	Solvent	Catalyst	Temp (°C)	Time (hr)	Product	Yield (%)
$CH_3(CH_2)_5OH$	Water[b]	Pt/C	Reflux	25	$CH_3(CH_2)_4CO_2H$	99
$CH_3(CH_2)_{11}OH$	*n*-Heptane	PtO_2	60	0.25	$CH_3(CH_2)_{10}CHO$	77
$CH_3(CH_2)_{11}OH$	*n*-Heptane	PtO_2	60	2	$CH_3(CH_2)_{10}CO_2H$	96
$CH_3(CH_2)_{13}OH$	*n*-Heptane	PtO_2	60	0.75	$CH_3(CH_2)_{12}CHO$	91
$CH_3(CH_2)_{15}OH$	*n*-Heptane	PtO_2	60	7	$CH_3(CH_2)_{14}CHO$	95
$HOCH_2CH_2OH$	Water[b]	Pt/C	Reflux	6	$HOCH_2CO_2H$	100
$HO(CH_2)_{10}OH$	*n*-Heptane	PtO_2	60	1.5	$OHC(CH_2)_8CHO$	54
$PhCH_2OH$	*n*-Heptane	PtO_2	60	1.0	PhCHO	78
$PhCH_2OH$	*n*-Heptane	Pt/C	Reflux	10	$PhCO_2H$	97
$CH_3(CH_2)_5CH(OH)CH_3$	*n*-Heptane	PtO_2	20	96	$CH_3(CH_2)_5COCH_3$	80
—OH	*n*-Heptane	PtO_2	20	0.75	=O	82
—OH	*n*-Heptane	PtO_2	20	1.5	=O	92
—OH	*n*-Heptane	PtO_2	17	1.0	=O	99

[a] From Heynes and Blazejewicz.[10]

[b] Alkaline solution.

Such a mechanism was proposed in the ruthenium-catalyzed oxidation of coordinated alcohols by molecular oxygen.[12] (Compare the catalytic oxidation of alcohols to aldehydes and ketones with dioxygen using catalysts consisting of combinations of ruthenium with either copper or iron.[13]) The formation of H_2O_2 observed during the reaction was consistent with a mechanism involving the Ru(II)/Ru(IV) couple:

Scheme VI:

$$\mathrm{Ru^{II}} + \mathrm{O_2} + \mathrm{{>}CHOH} \xrightarrow{\mathrm{H^+}} \mathrm{{>}CHORu^{IV}} + \mathrm{H_2O_2}$$

$$\mathrm{{>}CHORu^{IV}} \longrightarrow \mathrm{{>}C{=}O} + \mathrm{Ru^{II}} + \mathrm{H^+}$$

An analogous mechanism involving the Cu(I)/Cu(III) couple has been proposed[14] for the oxidation of galactose catalyzed by galactose oxidase (see Chapter 8).

Transition metal salts can also be used in conjunction with primary oxidants other than oxygen, as in the Ni(II)Br_2-catalyzed oxidation of primary and secondary alcohols to carbonyl compounds by benzoyl peroxide.[15] The use of NaOCl as the primary oxidant in conjunction with a porphyrinatomanganese(III) chloride catalyst, and a phase transfer catalyst in a mixture of CH_2Cl_2 and H_2O, has also been recently reported.[16] With this procedure, benzyl alcohol was smoothly oxidized to benzaldehyde in 94% yield at ambient temperature.

C. Catalytic Oxidation with Oxometal Reagents

Many stoichiometric procedures employing oxometal reagents [such as oxochromium(VI)[17] and oxovanadium(V)[18] compounds] for the oxidation of alcohols have recently been improved by the use of phase transfer catalysis to solubilize the oxometal reagent in an organic phase.[19] Alcohol oxidations with oxometal reagents probably all involve the basic transformation in eq 21.

$$\mathrm{{>}C(H){-}O{-}M^{n+}({=}O)_2} \longrightarrow \mathrm{{>}C{=}O} + \mathrm{O{=}M^{(n-2)+}{-}OH} \qquad (21)$$

In principle, it should be possible to utilize oxometal reagents as catalysts in conjunction with suitable primary oxidants which would regenerate the oxometal oxidant. For example, the selective, vapor-phase catalytic oxidation of *n*-propyl alcohol to propionaldehyde over V_2O_5 has been reported.[20]

Oxidation of *n*-propyl alcohol at 210°C over a V_2O_5 catalyst modified with an alkaline earth metal oxide (10 mol %) using a stoichiometric amount of dioxygen gave propionaldehyde in 94–99% selectivity.

The liquid-phase oxidation of secondary alcohols to the corresponding ketones was observed as a side reaction in the catalytic epoxidation of allylic alcohols[21] using an oxovanadium(V) catalyst and TBHP as the primary oxidant (see Chapter 3). The scope of this reaction as a general method for the catalytic oxidation of alcohols with TBHP has not been explored.

Ruthenium tetroxide is a useful reagent for the oxidation of secondary alcohols to ketones and of primary alcohols to the corresponding carboxylic acids.[3] In principle, RuO_4 can be used in catalytic amounts in conjunction with NaOCl as the primary oxidant under phase transfer conditions (compare the RuO_4-catalyzed cleavage of olefins). This possibility does not appear to have been explored. Aqueous sodium hypochlorite has been used, in the presence of a tetraalkylammonium salt as phase transfer catalyst, in two-phase reactions for the oxidation of primary and secondary alcohols to aldehydes and ketones in good yield.[22] Interestingly, ethyl acetate is the solvent of choice for this reaction. The use of NaOCl in conjunction with various oxometal catalysts should also provide synthetically useful methods for the mild, selective oxidation of alcohols.

II. ETHERS

Ethers are readily autoxidized to the corresponding α-hydroperoxy ethers.[1]

$$-\overset{|}{\underset{|}{C}}-O-\overset{H}{\underset{|}{C}}- \longrightarrow -\overset{|}{\underset{|}{C}}-O-\overset{O_2H}{\underset{|}{C}}- \quad (22)$$

These reactions do not, as yet, have any general synthetic utility.

Ruthenium oxide is a useful reagent for the mild, selective oxidation of ethers to esters,

$$\text{(n-}C_4H_9)_2O \xrightarrow[25°C,\ 5\ min]{[RuO_4],\ CCl_4} n\text{-}C_3H_7CO_2\text{-}n\text{-}C_4H_9 \quad (23)$$

100%

and to lactones, if the ether is cyclic,[3,23]

$$\text{tetrahydrofuran} \xrightarrow[25°C,\ 5\ min]{[RuO_4],\ CCl_4} \gamma\text{-butyrolactone} \quad (24)$$

100%

The use of catalytic amounts of RuO_4 and aqueous NaOCl or periodate as the primary oxidant in the presence of a phase transfer catalyst may provide an improved procedure for carrying out this reaction.[23a]

III. GLYCOLS

Glycols can be converted to ketones and aldehydes by oxidative carbon–carbon bond cleavage, i.e.,

$$R_2C(OH){-}CH(OH)R' \Longrightarrow R_2C{=}O + R'CHO \tag{25}$$

The aldehydes may be further oxidized to carboxylic acids under the reaction conditions. Hydroxylation of an olefin followed by oxidative cleavage of the glycol represents a two-step alternative to the direct oxidative cleavage of an olefin, as described in Chapter 9.

Several oxidants are known to cleave the carbon–carbon bond of glycols.[24,25] Stoichiometric oxidations with lead tetraacetate or periodate are generally the reagents of choice for laboratory-scale reactions, since they are effective under mild conditions.[24,25] By analogy with alcohol oxidations, catalytic amounts of transition metal salts or oxometal reagents, in conjunction with peroxides, NaOCl, or even O_2 as the primary oxidant, should provide synthetically useful procedures for the selective cleavage of glycols via a heterolytic pathway,

$$\begin{matrix}-\overset{|}{C}-O\\ |\\ -\underset{|}{C}-O\end{matrix}\!\!>\!M^{n+}(=O) \longrightarrow 2\,{>}C{=}O + O{=}M^{(n-2)+} \tag{26}$$

$$O{=}M^{(n-2)+} + -\underset{HO}{\overset{|}{C}}-\underset{OH}{\overset{|}{C}}- \xrightarrow{RO_2H} \begin{matrix}-\overset{|}{C}-O\\ |\\ -\underset{|}{C}-O\end{matrix}\!\!>\!M^{n+}(=O), \quad \text{etc.}$$

or a homolytic pathway,

$$-\underset{HO}{\overset{|}{C}}-\underset{OH}{\overset{|}{C}}- + M^{n+} \xrightarrow{-H^+} M^{(n-1)+} + -\underset{\cdot O}{\overset{|}{C}}-\underset{OH}{\overset{|}{C}}- \longrightarrow {>}C{=}O + {>}\dot{C}OH, \quad \text{etc.} \tag{27}$$

However, the various possibilities for catalytic oxidations do not appear to have been fully explored, although one example is the silver-catalyzed oxidative cleavage of glycols by potassium peroxysulfate.[26]

$$\begin{array}{cc} \text{HO} & \text{OH} \\ | & | \\ -\text{C}- & \text{C}- \\ | & | \end{array} + S_2O_8^{2-} \xrightarrow[-2\,H^+]{[Ag^+]} 2\ \rangle C{=}O + 2\,SO_4^{2-} \tag{28}$$

Nitric acid in the presence of vanadium(V) catalysts has been used for the oxidation of cyclohexane-1,2-diol to adipic acid.[27]

The selective cleavage of glycols by molecular oxygen in the presence of catalytic amounts of Co(II) salts has been reported[28] at 100°C in aprotic polar solvents. Depending on the reaction time, aldehydes or carboxylic acids are the major products. (See Chapter 5 for a discussion of the mechanism.)

It is conceivable that glycols can be oxidized stepwise to the corresponding 1,2-diketones by a route involving α-ketols as intermediates.

$$\text{R(H)C(OH)–C(OH)(H)R}' \xrightarrow{(a)} \text{R(H)C(OH)–C(=O)R}' \xrightarrow{(b)} \text{RC(=O)–C(=O)R}' \tag{29}$$

(When one of the α-carbon atoms is fully substituted, the α-ketol would not be further oxidized.) These reactions do not appear to have received much attention, but it is worth noting that Cu(II) salts oxidize α-ketols to the corresponding α-diketones [step b in reaction 29]. Catalytic amounts of Cu(II) can be used with ammonium nitrate as the primary oxidant.[29,30] A typical example is the oxidation of benzoin to benzil:

$$\text{PhCH(OH)–C(=O)Ph} \xrightarrow[\text{HOAc, reflux}]{[Cu(OAc)_2],\ NH_4NO_3} \text{PhC(=O)–C(=O)Ph} \tag{30}$$

The oxidation of glycols to 1,2-dicarbonyl compounds as well as carbon–carbon cleavage products by copper reagents has been recently reviewed.[31] Glycols have also been oxidized, without C–C bond cleavage, using molecular oxygen and a Pt/C catalyst under alkaline conditions.[10,11] Ethylene glycol, for example, affords glycolic acid (see Table II).

$$HOCH_2CH_2OH + O_2 \xrightarrow[\text{alkaline pH, reflux}]{[Pt/C]} \underset{100\%}{HOCH_2CO_2H} \tag{31}$$

IV. ALDEHYDES

Aldehydes undergo facile autoxidation, even at ambient temperatures. The propagation rate constants for the following chain process are listed in Table III.

TABLE III Rate Constants for Aldehyde Autoxidations[a]

Aldehyde	Temp (°C)	$k_p/(2k_t)^{1/2} \times 10^3$ ($M^{-1/2}$ sec$^{-1/2}$)	k_p (M^{-1} sec^{-1})	$2k_t \times 10^{-6}$ (M^{-1} sec^{-1})
Acetaldehyde	0	265	2700	104
Heptanal	0	390	3100	54
Octanal	0	470	3900	69
Decanal	5	450	2700	34
Cyclohexanecarboxaldehyde	0	440	1100	6.8
Pivalaldehyde	0	960	2500	6.6
Benzaldehyde	0	290	1200	1760
	5	130	1900	210

[a] From Howard.[2]

Scheme VII:

$$RCHO \xrightarrow{[\text{initiation}]} R\dot{C}O \quad (32)$$

$$R\dot{C}O + O_2 \longrightarrow RCO_3\cdot \quad (33)$$

$$RCO_3\cdot + RCHO \xrightarrow{k_p} RCO_3H + R\dot{C}O \quad (34)$$

The initially formed organic peracid is stable at low temperatures but is subject to a facile Baeyer–Villiger oxidation of the aldehyde.

$$RCO_3H + RCHO \longrightarrow 2\ RCO_2H \quad (35)$$

The oxidation of acetaldehyde is used for the manufacture of peracetic acid,[32] acetic anhydride,[33] and acetic acid.[34–36] For the latter, the batch autoxidation of acetaldehyde with air is carried out at 55°–65°C and 5 bar in acetic acid as solvent containing catalytic amounts of Mn(II) or Co(II) acetates. A continuous process has also been described[35] in which acetaldehyde conversions are 20–30% per pass and the selectivity is approximately 95%.

The metal catalyst initiates the oxidation via a one-electron process such as:

$$CH_3CHO + Mn^{III} \longrightarrow CH_3\dot{C}O + Mn^{II} + H^+ \quad (36)$$

The Mn(III) is subsequently regenerated by oxidation of the Mn(II) by peracetic acid or acetylperoxy radicals (see Chapter 3). In the process for the manufacture of acetic anhydride,[33] a mixture of Co(II) and Cu(II) acetates is employed as the catalyst. The key step in anhydride formation involves the oxidation of the intermediate acetyl radicals by Cu(II):

$$CH_3\dot{C}O + Cu^{II} \longrightarrow CH_3\overset{+}{C}O + Cu^{I} \tag{37}$$

$$CH_3\overset{+}{C}O + CH_3CO_2H \longrightarrow (CH_3CO)_2O + H^+ \tag{38}$$

Copper(II) is such an effective oxidant in reaction 37 that it competes successfully with molecular oxygen for the acyl radical. Copper(II) is unique in this respect, since other metal oxidants such as Mn(III) and Co(III) are much less effective.

The higher aldehydes can also be autoxidized to the corresponding carboxylic acids, providing a route for the preparation of higher fatty acids from olefins via hydroformylation and subsequent oxidation,[37] i.e.,

$$RCH{=}CH_2 \xrightarrow[\text{[catalyst]}]{CO/H_2} RCH_2CH_2CHO \xrightarrow{O_2} RCH_2CH_2COOH \tag{39}$$

A competing side reaction in the autoxidation of aldehydes is provided by the decarbonylation of the intermediate acyl radicals:

$$R\dot{C}O \longrightarrow R\cdot + CO \tag{40}$$

Decarbonylation is favored (relative to reaction 33) at high temperatures and low oxygen concentrations and when the alkyl moiety R is branched at the α position. Reaction 40 also appears to be promoted in the presence of metal catalysts such as Co and Mn. For example, the autoxidation of the branched aldehydes 2-ethylhexanal and 2-methylpentanal at 80°C in the presence of Co, Mn, or Ni catalysts produces only low yields (< 10%) of the corresponding carboxylic acids.[38] On the other hand, the straight autoxidation of isobutyraldehyde[39] or 2-ethylhexanal[40] at 30°C in the absence of metal catalysts affords the corresponding acids in high yield:

(2-ethylhexanal, CHO) $\xrightarrow{O_2}$ (2-ethylhexanoic acid, CO_2H) (41)

Catalyst	Temp (°C)	Selectivity (%)
None	30	86
Mn^{II}	30	63
Mn^{II}	10	83

Similarly, in the autoxidation of hydroformylated olefins, the aldehydes were rigorously purified to remove traces of metals (cobalt and rhodium catalysts are used in hydroformylations) prior to the oxidation step.[37] Aldehydes derived from C_5–C_{10} olefins gave the corresponding acids in high selectivities (~85%) even at relatively high temperatures (80°–95°C) after this treatment.

Schwab and co-workers[41] have described the selective autoxidation of hydroformylated oleate to 9(10)-carboxystearate at 20°C in the presence of 0.5% calcium naphthenate.

Scheme VIII:

$$CH_3(CH_2)_7CH{=}CH(CH_2)_7CO_2CH_3 \xrightarrow[\text{[Rh]}]{H_2/CO}$$

$$CH_3(CH_2)_7\underset{\substack{|\\ CHO}}{C}H(CH_2)_8CO_2CH_3 + CH_3(CH_2)_8\underset{\substack{|\\ CHO}}{C}H(CH_2)_7CO_2CH_3 \xrightarrow{O_2}$$

$$CH_3(CH_2)_7\underset{\substack{|\\ CO_2H}}{C}(CH_2)_8CO_2CH_3 + CH_3(CH_2)_8\underset{\substack{|\\ CO_2H}}{C}H(CH_2)_7CO_2CH_3$$

Methyl 9(10)-carboxystearates were obtained in 95% selectivity. In the presence of conventional autoxidation catalysts such as those containing Fe, Co, Mn, or Ce, lower selectivities (60–80%) were observed.

Aromatic aldehydes are also readily autoxidized to the corresponding benzoic acids. Although the reaction occurs in the absence of a catalyst, cobalt salts are often added to catalyze the reaction.[42,43] The benzoic acids are formed in virtually quantitative yields, since the competing decarbonylation in eq 40 is not a problem.

Autoxidations of α,β-unsaturated aldehydes such as acrolein[44,45] and methacrolein[46] are rather unselective and afford large amounts of polymeric material. However, Zawadzki and Ziolkowski[47] recently reported that acrolein can be selectively oxidized to acrylic acid using a $Co(acac)_3$ catalyst and acetaldehyde as promoter at 30°–40°C. The best results (93% selectivity at 32% conversion) were obtained in dioxane as solvent.

$$CH_2{=}CHCHO + \tfrac{1}{2}\,O_2 \xrightarrow[\text{dioxane}]{[Co(acac)_3]} CH_2{=}CHCO_2H \tag{42}$$

α,β-Unsaturated aldehydes can also be converted to the corresponding acids in good yield by autoxidation in the presence of copper catalysts. Methacrolein, for example, is converted to methacrylic acid in 96% yield using a $Cu(OAc)_2/Ni(OAc)_2$ catalyst in toluene or benzene solution at 30°C and 14 bar O_2 pressure.[48]

Copper catalysts are also involved in the interesting ammoxidation of aldehydes to the corresponding nitriles with a mixture of NH_3 and O_2.[49] For example, *p*-methoxybenzonitrile was obtained in 90% yield from the oxidation of anisaldehyde with a mixture of NH_3 and O_2 in the presence of NaOH and a catalytic amount of $CuCl_2 \cdot 2\,H_2O$ in methanol at 30°C. The reaction probably proceeds via the following sequence.

$$RCHO + NH_3 \longrightarrow RCH{=}NH + H_2O \tag{43}$$

$$RCH{=}NH + \tfrac{1}{2}\,O_2 \xrightarrow{[Cu^{II}]} RC{\equiv}N + H_2O \tag{44}$$

Aldehydes can also be oxidized to the corresponding carboxylic acids using stoichiometric amounts of inorganic oxidants, such as Cr(VI) compounds, permanganate, and MnO_2. However, since standard autoxidation procedures are usually satisfactory, the use of such stoichiometric reagents is limited to small-scale laboratory preparation.

Finally, we have already mentioned in Chapter 5 the various metal-catalyzed oxidative additions of aldehydes to olefins in which acyl radicals are intermediates.

Scheme IX:

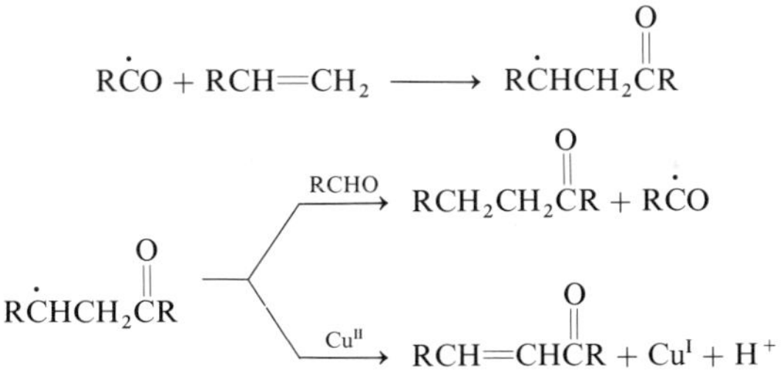

$$R\dot{C}O + RCH{=}CH_2 \longrightarrow R\dot{C}HCH_2\overset{\overset{\displaystyle O}{\|}}{C}R \tag{45}$$

$$R\dot{C}HCH_2\overset{\overset{\displaystyle O}{\|}}{C}R \xrightarrow{RCHO} RCH_2CH_2\overset{\overset{\displaystyle O}{\|}}{C}R + R\dot{C}O \tag{46}$$

$$R\dot{C}HCH_2\overset{\overset{\displaystyle O}{\|}}{C}R \xrightarrow{Cu^{II}} RCH{=}CH\overset{\overset{\displaystyle O}{\|}}{C}R + Cu^{I} + H^{+} \tag{47}$$

V. KETONES

A. Autoxidation

Ketones undergo autoxidation at the activated α-C—H bond to form an α-ketohydroperoxide as the primary product.[1]

Scheme X:

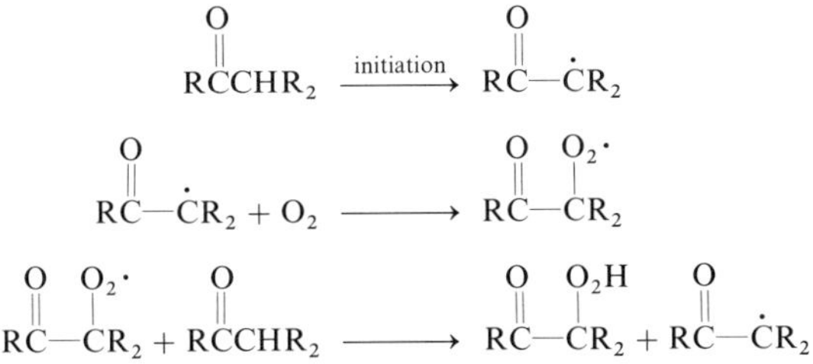

$$R\overset{\overset{\displaystyle O}{\|}}{C}CHR_2 \xrightarrow{\text{initiation}} R\overset{\overset{\displaystyle O}{\|}}{C}{-}\dot{C}R_2 \tag{48}$$

$$R\overset{\overset{\displaystyle O}{\|}}{C}{-}\dot{C}R_2 + O_2 \longrightarrow R\overset{\overset{\displaystyle O}{\|}}{C}{-}\overset{\overset{\displaystyle O_2\cdot}{|}}{C}R_2 \tag{49}$$

$$R\overset{\overset{\displaystyle O}{\|}}{C}{-}\overset{\overset{\displaystyle O_2\cdot}{|}}{C}R_2 + R\overset{\overset{\displaystyle O}{\|}}{C}CHR_2 \longrightarrow R\overset{\overset{\displaystyle O}{\|}}{C}{-}\overset{\overset{\displaystyle O_2H}{|}}{C}R_2 + R\overset{\overset{\displaystyle O}{\|}}{C}{-}\dot{C}R_2 \tag{50}$$

The α-ketohydroperoxide is thermally labile and decomposes to products resulting from C—C bond cleavage (*vide infra*). A few absolute rate constants for ketone autoxidations are listed in Table IV.

Ketones are also readily autoxidized at relatively low temperatures in the presence of stoichiometric amounts of KOH or NaOH.[50] The reaction proceeds best in polar aprotic solvents such as hexamethylphosphoric triamide, and carboxylic acids are formed in moderate to excellent yields. The decomposition of the α-ketohydroperoxide derived from cyclohexanone to adipic acid is illustrated in Scheme XI.

Scheme XI:

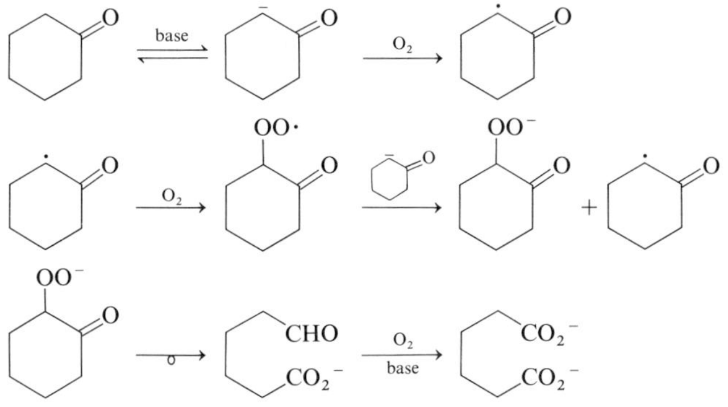

base O_2 $OO\cdot$ OO^- O_2 + OO^- CHO CO_2^- O_2 base CO_2^- CO_2^-

TABLE IV Absolute Rate Constants for Ketone Autoxidations at 30°C[a]

Ketone	$k_p/(2k_t)^{1/2} \times 10^3$ $(M^{-1/2}\ sec^{-1/2})$	k_p $(M^{-1}\ sec^{-1})$	$2k_t \times 10^{-6}$ $(M^{-1}\ sec^{-1})$
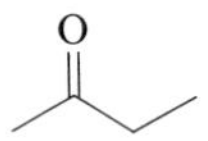	0.09	0.11	1.4
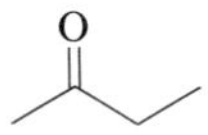	0.05	0.07	1.8
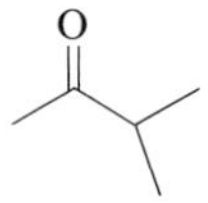	0.25	0.36	2.1
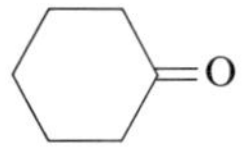	0.52	0.75	2.0

[a] From Howard.[2]

When the base-catalyzed autoxidation of cyclic ketones is carried out at low temperatures ($-20°C$), the major product is the corresponding α-diketone,[51] where $n = 0,1$.

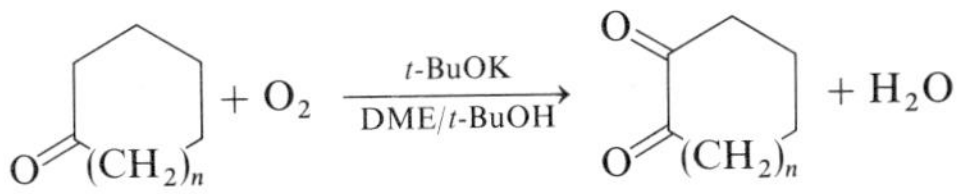

$$\text{cyclic ketone } (CH_2)_n + O_2 \xrightarrow[\text{DME}/t\text{-BuOH}]{t\text{-BuOK}} \alpha\text{-diketone } (CH_2)_n + H_2O \qquad (51)$$

The base-catalyzed autoxidation of ketones under phase transfer catalysis conditions may obviate the need for polar aprotic solvents (compare the autoxidation of *p*-nitrotoluene). Indeed, the micellar catalysis in the base-catalyzed autoxidation of diosphenol,

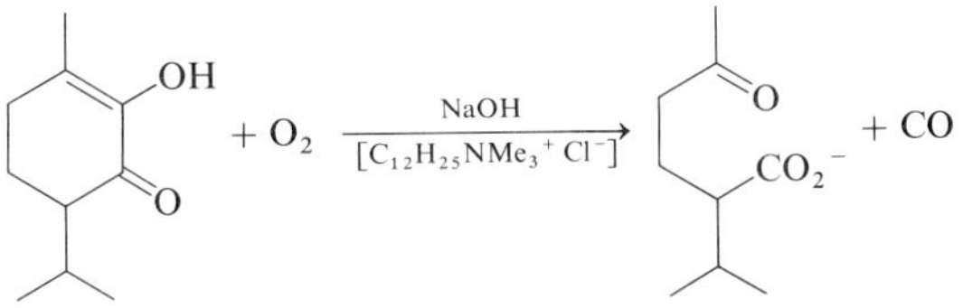

$$\text{diosphenol} + O_2 \xrightarrow[[C_{12}H_{25}NMe_3^+\,Cl^-]]{\text{NaOH}} \text{keto acid } CO_2^- + CO$$

has been shown to increase the yield of the keto acid from 47 to 76% yield.[51a]

B. Metal-Catalyzed Autoxidation

Ketones undergo facile metal-catalyzed autoxidation to give carboxylic acids resulting from C—C bond cleavage. Indeed, ketones are often added as promoters in the metal-catalyzed autoxidation of hydrocarbons. (See the oxidation of alkylaromatic compounds and *n*-butane in Chapters 10 and 11.) These reactions involve the direct oxidation of the ketone enolate by the metal oxidant, as described in Chapter 5. Thus, the rate of oxidation of acetophenone catalyzed by manganese(II) acetate in acetic acid and butyric acid at 130°C was equal to the rate of enolization under these conditions.[52]

Scheme XII:

$$PhCOCH_3 \overset{HOAc}{\rightleftharpoons} PhC(OH){=}CH_2 \xrightarrow{[Mn^{III}]} PhCOCH_2\cdot$$

$$PhCOCH_2\cdot \xrightarrow{O_2} PhCOCH_2OO\cdot \xrightarrow{[Mn^{III}]} PhCO_2H$$

Manganese salts are the catalysts of choice for the oxidation of acetophenone to benzoic acid and the conversion of cyclohexanone to adipic acid (see Chapter 11). For the oxidative condensation reaction of ketones with olefinic substrates see Chapter 5, Section IV.

C. Oxidation with Oxometal and Peroxometal Reagents

Ketones are oxidatively cleaved to carboxylic acids with stoichiometric amounts of permanganate and chromium(VI) compounds. Selenium dioxide is the reagent *par exellence* for the oxidation of ketones and aldehydes to 1,2-dicarbonyl compounds.[53,54] This reaction constitutes one of the more important synthetic applications of SeO_2.

$$RC(=O)-CH_2R' \xrightarrow{SeO_2} RC(=O)-C(=O)R' \qquad (52)$$

The reaction is usually carried out in acetic acid or alcohol solvents at 80°–100°C. A few examples are illustrated below:[55–57]

$$CH_3CHO \xrightarrow{SeO_2} \underset{90\%}{OHCCHO} \qquad (53)$$

$$PhCOCH_3 \xrightarrow{SeO_2} \underset{70\%}{PhCOCHO} \qquad (54)$$

$$\text{[bicyclic ketone]}{=}O \xrightarrow{SeO_2} \text{[bicyclic 1,2-dione]} \quad 90\% \qquad (55)$$

The mechanism probably involves the reaction of SeO_2 with the enolate (see Chapter 6). These reactions involve the use of stoichiometric amounts of SeO_2 and form colloidal selenium and organoselenium compounds as by-products. The use of catalytic amounts of SeO_2 in the presence of primary oxidants (e.g., peroxides) does not appear to have been fully investigated. Interestingly, the SeO_2-catalyzed oxidation of cyclic ketones with H_2O_2 in *tert*-butyl alcohol results in ring contraction:[58,59]

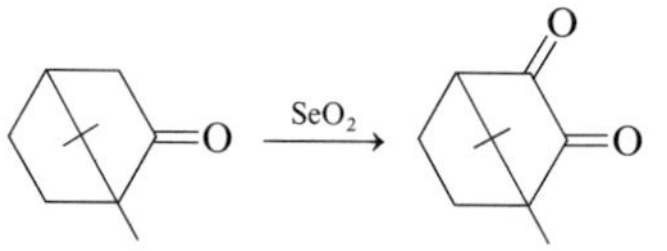

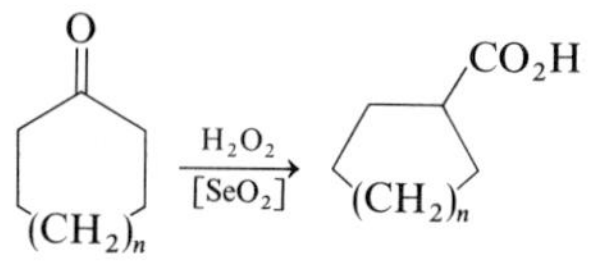

$$\text{cyclic ketone } (CH_2)_n \xrightarrow[\text{[SeO}_2]{H_2O_2} \text{ring-contracted } (CH_2)_n\text{–}CO_2H \qquad (56)$$

n	Yield (%)
7	32
2	34
1	32
0	23

The same reagent has also been used for the selective oxidation of acrolein to acrylic acid[60] (compare other methods for the oxidation of α,β-unsaturated aldehydes in the preceding section):

$$CH_2{=}CHCHO + H_2O_2 \xrightarrow{[SeO_2]} CH_2{=}CHCO_2H + H_2O \tag{57}$$

These reactions probably involve the peroxyselenium(IV) acid as the active oxidant:

$$H_2O_2 + SeO_2 \longrightarrow HO{-}Se({=}O){-}OOH \tag{58}$$

Another useful oxidative transformation of ketones is the Baeyer–Villiger reaction in which peracids produce the corresponding esters, or lactones in the case of cyclic ketones.[61]

$$R\overset{\overset{O}{\|}}{C}R + R'\overset{\overset{O}{\|}}{C}OOH \longrightarrow R\overset{\overset{O}{\|}}{C}OR + R'\overset{\overset{O}{\|}}{C}OH \tag{59}$$

A catalytic process for Baeyer–Villiger oxidations has recently been reported.[62,63] Arsonated polystyrenes, for example, catalyze the oxidation of ketones with H_2O_2 to esters or lactones.[62] Reactions were carried out under both biphasic (H_2O_2-miscible solvents) and triphasic (H_2O_2-immiscible solvents) conditions. The active oxidant in this system is the polymerbound peroxyarsonic acid.

Scheme XIII:

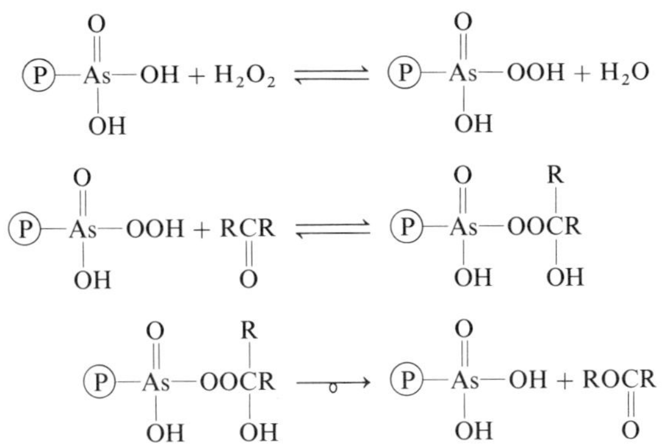

$$\text{Ⓟ}{-}As({=}O)(OH){-}OH + H_2O_2 \rightleftharpoons \text{Ⓟ}{-}As({=}O)(OH){-}OOH + H_2O$$

$$\text{Ⓟ}{-}As({=}O)(OH){-}OOH + R\overset{\overset{}{}}{C}({=}O)R \rightleftharpoons \text{Ⓟ}{-}As({=}O)(OH){-}OOC(R)(OH)R$$

$$\text{Ⓟ}{-}As({=}O)(OH){-}OOC(R)(OH)R \longrightarrow \text{Ⓟ}{-}As({=}O)(OH){-}OH + ROC({=}O)R$$

Mares and co-workers[63] also reported the Baeyer–Villiger oxidation of cyclic ketones with H_2O_2 and a catalyst consisting of peroxomolybdenum(VI) complexes stabilized by picolinato or pyridine-2,6-dicarboxylato ligands (see Chapter 4).

Scheme XIV:

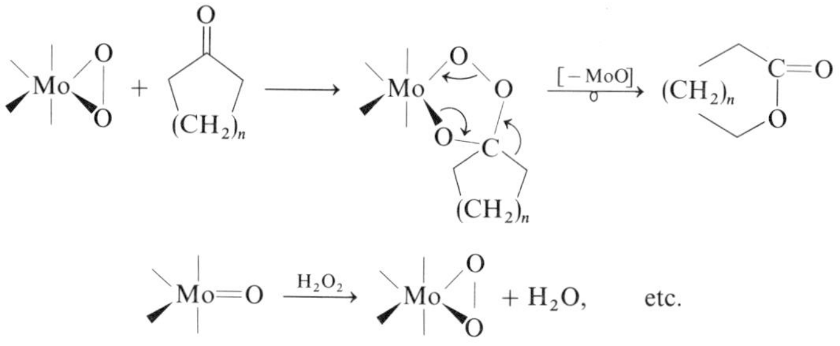

$$\text{Mo(O}_2) + (CH_2)_n\text{C=O} \longrightarrow \text{Mo–O–O–C(O)–}(CH_2)_n \xrightarrow{[-\text{MoO}]} (CH_2)_n\text{–C(=O)–O (lactone)}$$

$$\text{Mo=O} \xrightarrow{H_2O_2} \text{Mo(O}_2) + H_2O, \quad \text{etc.}$$

It should be noted, however, that both of these procedures appear to require the use of concentrated (90%) H_2O_2.

Finally, the selective oxidation of ketone enolates to α-ketols can be effected with a catalytic amount of OsO_4 and *N*-methylmorpholine oxide as the primary oxidant[64] (compare the olefin hydroxylation with the same reagent in Chapter 9).

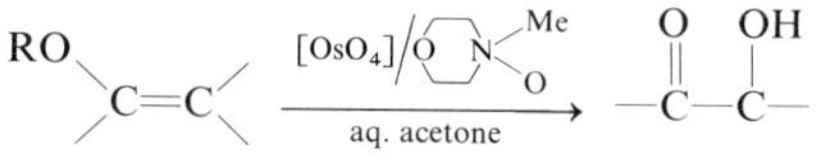

$$\text{RO}\!\!>\!C{=}C\!< \xrightarrow[\text{aq. acetone}]{[OsO_4]/\text{O}\langle\ \rangle\text{N(Me)}\to\text{O}} -\overset{\overset{O}{\|}}{C}-\overset{\overset{OH}{|}}{C}- \tag{60}$$

The same transformation can also be carried out using the molybdenum(VI)–peroxide reagents described in Chapter 4.

VI. PHENOLS

One of the characteristic chemical properties of phenols is their facile oxidation, which can be accomplished with almost any oxidant.[65–70] Not only are these reactions of synthetic importance, but they are also implicated in many biogenetic pathways.[67–70] (Enzymatic hydroxylations of phenols have been discussed in Chapter 8.) Because of their facile reaction with alkylperoxy radicals, phenols are used as inhibitors in free radical autoxidations. Autoxidation of phenols under neutral or basic[71] conditions generally affords a complex mixture of products. However, in the presence of certain metal catalysts, several synthetically useful oxidative transformations of phenols can be accomplished. These reactions probably all involve, as a key feature, the oxidation of phenolate to the corresponding phenoxy radical, e.g.,

$$ArOM^{n+} \longrightarrow ArO\cdot + M^{(n-1)+} \tag{61}$$

A. Copper-Catalyzed Oxidations

Depending on the nature of the copper catalyst and the reaction conditions, phenols undergo a variety of oxidative transformations in the presence of copper compounds, as summarized in Scheme XV,

Scheme XV:

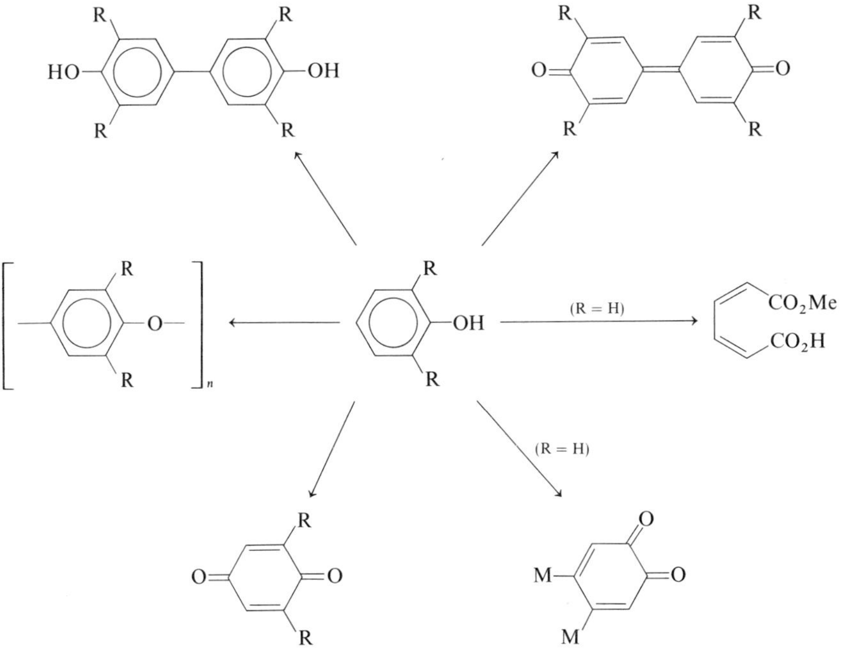

where M = morpholinyl group and R = alkyl group, except where noted.

Hay and co-workers[72–75] found that 2,6-disubstituted phenols are oxidized by molecular oxygen at room temperature in the presence of a cuprous salt and a tertiary amine (the catalyst is usually CuCl/pyridine) to give the corresponding polyphenylene ethers and/or diphenoquinones:

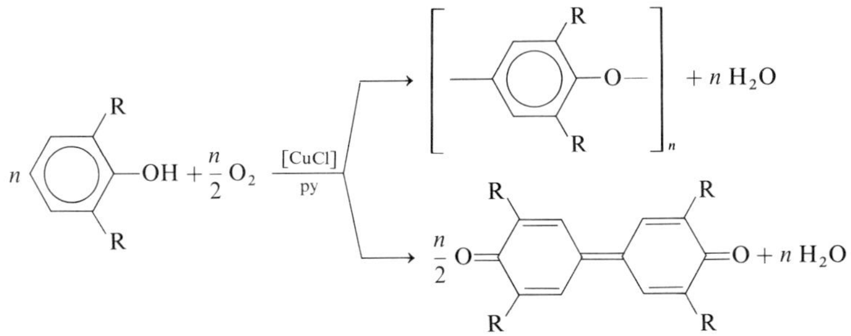

(62)

(63)

The ratio of carbon–oxygen (reaction 62) to carbon–carbon (reaction 63) coupling is determined by, *inter alia*, the size of the group R, the molar ratio of the amine to Cu(I), and the temperature. When R is a bulky group, such as *tert*-butyl, carbon–carbon coupling occurs and the diphenoquinone is the sole product.[72,73] With small substituents, as in 2,6-dimethylphenol, carbon–carbon coupling predominates at low catalyst concentrations and/or low molar ratios (1 : 1) of pyridine to Cu(I). Carbon–oxygen coupling is favored at high (10 : 1) molar ratios of pyridine to Cu(I).[74,75] Increasing the temperature or the use of sterically hindered amine ligands also favors carbon–carbon coupling.[75] The mechanism of oxidative coupling has been extensively studied.[73–84] Copper(II)–amine complexes act as catalysts only in the presence of a strong base such as KOH or NaOMe.[73,78] Although copper(II) hydroxide in pyridine is inactive, the addition of one equivalent of HCl produces an active catalyst.[73] It was concluded, therefore, that the active catalyst is a basic salt, e.g.,

$$(py)_2Cu^{II}Cl_2 \xrightarrow[-Cl^-]{HO^-} (py)_2Cu^{II}(OH)Cl \underset{}{\overset{-2\,py}{\rightleftharpoons}} (py)(HO)Cu^{II}(\mu\text{-}Cl)_2Cu^{II}(OH)(py) \qquad (64)$$

The same basic Cu(II) complex can be formed in the oxidation of the pyridine –copper(I) complex with dioxygen.[74] In methanol solution, the corresponding pyridine–copper(II) methoxide is formed. (See Chapter 4 for a discussion of the oxidation of CuCl in pyridine.)

$$2\,CuCl + \tfrac{1}{2}\,O_2 + 2\,ROH \xrightarrow[-H_2O]{2\,py} (py)(RO)Cu(\mu\text{-}Cl)_2Cu(OR)(py) \overset{2\,py}{\rightleftharpoons} 2\,(py)_2Cu(Cl)(OR)$$

where R = H, CH_3, Ar. The first step in the oxidative coupling reaction is generally accepted to be the formation of the phenoxy radical via one-electron oxidation of phenolate by Cu(II).

$$2,6\text{-}R_2C_6H_3\text{–}OCu^{II} \longrightarrow 2,6\text{-}R_2C_6H_3\text{–}O\cdot + Cu^{I} \qquad (65)$$

At low catalyst concentrations and/or molar ratio of pyridine to copper, the intermediate phenoxy radicals undergo dimerization by carbon–carbon coupling.

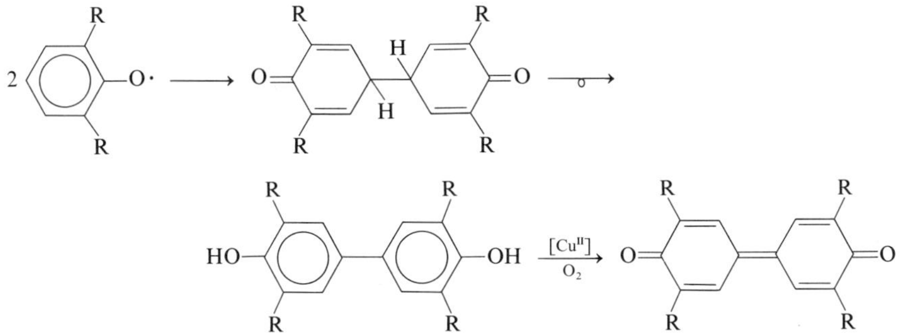

The intermediate dihydroxybiphenyl is further oxidized by Cu(II) and/or dioxygen to the diphenoquinone. It has recently been shown[85] that anaerobic oxidation of 2,6-dimethylphenol by stoichiometric amounts of a Cu(II)–phenylethylamine complex produces predominantly the dihydroxybiphenyl. At high catalyst concentrations and/or high molar ratios of pyridine to copper, the carbon–oxygen coupling results, probably from ligand transfer oxidation of the phenoxy radical by Cu(II) species.

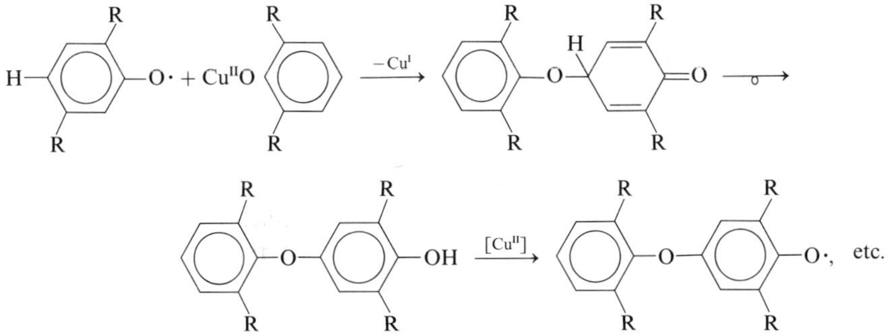

When the oxidation of alkyl-substituted phenols with stoichiometric amounts of copper(II) salts was carried out in an oxygen atmosphere (1–30 bar) in dimethylformamide as solvent, the corresponding *p*-benzoquinone was the major product.[86] Oxidation of 2,3,6-trimethylphenol, for example, afforded the vitamin E intermediate, 2,3,5-trimethyl-1,4-benzoquinone, in 96% yield.

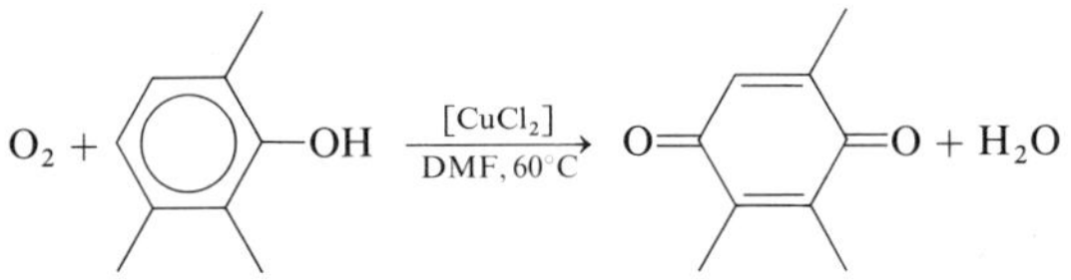

$$O_2 + \text{ArOH} \xrightarrow[\text{DMF, }60^\circ\text{C}]{[CuCl_2]} O{=}\text{Ar}{=}O + H_2O \qquad (66)$$

The first step in this reaction may be the chlorination of the phenol, which is known to occur in DMF in the absence of oxygen.[87]

Scheme XVI:

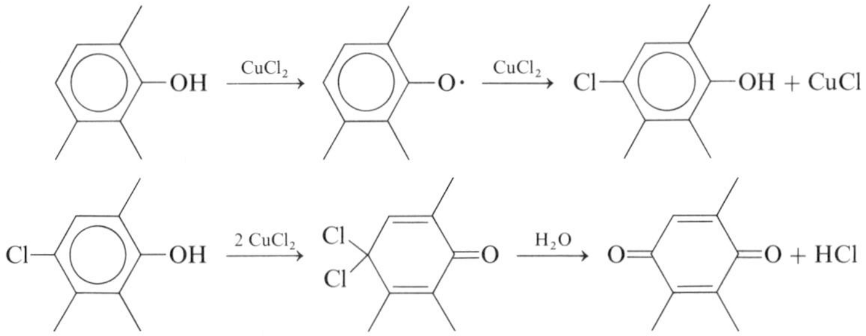

Alternatively, the benzoquinone could result from the autoxidation of the intermediate phenoxy radicals, followed by reaction with Cu(I):

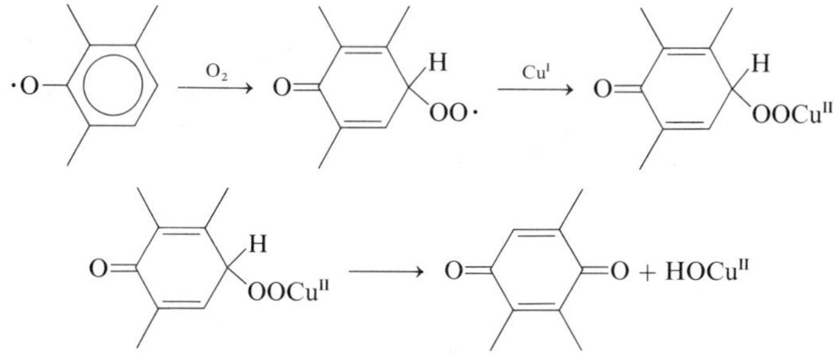

Indeed, both mechanisms may be operating, depending on the concentration of Cu(II), the oxygen pressure, and the presence of chloride. Thus, in support of the chlorination mechanism, substituted *p*-chlorophenols can be converted to the corresponding *p*-benzoquinones under the reaction conditions.[86] However, it was also found[86] that Cu(II) sulfate catalyzes the oxidation of phenols to the *p*-benzoquinone; i.e., the reaction proceeds in the absence of chloride. More recently, it has been shown[88] that phenol itself is selectively oxidized to 1,4-benzoquinone by reaction with dioxygen at high pressures (70 bar), in acetonitrile at 40°C, and in the presence of catalytic amounts of CuCl or $CuCl_2$.

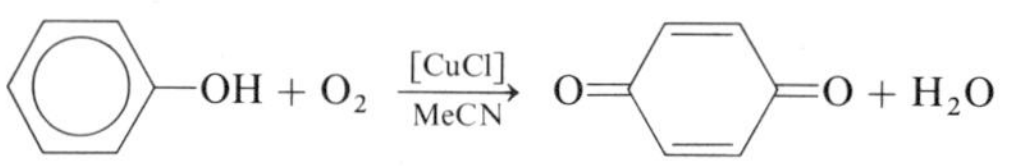

$$C_6H_5OH + O_2 \xrightarrow[\text{MeCN}]{[\text{CuCl}]} O{=}C_6H_4{=}O + H_2O \tag{67}$$

An 80% selectivity was achieved at 93% conversion. Other workers[89] obtained 1,4-benzoquinone in 60% selectivity at a 89% conversion of phenol by carrying out the oxidation with dioxygen (100 bar) in methanol at 70°C in the presence of a catalytic amount of $CuCl_2$ and a reducing agent (e.g., hydroquinone).

In contrast, the oxidation of phenol catalyzed by pyridine–Cu(I) in methanol affords the monomethyl ester of *cis,cis*-muconic acid by oxidative cleavage involving catechol and *o*-benzoquinone as intermediates[90] (see Chapter 4). The active oxidant is thought to be $py_2Cu(OMe)OH$.[90]

Scheme XVII:

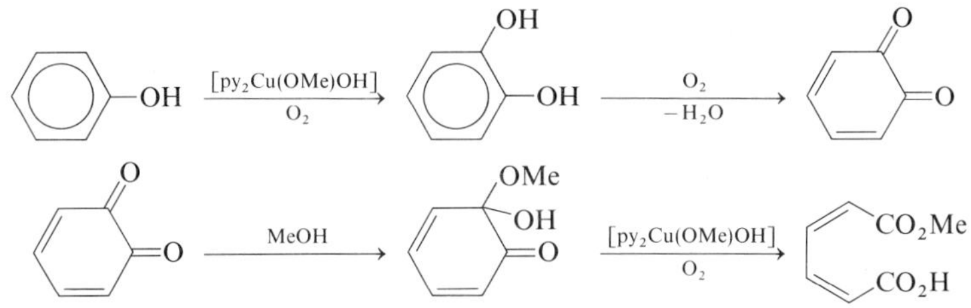

$$\text{C}_6\text{H}_5\text{OH} \xrightarrow[\text{O}_2]{[\text{py}_2\text{Cu(OMe)OH}]} \text{catechol} \xrightarrow[-\text{H}_2\text{O}]{\text{O}_2} o\text{-benzoquinone}$$

$$o\text{-benzoquinone} \xrightarrow{\text{MeOH}} \text{(OMe)(OH) adduct} \xrightarrow[\text{O}_2]{[\text{py}_2\text{Cu(OMe)OH}]} \text{MeO}_2\text{C-CH=CH-CH=CH-CO}_2\text{H}$$

Finally, the oxidation of phenol in the presence of a morpholine–copper(II) complex produces the dimorpholino-substituted *o*-benzoquinone,[91]

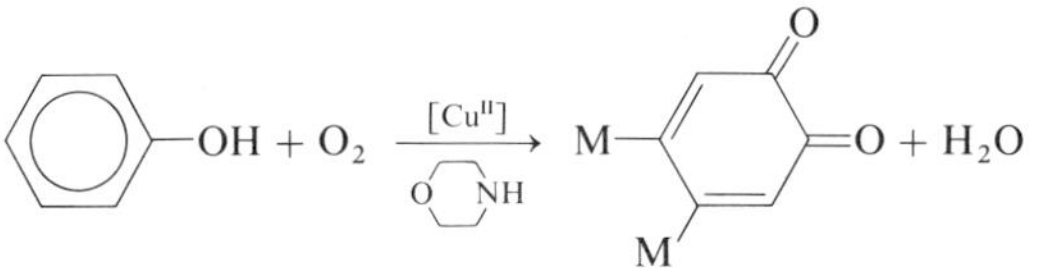

$$\text{C}_6\text{H}_5\text{OH} + \text{O}_2 \xrightarrow[\text{morpholine}]{[\text{Cu}^{\text{II}}]} \text{M}_2\text{-}o\text{-benzoquinone} + \text{H}_2\text{O} \qquad (68)$$

where M = morpholinyl. Apparently, in this case, the *o*-quinone intermediate is trapped as its morpholine adduct.

It is readily apparent from the above discussion that a diversity of phenol oxidative transformations can be accomplished in the presence of copper complexes. Many of these reactions are reminiscent of the biochemical reactions mediated by various copper- and iron-containing oxygenases (see Chapter 8). For example, the oxidative ring cleavage of 3,5-di-*tert*-butylcatechol by molecular oxygen is catalyzed by 2,2′-bipyridine–iron(II).[92]

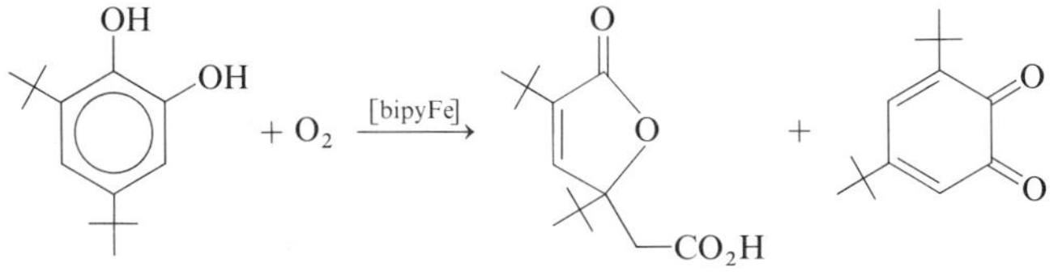

$$\text{3,5-di-}t\text{-butylcatechol} + \text{O}_2 \xrightarrow{[\text{bipyFe}]} \underset{\text{Minor}}{\text{lactone-CH}_2\text{CO}_2\text{H}} + \underset{\text{Major}}{\text{3,5-di-}t\text{-butyl-}o\text{-benzoquinone}}$$

This is the first example of the oxidative ring cleavage of phenols catalyzed by a simple iron complex, and it is of interest as a model for the dioxygenase, pyrocatechase, described in Chapter 8.

B. Cobalt(II)–Schiff Base Complexes as Catalysts

The oxidation of alkyl-substituted phenols can be effected by dioxygen in chloroform or methanol solutions in the presence of (Salen)Co(II) as catalyst (see Chapter 4 for the structure).[93] The corresponding *p*-benzoquinones are formed in moderate to good yields, e.g.,

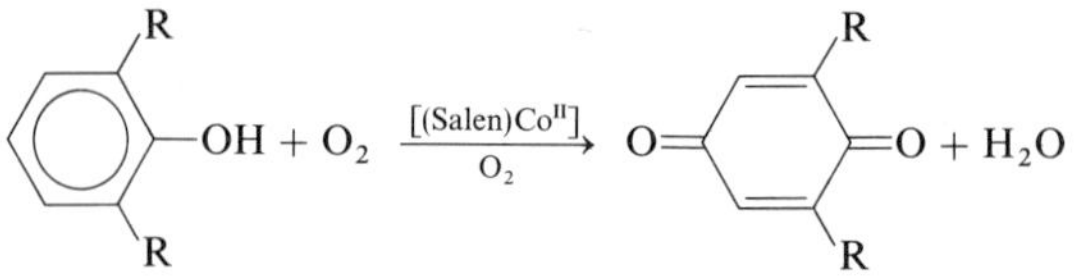

$$\text{2,6-}R_2C_6H_3OH + O_2 \xrightarrow[O_2]{[(\text{Salen})Co^{II}]} O{=}C_6H_2R_2{=}O + H_2O \qquad (69)$$

where R = CH_3, *tert*-Bu. This work was extended by Vogt and co-workers,[94] who showed that the (Salen)Co(II)-catalyzed autoxidations of phenols can give high yields of *p*-benzoquinones or diphenoquinones depending on the conditions (compare the copper-catalyzed oxidations described above).

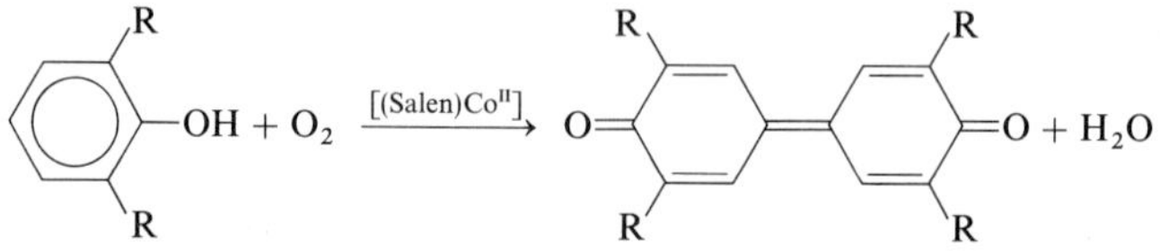

$$\text{2,6-}R_2C_6H_3OH + O_2 \xrightarrow{[(\text{Salen})Co^{II}]} O{=}C_6H_2R_2{=}C_6H_2R_2{=}O + H_2O \qquad (70)$$

The formation of benzoquinones is favored by high catalyst concentrations and low temperatures. Diphenoquinones are predominantly formed at low catalyst concentrations and high temperatures. From these results, it can be concluded that diphenoquinones result from the dimerization of intermediate phenoxy radicals, whereas the benzoquinones are derived from their reaction with the cobalt catalyst. (See Chapter 4 for a detailed discussion of the mechanism.)

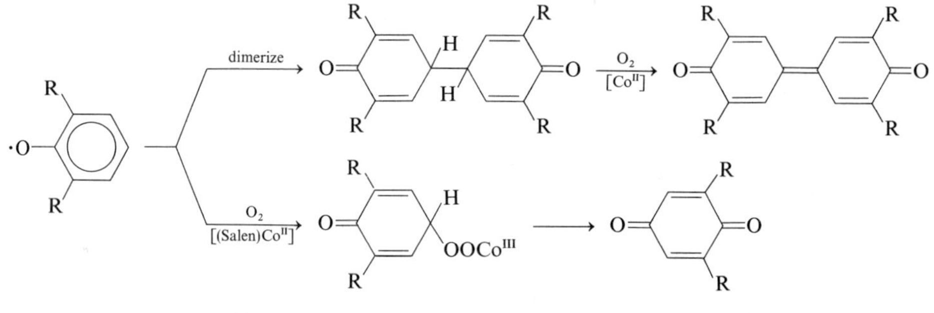

$$\cdot O\text{-}C_6H_3R_2 \xrightarrow{\text{dimerize}} O{=}C_6H_2R_2(H)\text{-}(H)C_6H_2R_2{=}O \xrightarrow[[Co^{II}]]{O_2} O{=}C_6H_2R_2{=}C_6H_2R_2{=}O$$

$$\cdot O\text{-}C_6H_3R_2 \xrightarrow[[(\text{Salen})Co^{II}]]{O_2} O{=}C_6H_2R_2(H)(OOCo^{III}) \longrightarrow O{=}C_6H_2R_2{=}O$$

The same group[95] also studied the effect of ring substituents (R and R′) in the Salen ligand on the rates and selectivities of these oxidations:

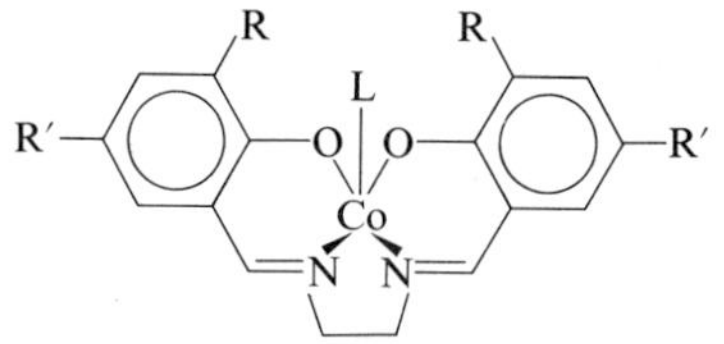

$$\text{(R,R'-substituted Salen)Co–L}$$

Electron-donating groups (e.g., methoxy) facilitated the formation of benzoquinones. Aquo-3-fluoro(Salen)cobalt(II) (R = F; R′ = H; L = H_2O) was found to be a good catalyst for the oxidation 2,6-dichlorophenol to the corresponding benzoquinone. [(Salen)Co(II) was ineffective with this unreactive phenol.] The reactivities of phenols toward oxidation increases significantly with the number of alkyl or other electron-donating substituents. Dialkyl-substituted phenols are smoothly oxidized, and monoalkylphenols are only moderately reactive. Phenol itself is oxidized very slowly. Subsequent studies have shown that higher rates and selectivities to *p*-benzoquinones can be obtained when these reactions are performed in dimethylformamide as solvent.[96–98] For example, 2,3,5-trimethylphenol was oxidized to 2,3,5-trimethyl-1,4-benzoquinone in 94% selectivity at 94% conversion, with little or no reaction at the free ortho position.[96b]

$$\text{2,3,5-trimethylphenol} + O_2 \xrightarrow[\text{DMF, 65°C}]{[(\text{Salen})\text{Co}^{\text{II}}]} \text{2,3,5-trimethyl-1,4-benzoquinone}$$

(Compare the copper-catalyzed oxidation in eq 66.) Examples of *p*-benzoquinone syntheses using various catalysts are compiled in Table V.

Alkyl-substituted phenols are preferentially oxidized to *p*-benzoquinones. However, suitably activated phenols containing a blocked para position can be selectively oxidized to the corresponding *o*-benzoquinone, e.g.,[99]

$$\text{2,4-di-}tert\text{-butylphenol} + O_2 \xrightarrow[\text{DMF, 60°C, 4 hr}]{[\text{py(Salen)Co}^{\text{II}}]} \text{3,5-di-}tert\text{-butyl-}o\text{-benzoquinone (91\%)} \quad (71)$$

When the phenol is blocked in both the ortho and para positions, selective formation of the corresponding *p*-quinol results in the presence of (Salpr)Co(II) (see Chapter 4 for structure), e.g.,[100]

$$\text{HO–C}_6\text{H}_2(t\text{-Bu})_2\text{–CH}_3 + O_2 \xrightarrow[\text{MeOH, 25°C}]{[(\text{Salpr})\text{Co}^{\text{II}}]} \text{2,6-di-}tert\text{-butyl-4-methyl-4-hydroxycyclohexa-2,5-dienone (96\%)} \quad (72)$$

Since the (Salen)Co(II)-catalyzed oxidations described above are usually carried out in the absence of water, it is noteworthy that 4-hydroxy-(Salen)cobalt(II) catalyzes the autoxidation of 2,6-di-*tert*-butylphenol to the *p*-benzoquinone in aqueous organic media under neutral or basic conditions.[100a]

TABLE V Autoxidation of Phenols to 1,4-Benzoquinones

Phenol	Catalyst[a]	Solvent	Temp (°C)	Pressure (bar)	Time (hr)	Phenol conversion (%)	Quinone selectivity (%)	Ref.
OH	A	DMF	25–45	1	0.75	100	92	96a
OH	A	DMF	65	1	2	94	94	96b
OH	A	DMF	65	1	2	87	90	96b
OH	A	DMF	65	1	2	35	90	96b
OH	A	DMF	65	1	2	99	90	96b

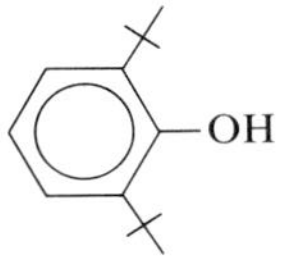 OH	A	DMF	20–45	4	0.5	100	99	97
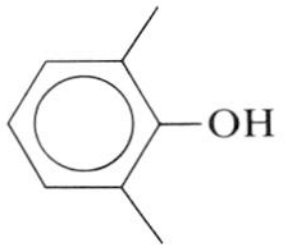 OH	A	DMF	20–45	4	0.75	100	~100	97
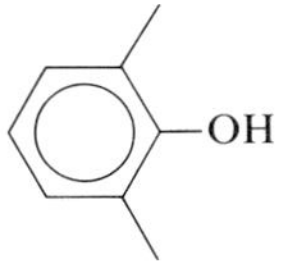 OH	A	$CHCl_3$	20	1	12	93	56	94
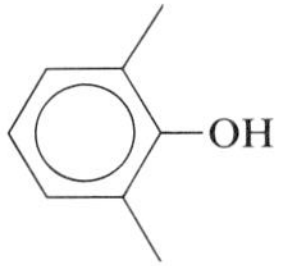 OH	B	DMF	22–28	4	5	100	100	97
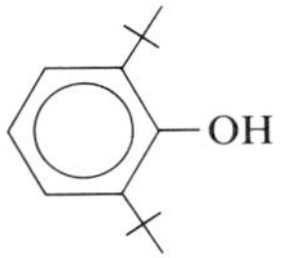 OH	B	DMF	22–27	4	3	100	73	97
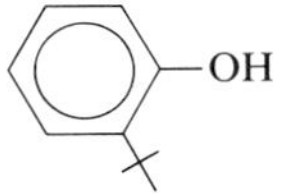 OH	C	DMF	20–35	4	1.5	46	96	97
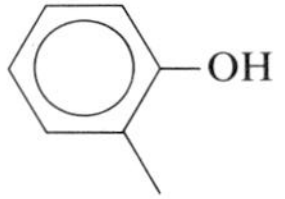 OH	C	DMF	22–40	4	3	26	92	97

(continued)

TABLE V (*continued*)

Phenol	Catalyst[a]	Solvent	Temp (°C)	Pressure (bar)	Time (hr)	Phenol conversion (%)	Quinone selectivity (%)	Ref.
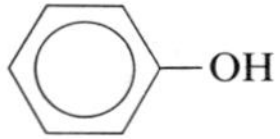 OH	C	DMF	68–73	6.5	2	23	~100	97
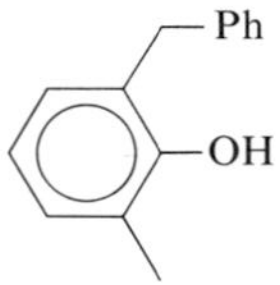 Ph, OH	C	$CHCl_3$	25	1	24	59	77	94,95
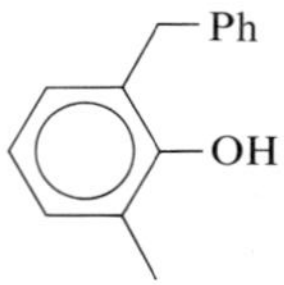 Ph, OH	D	$CHCl_3$	25	1	24	100	94	95
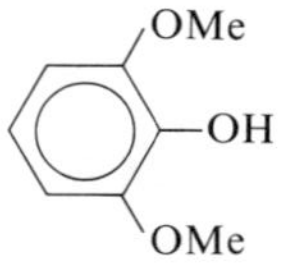 OMe, OH, OMe	A	DMF	30–50	1	4	86% yield		98
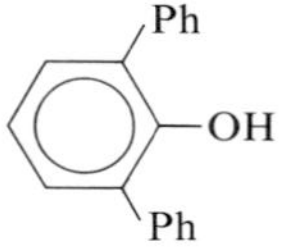 Ph, OH, Ph	A	DMF	30–50	1	4	91% yield		98

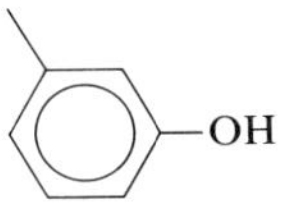 OH	E	DMK[b]	60	30	3	75% yield	86
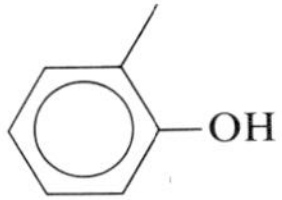 OH	E	DMF	60	30	3	41% yield	86
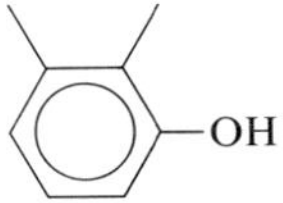 OH	E	DMK[b]	60	30	2.5	75% yield	86
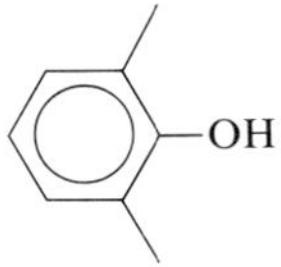 OH	E	DMF	60	30	2	77% yield	86
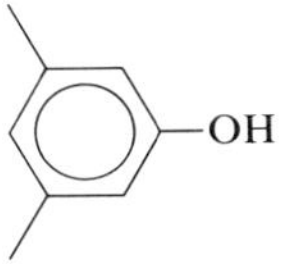 OH	E	DMK[b]	60	30	2	85% yield	86
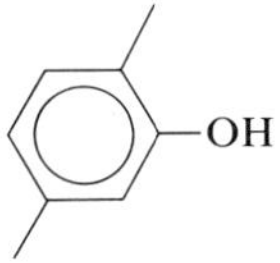 OH	E	DMK[b]	60	30	2.5	82% yield	86
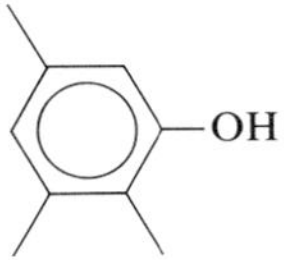 OH	E	DMK[b]	60	30	2.5	80% yield	86

(*continued*)

TABLE V (*continued*)

Phenol	Catalyst[a]	Solvent	Temp (°C)	Pressure (bar)	Time (hr)	Phenol conversion (%)	Quinone selectivity (%)	Ref.
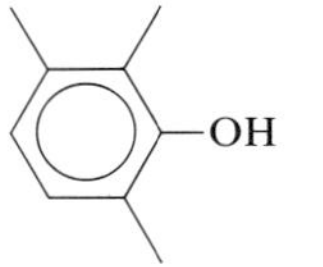 OH	E	DMF	60	1	2	96% yield		86
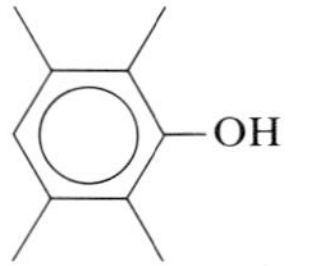 OH	E	ME[c]	60	1	3	99% yield		86
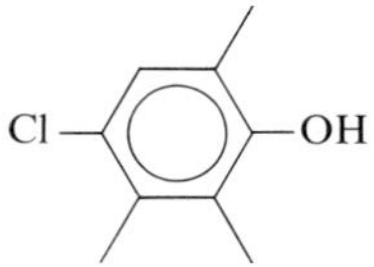 Cl, OH	E	DMF	60	1	3	99% yield		86
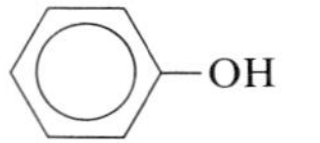 OH	F	MeCN	40	70	—	93	80	88

[a] A, (Salen)Co(II); B, phthalocyanine Co(II); C, (Salen)(py)Co(II); D, 3-methoxy(Salen)(py)Co(II) E, $CuCl_2$ (stochiometric amounts); F, CuCl.

[b] DMK, aqueous acetone.

[c] ME, Ethylene glycol monomethyl ether.

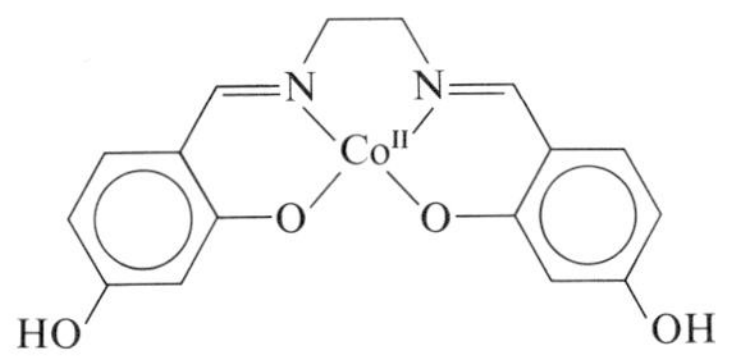

N N Co^{II} O O HO OH

An interesting recent development in this area is the use of polymer-supported Schiff base complexes of cobalt, which were shown to be as effective as their homogeneous counterparts for the oxidation of 2,6-dimethylphenol to the corresponding benzoquinone.[101] This catalyst system merits further investigation, particularly with regard to the stability of the catalyst to long-term usage. It could have important synthetic utility if the consumption or the recycling of expensive soluble cobalt complexes could be circumvented.

C. Oxidations with Hydrogen Peroxide

Hydroquinone is usually produced by the hydrogenation of *p*-benzoquinone, the latter being obtained by oxidation of aniline with MnO_2 in H_2SO_4. Recently, methods have been developed for the oxidation of phenol to benzoquinone (as described in the foregoing section). Alternatively, phenol can be hydroxylated directly to hydroquinone with hydrogen peroxide. Several processes have been reported for the hydroxylation of phenol to a mixture of hydroquinone and catechol (in a 1 : 2 molar ratio) using H_2O_2 in the presence of strong acid catalysts such as $HClO_4$ or HF.[102,103] A common feature of these processes is that an excess of phenol is used (10–30% conversion) to obtain high selectivities. These reactions involve hydroxylation of the aromatic ring by the putative hydroxyl cation, i.e.,

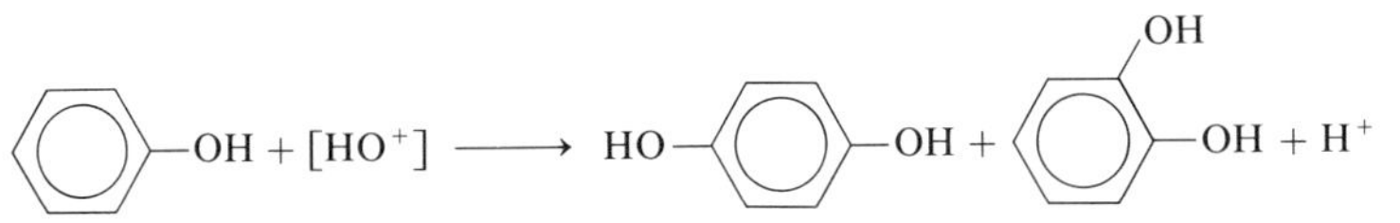

$$C_6H_5\text{–OH} + [HO^+] \longrightarrow HO\text{–}C_6H_4\text{–OH} + (OH)C_6H_4\text{–OH} + H^+ \qquad (73)$$

Phenol can also be hydroxylated with H_2O_2 in the presence of metal catalysts, such as copper(II),[104] chromium(III),[104] cobalt(II),[105] and ferrocenes,[106] to give hydroquinone and catechol in a molar ratio of about 2 : 1. Hydroxylation by $H_2O_2/Co(II)$, for example, affords dihydric phenols in approximately 90% yields.[105] These reactions presumably involve hydroxylation by the hydroxyl radical (see Chapter 3). In this context, it should be mentioned that hydroquinone can be prepared by the acid-catalyzed decomposition of the bishydroperoxide derived from 1,4-diisopropylbenzene[107] (compare the manufacture of phenol from cumene).

$$\text{(CH}_3)_2\text{CH–C}_6\text{H}_4\text{–CH(CH}_3)_2 + 2\,O_2 \longrightarrow \text{HOO–C(CH}_3)_2\text{–C}_6\text{H}_4\text{–C(CH}_3)_2\text{–OOH} \xrightarrow{[H^+]} \text{HO–C}_6\text{H}_4\text{–OH} + 2\,(CH_3)_2CO \quad (74)$$

Phenols can also be hydroxylated with alkaline persulfate (Elbs reaction).[108] Yields are, however, generally low (less than 50%). The reaction proceeds via a sulfate ester, probably formed via nucleophilic displacement:

$$^{-}\text{O–C}_6\text{H}_5 \xrightarrow{S_2O_8^{2-}} \text{O=C}_6\text{H}_4\text{(H)(OSO}_3^{-}) + SO_4^{2-}$$

$$\text{HO–C}_6\text{H}_4\text{–OSO}_3^{-} \xrightarrow{H_2O} \text{HO–C}_6\text{H}_4\text{–OH} + HSO_4^{-} \quad (75)$$

The selective oxidation of *p*-alkylphenols to dienone hydroperoxides with tetraperoxomolybdate (formed from molybdate and excess H_2O_2 in neutral aqueous media) has been described,[109] e.g.,

$$H_2O_2 + \text{HO–C}_6\text{H}_4\text{–CH}_2\text{CH}_2\text{COCH}_3 \xrightarrow[\text{MeOH, 25°C}]{[MoO_4^{2-}]} \text{O=C}_6\text{H}_4\text{(OOH)(CH}_2\text{CH}_2\text{COCH}_3) \quad (76)$$

D. Oxidation of Side Chains

Although phenolic oxidations invariably involve oxidation of the OH group and/or the aromatic ring (*vide supra*), an example has recently been reported of selective side-chain oxidation. Thus, oxygenation of substituted *p*-cresols in the presence of a large excess (10 : 1) of potassium *tert*-butoxide in dimethylformamide afforded the corresponding aldehydes,[110]

$$\text{2-R-6-R}'\text{-4-CH}_3\text{-C}_6\text{H}_2\text{OH} + O_2 \xrightarrow[\text{DMF, 25°C}]{\text{KO-}t\text{-Bu}} \text{2-R-6-R}'\text{-4-CHO-C}_6\text{H}_2\text{OH} \quad (77)$$

where R and R′ = H, CH_3, $(CH_3)_3C$, and CH_3O (see also Chapter 10, p. 326 ff.).

REFERENCES

1. E. T. Denisov, N. I. Mitskevich, and V. E. Agabekov, "Liquid-Phase Oxidation of Oxygen-Containing Compounds" (transl. by D. A. Paterson). Consultants Bureau, New York, 1977.

2. J. A. Howard, *Adv. Free-Radical Chem.* **4**, 49 (1972).
3. D. G. Lee, *in* "Oxidation" (R. L. Augustine, ed.), Vol. 1, pp. 56-81. Dekker, New York, 1969.

3a. E. J. Corey and G. Schmidt, *Tetrahedron Lett.* p. 399 (1979).

3b. S. J. Flatt, G. W. T. Fleet, and B. J. Taylor, *Synthesis* p. 534 (1978); J. Herscovici and K. Antonakis, *J. Chem. Soc., Chem. Commun.* p. 561 (1980).

4. W. R. Keeler, U.S. Patent 2,869,989 (1959) to Shell Development Co.
5. T. Hasuike and H. Matsuzawa, *Hydrocarbon Process.* **58**(2), 105 (1979).
6. A. L. Aleksandrov and E. T. Denisov, *Izv. Akad. Nauk SSSR, Ser. Khim.* **8**, 1652 (1969).
7. W. Brackman and C. J. Gaasbeek, *Recl. Trav. Chim. Pays-Bas* **85**, 242 (1966).
8. P. J. A. C. Camerman and J. D. V. Hanotier, Br. Patent 1,275,699 (1972).
9. T. F. Blackburn and J. Schwartz, *J. Chem. Soc., Chem. Commun.* p. 157 (1977); see also W. G. Lloyd, *J. Org. Chem.* **32**, 2816 (1967).
10. K. Heyns and L. Blazejewicz, *Tetrahedron* **9**, 67 (1960); see also K. Heyns and H. Paulsen, *in* "Newer Methods of Preparative Organic Chemistry" (W. Foerst, ed.), Vol. 2, pp. 303-335. Academic Press, New York, 1963.
11. I. I. Ioffe, Yu. T. Nikolaev, and M. S. Brodskii, *Kinet. Katal. U.S.S.R.* **1**, 112 (1960). See also W. Sauer, W. Werner, C. Dudeck and N. Petri, Eur. Patent App. 4,881 (1979) for silver catalysis.
12. B. S. Tovrog, S. E. Diamond, and F. Mares, *J. Am. Chem. Soc.* **101**, 5067 (1979); see also R. Tang, S. E. Diamond, N. Neary, and F. Mares, *J. Chem. Soc., Chem. Commun.* p. 562 (1978).
13. H. Mimoun, Ger. Patent 2,920,678 (1978) to Institut Français du Petrole.
14. G. A. Hamilton, P. K. Adolf, J. de Jersey, G. C. Dubois, G. A. Dyrkacz, and R. D. Libby, *J. Am. Chem. Soc.* **100**, 1899 (1978).
15. M. P. Doyle, W. J. Patrie, and S. B. Williams, *J. Org. Chem.* **44**, 2955 (1979).
16. I. Tabushi and N. Koga, *Tetrahedron Lett.* p. 3681 (1979).
17. K. B. Wiberg, *in* "Oxidation in Organic Chemistry" (K. B. Wiberg, ed.), Part A, p. 69. Academic Press, New York, 1965.
18. W. A. Waters and J. S. Littler, *in* "Oxidation in Organic Chemistry" (K. B. Wiberg, ed.), Part A, p. 186. Academic Press, New York, 1965.
19. See, for example, D. Pletcher and S. J. D. Tait, *Tetrahedron Lett.* p. 1601 (1978); D. Landini, F. Montanari, and F. Rolla, *Synthesis* p. 134 (1979).
20. Kh. M. Minachev, G. V. Antoshin, D. G. Klissurski, N. K. Guin, and N. Ts. Abadzhijeva, *React. Kinet. Catal. Lett.* **10**, 163 (1979).
21. T. Itoh, K. Jitsukawa, K. Kaneda, and S. Terunishi, *J. Am. Chem. Soc.* **101**, 159 (1979).
22. G. A. Lee and H. H. Freedman, *Tetrahedron Lett.* p. 1641 (1976); see also K. V. Una and S. Mayanna *J. Catal.* **61**, 165 (1980).
23. L. M. Berkowitz and P. N. Rylander, *J. Am. Chem. Soc.* **80**, 6682 (1958).

23a. A. B. Smith and R. M. Scarborough, *Synth. Commun.* **10**, 205 (1980).

24. A. S. Perlin, *in* "Oxidation" (R. L. Augustine, ed.), Vol. 1, p. 189. Dekker, New York, 1969.
25. C. A. Bunton, *in* "Oxidation in Organic Chemistry" (K. B. Wiberg, ed.), Part A, p. 367. Academic Press, New York, 1965.
26. F. P. Greenspan and H. M. Woodburn, *J. Am. Chem. Soc.* **76**, 6345 (1954); E. S. Huyser and L. G. Rose, *J. Org. Chem.* **37**, 851 (1972).
27. G. Gut, R. V. Falkenstein, and A. Guyer, *Chimia* **19**, 581 (1965).
28. G. de Vries and A. Schors, *Tetrahedron Lett.* p. 5689 (1968).
29. H. T. Clarke and E. E. Dreger, *Org. Synth., Col. Vol.* **1**, 87 (1932).
30. A. T. Blomquist and A. Goldstein, *Org. Synth.* **36**, 77 (1956); see also M. Weiss and M. Appel, *J. Am. Chem. Soc.* **70**, 3666 (1948).

31. W. G. Nigh, *in* "Oxidation in Organic Chemistry" (W. J. Trahanovsky, ed.), Part B, p. 1. Academic Press, New York, 1973.
32. J. A. John and F. J. Weymouth, *Chem. Ind.* (*London*) p. 62 (1962).
33. G. Benson, *Chem. Metall. Eng.* **47** (3), 150 (1940).
34. G. C. Allen and A. Aguilo, *Adv. Chem. Series* **76**, 363 (1968).
35. A. S. Hester and K. Kimmler, *Ind. Eng. Chem.* **51**, 1424 (1959).
36. G. H. Rieser and R. F. Smith, U.S. Patent 3,281,462 (1966).
37. K. Büchner and H. Tummes, U. S. Patent 3,047,599 (1962) to Ruhr Chemie.
38. P. Thuring and A. Peret, *Helv. Chim. Acta* **36**, 13 (1953).
39. A. Bouriot, Fr. Patent 1,532,460 (1968).
40. M. E. Ladhabhoy and M. M. Sharma, *J. Appl. Chem.* **20**, 274 (1970).
41. A. W. Schwab, E. N. Frankel, E. J. Dufek, and J. C. Cowan, *J. Am. Oil Chem. Soc.* **49**, 75 (1972); A. W. Schwab, *ibid.* **50**, 74 (1973).
42. C. E. H. Bawn and J. E. Jolley, *Proc. R. Soc. London, Sec. A* **236**, 297 (1956); C. E. H. Bawn, *Discuss.* Faraday Soc. **14**, 181 (1953).
43. F. Marta, E. Boga, and M. Matic, *Discuss. Faraday Soc.* **46**, 173 (1968).
44. Y. Ohkatsu, M. Takeda, T. Hara, and A. Misono, *Bul. Chem. Soc. Japan* **40**, 1413 (1967).
45. Y. Ohkatsu, T. Osa, and A. Misono, *Bul. Chem. Soc. Japan* **40**, 2111 (1967).
46. W. F. Brill and F. Lister, *J. Org. Chem.* **26**, 565 (1961).
47. M. Zawadzki and J. J. Ziolkowski, *React. Kinet. Catal. Lett.* **10**, 119 (1979).
48. J. M. Church and L. Lynn, *Ind. Eng. Chem.* **42**, 768 (1950).
49. W. Brackman and P. J. Smit, *Recl. Trav. Chim. Pays-Bas* **82**, 757 (1963).
50. T. J. Wallace, H. Pobiner, and A. Schriesheim, *J. Org. Chem.* **30**, 3768 (1965).
51. D. V. Rao, F. A. Stuber, and H. Ulrich, *J. Org. Chem.* **44**, 456 (1979).
51a. M. Utaka, S. Matsushita, H. Yamasaki and A. Takeda, *Tetrahedron. Lett.* **21**, 1063 (1980).
52. H. J. den Hertog and E. C. Kooyman, *J. Catal.* **6**, 357 (1966).
53. E. N. Trachtenberg, *in* "Oxidation" (R. L. Augustine, ed.), Vol. 1, p. 119. Dekker, New York, 1969.
54. N. Rabjohn, *Org. React.* **24**, 261 (1976).
55. H. L. Riley, J. F. Morley, and N. A. C. Friend, *J. Chem. Soc.* p. 1875 (1932).
56. H. A. Riley and A. R. Gray, *Org. Synth., Col. Vol.* **2**, 509 (1943).
57. J. Vene, *Bul. Soc. Sci. Bretagne* **19**, 14 (1946)[*Chem. Abstr.* **41**, 739h (1947)].
58. G. B. Payne and C. W. Smith, *J. Org. Chem.* **22**, 1680 (1957).
59. W. D. Dittmann, W. Kirchoff, and W. Stumpf, *Justus Liebigs* Ann. Chem. **681**, 30 (1965).
60. C. W. Smith and R. T. Holm, *J. Org. Chem.* **22**, 746 (1957).
61. C. H. Hassall, *Org. React.* **9**, 73 (1957).
62. S. E. Jacobson, F. Mares, and P. M. Zambri, *J. Am. Chem. Soc.* **101**, 6938 (1979).
63. S. E. Jacobson, R. Tang, and F. Mares, *J. Chem. Soc., Chem. Commun.* p. 888 (1978).
64. M. W. Johnson and J. P. McCormick, *178th Meet., Am. Chem. Soc.* Org. Chem. Div. Prepr., Pap. No. 66 (1979).
65. M. L. Mihailovic and Z. Cekovic, *in* "The Chemistry of the Hydroxyl Group" (S. Patai, .ed.), Part 1, p. 505. Wiley (Interscience), New York, 1971.
66. P. D. McDonald and G. A. Hamilton, *in* "Oxidation in Organic Chemistry" (W. S. Trahanovsky, ed.), Part B, p. 97. Academic Press, New York, 1973.
67. W. I. Taylor and A. R. Battersby, eds., "Oxidative Coupling of Phenols." Dekker, New York, 1967. See also B. Feringa and H. Wijnberg, *Bioorg. Chem.* **7**, 397 (1978).
68. T. Kametani and K. Fukumoto, *Synthesis* p. 657 (1972).
69. A. I. Scott, *Q. Rev., Chem. Soc.* **19**, 1 (1965).
70. H. Musso, *Angew. Chem.* **75**, 965 (1963).

71. G. Sosnovsky and H. Zaret, *in* "Organic Peroxides" (D. Swern, ed.), Vol. 1, p. 517. Wiley, New York, 1970.
72. A. S. Hay, H. S. Blanchard, G. F. Endres, and J. W. Eustance, *J. Am. Chem. Soc.* **81**, 6335 (1959); Br. Patent 930,993 (1959) to General Electric.
73. A. S. Hay, *J. Polym. Sci.* **58**, 581 (1962); *Adv. Polym. Sci.* **4**, 496 (1967).
74. H. Finkbeiner, A. S. Hay, H. S. Blanchard, and G. F. Endres, *J. Org. Chem.* **31**, 549 (1966).
75. G. F. Endres, A. S. Hay, and J. W. Eustance, *J. Org. Chem.* **28**, 1300 (1963).
76. G. D. Cooper, H. S. Blanchard, G. F. Endres, and H. Finkbeiner, *J. Am. Chem. Soc.* **87**, 3996 (1965).
77. G. F. Endres and J. Kwiatek, *J. Polym. Sci.* **58**, 593 (1962).
78. S. Tsuruya, T. Shirai, T. Kawamura, and T. Yonezawa, *Makromol. Chem.* **132**, 57 (1970).
79. C. C. Price and K. Nakaoka, *Macromolecules* **4**, 363 (1971).
80. J. Kresta, A. Tkáč, R. Přikryl, and L. Malik, *Makromol. Chem.* **176**, 157 (1975).
81. E. McNelis, *J. Org. Chem.* **31**, 1255 (1966).
82. W. J. Mijs, O. E. van Lohuizen, J. Bussink, and L. Vollbracht, *Tetrahedron* **23**, 2253 (1967).
83. E. Tsuchida, M. Kanedo, and H. Nishide, *Makromol. Chem.* **151**, 221, 235 (1972).
84. D. M. White, *Prepr., Div. Polym. Chem., Am. Chem. Soc.* **9**, 663 (1968).
85. B. Feringa and H. Wynberg, *Tetrahedron Lett.* p. 4447 (1977).
86. W. Brenner, Ger. Patent 2,221,624 (1972) to Hoffmann-La Roche.
87. E. M. Kosower and A. S. Wu, *J. Org. Chem.* **28**, 633 (1963).
88. E. L. Reilly, Br. Patent 1,511,813 (1978) to DuPont; see also P. Beltrame, P. L. Beltrame, and P. Carniti, *Ind. Eng. Chem. Prod. Res. Dev.* **18**(3), 208 (1979).
89. M. Constantine and M. Jouffret, Eur. Patent 0001199 (1979) to Rhone Poulenc.
90. M. M. Rogic, T. R. Demmin, and W. B. Hammond, *J. Am. Chem. Soc.* **98**, 7441 (1976); see also J. Tsuji and H. Takayanagi, *ibid.* **96**, 7349 (1974); J. Tsuji, H. Takayanagi, and I. Sakai, *Tetrahedron Lett.* p. 1245 (1975); J. Tsuji and H. Takayanagi, *ibid.* p. 1365 (1976); D. G. Brown, L. Beckmann, C. H. Ashby, G. C. Vogel, and J. T. Reinprecht, *ibid.* p. 1363 (1977).
91. W. Brackman and E. Havinga, *Recl. Trav. Chim. Pays-Bas* **74**, 937, 1021, 1070, 1100, 1107 (1955).
92. T. Funabiki and H. Sakamoto, *J. Chem. Soc., Chem. Commun.* p. 754 (1979).
93. H. M. van Dort and H. J. Geursen, *Recl. Trav. Chim. Pays-Bas* **86**, 520 (1967).
94. L. H. Vogt, J. G. Wirth, and H. L. Finkbeiner, *J. Org. Chem.* **34**, 273 (1969); see also A. McKillop and S. J. Ray, *Synthesis* p. 847 (1977), for analogous oxidations using Co(II) (acacen) as catalyst.
95. D. L. Tomaja, L. H. Vogt, and J. G. Wirth, *J. Org. Chem.* **35**, 2029 (1970).
96. (a) Br. Patent 1,268,653 (1972) to BASF.
    (b) A. J. de Jong and R. van Helden, Ger. Patent 2,460,665 and 2,517,870 (1975) to Shell.
97. V. Kothari and J. J. Tazuma, *J. Catal.* **41**, 180 (1976).
98. C. R. H. I. de Jonge, H. J. Hageman, G. Hoentjen, and W. J. Mijs, *Org. Synth.* **57**, 78 (1977).
99. R. A. Sheldon, to be published.
100. A. Nishinaga, K. Watanabe, and T. Matsuura, *Tetrahedron Lett.* p. 1291 (1974).
100a. T. J. Fullerton and S. P. Ahern, *Tetrahedron Lett.* p. 139 (1976).
101. R. S. Drago, J. Gaul, A. Zombek, and D. K. Straub, *J. Am. Chem. Soc.* **102,** 1033 (1980).
102. (a) C. Skopalik, K. Bauer, and R. Moeleken, Ger. Patent 2,138,735 (1973) to Haarman and Reimer.

(b) E. P. Bost, M. Costantini, M. Jouffret, and G. Lartigau, Ger. Patent 2,332,747 (1974) to Rhone-Poulenc.
(c) H. Seifert, H. Waldmann, W. Schwerdtel, and W. Swodenk, Ger. Patent 2,410,742, 2,410,758 (1975) to Bayer.
103. J. Varagnat, *Ind. Eng. Chem. Prod. Res. Dev.*, **15**(3), 212 (1976); see also J. P. Schirmann and S. Y. Delavarenne, "Hydrogen Peroxide in Organic Chemistry," pp. 92-123. Informations Chimie, Paris, 1979.
104. Br. Patent 1,332,420 (1973) to Mitsubishi.
105. P. Maggioni, Ger. Patent., 2,341,743 (1974) to Brichima.
106. P. Maggioni, Ger. Patent., 2,407,398 (1974) to Brichima.
107. J. Ewers, H. W. Voges, and G. Maleck, *Erdocl Kohle, Erdgas, Petrochem. Brennst.-Chem.* **28**, 34 (1975).
108. S. M. Sethna, *Chem. Rev.* **49**, 91 (1951).
109. Y. Hayaishi, S. Shioi, M. Togami, and T. Sakan, *Chem. Lett.* p. 651 (1973).
110. A. Nishinaga, T. Itahara, and T. Matsuura, *Angew. Chem. Int. Ed.* **14**, 356 (1975).

Chapter 13

Nitrogen, Sulfur, and Phosphorus Compounds

The autoxidations of nitrogen, sulfur, and phosphorus compounds tend to be complex, nonselective processes.[1] Indeed, sulfur[2]- and phosphorus[3]-containing compounds react so readily with alkylperoxy radicals that they are often used as inhibitors of autoxidations. In this chapter we shall be concerned primarily with the selective oxidations of these substrates by H_2O_2 and RO_2H in the presence of metal catalysts. These substrates react with peroxides via nucleophilic attack on the O—O bond. For example, the oxidation of tertiary amines to the corresponding amine oxides with organic peracids is a well-known reaction.[4]

$$R_3N \curvearrowright \underset{H}{O}-O\overset{O}{\overset{\|}{C}}R' \longrightarrow [R_3NOH^+ \ {}^-O_2CR'] \longrightarrow R_3NO + R'CO_2H \qquad (1)$$

Similar oxidations with the less electrophilic RO_2H and H_2O_2 are much slower. Oxidation with these reagents becomes facile, however, in the presence of high-valent transition metal catalysts such as Mo(VI), W(VI), V(V), and Ti(IV) and other Lewis acid catalysts such as Se(IV) and B(III), which promote heterolysis of peroxides.[5,6] The mechanisms of these reactions, which involve inorganic peracids or related compounds as putative intermediates, have been discussed in Chapter 3.

I. NITROGEN COMPOUNDS

Primary amines possessing an α-C—H bond are oxidized to oximes in 60–80% yield with hydrogen peroxide in the presence of the sodium salts of tungstic, molybdic, or vanadic acid.[7–9]

$$R_2CHNH_2 \xrightarrow[{[MoO_4^{2-},\ WO_4^{2-}]}]{H_2O_2} R_2C{=}NOH \quad (2)$$

This general reaction is of particular interest when applied to the conversion of cyclohexylamine to cyclohexanone oxime, an intermediate for nylon 6. More recent work[10,11] has shown that cyclohexylamine is selectively oxidized to cyclohexanone oxime with alkyl hydroperoxides in the presence of Mo, W, V, and Ti catalysts in hydrocarbon solvents at 80°–100°C. The reaction proceeds via the hydroxylamine, which becomes the major product at room temperature.[10]

$$C_6H_{11}NH_2 + RO_2H \xrightarrow[-ROH]{} C_6H_{11}NHOH \xrightarrow{RO_2H} C_6H_{10}{=}NOH + ROH + H_2O \quad (3)$$

Titanium catalysts, such as $(n\text{-BuO})_4Ti$, give the best results (selectivity > 90%).[10,11]

The catalytic oxidation of primary amines to imines by dioxygen in the presence of $RuCl_3$ has been reported.[12] For example, 2-aminohexane gave the corresponding imine (39%) together with 2-hexanone (31%), the product of imine hydrolysis.

$$CH_3(CH_2)_3CH(NH_2)CH_3 + O_2 \xrightarrow{[RuCl_3]} CH_3(CH_2)_3C({=}NH)CH_3 \xrightarrow[-NH_3]{H_2O} CH_3(CH_2)_3C({=}O)CH_3 \quad (4)$$

Benzylamine and *n*-butylamine afforded the corresponding nitriles, presumably via further oxidation of the intermediate imine.

$$RCH_2NH_2 \xrightarrow{[RuCl_3],\ O_2} RCH{=}NH \xrightarrow{[RuCl_3],\ O_2} RC{\equiv}N \quad (5)$$

By analogy with the oxidation of alcohols by Group VIII metal complexes (see Chapters 7 and 12), these reactions probably involve a β-hydride elimination mechanism:

$$>C(H)-N(H)-M^{n+}X \longrightarrow >C{=}N{-}H + HM^{n+}X \quad (6)$$

Similar oxidations can also be affected with oxometal reagents.[13] For example, in the $RuCl_3$ catalyzed oxidation of benzylamine to benzonitrile with peroxydisulfate,[13a] the active oxidant is the ruthenate anion $RuO_4{}^{2-}$. Potassium ferrate (K_2FeO_4) prepared by sodium hypochlorite oxidation of ferric nitrate, on the other hand, oxidizes primary amines to the corresponding aldehydes.[13b] These reactions presumably involve the initial formation of an imine intermediate via a cyclic transition state, followed by hydrolysis to the aldehyde in the case of $FeO_4{}^{2-}$, or by further oxidation to the nitrile in the case of $RuO_4{}^{2-}$.

$$\mathrm{R{-}C(H)(H){-}N{\cdots}M^{(n+2)+}{=}O} \xrightarrow{-(\mathrm{HOM}^{n+})} \mathrm{RCH{=}NH} \begin{cases} \xrightarrow{\mathrm{M{=}O}} \mathrm{RC{\equiv}N} \\ \xrightarrow{\mathrm{H_2O}} \mathrm{RCHO} \end{cases}$$

Primary amines are also oxidized to the corresponding nitriles with sodium hypochlorite under phase transfer condions with tetraalkylammonium salts in the absence of transition metals as catalyst.[13c]

Reaction of primary amines with stoichiometric amounts of $PdCl_2$ in the presence of carbon monoxide results in the formation of isocyanates via oxidative carbonylation.[14]

$$\mathrm{RNH_2 + CO + PdCl_2 \longrightarrow RNCO + Pd^0 + 2\,HCl} \qquad (7)$$

Catalytic oxidation is observed with the system Pd(II)/Cu(II)/O_2, resulting in the formation of ureas, presumably via an intermediate isocyanate,[15]

$$\mathrm{RNH_2 + CO \xrightarrow[O_2]{[Pd^{II}/Cu^{II}]} RNHCONHR} \qquad (8)$$

where R = Me, Bu, $PhCH_2$.

Primary aliphatic amines in which the NH_2 group is attached to a tertiary carbon atom are oxidized to the corresponding nitro compounds with H_2O_2 in the presence of sodium tungstate, presumably via the nitrosoalkane as intermediate.[7,9]

$$\mathrm{{-}\overset{|}{\underset{|}{C}}{-}NH_2 + 3\,H_2O_2 \xrightarrow{[Na_2WO_4]} {-}\overset{|}{\underset{|}{C}}{-}NO_2 + 4\,H_2O} \qquad (9)$$

The products of oxidation of aniline with TBHP depend on the particular metal catalyst used. In the presence of molybdenum or vanadium catalysts, the corresponding nitrobenzenes are formed.[16]

$$\mathrm{ArNH_2 + 3\,\mathit{t}\text{-}BuO_2H \xrightarrow{[Mo^{VI},\,V^{V}]} ArNO_2 + 3\,\mathit{t}\text{-}BuOH + H_2O} \qquad (10)$$

In the presence of titanium(IV) catalysts, on the other hand, the corresponding azoxy compounds are formed.[17]

$$2\,ArNH_2 + 3\,t\text{-}BuO_2H \xrightarrow{[Ti^{IV}]} ArN(\rightarrow O){=}NAr + 3\,t\text{-}BuOH + 2\,H_2O \qquad (11)$$

Secondary amines are oxidized by H_2O_2/Na_2WO_4 to the corresponding hydroxylamines.[7,9]

$$R_2NH + H_2O_2 \xrightarrow{[Na_2WO_4]} R_2NOH + H_2O \qquad (12)$$

The latter are, however, rather unstable and undergo further oxidation and/or condensation reactions. The corresponding reactions with alkyl hydroperoxides, which would be expected to be more selective, do not appear to have been studied.

The oxidation of cyclic secondary amines with TBHP in the presence of manganese(II) catalysts affords the corresponding imides in 65–85% yield via lactam intermediates.[18]

$$(CH_2)_n\langle CH_2, CH_2\rangle NH \xrightarrow[{[Mn^{II}]}]{TBHP} (CH_2)_n\langle CH_2, C{=}O\rangle NH \xrightarrow[{[Mn^{II}]}]{TBHP} (CH_2)_n\langle C{=}O, C{=}O\rangle NH \qquad (13)$$

In contrast to the reactions described above, reaction 13 almost certainly involves a homolytic mechanism (see Chapter 3).

A conversion of a secondary amine to an amide similar to that in eq 13 can be affected by sodium periodate in the presence of catalytic amounts of RuO_4,

$$\text{N-COR-pyrrolidine-2-}CO_2Me \xrightarrow[10\%\ aq\ NaIO_4]{[RuO_4]} \text{N-COR-5-oxopyrrolidine-2-}CO_2Me$$

in a two-phase system consisting of carbon tetrachloride or chloroform and water.[18a] [When $R = CH_3$, C_2H_5, cyclo-C_6H_{11} and $PhCH_2$, the yields of amide are 91, 80, 92 and 54%, respectively.] This transformation constitutes a key step in the chiral synthesis of L-glutamic acid from L-proline.

Secondary amines undergo oxidative carbonylation with CO/O_2 mixtures in the presence of copper salts as catalysts in methanol solvent at ambient temperature.[19]

$$2\,R_2NH + CO + \tfrac{1}{2}\,O_2 \xrightarrow{[Cu^I/Cu^{II}]} R_2NCONR_2 + H_2O \qquad (14)$$

The cyclic secondary amines piperidine and morpholine react particularly smoothly. The reaction was suggested to involve the following steps.

$$R_2NH + Cu^{II} \longrightarrow R_2N\cdot + Cu^{I} + H^+ \tag{15}$$

$$R_2N\cdot + (CO)_2Cu^{I} \longrightarrow R_2N{-}\underset{\underset{O}{\|}}{C}\cdot + Cu^{I}(CO) \tag{16}$$

$$R_2N{-}\underset{\underset{O}{\|}}{C}\cdot + Cu^{II} \longrightarrow R_2\overset{+}{N}{=}C{=}O + Cu^{I} \tag{17}$$

$$R_2\overset{+}{N}{=}C{=}O + R_2NH \longrightarrow R_2NCONR_2 + H^+ \tag{18}$$

Tertiary amines are readily oxidized to the corresponding amine oxides with TBHP in the presence of molybdenum and vanadium catalysts.[20–22] Vanadium compounds are the catalysts of choice.[20]

$$R_3N + t\text{-}BuO_2H \xrightarrow{[VO(acac)_2]} R_3NO + t\text{-}BuOH \tag{19}$$

Tolstikov and co-workers[23–25] used TAHP in the presence of molybdenum catalysts for the oxidation of a wide variety of nitrogen heterocycles to the corresponding *N*-oxides. Reaction rates and selectivities were significantly higher than those observed in the corresponding reactions with peracids. For example, acridine was oxidized quantitatively to its *N*-oxide in 2 hrs with TAHP/$MoCl_5$, whereas perbenzoic acid gives 50% conversion after several days.

$$\text{acridine} \xrightarrow[\text{[MoCl}_5\text{]}]{\text{TAHP}} \text{acridine } N\text{-oxide} \tag{20}$$

Interestingly, bipyridyl, phenanthroline, and 8-hydroxyquinoline were not oxidized, presumably because they bind too strongly to the catalyst, thus hindering complex formation with the hydroperoxide (see Chapter 3).

Imines (Schiff bases) are readily oxidized by TAHP to the corresponding oxaziridines, in 80–95% yield, with catalytic amounts of $Mo(CO)_6$ or $MoCl_5$.[23,24]

$$\begin{matrix}R' \\ R\end{matrix}\!\!>C{=}N{-}R'' \xrightarrow[\text{[Mo}^{VI}\text{]}]{\text{TAHP}} \begin{matrix}R' \\ R\end{matrix}\!\!>\overset{O}{\widehat{C{-}N}}{-}R'' \tag{21}$$

Oxaziridines containing aromatic and aliphatic constituents were formed smoothly, but attempts to prepare purely aromatic oxaziridines were unsuccessful. Thus, the oxidation of benzylidene aniline was accompanied by vigorous rearrangement of the intermediate oxaziridine to benzanilide.

Nitrosamines are oxidized to the corresponding nitramines with TAHP in the presence of molybdenum catalysts.[23]

$$\begin{matrix}R\\ \;\;\backslash\\ \;\;\;N{-}NO\\ \;\;/\\ R'\end{matrix} \xrightarrow[\text{[Mo}^{VI}]]{\text{TAHP}} \begin{matrix}R\\ \;\;\backslash\\ \;\;\;N{-}NO_2\\ \;\;/\\ R'\end{matrix} \qquad (22)$$

Nitroalkanes are converted to carbonyl compounds by reaction of the nitronate salt with TBHP in the presence of $VO(acac)_2$ as catalyst.[26]

$$RR'CHNO_2 \xrightarrow{t\text{-BuOK}} RR'C{=}\overset{+}{N}(O^-)_2 \xrightarrow[\text{[VO(acac)}_2]]{\text{TBHP}} \left[RR'C(OH)(NO_2)\right] \xrightarrow{-HNO_2} RR'C{=}O \qquad (23)$$

This method provides a synthetically useful alternative to the Nef reaction,[27] which requires strongly acidic conditions. It was used as a key step in the synthesis of a prostaglandin synthon.[26]

$$\text{7-}CH(OEt)_2\text{-5-}O_2N\text{-norbornene} \xrightarrow[\text{TBHP,[VO(acac)}_2]]{t\text{-BuOK}} \text{7-}CH(OEt)_2\text{-norbornen-5-one} \qquad (24)$$

II. SULFUR COMPOUNDS

Sulfides are oxidized to the corresponding sulfoxides with hydrogen peroxide[28] or alkyl hydroperoxides[29–34] in the presence of Mo, W, Ti and V catalysts. In the presence of excess hydroperoxide, further oxidation to the sulfone occurs.

$$R_2S \xrightarrow[\text{[Mo}^{VI}, V^{V}]]{R'O_2H} R_2SO \xrightarrow[\text{[Mo}^{VI}, V^{V}]]{R'O_2H} R_2SO_2 \qquad (25)$$

Sulfides are generally oxidized much faster than olefins. For example, with TBHP/$VO(acac)_2$ in ethanol at 25°C, the relative rates decreased in the order n-Bu_2S (100) > n-BuSPh (58) > n-Bu_2SO (1.7) > cyclohexene (0.2).[31] Unsaturated sulfides are selectively oxidized at the sulfur atom (see Chapter 3 for an example).[30]

The high rates of oxidation and virtually quantitative yields obtained under mild conditions emphasize the synthetic utility of the hydroperoxide–metal catalyst reagents for the conversion of sulfides to sulfoxides and sulfones. Tolstikov and co-workers,[33,34] for example, used TAHP/$MoCl_5$

for the oxidation of a wide variety of sulfides to sulfoxides and sulfones, many of which could not be selectively oxidized with organic peracids or other reagents.

When the oxidation of unsymmetric sulfides was carried out with TBHP/ $VO(acac)_2$ in a mixture of benzene and a chiral alcohol, such as (−)-menthol, as solvent, asymmetric induction was observed, although enantiomeric excesses were rather low (5–10%).[35]

The use of H_2O_2 in the presence of SeO_2[36] or $ArSeO_2H$[37] for the facile, selective oxidation of sulfides to sulfoxides has also been reported. The active oxidizing agents are presumably the perseleninic acids:

$$RSe(=O)OH + H_2O_2 \longrightarrow RSe(=O)OOH + H_2O \tag{26}$$

where R = Ar or HO

$$RSe(=O)OOH + R'_2S \longrightarrow RSe(=O)OH + R'_2SO \tag{27}$$

Sulfides also undergo oxidation with peroxide–metal catalyst reagents via homolytic pathways. For example, the reaction of *tert*-butyl peracetate with dialkyl sulfides in the presence of copper ions affords α-acetoxy derivatives without oxidation of the sulfur atom.[38–40] This is an example of the peroxy ester reaction (see Chapter 3).

$$t\text{-}BuO_2Ac + Cu^I \longrightarrow Cu^{II}OAc + t\text{-}BuO\cdot \tag{28}$$

$$t\text{-}BuO\cdot + RCH_2SR' \longrightarrow R\dot{C}HSR' + t\text{-}BuOH \tag{29}$$

$$R\dot{C}HSR' + Cu^{II}OAc \longrightarrow RCH(OAc)SR' + Cu^I \tag{30}$$

Thiols are readily oxidized to disulfides by dioxygen in basic media.[41] The reaction involves the intermediary of thiyl (RS·) radicals formed by one-electron oxidation of the thiolate anion (compare carbanion autoxidations).

$$RS^- + O_2 \longrightarrow RS\cdot + O_2^{\cdot -} \tag{31}$$

$$2\,RS\cdot \longrightarrow RSSR \tag{32}$$

The autoxidation of thiols is catalyzed by typical one-electron oxidants, e.g., Cu(II), Fe(III), Mn(III), and Co(III).[41] These catalysts are usually employed to enhance the rate of oxidation. A number of studies of the interactions of metal compounds with thiols have been carried out.[42–46] For example, stoichiometric oxidation of thiols with ferric octanoate proceeds via the following steps,[43] involving inner-sphere oxidation of RS^- by Fe(III).

$$XFe^{III} + RSH \longrightarrow RSFe^{III} + HX \tag{33}$$

$$RSFe^{III} \longrightarrow Fe^{II} + RS\cdot \tag{34}$$

The intermediate thiyl radicals could be intercepted by the addition of olefins, resulting in the formation of sulfides via the following chain propagation sequence [i.e., the Fe(III) acts as an initiator].

$$RS\cdot + R'CH{=}CH_2 \longrightarrow RSCH_2\dot{C}HR' \tag{35}$$

$$RSCH_2\dot{C}HR' + RSH \longrightarrow RSCH_2CH_2R' + RS\cdot \tag{36}$$

In contrast, when Mn(III) acetylacetonate was the stoichiometric oxidant, interception of thiyl radicals by added olefin was inefficient.[42] This was attributed to effective interception of the thiyl radicals by Mn(III):

$$RS\cdot + Mn^{III} \longrightarrow RS^+ + Mn^{II} \tag{37}$$

$$RS^+ + RSH \longrightarrow RSSR + H^+ \tag{38}$$

The cooxidation of thiols and olefins results in the formation of β-sulfinyl alcohols via rearrangement of intermediate β-thiohydroperoxides (see Chapter 2).[47]

$$RSH + \rangle C{=}C\langle + O_2 \longrightarrow RS{-}\overset{|}{\underset{|}{C}}{-}\overset{|}{\underset{|}{C}}{-}O_2H \longrightarrow \underset{\underset{O}{\|}}{RS}{-}\overset{|}{\underset{|}{C}}{-}\overset{|}{\underset{|}{C}}{-}OH \tag{39}$$

Cooxidation of thiols and acetylenes produces hemithioacetals of α-dicarbonyl compounds, e.g.,[48]

$$PhC{\equiv}CH + RSH + O_2 \longrightarrow Ph{-}\underset{\underset{O}{\|}}{C}{-}\underset{\underset{OH}{|}}{CH}{-}SR \tag{40}$$

The product is unstable and thermally decomposes to the α-dicarbonyl compound. This constitutes a method for the overall conversion of an acetylene to an α-dicarbonyl compound (see Chapter 9).

$$Ph\underset{\underset{O}{\|}}{C}{-}\underset{\underset{OH}{|}}{CH}{-}SR \longrightarrow Ph\underset{\underset{O}{\|}}{C}CHO + RSH \tag{41}$$

The oxidation of thiols with TBHP in the presence of catalysts that promote the heterolysis of peroxides, e.g., Mo(VI) and V(V), produces sulfonic acids, presumably via the corresponding sulfenic (RSOH) and sulfinic (RSO_2H) acids as intermediates:[49]

$$RSH + 3\ t\text{-}BuO_2H \xrightarrow{[Mo^{VI},\ V^{V}]} RSO_3H + 3\ t\text{-}BuOH \tag{42}$$

Hydrogen sulfide is oxidized by hexammineruthenium(III) to elemental sulfur in aqueous solution.[50]

$$2\,(NH_3)_6Ru^{3+} + H_2S \longrightarrow 2\,(NH_3)_6Ru^{2+} + 2H^+ + S \tag{43}$$

This oxidation coupled with the ready reoxidation of the ruthenium(II) species by dioxygen,[51] represents a potential for the catalytic coproduction of hydrogen peroxide according to the overall stoichiometry:

$$H_2S + O_2 \xrightarrow{[Ru]} S + H_2O_2$$

The first step in eq 43 probably proceeds via a one-equivalent oxidation to the sulfhydryl radical HS·, since alkyl mercaptors yield alkyl dissulfides under the same conditions.

III. PHOSPHORUS COMPOUNDS

Trivalent phosphorus compounds readily undergo autoxidation to give more stable pentavalent compounds.[52,53] Indeed, the autoxidation of trialkylphosphines can be so facile that they may even be pyrophoric. Reactions of trivalent phosphorus compounds with intermediate alkylperoxy radicals generally involve displacement and/or addition at the phosphorus atom.[54]

$$RO_2\cdot + R_3P \longrightarrow RO\cdot + R_3PO \tag{44a}$$

$$RO_2\cdot + R_3P \longrightarrow RO_2PR_2 + R\cdot \tag{44b}$$

These reactions have little synthetic utility. Triarylphosphines and trialkyl or triaryl phosphites are more stable toward oxidation. Smooth oxidation of these compounds has been accomplished, however, using TBHP in the presence of molybdenum and vanadium catalysts.[55,56]

$$Ar_3P + t\text{-}BuO_2H \xrightarrow{[Mo^{VI}, V^{V}]} Ar_3PO + t\text{-}BuOH \tag{45}$$

$$(RO)_3P + t\text{-}BuO_2H \xrightarrow{[Mo^{VI}, V^{V}]} (RO)_3PO + t\text{-}BuOH \tag{46}$$

REFERENCES

1. L. Horner, *in* "Autoxidation and Antioxidants" (W. O. Lundberg, ed.), p. 171. Wiley, New York, 1961.
2. P. Koelewijn and H. Berger, *Recl. Trav. Chim. Pays-Bas* **91**, 1275 (1972); **93**, 63 (1974).
3. K. J. Humphris and G. Scott, *Pure Appl. Chem.* **36**, 163 (1973); *J. Chem. Soc., Perkin II* p. 826 (1973).
4. A. R. Katritzky and J. M. Lagowski, "Chemistry of the Heterocyclic N-Oxides." Academic Press, New York, 1971.
5. G. A. Tolstikov, V. P. Yurev, and U. M. Dzhemilev, *Russ. Chem. Rev.* **44**, 319 (1975).

6. R. A. Sheldon, *Aspects of Homogeneous Catalysis* (R. Ugo, ed.), **4** 3, (1981).
7. P. Burckard, J. P. Fleury, and F. Weiss, *Bul. Soc. Chim. France* p. 2730 (1965).
8. K. Kahr, *Angew. Chem.* **72**, 135 (1960).
9. O. L. Lebedev and S. N. Kazarnovskii, *Zh. Obshch. Khim.* **30**, 1631 (1960).
10. J. L. Russell and J. Kollar, U.S. Patent 1,100,672 (1965) to Halcon International.
11. G. N. Koshel, M. I. Farberov, L. L. Zalygin, and G. A. Krushinskaya, *J. Appl. Chem. USSR* **44**, 885 (1971).
12. R. Tang, S. E. Diamond, N. Neary, and F. Mares, *J. Chem. Soc., Chem. Commun.* p. 562 (1978).
13. (a) M. Schroder and W. P. Griffith, *J. Chem. Soc. Chem. Commun.* p. 58 (1979).
    (b) R. J. Audette, J. W. Quail and P. J. Smith, *Tetrahedron Lett.* p. 279 (1971).
    (c) G. A. Lee and H. H. Freedman, *Tetrahedron Lett.* p. 1641 (1976).
14. E. W. Stern and M. L. Spector, *J. Org. Chem.* **26**, 3126 (1961).
15. Y. L. Sheludyakov, V. A. Golodov, and D. V. Sokolskii, *Dokl. Akad. Nauk SSSR* **249**(3), 658 (1979).
16. G. R. Howe and R. R. Hiatt, *J. Org. Chem.* **35**, 4007 (1970).
17. K. Kosswig, *Justus Liebigs Ann. Chem.* **749**, 206 (1971).
18. A. R. Doumaux and D. J. Trecker, *J. Org. Chem.* **35**, 2124 (1970).
18a. S. Yoshifuji, H. Matsumoto, K. Tanaka and Y. Nitta, Tetrahedron. Lett. **21**, 2963 (1980).
19. W. Brackman, *Discuss. Faraday Soc.* **46**, 122 (1968); see also J. Tsuji and N. Iwamoto, *J. Chem. Soc. Chem. Commun.* p. 380 (1966).
20. M. N. Sheng and J. G. Zajacek, *J. Org. Chem.* **33**, 588 (1968).
21. M. N. Sheng and J. G. Zajacek, *Org. Syn.* **50**, 56 (1970).
22. L. Kuhnen, *Chem. Ber.* **99**, 3384 (1966).
23. G. A. Tolstikov, U. M. Jemilev, V. P. Jurjev, F. B. Gershanov, and S. F. Rafikov, *Tetrahedron Lett.* p. 2807 (1971).
24. G. A. Tolstikov, U. M. Dzhemilev, and V. P. Yurev, *J. Org. Chem. USSR* **8**, 1200 (1971).
25. G. A. Tolstikov, U. M. Dzhemilev, V. P. Yurev, A. A. Pozdeeva, and F. G. Gerchikova, *J. Gen. Chem. USSR* **43**, 1350 (1973).
26. P. A. Bartlett, F. R. Green, and T. R. Webb, *Tetrahedron Lett.* p. 331 (1977).
27. W. E. Noland, *Chem. Rev.* **55**, 137 (1955).
28. H. S. Schultz, H. B. Freyermuth, and S. R. Buc, *J. Org. Chem.* **28**, 1140 (1963).
29. L. Kuhnen, *Angew. Chem.* **78**, 957 (1966).
30. V. F. List and L. Kuhnen, *Erdoel & Kohle*, *Erdgas*, *Petrochem* **20**, 192 (1967).
31. R. Curci, F. DiFuria, R. Testi, and G. Modena, *J. Chem. Soc.*, *Perkin II* p. 752 (1974).
32. R.Curci, F. DiFuria, and G. Modena, *J. Chem. Soc.*, *Perkin II* p. 576 (1977); S. Cenci, F. DiFuria, G. Modena, R. Curci, and J. O. Edwards, *ibid.* p. 979 (1978).
33. G. A. Tolstikov, U. M. Dzhemilev, N. N. Novitskaya, V. P. Yurev, and R. G. Kantyukova, *J. Gen. Chem. USSR* **41**, 1896 (1971).
34. G. A. Tolstikov, U. M. Dzhemilev, N. N. Novitskaya, and V. P. Yurev, *Bul. Acad. Sci. USSR*, *Div. Chem. Sci.* **21**, 2675 (1972).
35. F. DiFuria, G. Modena, and R. Curci, *Tetrahedron Lett.* p. 4637 (1976).
36. J. Drabowicz and M. Mikolajczyk, *Synthesis* p. 758 (1978).
37. H. J. Reich, F. Chow, and S. L. Peake, *Synthesis* p. 299 (1978).
38. S. O. Lawesson, C. Berglund, and S. Gronwall, *Acta Chem. Scand.* **15**, 249 (1961).
39. S. O. Lawesson and C. Berglund, *Acta Chem. Scand.* **15**, 36 (1961); C. Berglund and S. O. Lawesson, *ibid.* **16**, 773 (1962).
40. G. Sosnovsky, *J. Org. Chem.* **26**, 281 (1961); *Tetrahedron* **18**, 15, 903 (1962).
41. G. Capozzi and G. Modena, *in* "The Chemistry of the Thiol Group" (S. Patai, ed.), Part 2, p. 785. Wiley, New York, 1974.

42. T. Nakaya, H. Arabori, and M. Imoto, *Bul. Chem. Soc. Japan* **43**, 1888 (1970).
43. T. J. Wallace, *J. Org. Chem.* **31**, 3071 (1966).
44. C. F. Cullis and D. L. Trimm, *Discuss. Faraday Soc.* **46**, 144 (1968).
45. C. J. Swan and D. L. Trimm, *Adv. Chem. Series* **76**, 182 (1968).
46. J. D. Hopton, C. J. Swan, and D. L. Trimm, *Adv. Chem. Series* **75**, 216 (1968).
47. M. S. Kharasch, W. Nudenberg, and G. J. Mantell, *J. Org. Chem.* **16**, 524 (1951).
48. K. Griesbaum, A. A. Oswald, and B. E. Hudson, *J. Am. Chem. Soc.* **85**, 1969 (1963).
49. M. N. Sheng and J. G. Zajacek, U.S. Patent 3,670,002 (1972) to Atlantic Richfield.
50. S. E. Diamond, B. S. Tovrog and F. Mares, *J. Am. Chem. Soc.* **102**, 5908 (1980).
51. D. M. Stanbury, O. Haas and H. Taube, *Inorg. Chem.* **19**, 518 (1980).
52. S. A. Buckler, *J. Am. Chem. Soc.* **84**, 3093 (1962).
53. M. B. Floyd and C. B. Boozer, *J. Am. Chem. Soc.* **85**, 984 (1963).
54. G. A. Razuvaev, V. A. Shushunov, V. A. Dodonov, and T. G. Brilkina, *in* "Organic Peroxides" (D. Swern, ed.), Vol. 3, p. 141. Wiley (Interscience), New York, 1972.
55. R. Hiatt and C. McColeman, *Can. J. Chem.* **49**, 1712 (1971).
56. D. G. Pobedimskii, E. G. Chebotareva, S. A. Nasybullin, P. A. Kirpchnikov, and A. L. Buchachenko, *Proc. Acad. Sci. USSR, Phys. Chem. Sect.* **220**, 59 (1975).

ADDITIONAL READING

J. P. Schirmann and S. Y. Delavarenne, "Hydrogen Peroxide in Organic Chemistry." Informations Chimie, Paris, 1979.

G. Sosnovsky and D. J. Rawlinson, *in* "Organic Peroxides" (D. Swern, ed.), Vol. 1, p. 585. Wiley (Interscience), New York, 1970.

G. Sosnovsky and D. J. Rawlinson, *in* "Organic Peroxides" (D. Swern, ed.), Vol. 2, p. 153. Wiley (Interscience), New York, 1971.

G. Sosnovsky, *in* "Organic Peroxides", (D. Swern, ed.), Vol. 2, p. 269. Wiley (Interscience), New York, 1971.

Index

D

E

F

I

J

K

Q

R

T

W

X

Y

Z